D0849381

Handbook of Biochemistry and Molecular Biology

Handbook of Biochemistry and Molecular Biology

3rd Edition

Proteins — Volume I

EDITOR

Gerald D. Fasman, Ph. D.

Rosenfield Professor of Biochemistry
Graduate Department of Biochemistry
Brandeis University
Waltham, Massachusetts

18901 Cranwood Parkway · Cleveland, Ohio 44128

Library of Congress Cataloging in Publication Data

Main entry under title:

Handbook of biochemistry and molecular biology.

Previous editions published under title: Handbook of biochemistry.
Includes bibliographies and indexes.
CONTENTS: A. Proteins. 2. v.–B. Nucleic acids. 2. v.–C. Lipids, carbohydrates, steroids.
1. Biological chemistry – Handbooks, manuals, etc. 2. Molecular biology – Handbooks, manuals, etc. I. Fasman, Gerald D. II. Sober, Herbert Alexander, 1918- [DNLM: 1. Nucleic acids – Tables. 2. Alkaloids – Tables. 3. Carbohydrates – Tables. 4. Lipids – Tables. 5. Steroids – Tables. QU16]

QP514.2.H34 1975 574.1'92 75-29514
ISBN 0-87819-504-1(v.1)

This book presents data obtained from authentic and highly regarded sources. Reprinted material is quoted with permission, and sources are indicated. A wide variety of references are listed. Every reasonable effort has been made to give reliable data and information but the editor and the publisher cannot assume responsibility for the validity of all materials or for the consequences of their use.

International Standard Book Number (ISBN)

Complete Set 0-87819-503-3
Proteins, Volume I 0-87819-504-1

Library of Congress Card No. 75-29514

Handbook of Biochemistry and Molecular Biology

3rd Edition

Proteins

Volume I

Editor
Gerald D. Fasman, Ph. D.

Rosenfield Professor of Biochemistry
Graduate Department of Biochemistry
Brandeis University
Waltham, Massachusetts

The following is a list of the four major sections of the *Handbook*, each consisting of one or more volumes

Proteins – Amino Acids, Peptides, Polypeptides, and Proteins

Nucleic, Acids – Purines, Pyrimidines, Nucleotides, Oligonucleotides, tRNA, DNA, RNA

Lipids, Carbohydrates, Steroids

Physical and Chemical Data, Miscellaneous – Ion Exchange, Chromatography, Buffers, Miscellaneous, e.g., Vitamins

ADVISORY BOARD (continued)

ADVISORY BOARD (continued)

CONTRIBUTORS

Mary G. Ampola
Clinical Unit
Tufts-New England Medical Center
New England Medical Center Hospitals
Boston, Massachusetts 02111

Charles C. Bigelow
Department of Biochemistry
Memorial University of Newfoundland
St. John's, Newfoundland
Canada A1C 5S7

M. Channon
Department of Chemistry
Memorial University of Newfoundland
St. John's, Newfoundland
Canada A1C 5S7

Waldo E. Cohn
Biology Division
Oak Ridge National Laboratory
Oak Ridge, Tennessee 37830

H. Joseph Goren
Division of Medical Biochemistry
Faculty of Medicine
The University of Calgary
Calgary, Alberta
Canada T2N 1N4

Walter B. Gratzer
MRC Neurobiology Unit
Department of Biophysics
King's College
London WC2B 5RL
England

Erhard Gross
National Institute of Child Health
and Human Development
Laboratory of Biomedical Sciences
National Institutes of Health
Bethesda, Maryland 20014

Paul B. Hamilton
Alfred I. DuPont Institute
Wilmington, Delaware 19899

Leonard Katzin
Chemistry Division
Argonne National Laboratory
Argonne, Illinois 60439

M. C. Khosla
Cleveland Clinic
Cleveland, Ohio 44106

Garland R. Marshall
Department of Physiology, Biophysics,
and Biological Chemistry
Washington University Medical School
St. Louis, Missouri 63110

Ruth S. McDiarmid
Laboratory of Physical Biology
National Institute of Arthritis and
Metabolic Diseases
National Institutes of Health
Bethesda, Maryland 20014

R. B. Merrifield
The Rockefeller University
New York, New York 10021

Elemer Mihalyi
Laboratory of Biochemistry
National Heart and Lung Institute
National Institutes of Health
Bethesda, Maryland 20014

Elizabeth D. Mooz
Department of Chemistry
Bowdoin College
Brunswick, Maine 04011

Vivian E. Shih
The Massachusetts General Hospital
Boston, Massachusetts 02114

Robert F. Steiner
Department of Chemistry
University of Maryland
Baltimore County
Baltimore, Maryland 21228

Arieh Yaron
Department of Biophysics
The Weizmann Institute of Science
Rehovot, Israel

PREFACE

The rapid pace at which new data is currently accumulated in science presents one of the significant problems of today – the problem of rapid retrieval of information. The fields of biochemistry and molecular biology are two areas in which the information explosion is manifest. Such data is of interest in the disciplines of medicine, modern biology, genetics, immunology, biophysics, etc., to name but a few related areas. It was this need which first prompted CRC Press, with Dr. Herbert A. Sober as Editor, to publish the first two editions of a modern *Handbook of Biochemistry,* which made available unique, in depth compilations of critically evaluated data to graduate students, post-doctoral fellows, and research workers in selected areas of biochemistry.

This third edition of the *Handbook* demonstrates the wealth of new information which has become available since 1970. The title has been changed to include molecular biology; as the fields of biochemistry and molecular biology exist today, it becomes more difficult to differentiate between them. As a result of this philosophy, this edition has been greatly expanded. Also, previous data has been revised and obsolete material has been eliminated. As before, however, all areas of interest have not been covered in this edition. Elementary data, readily available elsewhere, has not been included. We have attempted to stress the areas of today's principal research frontiers and consequently certain areas of important biochemical interest are relatively neglected, but hopefully not totally ignored.

This third edition is over double the size of the second edition. Tables used from the second edition without change are so marked, but their number is small. Most of the tables from the second edition have been extensively revised, and over half of the data is new material. In addition, a far more extensive index has been compiled to facilitate the use of the Handbook. To make more facile use of the Handbook because of the increased size, it has been divided into four sections. Each section will have one or more volumes. The four sections are titled:

Proteins – Amino Acids, Peptides, Polypeptides, and Proteins

Nucleic Acids – Purines, Pyrimidines, Nucleotides, Oligonucleotides, tRNA, DNA, RNA

Lipids, Carbohydrates, Steroids

Physical and Chemical Data, Miscellaneous – Ion Exchange, Chromatography, Buffers, Miscellaneous, e.g., Vitamins

By means of this division of the data, we can continuously update the *Handbook* by publishing new data as they become available.

The Editor wishes to thank the numerous contributors, Dr. Herbert A. Sober, who assisted the Editor generously, and the Advisory Board for their counsel and cooperation. Without their efforts this edition would not have been possible. Special acknowledgments are due to the editorial staff of CRC Press, Inc., particularly Ms. Susan Cubar Benovich, Ms. Sandy Pearlman, and Mrs. Gayle Tavens, for their perspicacity and invaluable assistance in the editing of the manuscript. The editor alone, however, is responsible for the scope and the organization of the tables.

We invite comments and criticisms regarding format and selection of subject matter, as well as specific suggestions for new data (and their sources) which might be included in subsequent editions. We hope that errors and omissions in the data that appear in the Handbook will be brought to the attention of the Editor and the publisher.

Gerald D. Fasman
Editor
August 1975

PREFACE TO AMINO ACIDS, PEPTIDES, POLYPEPTIDES, AND PROTEINS, VOLUME I

The section of the *Handbook of Biochemistry and Molecular Biology* on Amino acids, Peptides, Polypeptides and Proteins is divided into three volumes. The first volume contains information relating to the naturally occurring amino acids, α, β-unsaturated amino acids, amino acid antagonists, and α-keto analogues of amino acids.

Data on ultraviolet absorption, fluorescence, optical rotatory dispersion and circular dichroism of amino acids and their derivatives are contained herein.

Relevant data on peptide synthesize is included: Amino acid derivatives, preparation of sequential polypeptides, solid state synthesize, and poly-α-amino acids.

The second and third volumes will contain material mainly on proteins.

Although the data, for which the editor alone is responsible, are far from complete, it is hoped these volumes will be of assistance to those working in the field of biochemistry and molecular biology.

Gerald D. Fasman
Editor
January 1976

THE EDITOR

Gerald D. Fasman, Ph.D., is the Rosenfield Professor of Biochemistry, Graduate Department of Chemistry, Brandeis University, Waltham, Massachusetts.

Dr. Fasman graduated from the University of Alberta in 1948 with a B.S. Honors Degree in Chemistry, and he received his Ph.D. in Organic Chemistry in 1952 from the California Institute of Technology, Pasadena, California. Dr. Fasman did postdoctoral studies at Cambridge University, England, Eidg. Technische Hochschule, Zurich, Switzerland, and the Weizmann Institute of Science, Rehovoth, Israel. Prior to moving to Brandeis University, he spent several years at the Children's Cancer Research Foundation at the Harvard Medical School. He has been an Established Investigator of the American Heart Association, a National Science Foundation Senior Postdoctoral Fellow in Japan, and recently was a John Simon Guggenheim Fellow.

Dr. Fasman is a member of the American Chemical Society, a Fellow of the American Association for the Advancement of Science, Sigma Xi, The Biophysical Society, American Society of Biological Chemists, The Chemical Society (London), the New York Academy of Science, and a Fellow of the American Institute of Chemists. He has published 180 research papers.

The Editor and CRC Press, Inc. would like to dedicate this third edition to the memory of Eva K. and Herbert A. Sober. Their pioneering work on the development of the Handbook is acknowledged with sincere appreciation.

TABLE OF CONTENTS

Nomenclature

BIOCHEMICAL NOMENCLATURE

This synopsis of the recommendations of the IUPAC-IUB Commission on Biochemical Nomenclature (CBN) was prepared by Waldo E. Cohn, Director, NAS-NRC Office of Biochemical Nomenclature (OBN, located at Biology Division, Oak Ridge National Laboratory, Oak Ridge, TN 37830), from whom reprints of the CBN publications listed below and on which the synopsis is based are available.

The synopsis is divided into three sections: Abbreviations, symbols, and trivial names. Each section contains material drawn from the documents (A1 to C1, inclusive) listed below, which deal with the subjects named.

Additions consonant with the CBN Recommendations have been made by OBN throughout the synopsis.

RULES AND RECOMMENDATIONS AFFECTING BIOCHEMICAL NOMENCLATURE AND PLACES OF PUBLICATION (AS OF FEBRUARY 1975)

I. IUPAC-IUB Commission on Biochemical Nomenclature
- A1. Abbreviations and Symbols [General; Section 5 replaced by A6]
- A2. Abbreviated Designation of Amino-acid Derivatives and Peptides (1965) [Revised 1971; Expands Section 2 of A1]
- A3. Synthetic Modifications of Natural Peptides (1966) [Revised 1972]
- A4. Synthetic Polypeptides (Polymerized Amino Acids) (1967) [Revised 1971]
- A5. A One-letter Notation for Amino-acid Sequences (1968)
- A6. Nucleic Acids, Polynucleotides, and their Constituents (1970)

- B1. (Nomenclature of Vitamins, Coenzymes, and Related Compounds)
  - a. Miscellaneous [A, B's, C, D's, tocols, niacins; see B2 and B3]
  - b. Quinones with Isoprenoid Side-chains: E, K, Q [Revised 1973]
  - c. Folic Acid and Related Compounds
  - d. Corrinoids: B-12's [Revised 1973]
- B2. Vitamins B-6 and Related Compounds [Revised 1973]
- B3. Tocopherols (1973)

- C1. Nomenclature of Lipids (1967) [Amended 1970; see also II, 2]
- C2. Nomenclature of α-Amino Acids (1974) [See also II, 5]

- D1. Conformation of Polypeptide Chains (1970) [See also III, 2]

- E1. Enzyme Nomenclature (1972)[a] [Elsevier (in paperback); Replaces 1965 edition.]
- E2. Multiple Forms of Enzymes (1971) [Chapter 3 of E1]
- E3. Nomenclature of Iron-sulfur Proteins (1973) [Chapter 6.5 of E1]
- E4. Nomenclature of Peptide Hormones (1974)

II. Documents Jointly Authored by CBN and CNOC [See III]
1. Nomenclature of Cyclitols (1968) [Revised 1973]
2. Nomenclature of Steroids (1968) [Amended 1971; Revised 1972]
3. Nomenclature of Carbohydrates-I (1969)
4. Nomenclature of Carotenoids (1972) [Revised 1975]
5. Nomenclature of α-Amino Acids (1974) [Listed under I, C2 in the following table]

III. IUPAC Commission on the Nomenclature of Organic Chemistry (CNOC)
1. Section A (Hydrocarbons), Section B (Heterocyclics): *J. Am. Chem. Soc.*, 82, 5545;[a] Section C (Groups containing N, Hal, S, Se/Te): *Pure Appl. Chem.*, 11, Nos. 1–2[a] [A, B, and C Revised 1969:[a] Butterworth's, London (1971)]
2. Section E (Stereochemistry):[b] *J. Org. Chem.*, 35, 2489 (1970); *Biochim. Biophys. Acta*, 208, 1 (1970); *Eur. J. Biochem.*, 18, 151 (1970) [See also I, D1]

[a]No reprints available from OBN; order from publisher.
[b]Reprints available from OBN (in addition to all in IA to ID and II).

RULES AND RECOMMENDATIONS AFFECTING BIOCHEMICAL NOMENCLATURE AND PLACES OF PUBLICATION (AS OF FEBRUARY 1975)(continued)

IV. Physiochemical Quantities and Units (IUPAC)[a] *J. Am. Chem. Soc.*, 82, 5517 (1960) [Revised 1970: *Pure Appl. Chem.*, 21, 1 (1970)]

V. Nomenclature of Inorganic Chemistry (IUPAC) *J. Am. Chem. Soc.*, 82, 5523[a] [Revised 1971: *Pure Appl. Chem.*, 28, No. 1 (1971)][a]

VI. Drugs and Related Compounds or Preparations

1. U.S. Adopted Names (USAN) No. 10 (1972) and Supplement [U.S. Pharmacopeial Convention, Inc., 12601 Twinbrook Parkway, Rockville, Md.]
2. International Nonproprietary Names (INN) [WHO, Geneva]

CBN RECOMMENDATIONS APPEAR IN THE FOLLOWING PLACES[a]

	Arch. Biochem. Biophys.	*Biochem. J.*	*Biochemistry*	*Biochim. Biophys. Acta*	*Eur. J. Biochem.*	*J. Biol. Chem.*	*Pure Appl. Chem.*[b]	*Biochimie (Bull. Soc.)*[c]	*Molek. Biol.*[d]	*Z. Phys. Chem.*[e]
A1[f]	136,1	101,1	5,1445		1,259	241,527		50,3	1,872	348,245
A2(Revised)	150,1(R)	126,773(R)	11,1726(R)	263,205(R)	27,201(R)	247,977(R)	40,(R)	49,121*	2,282*	348,256*
A3(Revised)	121,6*	104,17*	6,362*	133,1*	1,379*	242,555*	31,649(R)	49,325*	2,466*	348,262*
A4(Revised)[g]	151,597(R)	127,753(R)	11,942(R)	278,211(R)	26,301(R)	247,323(R)	33,439(R)	51,205*	5,492(R)	349,1013*
A5	125(3),i	113,1	7,2703	168,6	5,151	243,3557	31,641	50,1577	3,473	350,793
A6[h]	145,425	120,449	9,4022	247,1	15,203	245,5171	40,		6,167	351,1055
B1*	118,505	102,15		107,1(a–c)	2,1	241,2987		49,331		348,266
B1b(Revised)	165.1(R)	147,15(R)		387,397(R)	53,15(R)		38,439			
Bld(Revised)	161(2),iii(R)	147,1(R)	13,1555(R)		45,7(R)					
B2(Revised)	162,1(R)	137,417(R)	13,1056(R)	354,155(R)	40,325(R)	245,4229*	33,447(R)			351,1165*
B3(Revised)	165,6(R)	147,11(R)			46,217(R)					
Cl[f]	123,409	105,897	6,3287	152,1	2,127	242,4845		50,1363	2,784	350,279
Amendments		116(5)		202,404	12,1	245,1511				
C2			14,449		53,1					
D1[i]	145,405	121,577	9,3471	229,1	17,193	245,6489			7,289	
E2	147,1	126,769	10,4825	258,1	24,1	246,6127		54,123		353,852
E3	160,355	135,5	12,3582	310,295	35,1	248,5907				
E4		151,1	14,2559			250,3215				
II,I(Revised)	128,269*	112,17*		165,1*	5,1*	243,5809* }	37,285(R)	51,3*		350,523*
II,2[f]	136,13	113,5	8,2227	164,453	10,1	}	31,285(R)	51,819		351,663
Amendments	147,4	127,613	10,4994	248,387	25,2					
II,3		125,673	10,3983	244,223	21,455	247,613				
II,4		127,741	10,4827	286,217	25,397	247,2633				
Amendments		151,507	14,1803							

[a]Reprints available from OBN.
[b]No reprints available from OBN; order from publisher.
[c]In French.
[d]In Russian.
[e]In German.
[f]Also in other journals.
[g]Also in *Biopolymers*, 11, 321.
[h]*J. Mol. Biol.*, 55, 299.
[i]*J. Mol. Biol.*, 52, 1.

*First, unrevised version.
(R) = revised version.

ABBREVIATIONS

Abbreviations are distinguished from **symbols** as follows (taken from Reference A1):

a. **Symbols**, for monomeric units in macromolecules, are used to make up abbreviated structural formulas (e.g., Gly-Val-Thr for the tripeptide glycylvalylthreonine) and can be made fairly systematic.

b. **Abbreviations** for semi-systematic or trivial names (e.g., ATP for adenosine triphosphate; FAD for flavinadenine dinucleotide) are generally formed of three or four capital letters, chosen for brevity rather than for system. It is the indiscriminate coining and use of such abbreviations that has aroused objections to the use of abbreviations in general.

[Abbreviations are thus distinguished from symbols in that they (a) are for semi-systematic or trivial names, (b) are brief rather than systematic, (c) are usually formed from three or four capital letters, and (d) are not used – as are symbols – as units of larger structures. ATP, FAD, etc., are abbreviations. Gly, Ser, Ado, Glc, etc., are symbols (as are Na, K, Ca, O, S, etc.); they are sometimes useful as abbreviations in figures, tables, etc., where space is limited, but are usually not permitted in text. The use of abbreviations is permitted when necessary but is never required.]

1. Nucleotides (N = A, C, G, I, O, T, U, X, ψ – see Symbols)

NMP	Nucleoside 5′-phosphate
NDP	Nucleoside 5′-di(or pyro)phosphate
NTP	Nucleoside 5′-triphosphate

Prefix d indicates deoxy.

2. Coenzymes, vitamins

CoA(or CoASH)	Coenzyme A
CoASAc	Acetyl Coenzyme A
DPN[a]	Diphosphopyridine nucleotide
FAD	Flavin-adenine dinucleotide
FMN	Riboflavin 5′-phosphate
GSH	Glutathione
GSSG	Oxidized glutathione
NAD[b]	Nicotinamide-adenine dinucleotide (cozymase, Coenzyme I, diphosphopyridine nucleotide)
NADP[b]	Nicotinamide-adenine dinucleotide phosphate (Coenzyme II, triphosphopyridine nucleotide)
NMN	Nicotinamide mononucleotide
TPN[c]	Triphosphopyridine nucleotide

3. Miscellaneous

ACTH	Adrenocorticotropin, adrenocorticotropic hormone, or corticotropin
CM-cellulose	*O*-(Carboxymethyl)cellulose
DEAE-cellulose	*O*-(Diethylaminoethyl)cellulose
DDT	1,1,1-Trichloro-2,2-bis(*p*-chlorophenyl)ethane
EDTA	Ethylenediaminetetraacetate
Hb,HbCO,HbO_2	Hemoglobin, carbon monoxide hemoglobin, oxyhemoglobin
P_i	Inorganic orthophosphate

[a]Replaced by NAD (also DPN^+ by NAD^+, DPNH by NADH).
[b]Generic term; oxidized and reduced forms are NAD^+, NADH ($NADP^+$, NADPH).
[c]Replaced by NADP (also TPN^+ by $NADP^+$, TPNH by NADPH).

PP_i	Inorganic pyrophosphate
TEAE-cellulose	*O*-(Triethylaminoethyl)cellulose
Tris	Tris(hydroxymethyl)aminomethan (2-amino-2-hydroxymethylpropane-1,3-diol)

4. Nucleic Acids

DNA, RNA	Deoxyribonucleic acid, ribonucleic acid (or -nucleate)
hnRNA	Heterogeneous RNA
mtDNA	Mitochondrial DNA
cRNA	Complementary RNA
mRNA	Messenger RNA
nRNA	Nuclear RNA
rRNA	Ribosomal RNA
tRNA	Transfer RNA (generic term; sRNA should not be used for this or any other purpose)
$tRNA^{Ala}$	Alanine tRNA; $tRNA_1^{Ala}$, $tRNA_2^{Ala}$: isoacceptor alanine tRNA's
AA-tRNA	Aminoacyl-tRNA; aminoacylated tRNA; "charged" tRNA (generic term)
Ala-tRNA or Ala-$tRNA^{Ala}$	Alanyl-tRNA
$tRNA^{Met}$	Methionine tRNA (not enzymically formylatable)
$tRNA^{fMet}$ or $tRNA_f^{Met}$	Methionine tRNA, enzymically formylatable to . . .
fMet-tRNA	Formylmethionyl-tRNA (small f, to distinguish from fluorine F)

SYMBOLS

Symbols are distinguished from abbreviations in that they are designed to represent specific parts of larger molecules, just as the symbols for the elements are used in depicting molecules, and are thus rather systematic in construction and use. Symbols are not designed to be used as abbreviations and should not be used as such in text, but they may often serve this purpose when space is limited (as in a figure or table). Symbols are always written with a single capital letter, all subsequent letters being lower-case (e.g., Ca, Cl, Me, Ac, Gly, Rib, Ado), regardless of their position in a sequence, a sentence, or as a superscript or subscript.

Some abbreviations expressed in symbols (see also Section II F below), as examples of the use of symbols:

Dimethylsulfoxide	Me_2SO [a]
Tetranitromethane	$(NO_2)_4C$ [b]
Guanidine hydrochloride	Gdn · HCl [c]
Guanidinium chloride	GdmCl
Cetyltrimethylammonium bromide	$CtMe_3NBr$ [d]
Ethyl methanesulfonate	$MeSO_3Et$
Methylnitronitrosoguanidine	MeN_2O_3Gdn
-nitrosourea	-Nur [e]
-nitrosamine	-Nam [f]
-fluorene	-Fln
Aminofluorene	NH_2Fln
Acetylaminofluorene	AcNHFln [g]
Acetoxyacetylaminofluorene	Ac(AcO)NFln
N-Acetylneuraminic acid	AcNeu [h]

[a]Replaces DMSO.
[b]Replaces TNM.
[c]Replaces Gu, Gd, and G.
[d]Replaces CTAB (similarly for other ammonium compounds).
[e]Replaces NU.
[f]Replaces NA.
[g]Replaces AAF.
[h]Not NANA.

I. Phosphorylated Compounds (Reference A1)

$-PO_3H_2$ (or its ions)	*-P*(or *P*-) ("p" in Nucleic Acids; see IV)
$-PO_2H$-(or its ion)	*-P*-(hyphen in Nucleic Acids; see IV)
$-PO_2H-PO_3H_2$ (or ions)	*-P-P* or *-PP* or *PP*- (cf. PP_i in Abbreviations)

Examples:[a]

Glucose 6-phosphate	Glucose-6-*P* (or Glc-6-*p*; see II below).
Phosphenolpyruvate (pyruvenol phosphate)	*P-enol*Pyruvate or *enol*Pyruvate-*P* or *e* Prv-*P*[b]
Fructose 1,6-bisphosphate (not di)	Fructose-1,6-P_2 (or Fru-1,6-P_2 ; see II below).
Creatine phosphate	Creatine-*P*
Phosphocreatine	*P*-Creatine

[a]Note that symbols are hyphenated even where names are not.
[b]Recommended by OBN.

II. Peptides and Proteins (References A1–A5)

A. Symbols (Reference A2–A5)

1. Common amino acids

Name	Symbol: Three-letter[a]	Symbol: One-letter[b]	Name	Symbol: Three-letter[a]	Symbol: One-letter[b]
Alanine	Ala	A	Lysine[e]	Lys	K
Arginine	Arg	R	Methionine	Met	M
Asparagine	Asn [c,d,e]	N	Phenylalanine	Phe	F
Aspartic acid	Asp [d,e]	D	Proline	Pro	P
Cysteine	Cys [e]	C	Serine[e]	Ser	S
Glutamic acid	Glu [f]	E	Threonine[e]	Thr	T
Glutamine	Gln [e,f,g]	Q	Tryptophan[e]	Trp	W
Glycine	Gly	G	Tyrosine[e]	Tyr	Y
Histidine	His [e]	H	Valine	Val	V
Isoleucine	Ile	I	Unknown or "other"	AA [h]	X
Leucine	Leu	L			

[a]One capital, two small letters, at all times.
[b]For special uses and with special conventions; see III, *I* following.
[c]Or Asp (NH_2); see Footnotes e and g.
[d]Uncertainty as between Asp and Asn may be designated by Asx (or B).
[e]Substitution on a functional group may be indicated, as shown in Footnotes c and g, by parenthesis following the symbol, e.g., Cys (Cme), Ser (*P*); see C 2 below.
[f]Uncertainty, as between Glu and Gln, may be designated by Glx (or Z); pyroglutamate is pGlu or <Glu, not PCA.
[g]Or Glu (NH_2); see Footnotes c and e.
[h]See Abbreviations, Part 4 (AA-tRNA).

2. Less common amino acids

Name	Symbol
β-Alanine	βAla
Alloisoleucine	aIle
2-Aminoadipic acid	Aad
3-Aminoadipic acid	βAad
2-Aminobutyric acid	Abu
6-Aminocaproic acid[a]	ϵAhx [a]
2-Aminopimelic acid	Apm
2,4-Diaminobutyric acid	A_2bu [b]
2,2′-Diaminopimelic	A_2pm [b]
2,3-Diaminopropionic acid	A_2pr [b]
N-Ethylglycine, etc.	EtGly, etc.
Hydroxylysine	Hyl
allo-Hydroxylysine	*a*Hyl
3-Hydroxyproline	3Hyp [c]
4-Hydroxyproline	4Hyp[c]
N-Methylglycine (sarcosine)	MeGly or Sar
N-Methylisoleucine	MeIle
N-Methylvaline, etc.	MeVal, etc.
Norleucine	Nle
Norvaline	Nva
Ornithine	Orn

[a]6-Aminohexanoic acid is preferred; see Reference C2.
[b]The use of D (for di), T (for tri or tetra), etc., is undesirable. Hence, in this context, A_2 for diamino is recommended (cf. II F below).
[c]Or Pro(PH) for hydroxyproline.

B. Sequence, Direction, and Bonding (Reference A2)

→ peptide bond, originating in peptide -CO-
– peptide bond, originating in peptide -CO- of residue at left.
, separates symbols in unknown sequence (the entire unknown sequence is enclosed in parentheses).
| bond originating in first letter of symbol of a residue having a substituted functional group (-SH, 3- or 4-COOH, -OH, 6-NH_2, etc., or the remaining H of a peptide bond; see C2 below).

C. Substitution (Reference A2)

Groups substituted for hydrogen or for hydroxyl may be indicated either by their structural formulae, or by symbols, or by combinations of both, e.g.,

Benzoylglycine (hippuric acid)	PhCo-Gly or C_6H_5CO-Gly or Bz-Gly
Glycine methyl ester	Gly-OCH_3 or Gly-OMe
Trifluoroacetylglycine	CF_3CO-Gly

1. In α-NH_2 or α-COOH groups: horizontal dash (hyphen) to left or right, respectively.

N-Acetylglycine	Ac-Gly
Glycine ethyl ester	Gly-OEt
N^2-Acetyllysine	Ac-Lys
Serine methyl ester	Ser-OMe
O^1-Ethyl *N*-acetylglutamate	Ac-Glu-OEt
Isoglutamine	Glu-NH_2
N-Tosylphenylalanyl chloromethyl ketone (TPCK)	Tos-Phe CH_2Cl
N-Ethyl-*N*-methylglycine	Et-(Me-)Gly or Et/Me>Gly

2. On functional group: Vertical bond (see B above) or parentheses (see II A1 above, Footnote e).

4-Methyl aspartate	Asp (OMe above) or Asp (OMe below) or Asp(OMe)	*O*-Methyltyrosine	Tyr (Me above) or Tyr (Me below) or Tyr (Me)
N^6-Acetyllysine	Lys (Ac above) or Lys (Ac below) or Lys(Ac)	*S*-Ethylcyteine	Cys (Et above) or Cys (Et below) or Cys(Et)
O-Phosphoserine	Ser (*P* above) or Ser (*P* below) or Ser( )	N^τ-Methylhistidine* (see 3.3) (*tele*methylhistidine)	His (τ Me above) or His (τ Me below) or His(τMe)
		similarly for N^π substitution **(prosmethylhistidine)**	

N-Glycylsarcosine: Gly —┬— Gly (Me below) or Gly-(Me-)Gly or Gly-Sar

N-Glycyl-*N*-acetylglycine: Gly —┴— Gly (Ac above) or Gly-(Ac-)Gly or Ac, Gly >Gly

N,N-Diglycylglycine: Gly —┘ Gly —┴— Gly or Gly_2 > Gly or Gly, Gly >Gly

***The prolonged and well-entrenched ambiguity in the nomenclature of the *N*-1 being the biochemist's *N*-3 and *vice versa*) led to a new trivial system for designating these substances: The imidazole N *nearer* the alanine residue is designated *pros* (symbol π) and the one *far*ther *tele* (symbol τ), to give the following names and symbols: *pros*methylhistidine or π-methylhistidine, His(πMe); *tele*methylhistidine or τ-methylhistidine, His(τMe).**

D. Polypeptides: Follow Rules for Substitution (C above) (Reference Az)

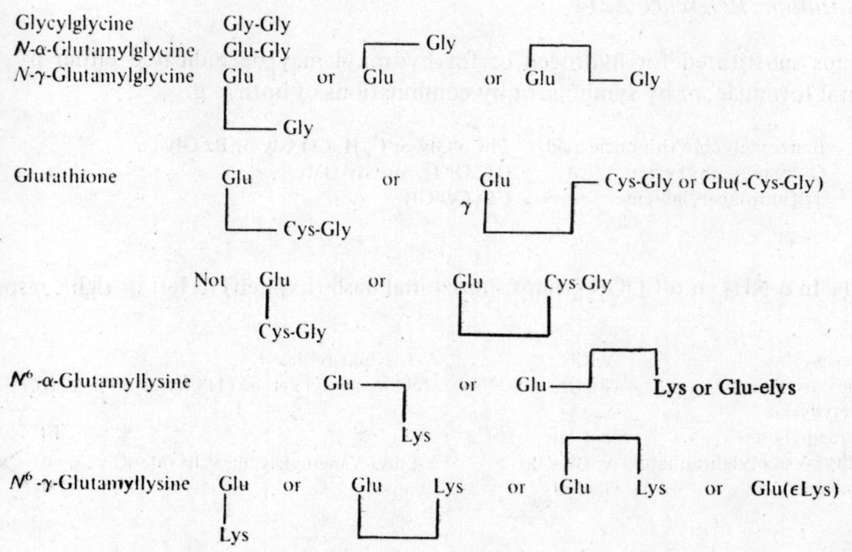

Glycyllysylglycine dihydrochloride $^{+}H_2$–Gly-Lys-Gly–OH · 2HCl

Its N^6-formylderivative Gly-Lys-Gly (with CHO on Lys) or Gly-Lys(CHO)-Gly

E. Cyclic Polypeptides (Reference A2)

1. Homodetic: Gramicidin *S*

cyclo-Val-Orn-Leu-DPhe-Pro-Val-Orn-Leu-DPhe-Pro

Val-Orn-Leu-DPhe-Pro-Val-Orn-Leu-DPhe-Pro (cyclic)

→Val → Orn → Leu — DPhe → Pro→
└Pro ← DPhe ← Leu — Orn ← Val←┘

2. Heterodetic:

Oxytocin Cys-Tyr-Ile-Asn-Gln-Cys-Pro-Leu-Gly-NH_2 (Cys–Cys bridged)

Cyclic ester of threonylglycylglycylglycine

Thr-Gly-Gly-Gly (cyclic) or H–Thr-Gly-Gly-Gly (cyclic)

F. Substituents (Reference A2)

1. NH_2 protecting groups of the urethan type (partial list)

Benzyloxycarbonyl-	Z- or Cbz-[a]	*p*-Methoxyphenylazobenzyloxycarbonyl-	Mz-
p-Nitrobenzyloxycarbonyl-	Z(NO_2)-	*p*-Phenylazobenzyloxycarbonyl-	Pz-
p-Bromobenzyloxycarbonyl-	Z(Br)-	*t*-Butoxycarbonyl-	Boc-[b] or Bu^tOCO-
p-Methoxybenzyloxycarbonyl-	Z(OMe)-	Cyclopentyloxycarbonyl-	Poc- or cPeOCO-

2. Other N- protecting groups (partial list)

Acetyl-	Ac-	Maleoyl- (–OC–CH=CH–CO–)	Mal-[e] or Mal<
Benzoyl-(C_6H_5CO-)	PhCO- or Bz-	Maleyl- (HOOC–CH=CH–CO–)	Mal-
Benzyl- ($C_6H_5CH_2$-)	$PhCH_2$- or Bzl	Methylthiocarbamoyl-	MeNHCS-[a] or Mtc-[f]
Benzylthiomethyl-	$PhSCH_2$- or Btm-	*o*-Nitrophenylthio-	Nps-
Carbamoyl-	NH_2CO- (preferred to Cbm)	Phenylthiocarbamoyl-	PhNHCS-[a] or Ptc-[f]
		Phthaloyl-	–Pht- or Pht<
1-Carboxy-2-nitrophenyl-5-thio-	Nbs-	Phthalyl-	Pht-
3-Carboxypropionyl- ($HOOC–CH_2–CH_2–CO–$)	Suc-	Succinyl- ($–OC–CH_2–CH_2–CO–$)	–Suc- or Suc<
		Tetrahydropyranyl-	H_4 pyran-[c,g]

Dansyl-(5-dimethylamino-naphthalene-inonaphthalene-1-sulfonyl)	Dns-[c] or dansyl[a]	Tosyl- (*p*-tolylsulfonyl)	Tos- or tosyl
Dinitrophenyl-	N_2ph-[a,c] or Dnp	Trifluoroacetyl-	CF_3CO- or F_3Ac-[a,h]
Formyl-	HCO-[d] or CHO-	Trityl- (triphenylmethyl)	Ph_3C-[a,i] or Trt-
p-Iodophenylsulfonyl (pipsyl)	Ips or pipsyl		

3. Substituents at carboxyl group

Benzyloxy- (benzyl ester)	$-OCH_2Ph$ or $-OBzl$	*p*-Nitrophenoxy- (*p*-nitrophenyl ester)	–ONph
Cyanomethoxy- (cyanomethyl ester)	$-OCH_2CN$ or –OMeCN	*p*-Nitrophenylthio-	–SNph
Diphenylmethoxy- (benzhydryl ester)	$-OCHPh_2$ or –OBzh	Phenylthio- (phenylthiolester)	–SPh
Ethoxy- (ethyl ester)	–OEt	1-Piperidino-oxy-	–OPip
Methoxy- (methyl ester)	–OMe	8-Quinolyloxy-	–OQu
		Succinimido-oxy-	–ONSuc
		Tertiary butoxy- (*t*-butyl ester)	$-OBu^t$

[a]Preferred.
[b]Not BOC or *t*BOC.
[c]The use of D for di and T for tri (or tetra) is discouraged. Recognized symbols with numerical subscripts are recommended.
[d]fMet is approved for formylmethionine.
[e]MalNEt is recommended for *N*-ethylmaleimide (not NEM).
[f]Mtc and Ptc have been used to denote methyl- and phenylthiohydantoins (e.g., Ptc-Leu). Since this incorrectly implies the substitution of an amino acid by a "phenyl (or methyl) thiohydantoyl" group, the correct representation, CS-Leu-NPh or PhNCS-Leu or, in text, Leu>PhNCS, is recommended.
[g]Not THP or Thp (see Footnote c).
[h]Not TFA (see Footnote c).
[i]Or trityl.

4. Other substituents (and reagents)

2-Aminoethyl-[a]	$-(CH_2)_2NH_2$ or Aet-[a,b]	Chloroethylamine	$Cl(CH_2)_2NH_2$ or AetCl
Carbamoylmethyl-	$-CH_2CONH_2$ or Ncm-	Chloroacetamide	$ClCH_2CONH_2$ or NcmCl
Carboxymethyl-	$-CH_2CO_2H$ or Cxm-	Chloroacetic acid	$ClCH_2CO_2H$ or CxmCl
p-Carboxyphenylmercuri-	-HgBzOH	*p*-Chloromercuribenzoate	$ClHgBzO^{-}$[c]
1-Carboxy-2-mitrophenyl-5-thio-nitrophenyl-5-thio-	Nbs-	5,5′-Dithiobis(2-nitrobenzoic acid) (2-nitrobenzoic acid)	Nbs_2[d]
Diazoacetyl-	N_2CHCO- or N_2<Ac-		
-Diisopropylphosphor	iPr_2P-[e]	Diisopropylfluorophosphate	iPr_2P-F[f]
Dinitrophenyl-	N_2ph[g]	Fluorodinitrobenzene	N_2ph-F[h]
Hydroxyethyl-	$-(CH_2)_2OH$ or HOEt-	Ethylene oxide	$(CH_2)_2O$ or Et>O
		N-Ethylmaleimide	MalNEt[i]
		Tetrahydrofuran	H_4furan[j]
		Tosyllysyl chloromethyl ketone	$Tos\text{-}LysCH_2Cl$[k]

		Tosylarginine methyl ester	Tos-ArgOMe [l]
Trifluoroacetyl-	F_3Ac-[m]	Trifluoroacetic acid	F_3AcOH
Trimethylsilyl-	Me_3Si-[n]	Tetramethylsilane	Me_4Si [n]

[a]For -ethylamine, -Etn; for -ethanolamine (see Lipids), -OEtn.
[b]Not AET.
[c]Replaces PCMB, *p*CMB, and CMB.
[d]Replaces DTNB.
[e]Replaces, DIP and Dip.
[f]Replaces DPF, DFP, DIPF, etc.
[g]Replaces DNP and Dnp.
[h]Replaces FDNB.
[i]Replaces NEM.
[j]Replaces THF. Similarly, H_4 folate.
[k]Replaces TLCK (similarly for TPCK, etc.).
[l]Replaces TAME (similarly for other N-substituted amino-acid esters. See C1 above).
[m]Replaces TFA.
[n]Not TMS- or TMS. Similarly, Me_2SO, not DMSO; NAc_3, not NTA.

G. Polymerized Amino Acids (Synthetic Polypeptides) (Reference A4)

1. Linear polymers (only normal peptide links are involved).

a. Homopolymer: polylysine; poly(Lys) or $(Lys)_n$ (n may be replaced by a number).

b. Copolymer, alternating sequence: poly(alanine-lysine); poly(Ala-Lys) or $(Ala\text{-}Lys)_n$.

c. Copolymer, random sequence, composition unspecified: poly(alanine, lysine); poly(Ala,Lys) or $(Ala,Lys)_n$.

d. Copolymer, random sequence, molar percentages (Σ = 100%) known: poly($DLGlu^{56}Lys^{38}DTyr^{6}$) or $(DLGlu^{56}Lys^{38}DTyr^{6})_n$ (only lysine is L).

e. Block polymer of poly(Glu) linked via α-COOH to α-NH_2 of poly(Lys): poly(Glu^{56})-poly(Lys^{44}) or $(Glu^{56})_n$-$(Lys^{44})_n$.

f. Block polymer, a repeating series of the known sequence Glu-Lys-Lys-Tyr: poly(Glu-Lys_2-Tyr) or $(Glu\text{-}Lys_2\text{-}Tyr)_n$.

g. Block polymer of two repeating series: poly $(Glu\text{-}Lys)^{25}$-poly$(Ala\text{-}Tyr_2\text{-}Glu)^{12.5}$ or $(Glu\text{-}Lys)_n^{25}$-$(Ala\text{-}Tyr_2\text{-}Glu)_n^{12.5}$ (molar percentages = 100).

2. Branched graft polymers (functional groups are involved).

a. Main chain is a repeating sequence (see 1f above), sidechain is of random sequence, connection is from ϵ-NH_2 of a lysine to an unknown group in the sidechain

poly($Asp^{30}Glu^{50}$) or $(Asp^{30}Glu^{50})_n$
| |
poly$(Glu\text{-}Lys_2\text{-}Tyr)^5$ $(Glu\text{-}Lys_2\text{-}Tyr)_n^5$

or

poly$(Glu\text{-}Lys_2\text{-}Tyr)^5$ ⌐—⌐ poly($Asp^{30}Glu^{50}$) or $(Glu\text{-}Lys_2\text{-}Tyr)_n^5$ ⌐—⌐ $(Asp^{30}Glu^{50})_n$

b. Main chain of unknown sequence, linked via ϵ-NH_2 of a lysine to the α-COOH of an L-tyrosine in the sidechain, which is a block polymer (no analytical data):

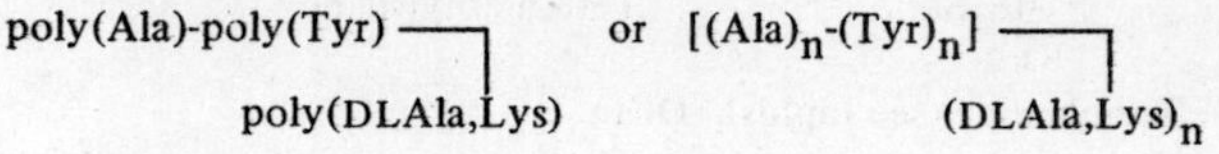

poly(Ala)-poly(Tyr) ─┐ poly(DLAla,Lys) or [(Ala)$_n$-(Tyr)$_n$] ─┐ (DLAla,Lys)$_n$

or

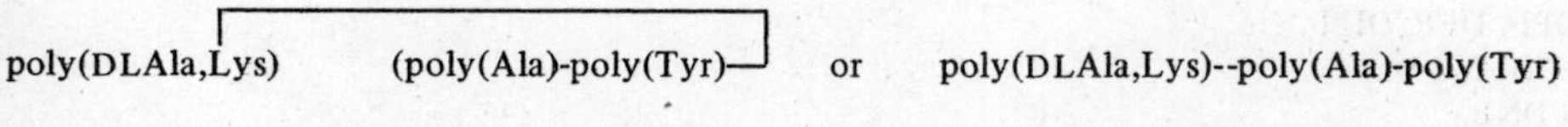

poly(DLAla,Lys) (poly(Ala)-poly(Tyr)) [bracket linking first and last] or poly(DLAla,Lys)--poly(Ala)-poly(Tyr)

Note: The points of attachment of Lys and Tyr cannot be specified in the last example. This system, depending on double hyphens to express functional group involvement, is not recommended.

c. Two linear copolymers of unknown sequence, triply linked between ϵ-NH_2 groups of lysines and γ-COOH residues of glutamates:

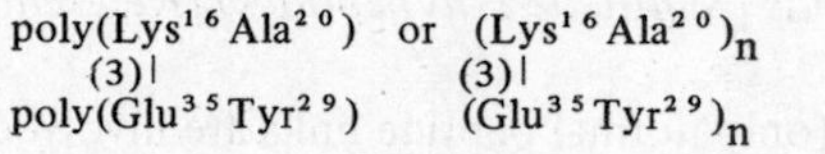

poly(Lys^{16} Ala^{20}) (3)| poly(Glu^{35} Tyr^{29}) or (Lys^{16} Ala^{20})$_n$ (3)| (Glu^{35} Tyr^{29})$_n$

or

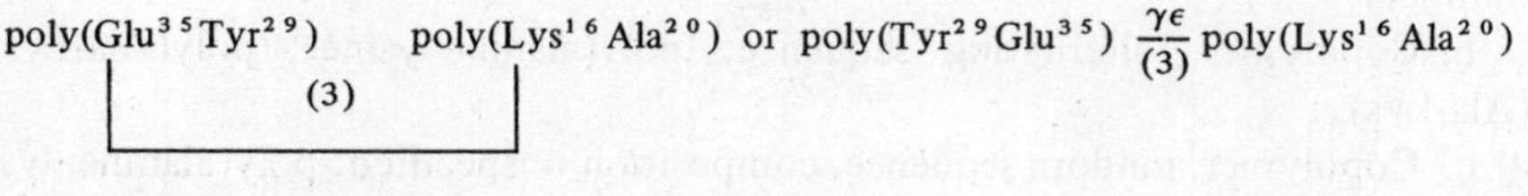

poly(Glu^{35} Tyr^{29}) poly(Lys^{16} Ala^{20}) [linked by (3)] or poly(Tyr^{29} Glu^{35}) $\frac{\gamma\epsilon}{(3)}$ poly(Lys^{16} Ala^{20})

(See comment under b with respect to last example).

d. Linear, random-sequence chain attached via terminal α-COOH group (of either Tyr or Glu) to NH_2 terminal of poly(DLalanine) in turn connected, via terminal COOH, to ϵ-NH_2 groups(s) of poly(L-lysine) (no analytical data):

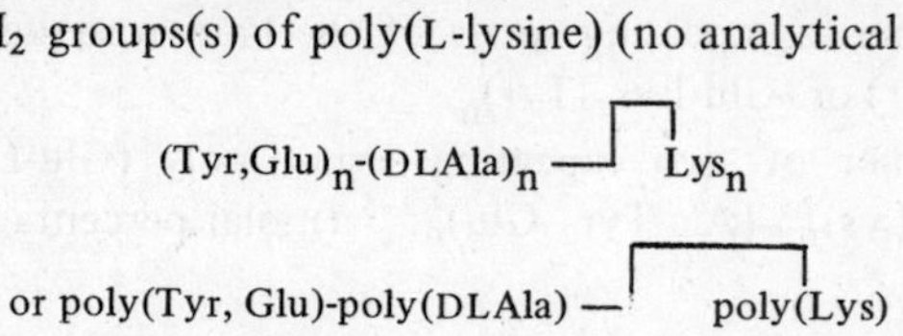

(Tyr,Glu)$_n$-(DLAla)$_n$ ─┘ Lys$_n$

or poly(Tyr, Glu)-poly(DLAla) ─┘ poly(Lys)

H. Synthetic Modifications of Natural Peptides (Reference A3)

Modification and Name	Abbreviation
1. Replacement:	
a. of 8th residue in vasopressin by citrulline:	
[8-Citrulline] vasopressin	[Cit^8] vasopressin
b. at 5 and 7 positions in hypertensin II:	
[5-Isoleucine, 7-alanine] hypertensin II	[Ile^5, Ala^7] hypertensin II
2. Extension of X by valyl residue, at N and C terminals:	
valyl-X	Val-X
X(yl)-valine	X(yl)-Val
3. Insertion of tyrosine residue between 4th and 5th residue	
4a-endo-tyrosine-hypertensin II	endo-Tyr^{4a}-hypertensin II
4. Removal of proline from position 7 in oxytocin:	
des-7-proline-oxytocin	des-Pro^7-oxytocin

5. Substitution of valine on ϵ-nitrogen of a lysine at position 2 in peptide X:
   $N^{\epsilon 2}$-valyl-X $N^{\epsilon 2}$-Val-X
6. Substitution of valine on γ-carboxyl of glutamate at position 3 in peptide X:
   $C^{\gamma 3}$-X(yl)-valine $C^{\gamma 3}$-X(yl)-Val
7. Fragments, or partial sequences: fragments from α-MSH.
   Ac-Ser-Tyr-Ser-Met-Glu-His-Phe-Arg-Trp-Gly-Lys-Pro-Val-NH_2 α-MSH
   1 2 3 4 5 6 7 8 9 10 11 12 13

   Met-Glu-His-Phe-Arg-Trp-Gly α-MSH-(4–10)-heptapeptide
   6 10

   His-Phe-Arg-Lys-Pro-Val-NH_2 α-MSH-(6–8)-(11–13)-hexapeptide amide
   6 8 11 13

I. One-letter Notation[a] (Reference A5)

1. Symbols: see II.A.1 above (NH_2 terminal at left, COOH terminal at right).
2. Known sequence: space[b] between symbols.
3. Unknown sequence: comma[c] between symbols, parentheses[d] enclosing.
4. Adjacent unknown sequences: = replaces)(.
5. Uncertainty as to sequence or terminus: / (see examples b and c).

Examples: a. (Ala, Cys, Asp) (Arg, Ser) (Gly, His, Ile) Lys-Leu-Met-Asn-Pro-Gln
becomes (A, C, D = R, S = G, H, I) K L M N P Q

b. A C D E F G H I K L M N P Q
c. (A. C. D = R, S = G. H. I) K L/ M N/ P Q/

In c., the tripeptides A . C . D and G . H . I are not of known sequence, but are inferred by analogy with the known peptide b.; the inference is expressed by periods instead of commas. The comma between R and S indicates that no inference as to sequence can be drawn for this dipeptide. The internal slashes indicate that no connection between L and M, and N and P, has been proven, although KL, MN and PQ are each of known internal sequence. The final slash indicates that Q has not been proven to be the COOH terminal residue of the entire peptide, although it is the terminus of the PQ dipeptide.

[a]For display of very long sequences or computer use only.
[b]In place of hyphen in three-letter system. Spaces must be equal to characters, as in typing. So must commas, dots, and all other symbols.
[c]As in three-letter system; becomes a dot (period) when sequence is inferred but not demonstrated (see example c).
[d]The double symbol,)(, is replaced by = (see 4) to preserve equal spacing.

NOMENCLATURE OF LABELED COMPOUNDS

The statement below was adopted by the IUB Commission of Editors of Biochemical Journals* (CEBJ) and appears, in the same or in similar form, in the Instructions to Authors of their journals. This system originated with the Chemical Society (London) and was subsequently adopted by the American Chemical Society (Handbook for Authors, 1967). It was adopted by CEBJ in 1971 and is the only system currently permitted in the pages of their journals.

ISOTOPICALLY LABELED COMPOUNDS

The symbol for the isotope introduced is placed in *square* brackets directly attached to the front of the name (word), as in [^{14}C]urea. When more than one position in a substance is labeled by means of the same isotope and the positions are not indicated (as below), the number of labeled positions is added as a right-hand subscript, as in [$^{14}C_2$]glycollic acid. The symbol "U" indicates uniform and "G" general labeling, e.g., [U-^{14}C]glucose (where the ^{14}C is uniformly distributed among all six positions) and [G-^{14}C]glucose (where the ^{14}C is distributed among all six positions, but not necessarily uniformly); in the latter case it is often sufficient to write simply "[^{14}C]glucose."

The isotopic prefix precedes that part of the name to which it refers, as in sodium [^{14}C]formate, iodo[$^{14}C_2$]acetic acid, 1-amino[^{14}C]methylcyclopentanol ($H_2N{-}^{14}CH_2{-}C_5H_8{-}OH$), α-naphth[^{14}C]oic acid ($C_{10}H{-}^{14}CO_2H$), 2-acetamido-7-[^{131}I]iodofluorene, fructose 1,6-[1-^{32}P]diphosphate, D-[^{14}C]glucose, 2*H*-[2-^{2}H]pyran, *S*-[8-^{14}C]adenosyl[^{35}S]methionine. Terms such as "^{131}I-labeled albumin" should not be contracted to "[^{131}I]albumin" (since native albumin does not contain iodine), and "^{14}C-labeled amino acids" should similarly not be written as "[^{14}C]amino acids" (since there is no carbon in the amino group).

When isotopes of more than one element are introduced, their symbols are arranged in alphabetical order, including ^{2}H and ^{3}H for deuterium and tritium, respectively.

When not sufficiently distinguished by the foregoing means, the positions of isotopic labeling are indicated by Arabic numerals, Greek letters, or prefixes (as appropriate), placed within the square brackets and before the symbol of the element concerned, to which they are attached by a hyphen; examples are [1-^{2}H]ethanol ($CH_3{-}C^2H_2{-}OH$), [1-^{14}C]aniline, L-[2-^{14}C]leucine (or L-[α-^{14}C]-leucine), [*carboxy*-^{14}C]leucine, [*Me*-^{14}C]isoleucine, [2,3-^{14}C]maleic anhydride, [6,7-^{14}C]xanthopterin, [3,4-^{13}C,^{35}S]-methionine, [2-^{13}C; 1-^{14}C]acetaldehyde, [3-^{14}C; 2,3-^{2}H; ^{15}N]serine.

The same rules apply when the labeled compound is designated by a standard abbreviation or symbol, other than the atomic symbol, e.g. [γ-^{32}P]ATP.

For simple molecules, however, it is often sufficient to indicate the labeling by writing the chemical formulae, e.g. $^{14}CO_2$, $H_2{}^{18}O$, $^{2}H_2O$ (not D_2O), $H_2{}^{35}SO_4$, with the prefix superscripts attached to the proper atomic symbols in the formulae. The square brackets are not to be used in these circumstances, nor when the isotopic symbol is attached to a word that is not a chemical name, abbreviation or symbol (e.g. ^{131}I-labeled).

*CEBJ consists of the Editors-in-Chief of the following journals: *Archives of Biochemistry and Biophysics, Biochemical Journal, Biochemistry, Biochimica et Biophysica Acta, Biochimie, European Journal of Biochemistry, Hoppe-Seyler's Zeitschrift für Physiologische Chemie, Journal of Biochemistry, Journal of Biological Chemistry, Journal of Molecular Biology, and Molekulyarnaya Biologiya;* corresponding members include *Proceedings of the National Academy of Sciences* (U.S.A.) and approximately 40 others.

THE CITATION OF BIBLIOGRAPHIC REFERENCES IN BIOCHEMICAL JOURNALS RECOMMENDATIONS (1971)*

IUB Commission of Editors of Biochemical Journals (CEBJ)

These Recommendations were reviewed by the Commission in August 1972, when it was decided to publish them.

PREAMBLE

Two basic systems for the citation of references are used at present. The so-called Harvard System (where names of authors and the date are cited in the text, and the reference list is in alphabetical order) and the Numbering System (where numbers, but not necessarily names of authors, are cited in the text, and the reference list is in order of citation in the text). Several ways of quoting references in the list are in current use.

The Commission is of the opinion, arrived at as a result of much consultation between many senior editors, that it is unlikely that all journals would accept a recommendation to use either the Harvard or the Numbering System to the exclusion of the other. It believes, however, that most biochemists will accept the need for, and indeed welcome, a substantial degree of unification of practices, there being no strong case for the individuality of each journal on this issue. Accordingly, the Commission makes the following Recommendations to all biochemical journals; the reasons for some of them are given. The Recommendations deal first with the way in which references should be cited in the list; the proposal is suitable for journals adopting either the Harvard or the Numbering System. Secondly, there are Recommendations about the way in which each of these systems is used. Thirdly, abbreviations for titles of journals and a few other points are considered. Implementation of the Recommendations would mean that any very small differences between journals in their practices would be of the type that can be attended to at the redactory stage of preparation for press. The Commission recognizes that it cannot deal with a number of smaller problems concerning citations that arise from time to time.

RECOMMENDATIONS

1. Citations of References in the List of References Should Be as Follows

Braun, A., Brown, B. & LeBrun, C. (1971) *Journal*, 11, 111–113.

Notes: (a) This form can be used by both systems.

(b) Journals using the Numbering System should arrange the references in numerical order beside the number (which can be italicized or in brackets according to the house custom of the journal).

(c) Journals using the Harvard System should arrange the references in alphabetical order, whatever the language, except in certain situations (see Recommendation 4a below).

(d) This recommendation incorporates the following points:

i. Initials after surnames (full first names are not given in the list).

ii. The use of the symbol "&" is recommended if at all possible because of its widespread usage and the fact that it is independent of the language. No comma before "&."

*From IUB Commission of Editors of Biochemical Journals (CEBJ), *J. Biol. Chem.*, 248(21), 7279–7280 (1973). With permission.

iii. Year in parentheses (this follows immediately after the authors' names because it is essential to the Harvard System).

iv. Journal title (abbreviated). This can be in italics according to house practice (see Recommendation 7 below concerning journal title abbreviations).

v. Volume number. This can be in heavy type or italics according to house practice.

vi. A few journals do not have volume numbers in which case the page numbers should follow immediately after the abbreviated journal title.

If it is necessary to quote both a volume and a part number, the reference should read: Brown, B. (1971) *Journal,* 11, pt 1, 121–123.

vii. First and last pages should be given. The Commission decided to make this Recommendation mainly on the basis of evidence that the additional information provided by quoting the last page was being required increasingly in many types of library and information retrieval services. Citation of the last page (as well as the first) has been requested for some time by the secondary and abstracting journals. Citation of both first and last pages is also an aid in the prevention of errors.

viii. The number of stops and commas is kept as small as possible.

(e) Authors' names and the abbreviated name of the journal when repeated in the next reference should be spelled out in full; ibid. and similar terms should not be used.

(f) Recommendations of the IUPAC-IUB Commission on Biochemical Nomenclature (CBN) and similar documents should be referred to as: Commission on Biochemical Nomenclature (1970) followed by a journal reference.

(g) Junior should be abbreviated to "Jr," not "jun."

2. Numbering System in the Text

The use of authors' names is permissible as authors wish; only the initial letter of the name should be in capital type. Numbers can be inserted in parentheses or as superscripts according to house custom. The printing of references at the foot of the page on which they are first quoted is considered to be helpful with the Numbering System but is not part of the Recommendation because the extra cost is generally considered to be prohibitive.

3.Harvard System in the Text

For multi-author papers, it is recommended that:

a. Not more than two authors to be named either on the first or any subsequent occasion;

b. et al. should be used for three or more authors on every occasion;

c. Each name to have the initial letter in capital type only.

Examples (Harvard System style):

Braun et al. (1969) did some work that was confirmed by LeBrun (1970).

These results (Braun et al., 1969; LeBrun, 1970) have been discussed by Brown & Braun (1971).

The same Recommendation (without the year) applies when authors are quoted in the text in the Numbering System.

4. Harvard System in the List of References

a. A special problem arises in the list when there are several papers by, e.g., Green et al. in the same or over several years. While the list could be in strict alphabetical order of the full reference, the reader will find no clue in the text to the alphabetical status of the names of the second and subsequent authors (see Recommendations 3a and 3b). It is therefore recommended that all the papers by Green et al. (that is by Green and more than one co-author) should be arranged, irrespective of the names of the other authors, in chronological order (over many years if necessary) and designate tham a, b, c, etc.

Examples:

Green, G. (1970) etc.
Green, G. & Brown, B. (1971) etc.
Green, G. & White, W. (1969) etc.
Green, G., White, W. & Black, B. (1968a) etc. — sequence governed by order or date of publication, as far as can be ascertained.
Green, G., Brown, B. & Black, B. (1968b) etc.
Green, G., White, W., Black, B. & Brown, B. (1969) etc.
Green, G., Black, B. & Brown, B. (1970) etc.

b. Names beginning with "Mc" should be listed under "Mc" and not under "Mac," to decide alphabetical order.

c. Names beginning with "De," "Van," or "von," etc. should be arranged under D or V/v, etc.

5. Reference to Books

These should appear in text like any reference to a journal paper. The reference in the list should read: Brown, B. & Braun, A. (1971) in *Book Title* (LeBrun, C., ed.), pp. 1–20, Publisher, Town.

Notes:

a. If a volume number has to be quoted, this would appear before the pp. as, e.g., "vol. 2," with the number in Arabic numerals (even when Roman numerals are printed on the cover of the book).

b. Where an author wishes to refer to a specific page within a book reference, this should be given in the text.

Example (in text): ". . . discussed on p. 21 of Braun et al.(1971)."

6. Other Forms of References

a. *In the press.* It is recommended that (i) this should mean that the paper has been finally accepted by a journal, (ii) it is quoted in the text (both systems) just as any other paper, (iii) the year quoted should be the best estimate revised if necessary at proof stage, and (iv) the full citation in the list to read: Braun, A. & Brown, B. (1971) *Journal,* in the press.

b. *Submitted for publication* should be used in a typescript only when it is reasonable to expect that it will be possible to alter the quotation to a final form at a stage before publication; if such alteration cannot be made then the name of the journal involved should be stated.

c. The use of *in preparation* and *private communication* should not be allowed because they have no real value.

d. *Personal communication* and *unpublished work* should be permitted in the text only, i.e., not in the list of references. Editors may require to see written evidence of the former.

7. Abbreviations for Journal Titles

Most biochemical journals use the *Chemical Abstract** system but a few use the World List, 4th Edition. The Commission noted that the latest information available (International List of Periodical Title Word Abbreviations prepared for the UNISIST/ ICSU-AB Working Group on Bibliographical Descriptions) suggests that the abbreviations that will be recommended finally by ICSU will be very similar to those now used by *Chemical Abstracts.*

Believing that complete uniformity on this issue is highly desirable now and estimating that it may be a few more years before ICSU finally reports, the Commission recommends that all biochemical journals should now use the *Chemical Abstracts* (American Chemical Society) system. The Commission believes that any changes that will be required when ICSU eventually issues recommendations on this point will be comparatively minor ones.

8. Implementation of these Recommendations

The Commission at its meeting in Menton, May 7 to 8, 1971, has taken the view that the degree of uniformity envisaged in the Recommendations is highly desirable and therefore further recommends to all biochemical journals that the changes required should be made as soon as possible. The Commission recognizes that all journals will have to make some changes (in most cases these are minor) from their present established practices to implement these Recommendations in full. It considers that the possible objections of difficulties even for a commercial publisher with an established "house style" are outweighed by the advantage that conformity of style in the citation of references will prove to the authors, editors, and readers upon whom all journals depend for their existence.

*The journal-title abbreviations in *Biological Abstracts* are essentially the same in *Chemical Abstracts.* A *List of Serials with Title Abbreviations* is available from BioSciences Information Service of Biological Abstracts, 2100 Arch Street, Philadelphia, PA 19103.

IUPAC TENTATIVE RULES FOR THE NOMENCLATURE OF ORGANIC CHEMISTRY SECTION E. FUNDAMENTAL STEREOCHEMISTRY*

International Union of Pure and Applied Chemistry

INTRODUCTION

This Section of the IUPAC Rules for Nomenclature of Organic Chemistry differs from previous Sections in that it is here necessary to legislate for words that describe concepts as well as for names of compounds.

At the present time, concepts in stereochemistry (that is, chemistry in three-dimensional space) are in the process of rapid expansion, not merely in organic chemistry, but also in biochemistry, inorganic chemistry, and macromolecular chemistry. The aspects of interest for one area of chemistry often differ from those for another, even in respect to the same phenomenon. This rapid evolution and the variety of interests have led to development of specialized vocabularies and definitions that sometimes differ from one group of specialists to another, sometimes even within one area of chemistry.

The Commission on the Nomenclature of Organic Chemistry does not, however, consider it practical to cover all aspects of stereochemistry in this Section E. Instead, it has two objects in view: To prescribe, for basic concepts, terms that may provide a common language in all areas of stereochemistry; and to define the ways in which these terms may, so far as necessary, be incorporated into the names of individual compounds. The Commission recognizes that specialized nomenclatures are required for local fields; in some cases, such as carbohydrates, amino acids, peptides and proteins, and steroids, international rules already exist; for other fields, study is in progress by specialists in Commissions or Subcommittees; and further problems doubtless await identification. The Commission believes that consultations will be needed in many cases between different groups within IUPAC and IUB if the needs of the specialists are to be met without confusion and contradiction between the various groups.

The Rules in this Section deal only with Fundamental Stereochemistry, that is, the main principles. Many of these Rules do little more than codify existing practice, often of long standing; however, others extend old principles to wider fields, and yet others deal with nomenclature that is still subject to controversy.

Rule E-0

The stereochemistry of a compound is denoted by an affix or affixes to the name that does not prescribe the stereochemistry; such affixes, being additional, do not change the name or the numbering of the compound. Thus, enantiomers, diastereoisomers, and *cis–trans* isomers receive names that are distinguished only by means of different stereochemical affixes. The only exceptions are those trivial names that have stereochemical implications (for example, fumaric acid, cholesterol).

Note: In some cases (see Rules E-2.23 and E-3.1) stereochemical relations may be used to decide between alternative numberings that are otherwise permissible.

E-1. Types of Isomerism

E-1.1. The following nonstereochemical terms are relevant to the stereochemical nomenclature given in the Rules that follow.

*From *IUPAC Inf. Bull. Append. Tentative Nomencl. Sym. Units Stand.*, No. 35, August 1974, pp. 36–80. With permission.

(a) The term structure may be used in connection with any aspect of the organization of matter.

Hence: structural (adjectival)

(b) Compounds that have identical molecular formulas but differ in the nature or sequence of bonding of their atoms or in arrangement of their atoms in space are termed isomers.

Hence: isomeric (adjectival)
isomerism (phenomenological)

Examples:

$H_3C — O — CH_3$ is an isomer of $H_3C — CH_2 — OH$

```
H3C       CH3                      H        CH3
   \     /                          \      /
    C = C        is an isomer of     C = C
   /     \                          /      \
  H       H                      H3C        H
```

(In this and other Rules a broken line denotes a bond projecting behind the plane of the paper, and a thickened line denotes a bond projecting in front of the plane of the paper. In such cases a line of normal thickness denotes a bond lying in the plane of the paper.)

(c) The constitution of a compound of given molecular formula defines the nature and sequence of bonding of the atoms. Isomers differing in constitution are termed constitutional isomers.

Hence: constitutionally isomeric (adjectival)
constitutional isomerism (phenomenological)

Example:

$H_3C–O–CH_3$ is a constitutional isomer of $H_3C–CH_2–OH$.

Note: Use of the term "structural" with the above connotation is abandoned as insufficiently specific.

E-1.2. Isomers are termed stereoisomers when they differ only in the arrangement of their atoms in space.

Hence: stereoisomeric (adjectival)
stereoisomerism (phenomenological)

Examples:

```
H3C       CH3                         H3C       H
   \     /                               \     /
    C = C        is a stereoisomer of     C = C
   /     \                               /     \
  H       H                             H       CH3
```

```
     CHO                                  CHO
  H   |                              HO   |
     C           is a stereoisomer of      C
HO      CH2OH                         H      CH2OH
```

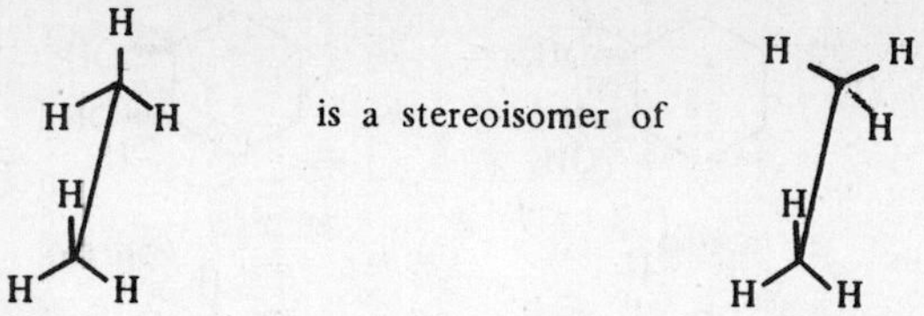

is a stereoisomer of

E-1.3. Stereoisomers are termed *cis–trans* isomers when they differ only in the positions of atoms relative to a specified plane in cases where these atoms are, or are considered as if they were, parts of a rigid structure.

Hence: *cis–trans* isomeric (adjectival)
cis–trans isomerism (phenomenological)

Examples:

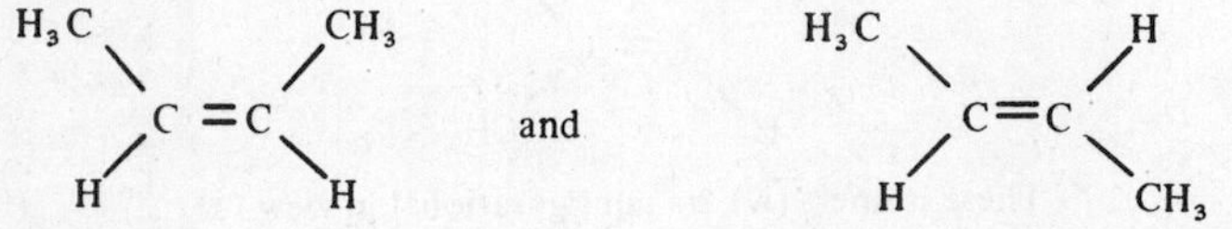

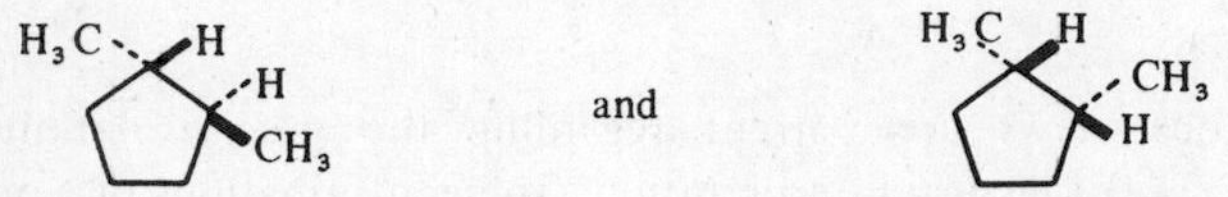

and

and

E-1.4. Various views are current regarding the precise definition of the term "configuration." (a) Classical interpretation: The configuration of a molecule of defined constitution is the arrangement of its atoms in space without regard to arrangements that differ only as after rotation about one or more single bonds. (b) This definition is now usually limited so that no regard is paid also to rotation about π bonds or bonds of partial order between one and two. (c) A third view limits the definition further so that no regard is paid to rotation about bonds of any order, including double bonds.

Molecules differing in configuration are termed configurational isomers.

Hence: configurational isomerism

Notes: (1) Contrast conformation (Rule E-1.5). (2) The phrase "differ only as after rotation" is intended to make the definition independent of any difficulty of rotation, in particular independent of steric hindrance to rotation. (3) For a brief discussion of views (a) to (c), see Appendix 1. It is hoped that a definite consensus of opinion will be established before these Rules are made "Definitive."

Examples: The following pairs of compounds differ in configuration:

(i)

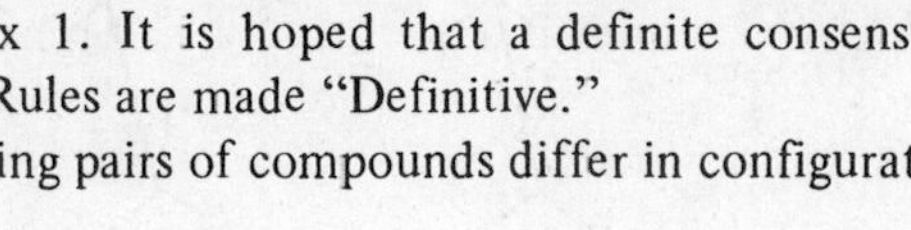

(ii)

(iii)

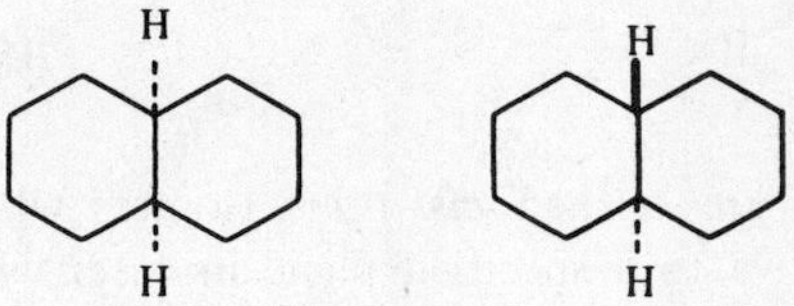

(iv)

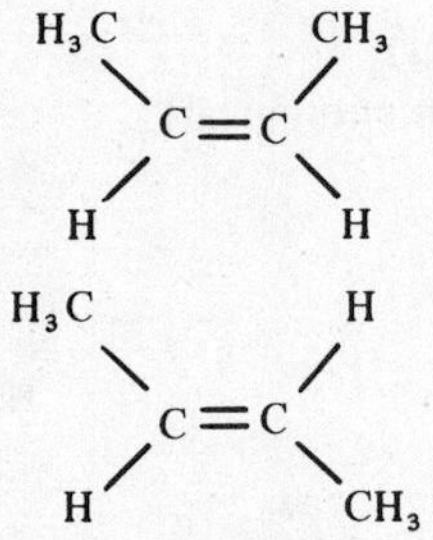

H_3C CH_3 C=C H H

H_3C H C=C H CH_3

These isomers (iv) are configurational in view (a) or (b) but are conformational (see Rule E-1.5) in view (c)

E-1.5. Various views are current regarding the precise definition of the term "conformation." (a) Classical interpretation: The conformations of a molecule of defined configuration are the various arrangements of its atoms in space that differ only as after rotation about single bonds. (b) This is usually now extended to include rotation about π bonds or bonds of partial order between one and two. (c) A third view extends the definition further to include also, rotation about bonds of any order, including double bonds.

Molecules differing in conformation are termed conformational isomers.

Hence: conformational isomerism

Notes: All the Notes to Rule E-1.4 apply also to E-1.5.

Examples: Each of the following pairs of formulas represents a compound in the same configuration but in different conformations.

(a, b, c)

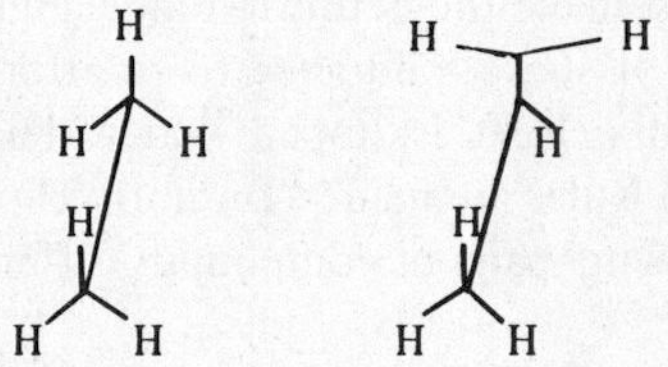

(a, b, c)

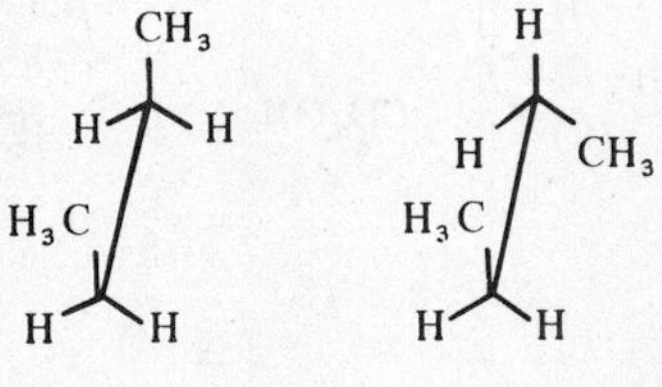

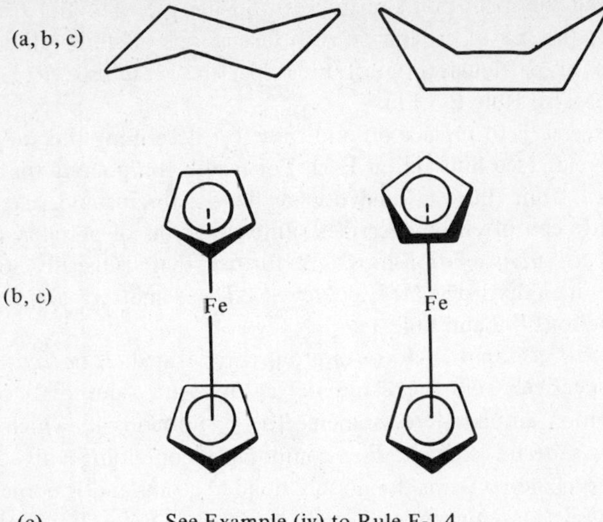

(c) See Example (iv) to Rule E-1.4.

E-1.6. The terms relative stereochemistry and relative configuration are used with reference to the positions of various atoms in a compound relative to one another, especially, but not only, when the actual positions in space (absolute configuration) are unknown.

E-1.7. The terms absolute stereochemistry and absolute configuration are used with reference to the known actual positions of the atoms of a molecule in space.*

E-2. *cis–trans* Isomerism†

Preamble. The prefixes *cis* and *trans* have long been used for describing the relative positions of atoms or groups attached to nonterminal doubly bonded atoms of a chain or attached to a ring that is considered as planar. This practice has been codified for hydrocarbons by IUPAC.** There has, however, not been agreement on how to assign *cis* or *trans* at terminal double bonds of chains or at double bonds joining a chain to a ring. An obvious solution was to use *cis* and *trans* where doubly bonded atoms formed the backbone and were nonterminal and to enlist the sequence-rule preferences to decide other cases; however, since the two methods, when generally applied, do not always produce analogous results, it would then be necessary to use different symbols for the two procedures. A study of this combination showed that both types of symbols would often be required in one name and, moreover, it seemed wrong in principle to use two symbolisms for essentially the same phenomenon. Thus it seemed to the Commission wise to use only the sequence-rule system, since this alone was applicable to all cases. The same decision was taken independently by Chemical Abstracts Service who introduced *Z* and *E* to correspond more conveniently to *seqcis* and *seqtrans* of the sequence rule.

It is recommended in the Rules below that these designations *Z* and *E* based on the sequences rule shall be used in names of compounds, but *Z* and *E* do not always correspond to the classical *cis* and *trans* which show the steric relations of like or similar

*Determination of absolute configuration became possible through work by Bijvoet, J. M., Peerdeman, A. F., and van Bommel, A. J., *Nature,* 168, 271 (1951); cf. Bijvoet, J. M., *Proc. Kon. Ned. Akad. Wetensch.*, 52, 313 (1949).

†These Rules supersede the Tentative Rules for olefinic hydrocarbons published in the Comptes rendus of the 16th IUPAC Conference, New York, N.Y., 1951, pp. 102–103.

**Blackwood, J. E., Gladys, C. L., Loening, K. L., Petrarca, A. E., and Rush, J. E., *J. Amer. Chem. Soc.*, 90, 509 (1968); Blackwood, J. E., Gladys, C. L., Petrarca, A. E., Powell, W. H., and Rush, J. E., *J. Chem. Doc.*, 8, 30 (1968).

groups that are often the main point of interest. So the use of *Z* and *E* in names is not intended to hamper the use of *cis* and *trans* in discussions of steric relations of a generic type or of groups of particular interest in a specified case (see Rule E-2.1 and its Examples and Notes, also Rule E-5.11).

It is also not necessary to replace *cis* and *trans* for describing the stereochemistry of substituted monocycles (see Subsection E-3). For cyclic compounds the main problems are usually different from those around double bonds; for instance, steric relations of substitutents on rings can often be described either in terms of chirality (see Subsection E-5) or in terms of *cis–trans* relationships, and, further, there is usually no single relevant plane of reference in a hydrogenated polycycle. These matters are discussed in the Preambles to Subsections E-3 and E-4.

E-2.1. *Definition of cis–trans.* Atoms or groups are termed *cis* or *trans* to one another when they lie respectively on the same or on opposite sides of a reference plane identifiable as common among stereoisomers. The compounds in which such relations occur are termed *cis–trans* isomers. For compounds containing only doubly bonded atoms, the reference plane contains the doubly bonded atoms and is perpendicular to the plane containing these atoms and those directly attached to them. For cyclic compounds, the reference plane is that in which the ring skeleton lies or to which it approximates. When qualifying another word or a locant, *cis* or *trans* is followed by a hyphen. When added to a structural formula, *cis* may be abbreviated to *c*, and *trans* to *t* (see also Rule E-3.3).

Examples: (Rectangles here denote the reference planes and are considered to lie in the plane of the paper.)

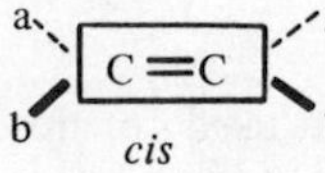

a, a, C=C, b, b — *cis*

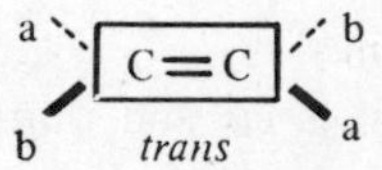

a, b, C=C, b, a — *trans*

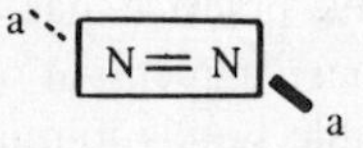

a, N=N, a — *trans*

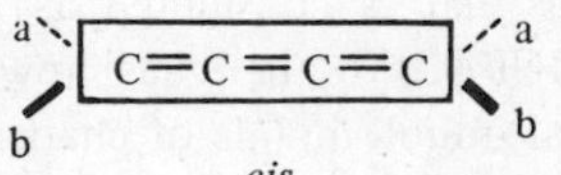

a, a, C=C=C=C, b, b — *cis*

The groups or atoms a,a are the pair selected for designation but are not necessarily identical; b,b are also not necessarily identical but must be different from a,a.

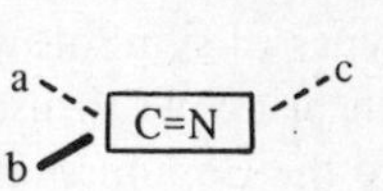

a, b, C=N, c — *cis* or *trans* according as a or b is taken as basis of comparison

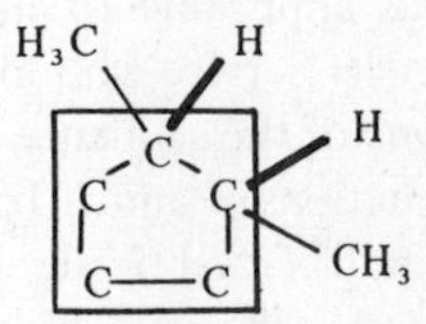

H_3C, H, C, C, C, C, C, H, CH_3 — *cis*

a, b, C=C, CH_2—CH_2, CH_2—CH_2, C=C, a, b — *trans*

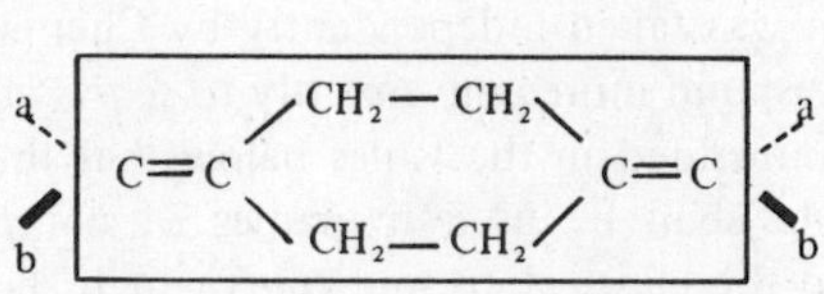

Notes: The formulas above are drawn with the reference plane in the plane of the paper, but for doubly bonded compounds it is customary to draw the formulas so that this plane is perpendicular to that of the paper; atoms attached directly to the doubly bonded atoms then lie in the plane of the paper and the formulas appear as, for instance

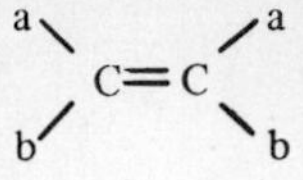

$$\begin{array}{ccc} a & & a \\ & C{=}C & \\ b & & b \end{array}$$

cis

Cyclic structures, however, are customarily drawn with the ring atoms in the plane of the paper, as above. However, care is needed for complex cases, such as

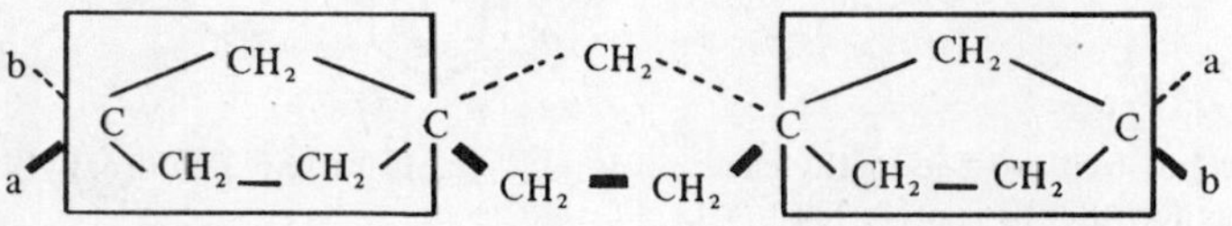

[Structure: two cyclopentane rings (each drawn as C with CH_2, CH_2—CH_2) bearing groups a and b, joined through a central five-membered ring containing CH_2 and CH_2—CH_2]

The central five-membered ring lies (approximately) in a plane perpendicular to the plane of the paper. The two a groups are *trans* to one another; so are the b groups; the outer cyclopentane rings are *cis* to one another with respect to the plane of the central ring. *cis* or *trans* (or *Z* or *E*; see Rule E-2.21) may also be used in cases involving a partial bond order when a limiting structure is of sufficient importance to impose rigidity around the bond of partial order. An example is

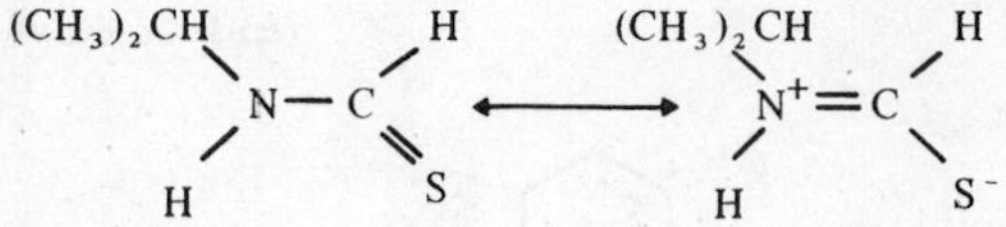

$$(CH_3)_2CH(H)N{-}C(H){=}S \longleftrightarrow (CH_3)_2CH(H)N^{+}{=}C(H)S^{-}$$

trans (or E)

E-2.2. *cis-trans Isomerism around Double Bonds.*

E-2.21. In names of compounds steric relations around one or more double bonds are designated by affixes *Z* and/or *E*, assigned as follows. The sequence-rule-preferred* atom or group attached to one of a doubly bonded pair of atoms is compared with the sequence-rule-preferred atom or group attached to the other of that doubly bonded pair of atoms; if the selected pair are on the same side of the reference plane (see Rule 2.1) an italic capital letter *Z* prefix is used; if the selected pair are on opposite sides an italic capital letter *E* prefix is used.† These prefixes, placed in parentheses and followed by a hyphen, normally precede the whole name; if the molecule contains several double bonds, then each prefix is immediately preceded by the lower or less primed locant of the relevant double bond.

Examples:

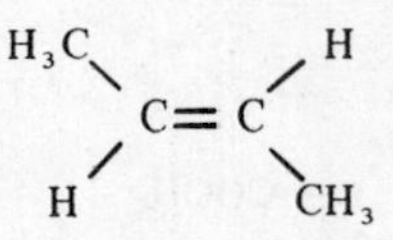

$H_3C(H)C{=}C(H)CH_3$	$H_3C(H)C{=}C(COOH)CH_3$	$H_3C(H)C{=}C(CH_3)COOH$
(*E*)-2-Butene	(*Z*)-2-Methyl-2-butenoic acid** *or* (*Z*)-2-methylisocrotonic acid (see Exceptions below)	(*E*)-2-Methyl-2-butenoic acid†† *or* (*E*)-2-Methylcrotonic acid (see Exceptions below)

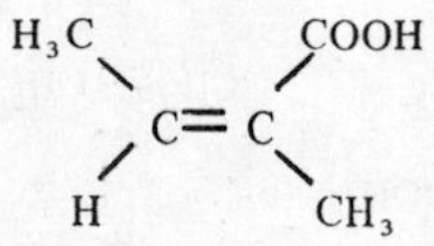

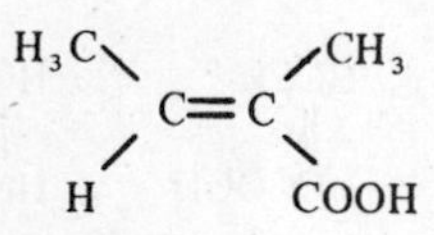

*For sequence-rule preferences see Appendix 2.

†These prefixes may be rationalized as from the German *zusammen* (together) and *entgegen* (opposite).

**The name angelic acid is abandoned because it has been associated with the designation *trans* with reference to the methyl groups.

††The name tiglic acid is abandoned because it has been associated with the designation *cis* with reference to the methyl groups.

(*Z*)-3-Chloroacrylonitrile

(*E*)-2,3-Dichloroacrylonitrile

(*Z*)-1,2-Dibromo-1-chloro-2-iodoethylene
(By the sequence rule, Br is preferred to Cl, but I to Br)

(*E*)-(3-Bromo-3-chloroallyl)benzene

(*E*)-Cyclooctene

(*E*)-1-*sec*-Butylideneindene

(*Z*)-1-Chloro-2-ethylidene-2*H*-indene

(*E*)-1,1′-Biindenylidene

(*E*)-Azobenzene

Exceptions to Rule E-2.21. The following are examples of accepted trivial names in which the stereochemistry is prescribed by the name and is not cited by a prefix.

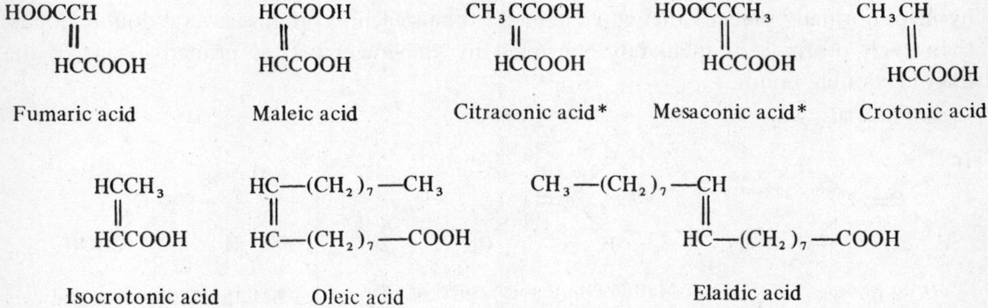

Fumaric acid   Maleic acid   Citraconic acid*   Mesaconic acid*   Crotonic acid

Isocrotonic acid   Oleic acid   Elaidic acid

E-2.22 (*Alternative to Part of E-2.21*). (a) When more than one series of locants starting from unity is required to designate the double bonds in a molecule, or when the name consists of two words, the *Z* and *E* prefixes together with their appropriate locants may be placed before that part of the name where ambiguity is most effectively removed.

(b) [Alternative to (a)] When several *Z* or *E* prefixes are required they are arranged in

*Systematic names are recommended for derivatives of these compounds formed by substitution on carbon.

order as follows: Of the four atoms or groups attached to each doubly bonded pair of atoms, that one preferred by the sequence rule is selected; the single atoms or groups thus selected are then arranged in their sequence rule order (determined in respect of their position in the whole molecule), and the prefixes *Z* and/or *E* for the respective double bonds are placed in that order, but *without* their locants.

Note: In method (a) the final choice is left to an author or editor because of the variety of cases met and because the problems are not always the same in different languages. The presence of the locants usually eases translation from the name to a formula, but this method (a) may involve the logical difficulty explained for the third example below. Method (b) always gives a single unambiguous order and is not subject to the logical difficulty just mentioned, but translation from the name to the formula is harder than for method (a). Method (a) may be more suitable for cursive text, and method (b) for compendia. If method (b) is used it should be used whenever more than one double bond is involved, but method (a) is to be used only under the special conditions detailed in the rule.

Examples:

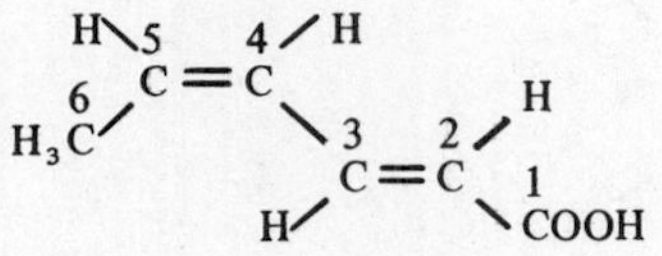

(a) (2*E*,4*Z*)-2,4-Hexadienoic acid
(b) (*E*,*Z*)-2,4-Hexadienoic acid

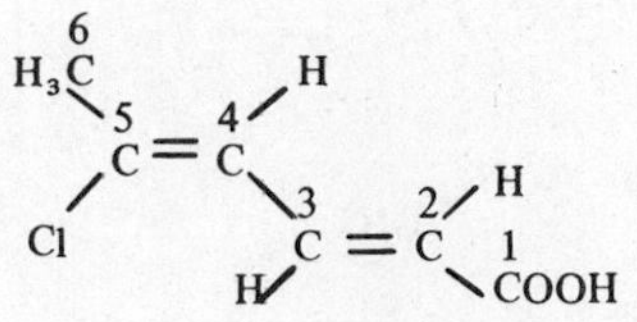

(a) (2*E*,4*Z*)-5-Chloro-2,4-hexadienoic acid
(b) (*Z*,*E*)5-Chloro-2,4-hexadienoic acid

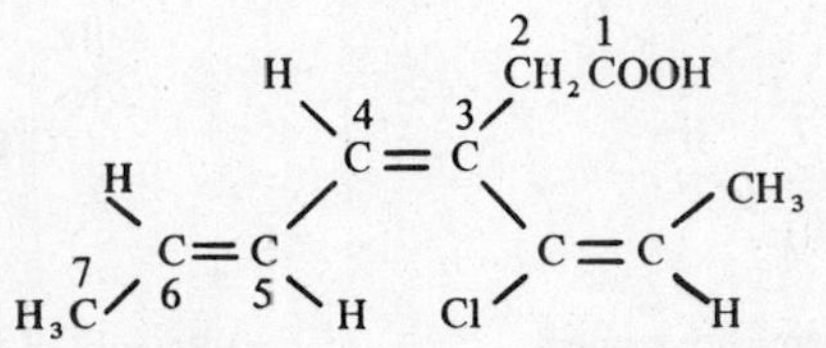

(a) 3-[(*E*)-1-Chloropropenyl]-(3*Z*,5*E*)-3,5-heptadienoic acid
(b) (*E*,*Z*,*E*)-3-(1-Chloropropenyl)-3,5-heptadienoic acid

[The last example shows the disadvantages of both methods. In method (a) there is a fault of logic, namely, the 3*Z*,5*E* are not the property of the unsubstituted heptadienoic acid chain, but the 3*Z* arises only because of the side chain that is cited before the 3*Z*,5*E*. In method (b) it is some trouble to assign the *E*,*Z*,*E* to the correct double bonds.]

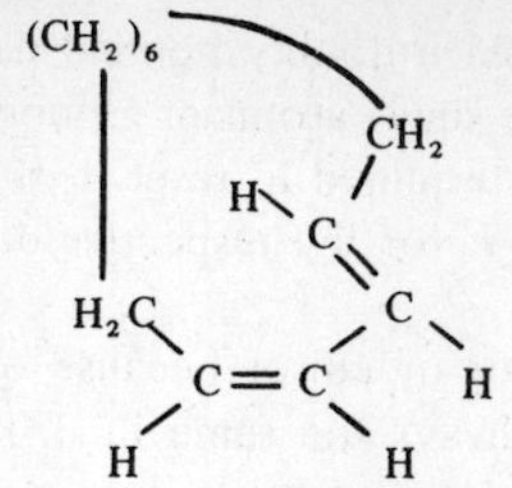

(a) (1*Z*,3*E*)-1,3-Cyclododecadiene
(b) (*Z*,*E*)-1,3-Cyclododecadiene

[The lower locant is assigned to the *Z* double bond.]

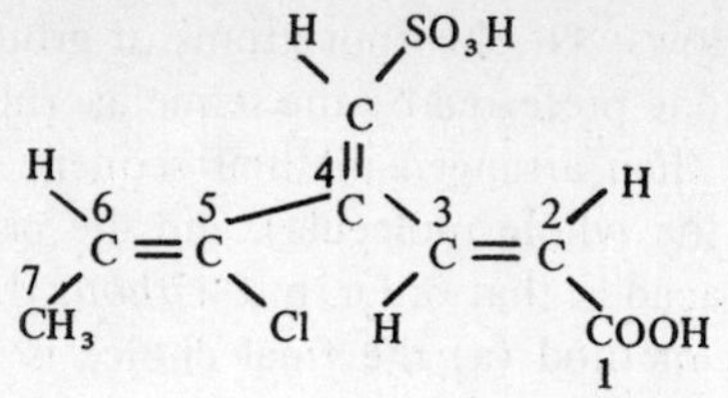

(a) 5-Chloro-4-(*E*-sulfomethylene)-(2*E*,5*Z*)-2,5-heptadienoic acid
(b) (*Z*,*E*,*E*)-5-Chloro-4-(sulfomethylene)-2,5-heptadienoic acid

[In application of the sequence rule, the relation of the SO_3H to CCl (rather than to C-3), and of the CH_3 to Cl, are decisive.]

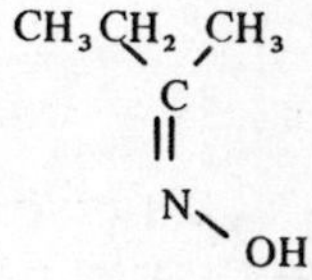

(a) Butanone (*E*)-oxime*
(b) (*E*)-Butanone oxime

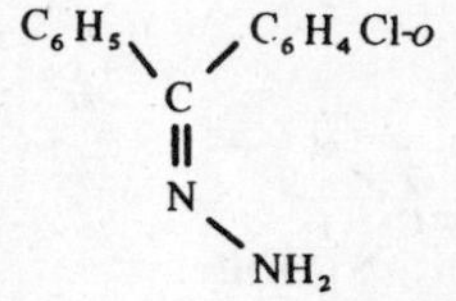

(a) 2-Chlorobenzophenone (*Z*)-hydrazone
(b) (*Z*)-2-Chlorobenzophenone hydrazone

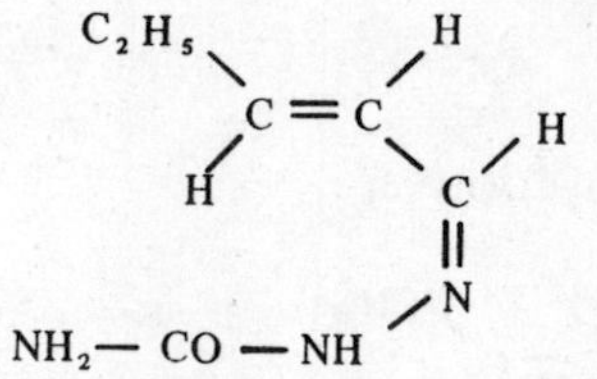

(a) (*E*)-2-Péntenal (*Z*)-semicarbazone
(b) (*Z*,*E*)-2-Pentenal semicarbazone

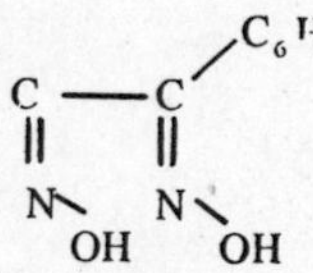

(a) Benzil (*Z*,*E*)-dioxime
(b) (*Z*,*E*)-Benzil dioxime

E-2.23. When Rule C-13.1 or E-2.22(b) permits alternatives, preference for lower locants and for inclusion in the principal chain is allotted as follows, in the order stated, so far as necessary: *Z* over *E* groups; *cis* over *trans* cyclic groups; *R* over *S* groups (also *r* over *s*, etc., as in the sequence rule); if the nature of these groups is not decisive, then the lower locant for such a preferred group at the first point of difference.

Examples:

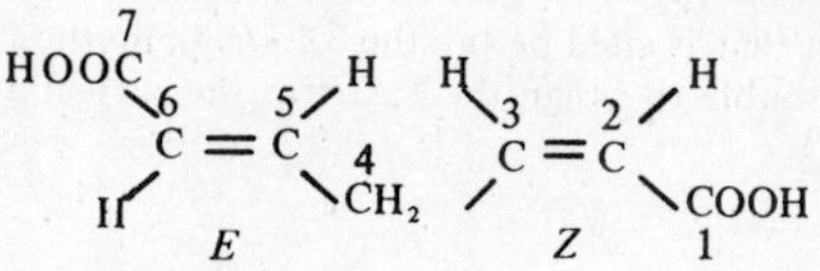

(a) (2*Z*,5*E*)-2,5-Heptadienedioic acid
(b) (*E*,*Z*)-2,5-Heptadienedioic acid

[The lower numbers are assigned to the *Z*double bond.]

*The terms *syn, anti*, and *amphi* are abandoned for such compounds.

```
H\1   2/H
   C=C           5
  /  Z  \3   2  /CH3
Cl        C=C
         /    \H
H\  E   /
  C=C
 /    \
Cl     H
```

(a) 1-Chloro-3-[2-chloro-(*E*)-vinyl]-(1*Z*,3*Z*)-1,3-pentadiene
(b) (*E*,*Z*,*Z*)-1-Chloro-3-(2-chlorovinyl)-1,3-pentadiene

[According to Rule C-13.1 the principal chain must include the C=C–CH_3 group because this gives lower numbers to the double bonds (1,3 rather than 1,4); then the Cl-containing *Z* group is chosen for the remainder of the principal chain in accord with Rule E-2.23.]

```
   6    5     4                 1
(R)-CH2 CH2 CH(CH3)           COOH
                  \3       2/
                    C  =  C
                  /         \
(S)-CH3 CH2 CH(CH3)           H
```

(a,b) (*Z*)-(4*R*)-3-[(*S*)-*sec*-Butyl]-4-methyl-2-hexenoic acid

[The principal chain is chosen to include the (*R*)-group, and the prefix *Z* refers to the (*R*)-group.]

E-3. Relative Stereochemistry of Substituents in Monocyclic Compounds†

Preamble. The prefixes *cis* and *trans* are commonly used to designate the positions of substituents on rings relative to one another; when the ring is, or is considered to be, rigidly planar or approximately so and is placed horizontally, these prefixes define which groups are above and which below the (approximate) plane of the ring. This differentiation is often important, so this classical terminology is retained in Subsection E-3; since the difficulties inherent in end groups do not arise for cyclic compounds, it is unnecessary to resort to the less immediately informative *E*/*Z* symbolism.

When the *cis*–*trans* designation of substituents is applied, rings are considered in their most extended form; reentrant angles are not permitted; for example

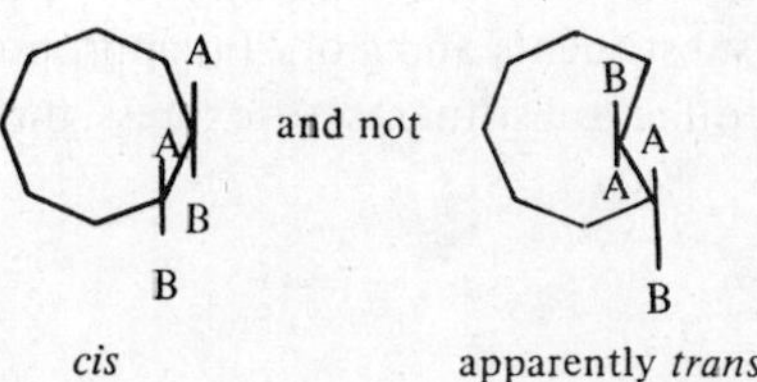

cis apparently *trans*

The absolute stereochemistry of optically active or racemic derivatives of monocyclic compounds is described by the sequence-rule procedure (see Rule E-5.9 and Appendix 2). The relative stereochemistry may be described by a modification of sequence-rule symbolism as set out in Rule E-5.10. If either of these procedures is adopted, it is then superfluous to use also *cis* or *trans* in the names of individual compounds.

†Formulas in Examples to this Rule denote relative (not absolute) configurations.

E-3.1. When alternative numberings of the ring are permissible according to the Rules of Section C, that numbering is chosen which gives a *cis* attachment at the first point of difference; if that is not decisive, the criteria of Rule E-2.23 are applied. The prefixes *cis* and *trans* may be abbreviated to *c* and *t*, respectively, in names of compounds when more than one such designation is required.

Examples:

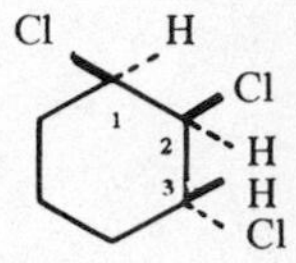

1,*c*-2,*t*-3-Trichlorocyclohexane

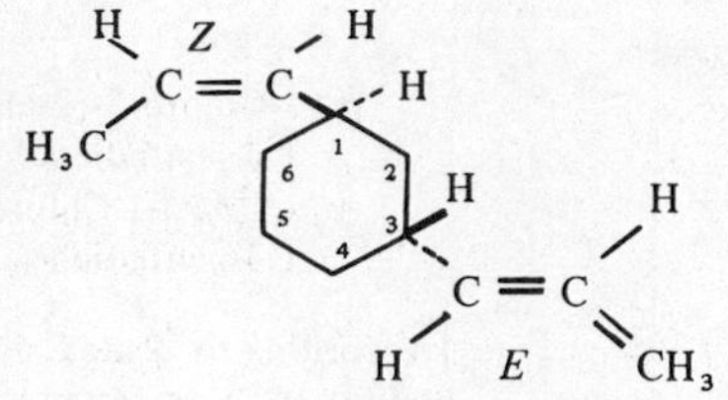

1-(*Z*)-Propenyl-*trans*-3-(*E*)-propenylcyclohexane

E-3.2. When one substituent and one hydrogen atom are attached at each of two positions of a monocycle, the steric relations of the two substituents are expressed as *cis* or *trans*, followed by a hyphen and placed before the name of the compound.

Examples:

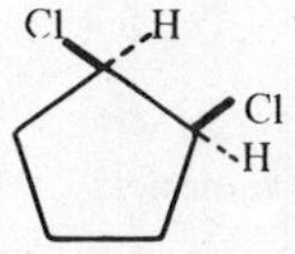

cis-1,2-Dichlorocyclopentane

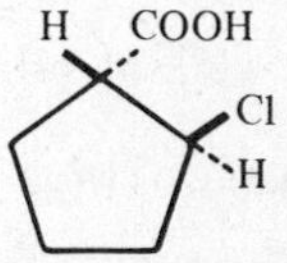

trans-2-Chloro-1-cyclopentanecarboxylic acid

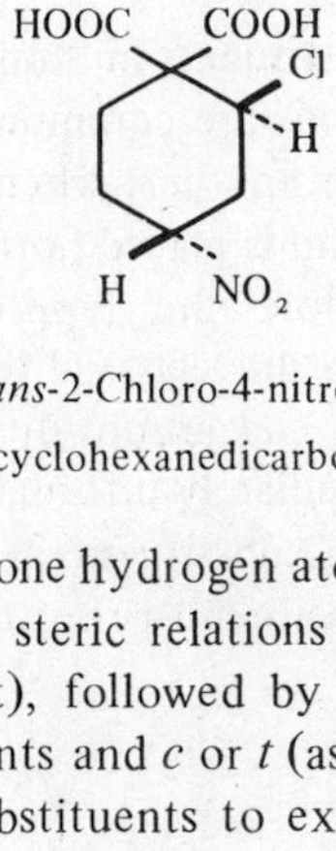

trans-2-Chloro-4-nitro-1,
1-cyclohexanedicarboxylic acid

E-3.3. When one substituent and one hydrogen atom are attached at each of more than two positions of a monocycle, the steric relations of the substituents are expressed by adding *r* (for *reference* substituent), followed by a hyphen, before the locant of the lowest numbered of these substituents and *c* or *t* (as appropriate), followed by a hyphen, before the locants of the other substituents to express their relation to the reference substituent.

Examples:

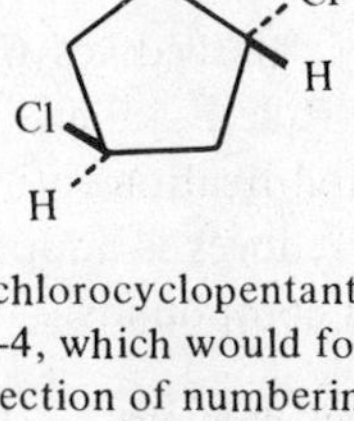

r-1,*t*-2,*c*-4-Trichlorocyclopentant
(not *r*-1, *t*-2, *t*-4, which would follow from the alternative direction of numbering; see Rule E-3.1)

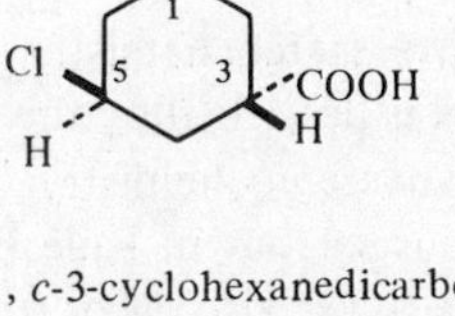

t-5-Chloro-*r*-1, *c*-3-cyclohexanedicarboxylic acid

E-3.4. When two different substituents are attached at the same position of a monocycle, then the lowest numbered substituent named as suffix is selected for designation as reference group in accordance with Rule E-3.2 or E-3.3; or, if none of the substituents is named as suffix, then of the lowest numbered pair that one preferred by the sequence rule is selected as reference group; and the relation of the sequence-rule preferred group at each other position, relative to the reference group, is cited as *c* or *t* (as appropriate).

Examples:

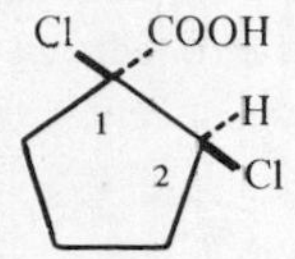

1,*t*-2-Dichloro-*r*-1-cyclopentanecarboxylic acid

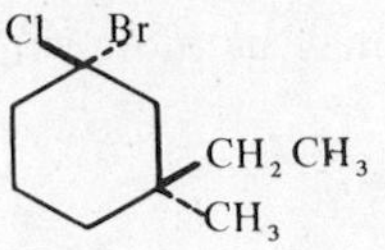

r-1-Bromo-1-chloro-*t*-3-ethyl-3-methylcyclohexane (alphabetical order of prefixes)

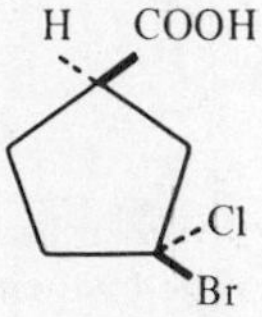

c-3-Bromo-3-chloro-*r*-1-cyclopentanecarboxylic acid

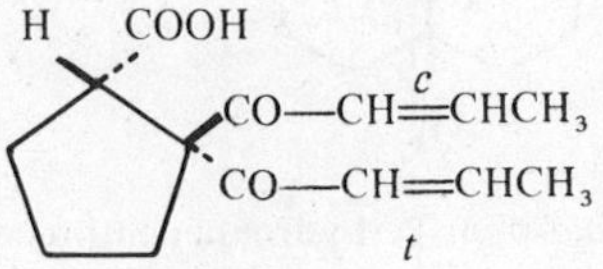

2-Crotonoyl-*t*-2-isocrotonoyl-*r*-1-cyclopentanecarboxylic acid

E-4. Fused Rings

Preamble. In simple cases the relative stereochemistry of substituted fused-ring systems can be designated by the methods used for monocycles. For the absolute stereochemistry of optically active and racemic compounds the sequence-rule procedure can be used in all cases (see Rule E-5.9 and Appendix 2), and for related relative stereochemistry the procedure of Rule E-5.10 can be applied. Sequence-rule methods are, however, not descriptive of geometrical shape for other than quite simple cases. There is as yet no generally acceptable system for designating in an immediately interpretable manner the stereochemistry of polycyclic bridged ring compounds (for instance, the *endo–exo* nomenclature, which should solve one set of problems, has been used in different ways). These and related problems (e.g., cyclophanes, catenanes) will be considered in a later document.

E-4.1. Steric relations at saturated bridgeheads common to two rings are denoted by *cis* or *trans*, followed by a hyphen and placed before the name of the ring system, according to the relative positions of the exocyclic atoms or groups attached to the bridgeheads. Such rings are said to be *cis* fused or *trans* fused.

Examples:

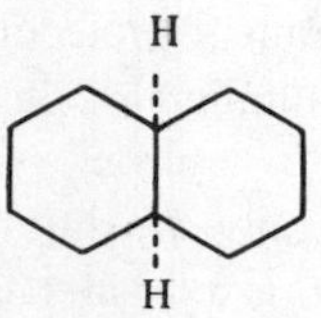

cis-Decalin

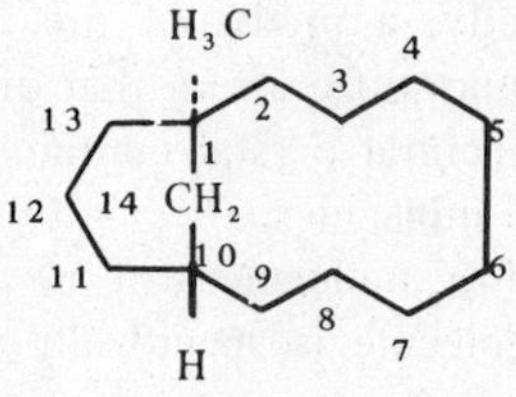

1-Methyl-*trans*-bicyclo[8.3.1]tetradecane

E-4.2. Steric relations at more than one pair of saturated bridgeheads in a polycyclic compound are denoted by *cis* or *trans*, each followed by a hyphen and, when necessary, the corresponding locant of the lower numbered bridgehead and a second hyphen, all placed before the name of the ring system. Steric relations between the nearest atoms* of *cis*- or *trans*-bridgehead pairs may be described by affixes *cisoid* or *transoid*, followed by a hyphen and, when necessary, the corresponding locants and a second hyphen, the whole placed between the designations of the *cis*- or *trans*-ring junctions concerned. When a choice remains among nearest atoms, the pair containing the lower numbered atom is selected; *cis* and *trans* are not abbreviated in such cases. In complex cases, however, designation may be more simply effected by the sequence-rule procedure (see Appendix 2).

Examples:

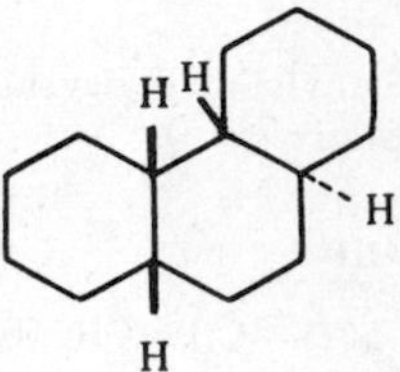

cis-cisoid-trans-Perhydrophenanthrene

cis-cisoid-4a, 10a-*trans*-Perhydroanthracene
or rel-(4a*R*, 8a*S*, 9a*S*, 10a*S*)-Perhydroanthracene†

trans-3a-*cisoid*-3a, 4a-*cis* -4a-Perhydrobenz[*f*] indene
or rel-(3a*R*, 4a*S*, 8a*R*, 9a*R*)-Perhydrobenz[*f*] indene

E-5. Chirality

E-5.1. The property of nonidentity of an object with its mirror image is termed chirality. An object, such as a molecule in a given configuration or conformation, is termed chiral when it is not identical with its mirror image; it is termed achiral when it is identical with its mirror image.

Notes: (1) Chirality is equivalent to handedness, the term being derived from the Greek Χειρ = hand.

(2) All chiral molecules are molecules of optically active compounds, and molecules of all optically active compounds are chiral. There is a 1:1 correspondence between chirality and optical activity.

(3) In organic chemistry the discussion of chirality usually concerns the individual molecule or, more strictly, a model of the individual molecule. The chirality of an assembly of molecules may differ from that of the component molecules, as in a chiral quartz crystal or in an achiral crystal containing equal numbers of dextrorotatory and levorotatory tartaric acid molecules.

(4) The chirality of a molecule can be discussed only if the configuration or conformation of the molecule is specifically defined or is considered as defined by

*The term "nearest atoms" denotes those linked together through the smallest number of atoms, irrespective of actual separation in space. For instance, in the second Example to this Rule, the atom 4a is "nearer" to 10a than to 8a.

†For the designation *rel*, see Rule E-5.10.

common usage. In such discussions structures are treated as if they were (at least temporarily) rigid. For instance, ethane is configurationally achiral although many of its conformations, such as (A), are chiral; in fact, a configuration of a mobile molecule is chiral only if all its possible conformations are chiral; and conformations of ethane such as (B) and (C) are achiral.

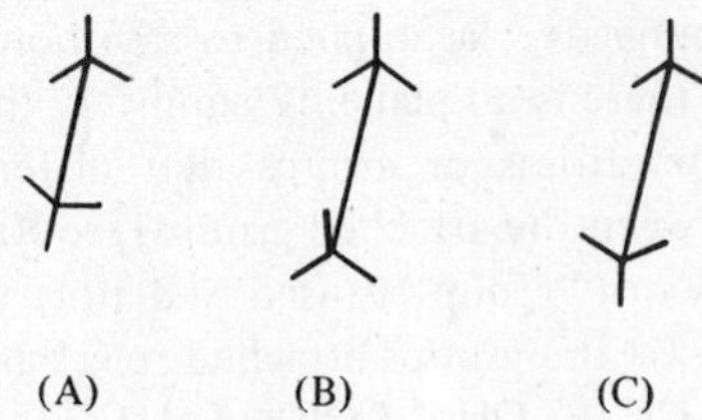

(A) (B) (C)

Examples:

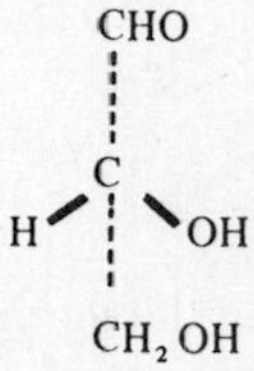

CHO / C / H OH / CH_2OH (D)

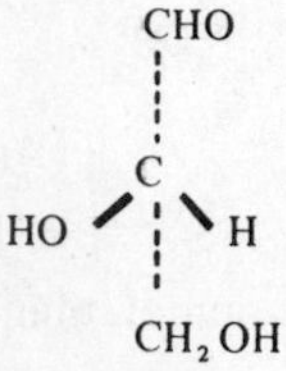

CHO / C / HO H / CH_2OH (E)

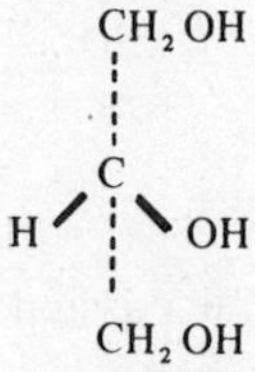

CH_2OH / C / H OH / CH_2OH (F)

(D) and (E) are mirror images and are not identical, not being superposable. They represent chiral molecules. They represent (D) dextrorotatory and (E) levorotatory glyceraldehyde.

(F) is identical with its mirror image. It represents an achiral molecule, namely, a molecule of *1,2,3*-propanetriol (glycerol).

E-5.2. The term asymmetry denotes absence of any symmetry. An object, such as a molecule in a given configuration or conformation, is termed asymmetric if it has no element of symmetry.

Notes: (1) All asymmetric molecules are chiral, and all compounds composed of them are therefore optically active; however, not all chiral molecules are asymmetric since some molecules having axes of rotation are chiral.

(2) Notes (3) and (4) to Rule E-5.1 apply also in discussions of asymmetry.

Examples:

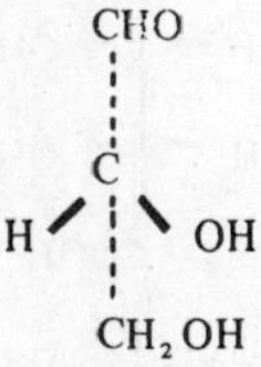

CHO / C / H OH / CH_2OH

has no element of symmetry and represents a molecule of an optically active compound.

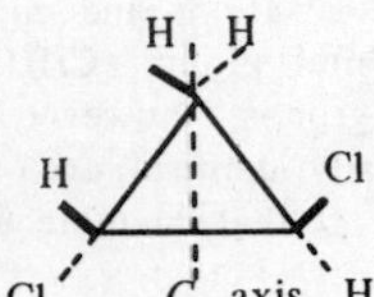

has a C_2 axis of rotation; it is chiral although not asymmetric, and is therefore a molecule of an optically active compound.

E-5.3. (a) An asymmetric atom is one that is tetrahedrally bonded to four different atoms or groups, none of the groups being the mirror image of any of the others.

(b) An asymmetric atom may be said to be at a chiral center since it lies at the center of a chiral tetrahedral structure. In a general sense, the term "chiral center" is not restricted to tetrahedral structures; the structure may, for instance, be based on an octahedron or tetragonal pyramid.

(c) When the atom by which a group is attached to the remainder of a molecule lies at a chiral center, the group may be termed a chiral group.

Notes: (1) The term "asymmetric," as applied to a carbon atom in rule E-5.3 (a), was chosen by van't Hoff because there is no plane of symmetry through a tetrahedron whose corners are occupied by four atoms or groups that differ in scalar properties. For differences of vector sense between the attached groups, see Rule E-5.8.

(2) In Subsection E-5 the word "group" is used to denote the series of atoms attached to one bond. For instance, in (i) the groups attached to C* are $-CH_3$, $-OH$, $-CH_2CH_3$, and $-COOH$; in (ii) they are $-CH_3$, $-OH$, $-COCH_2CH_2CH_2$, and $-CH_2CH_2CH_2CO$.

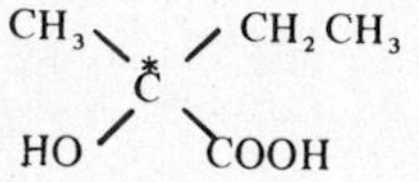

CH_3, CH_2CH_3, C*, HO, COOH

(i)

CH_2—CO, C*, CH_3, OH, CH_2—CH_2

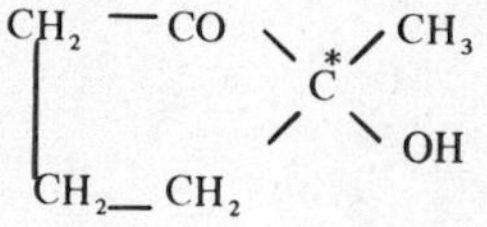

(ii)

(3) For the chiral axis and chiral plane (which are less common than the chiral center), see Appendix 2.

(4) There may be more than one chiral center in a molecule and these centers may be identical, or structurally different, or structurally identical but of opposite chirality; however, the presence of an equal number of structurally identical chiral groups of opposite chirality, and no other chiral group, leads to an achiral molecule. These statements apply also to chiral axes and chiral planes. Identification of the sites and natures of the various factors involved is essential if the overall chirality of a molecule is to be understood.

(5) Although the term "chiral group" is convenient for use in discussions it should be remembered that chirality attaches to molecules and not to groups or atoms. For instance, although the *sec*-butyl group may be termed chiral in dextrorotatory 2-*sec*-butyl-naphthalene, it is not chiral in the achiral compound $(CH_3CH_2)(CH_3)CH–CH_3$.

Examples:

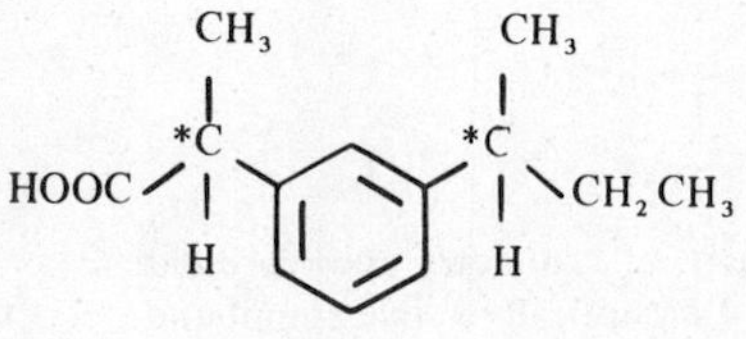

CH_3, *C, HOOC, H, CH_3, *C, CH_2CH_3, H

In this chiral compound there are two asymmetric carbon atoms, marked C*, each lying at a chiral center. These atoms form part of different chiral groups, namely, $-CH(CH_3)$-COOH and $-CH(CH_3)CH_2CH_3$

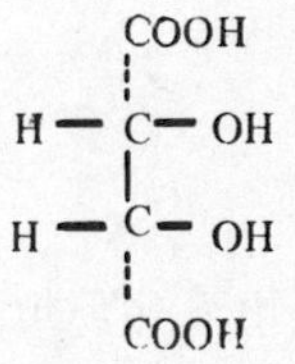

COOH, H—C—OH, H—C—OH, COOH

In this molecule (*meso*-tartaric acid) the two central carbon atoms are asymmetric atoms and each is part of a chiral group –CH(OH)COOH. These groups, however, although structurally identical, are of opposite chirality, so that the molecule is achiral.

E-5.4. Molecules that are mirror images of one another are termed enantiomers and may be said to be enantiomeric. Chiral groups that are mirror images of one another are termed enantiomeric groups.

Hence: enantiomerism (phenomenological)

Note: Although the adjective enantiomeric may be applied to groups, enantiomerism strictly applies only to molecules [see Note (5) to Rule E-5.3].

Examples: The following pairs of molecules are enantiomeric.

(i)

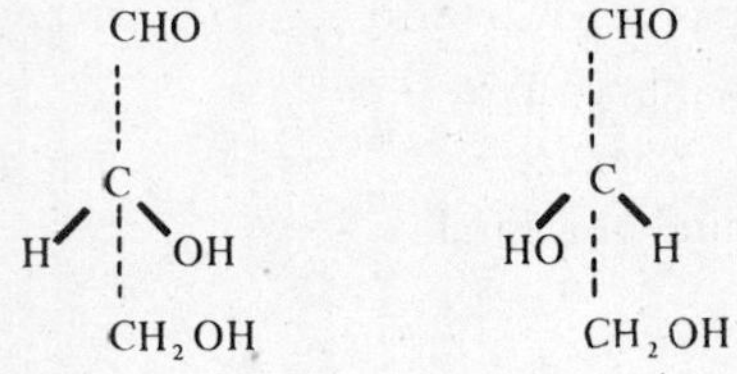

(ii)

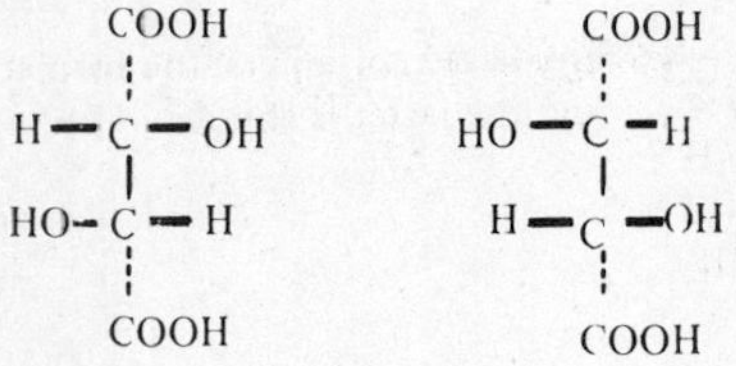

(iii)

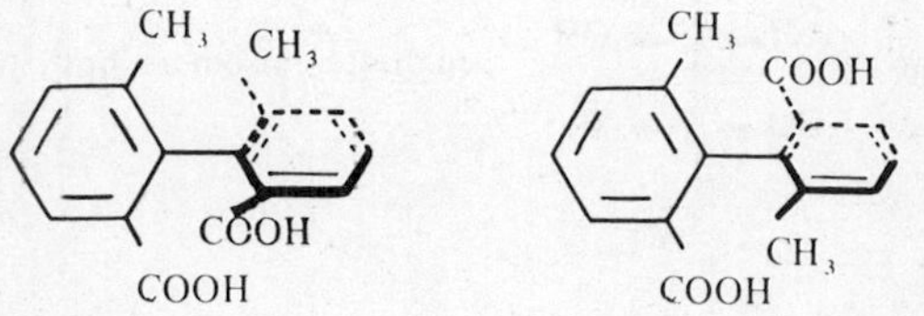

(iv)

Cyclooctene

(v)

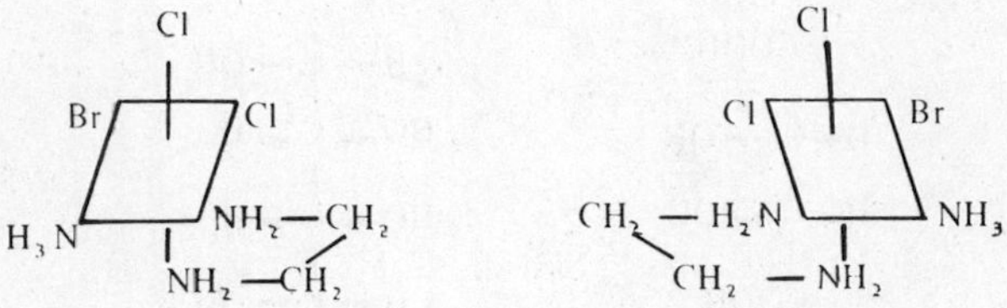

(vi)

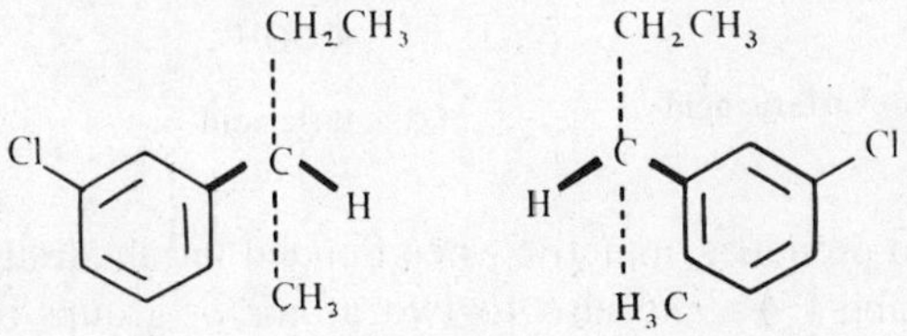

The *sec*-butyl groups in (vi) are enantiomeric.

E-5.5. When equal amounts of enantiomeric molecules are present together, the product is termed racemic, independently of whether it is crystalline, liquid, or gaseous. A homogeneous solid phase composed of equimolar amounts of enantiomeric molecules is termed a racemic compound. A mixture of equimolar amounts of enantiomeric molecules present as separate solid phases is termed a racemic mixture. Any homogeneous solid containing equimolar amounts of enantiomeric molecules is termed a racemate.

Examples: The mixture of two kinds of crystal (mirror-image forms) that separate below 28° from an aqueous solution containing equal amounts of dextrorotatory and levorotatory sodium ammonium tartrate is a racemic mixture.

The symmetrical crystals that separate from such a solution above 28°, each containing equal amounts of the two salts, provide a racemic compound.

E-5.6. Stereoisomers that are not enantiomeric are termed diastereoisomers.

Hence: diastereoisomeric (adjectival)
diastereoisomerism (phenomenological)

Note: Diastereoisomers may be chiral or achiral.
Examples:

```
   COOH               COOH
    |                  |
H — C — OH         H — C — OH
    |        and       |
H — C — OH        HO — C — H
    |                  |
   COOH               COOH
```

are diastereoisomers; the former is achiral, and the latter is chiral.

```
   COOH               COOH
    |                  |
H — C — OH         H — C — OH
    |        and       |
H — C — OH        HO — C — H
    |                  |
   CH3                CH3
```

are diastereoisomers; both are chiral.

E-5.7. A compound whose individual molecules contain equal numbers of enantiomeric groups, identically linked, but no other chiral group, is termed a *meso* compound.
Example:

```
                       COOH
                        |
   COOH             H — C — OH
    |                   |
H — C — OH         HO — C — H
    |                   |
H — C — OH         HO — C — H
    |                   |
   COOH             H — C — OH
                        |
                       COOH
```

meso-Tartaric acid Galactaric acid

E-5.8. An atom is termed pseudoasymmetric when bonded tetrahedrally to one pair of enantiomeric groups (+)-a and (−)-a and also to two atoms or groups b and c that are different from group a, different from each other, and not enantiomeric with each other.
Examples:

```
       H    H    H                    H    H    H
       |    |    |                    |    |    |
HOOC — C — C* — C — COOH       HOOC — C — C* — C — COOH
       |    |    |                    |    |    |
      HO   OH   OH                   HO   CH2  OH
                                           |
            (A)                   CH3 — C — CH2CH3      (B)
                                           |
                                           H
```

C* are pseudoasymmetric

Notes: (1) The orientation, in space, of the atoms around a pseudoasymmetric atoms is not reversed on reflection; for a chiral atom (see Note to Rule E-5.3) this orientation is always reversed.

(2) Molecules containing pseudoasymmetric atoms may be achiral or chiral. If ligands b and c are both achiral, the molecule is achiral as in the first example to this Rule. If either or both of the nonenantiomeric ligands b and c are chiral, the molecule is chiral, as in the second example to this Rule, that is the molecule is not identical with its mirror image. A molecule (i) is also chiral if b and c are enantiomeric, that is, if the molecule can be symbolized as (ii), but then, by definition, it does not contain a pseudoasymmetric atom.

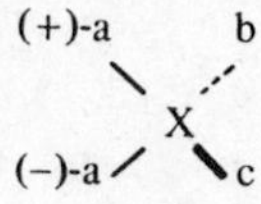

(+)-a b X (−)-a c

(i)

(+)-a b-(+) X (−)-a b-(−)

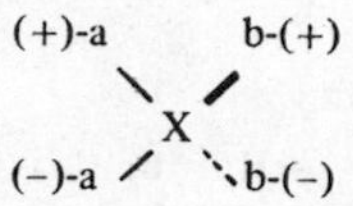

(ii)

(3) Compounds differing at a pseudoasymmetric atom belong to the larger class of diastereoisomers.

(4) In example (A), interchange of H and OH on C* gives a different achiral compound, which is an achiral diastereoisomer of (A) (see Rule E-5.6). In example (B), diastereoisomers are produced by inversion at C* or ◇C, giving in all four diastereoisomers, all chiral because of the $-CH(CH_3)CH_2CH_3$ group.

E-5.9. Names of chiral compounds whose absolute configuration is known are differentiated by prefixes *R*, *S*, etc., assigned by the sequence-rule procedure (see Appendix 2), preceded when necessary by the appropriate locants.

Examples:

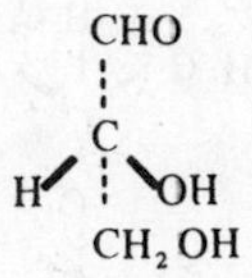

CHO C H OH CH_2OH

(*R*)-Glyceraldehyde

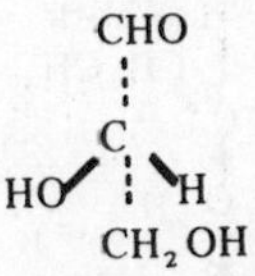

CHO C HO H CH_2OH

(*S*)-Glyceraldehyde

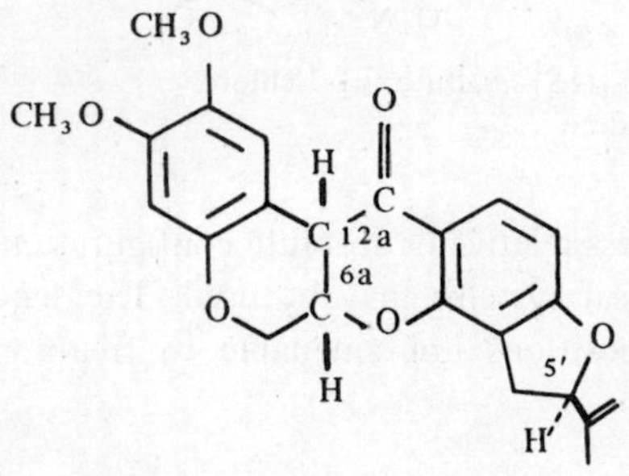

CH_3O CH_3O O H C C 12a 6a O C O H O 5′ H′

(6a*S*,12*S*,5′*R*)-Rotenone

C_6H_5—S—CH_3 O

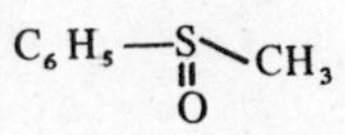

Methyl phenyl (*R*)-sulfoxide

E-5.10. (a) Names of compounds containing chiral centers, of which the relative but not the absolute configuration is known, are differentiated by prefixes *R**, *S** (spoken R star, S star), preceded when necessary by the appropriate locants, these prefixes being assigned by the sequence-rule procedure (see Appendix 2) on the arbitrary assumption that the prefix first cited is *R*.

(b) In complex cases the stars may be omitted and, instead, the whole name is prefixed by *rel* (for *relative*).

(c) When only relative configuration is known, enantiomers are distinguished by a prefix (+) or (–), referring to the direction of rotation of plane-polarized light passing through them (wavelength, temperature, solvent, and/or concentration should also be specified, particularly when known to affect the sign).

(d) When a substituent of known absolute chirality is introduced into a compound of which only the relative configuration is known, then starred symbols R^*, S^* are used and not the prefix *rel.*

Note: This Rule does not form part of the procedure formulated in the sequence-rule papers by Cahn, Ingold, and Prelog (see Appendix 2).

Examples:

(a) or

($1R^*$, $3S^*$,)-1-Bromo-3-chlorocyclohexane

(b) or

rel-($1R$,$3R$,$5R$)-1-Bromo-3-chloro-5-nitrocyclohexane

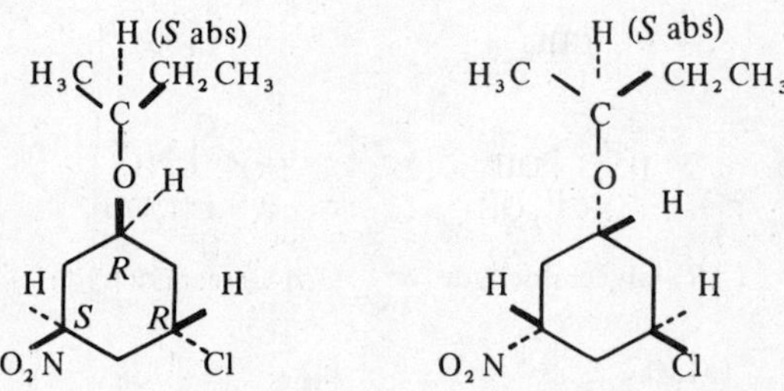

($1R^*$,$3R^*$,$5S^*$)-[($1S$)-*sec*-Butoxy]-3-chloro-5-nitrocyclohexane

E-5.11. When it is desired to express relative or absolute configuration with respect to a class of compounds, specialized local systems may be used. The sequence rule may, however, be used additionally for positions not amenable to treatment by the local system.

Examples:

gluco, arabino, etc., combined when necessary with D or L, for carbohydrates and their derivatives [see IUPAC-IUB Tentative Rules for Carbohydrate Nomenclature; see also *J. Org. Chem.*, 28, 281 (1963)].

D, L for amino acids and peptides [see Comptes rendus of the 16th IUPAC Conference, New York, N.Y., 1951., pp. 107–108; also published in *Chem. Eng. News,* 30, 4522 (1952)].

D, L, and a series of other prefixes and trivial names for cyclitols and their derivatives [see IUPAC-IUB Tentative Rules for the Nomenclature of Cyclitols, 1967, *IUPAC Inf. Bull.*, No. 32, 51 (1968); also published in *J. Biol. Chem.*, 243, 5809 (1968)].

α, β, and a series of trivial names for steroids and related compounds [see IUPAC-IUB Revised Tentative Rules for the Nomenclature of Steroids, 1967, *IUPAC Inf. Bull.*, No. 33, 23 (1968); also published in *J. Org. Chem.*, 34, 1517 (1969)].

The α, β system for steroids can be extended to other classes of compounds such as terpenes and alkaloids when their absolute configurations are known; it can also be combined with stars or the use of the prefix *rel* when only the relative configurations are known.

In spite of the Rules of Subsection E-2, *cis* and *trans* are used when the arrangement of the atoms constituting an unsaturated backbone is the most important factor, as, for instance, in polymer chemistry and for carotenoids. When a series of double bonds of the same stereochemistry occurs in a backbone, the prefix all-*cis* or all-*trans* may be used.

E-5.12. (a) An achiral object having at least one pair of features that can be distinguished only by reference to a chiral object or to a chiral reference frame is said to be prochiral, and the property of having such a pair of features is termed prochirality. A consequence is that, if one of the paired features of a prochiral object is considered to differ from the other, the resultant object is chiral.

(b) In a molecule an achiral center or atom is said to be prochiral if it would be held to be chiral when two attached atoms or groups, that taken in isolation are indistinguishable, are considered to differ.

Notes: (1) For a tetrahedrally bonded atom this requires a structure Xaabc (where none of the groups a, b, or c is the enantiomer of another).

(2) For a fuller exploration of this concept, which is of particular importance to biochemists and spectroscopists, and for its extension to axes, planes, and unsaturated compounds, see Hanson, K. R., *J. Am. Chem. Soc.*, 88, 2731 (1966).

Examples:

```
      CH3              CHO
       :                :
                    H—C—OH
  H—C—H                 |
       :            H—C—H
      OH                :
                       OH

      (A)              (B)
```

In both examples (A) and (B), the methylene carbon atom is prochiral; in both cases it would be held to be at a chiral center if one of the methylene hydrogen atoms were considered to differ from the other. An actual replacement of one of these protium atoms by, say, deuterium would produce an actual chiral center at the methylene carbon atom; as a result, compound (A) would become chiral, and compound (B) would be converted into one of two diastereoisomers.

E-5.13. Of the identical pair of atoms or groups in a prochiral compound, that one which leads to an (*R*) compound when considered to be preferred to the other by the sequence rule (without change in priority with respect to other ligands) is termed *pro-R*, and the other is termed *pro-S*.

Example:

CHO
|
H^1 --- C — OH
/
H^2

H^1 is *pro-R*.
H^2 is *pro-S*.

E-6. Conformations

E-6.1. A molecule in a conformation into which its atoms return spontaneously after small displacements is termed a conformer.

Examples:

CH_3 / H H / CH_3 H / H CH_3 / H H / H H / CH_3 are different conformers.

E-6.2. (a) When, in a six-membered saturated ring compound, atoms in relative positions 1, 2, 4, and 5 lie in one plane, the molecule is described as in the chair or boat conformation according as the other two atoms lie, respectively, on opposite sides or on the same side of that plane.

Examples:

Chair Boat

Note: These and similar representations are idealized, minor divergences being neglected.

(b) A molecule of a monounsaturated six-membered ring compound is described as being in the half-chair or half-boat conformation according as the atoms not directly bound to the doubly bonded atoms lie, respectively, on opposite sides or on the same side of the plane containing the other four (adjacent) atoms.

Examples:

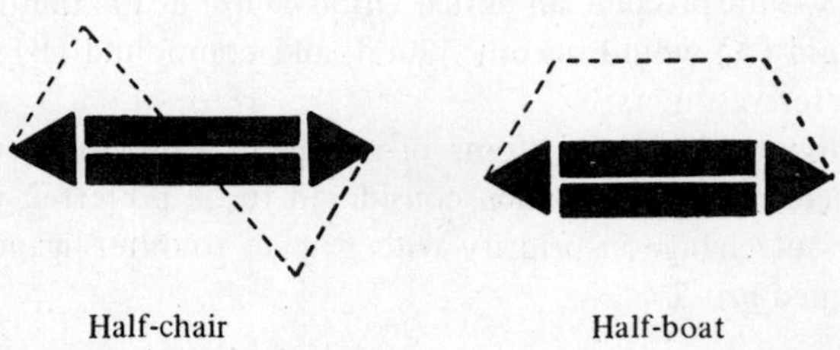

Half-chair Half-boat

(c) A median conformation through which one boat form passes during conversion into the other boat form is termed a twist conformation. Similar twist conformations are involved in conversion of a chair into a boat form or vice versa.

Examples:

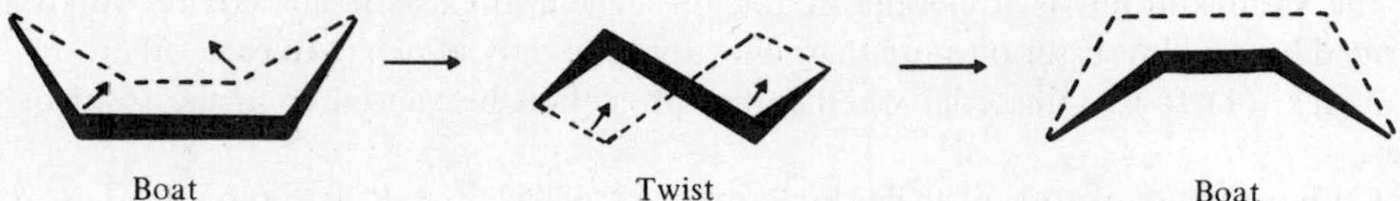

E-6.3. (a) Bonds to a tetrahedral atom in a six-membered ring are termed equatorial or axial according as they or their projections make a small or a large angle, respectively, with the plane containing a majority of the ring atoms.* Atoms or groups attached to such bonds are also said to be equatorial or axial, respectively.

Notes: (1) See, however, pseudoequatorial and pseudoaxial [Rule E-6.3(b)]. (2) The terms equatorial and axial may be abbreviated to e and a when attached to formulas; these abbreviations may also be used in names of compounds and are there placed in parentheses after the appropriate locants, for example, 1(e)-bromo-4(a)-chlorocyclohexane.

Examples:

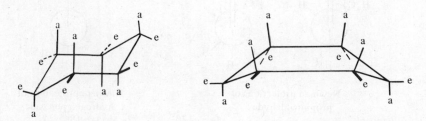

(b) Bonds from atoms directly attached to the doubly bonded atoms in a monounsaturated six-membered ring are termed pseudoequatorial or pseudoaxial according as the angles that they make with the plane containing the majority of the ring atoms approximate those made by, respectively, equatorial or axial bonds from a saturated six-membered ring. Pseudoequatorial and pseudoaxial may be abbreviated to e′ and a′, respectively, when attached to formulas; these abbreviations may also be used in names, then being placed in parentheses after the appropriate locants.

Example:

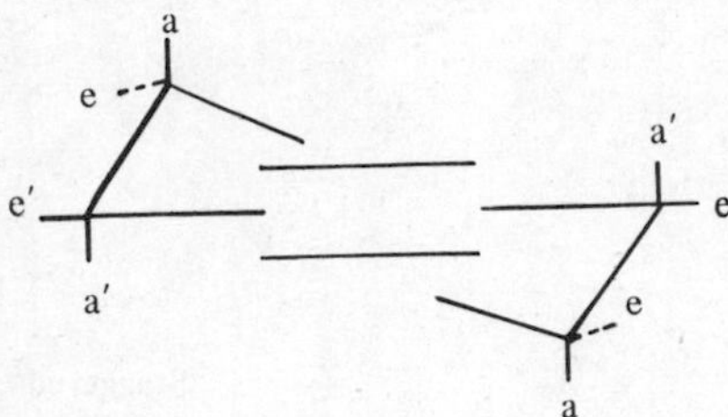

E-6.4. Torsion angle: In an assembly of attached atoms X–A–B–Y, where neither X nor Y is collinear with A and B, the smaller angle subtended by the bonds X–A and Y–B in a plane projection obtained by viewing the assembly along the axis A–B is termed the

*The terms axial, equatorial, pseudoaxial, and pseudoequatorial [see Rule E-6.3(b)] may be used also in connection with other than six-membered rings if, but only if, their interpretation is then still beyond dispute.

torsion angle (denoted by the Greek lower case letter theta θ or omega ω). The torsion angle is considered positive or negative according as the bond to the front atom X or Y requires rotation to the right or left, respectively, in order that its direction may coincide with that of the bond to the rear selected atom Y or X. The multiplicity of the bonding of the various atoms is irrelevant. A torsion angle also exists if the axis for rotation is formed by a collinear set of more than two atoms directly attached to each other.

Notes: (1) It is immaterial whether the projection be viewed from the front or the rear.

(2) For the use of torsion angles in describing molecules see Rule E-6.6.

Examples: (For construction of Newman projections, as here, see Rule E-7.2.)

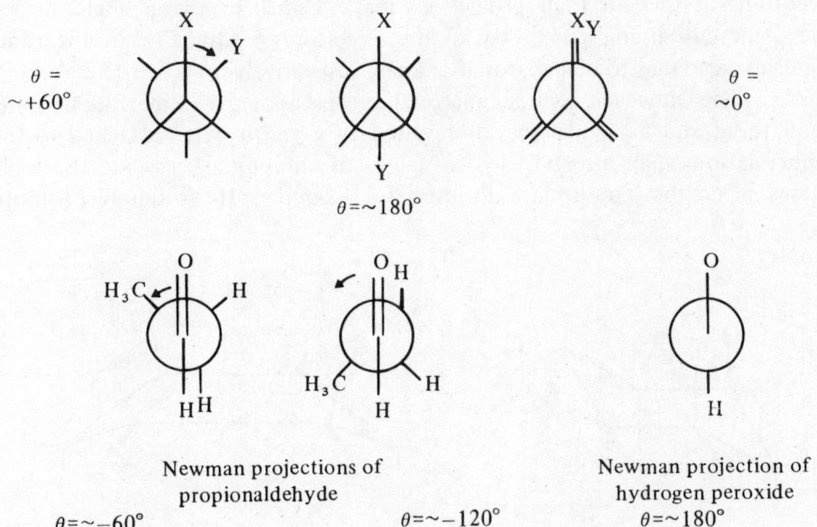

Newman projections of propionaldehyde

Newman projection of hydrogen peroxide θ=~180°

E-6.5. If two atoms or groups attached at opposite ends of a bond appear one directly behind the other when the molecule is viewed along this bond, these atoms or groups are described as eclipsed, and that portion of the molecule is described as being in the eclipsed conformation. If not eclipsed, the atoms or groups and the conformation may be described as staggered.

Examples:

Eclipsed conformation.
The pairs a/a′, b/b′, and c/c′ are eclipsed.

Staggered conformation.
All the attached groups are staggered.

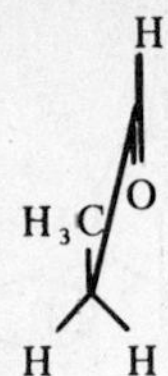

Projection of CH_3CH_2CHO.
The CH_3 and the H of the CHO are eclipsed.
The O and H's of CH_2 in CH_2CH_3 are staggered.

E-6.6. Conformations are described as synperiplanar (*sp*), synclinal (*sc*), anticlinal (*ac*), or antiperiplanar (*ap*) according as the torsion angle is within ±30° of 0°, ±60°, ±120°, or ±180°, respectively; the letters in parentheses are the corresponding abbreviations. Atoms or groups are selected from each set to define the torsion angle according to the following criteria: (1) if all the atoms or groups of a set are different, that one of each set that is preferred by the sequence rule; (2) if one of a set is unique, that one; or (3) if all of a set are identical, that one which provides the smallest torsion angle.
Examples:

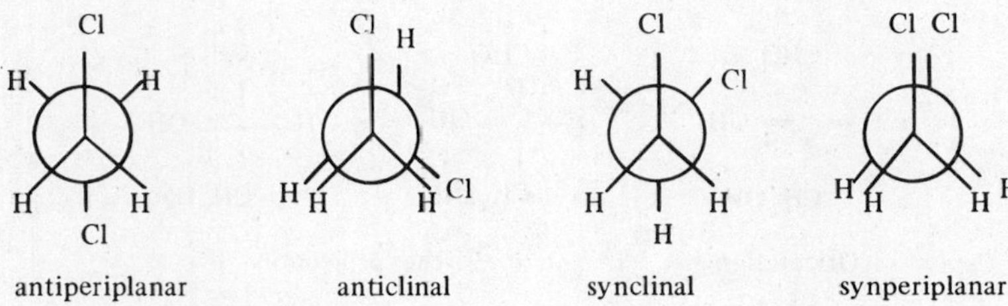

In the above conformations, all $CH_2Cl–CH_2Cl$, the two Cl atoms decide the torsion angle.

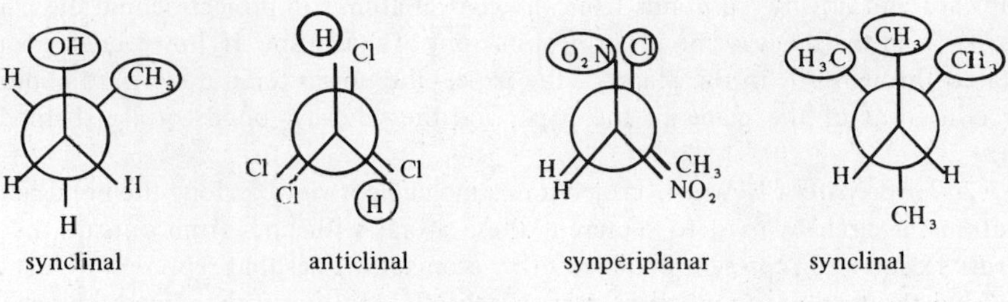

Criterion for:				
rear atom	2	2	1	3
front atom	2	2	1	2

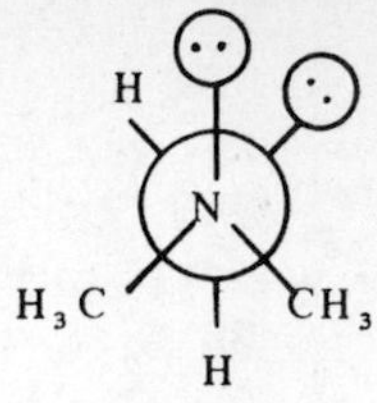

H, H, H_3C, CH_3, N, H, Cl, H, H, CH_3, O, H_3C, CH, O, H_2N, H

$(CH_3)_2N–NH_2$ synclinal*	$CH_3CH_2–COCl$ anticlinal	$(CH_3)_2CH–CONH_2$ antiperiplanar

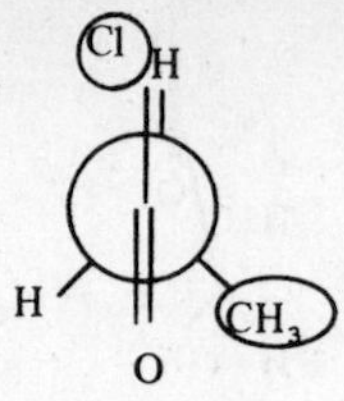

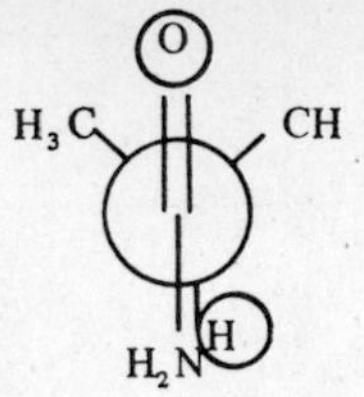

Criterion for:			
rear atom	2	2	2
front atom	2	1	1

E-7. Stereoformulas

E-7.1. In a Fischer projection the atoms or groups attached to a tetrahedral center are projected on to the plane of the paper from such an orientation that atoms or groups appearing above or below the central atom lie behind the plane of the paper and those appearing to left and right of the central atom lie in front of the plane of the paper, and that the principal chain appears vertical with the lowest numbered chain member at the top.

Examples:

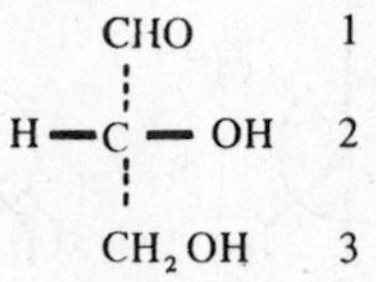

```
      CHO       1        CHO                 CHO
       ⋮                  |                   |
  H —C— OH      2    H—C—OH     or   H————————OH
       ⋮                  |                   |
     CH2OH      3       CH2OH               CH2OH

   Orientation              Fischer projection
```

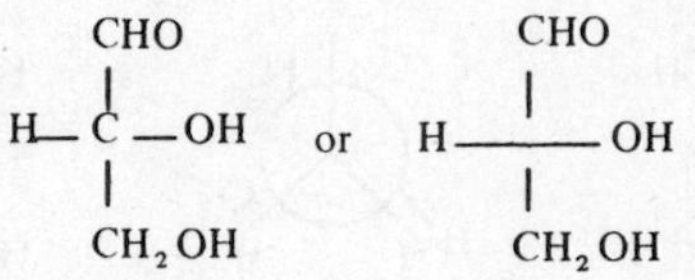

Notes: (1) The first of the two types of Fischer projection should be used whenever convenient.

(2) If a Fischer projection formula is rotated through 180° in the plane of the paper, the upward and downward bonds from the central atom still project behind the plane of the paper, and the sideways bonds project in front of that plane. If, however, the formula is rotated through 90° in the plane of the paper, the upward and downward bonds now project in front of the plane of the paper and the sideways bonds project behind that plane.

E-7.2. To prepare a Newman projection, a molecule is viewed along the bond between two atoms; a circle is used to represent these atoms with lines from outside the circle toward its center to represent bonds to other atoms; the lines that represent bonds to the nearer and the further atom end at, respectively, the center and the circumference of the circle. When two such bonds would be coincident in the projection, they are drawn at a small angle to each other.†

*The lone pair of electrons (represented by two dots) on the nitrogen atoms are the unique substituents that decide the description of the conformation (these are the "phantom atoms" of the sequence-rule symbolism).

†Cf. Newman, M. S., *Rec. Chem. Progr.*, 13, 111 (1952); *J. Chem. Educ.*, 33, 344 (1955); *Steric Effects in Organic Chemistry*, John Wiley & Sons, New York, 1956, 5.

Examples:

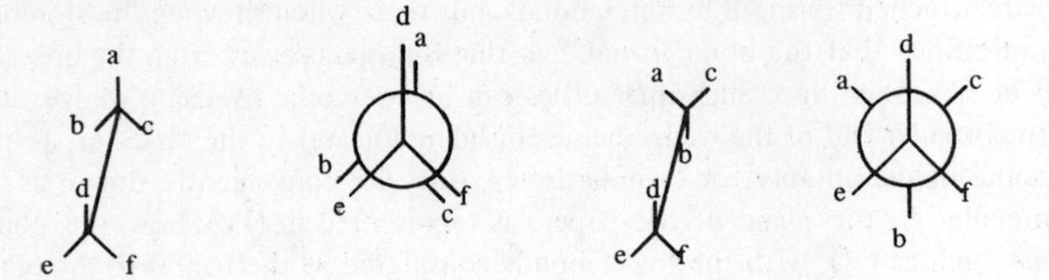

Perspective Newman projection Perspective Newman projection

E-7.3. *General Note*; Formulas that display stereochemistry should be prepared with extra care so as to be unambiguous and, whenever possible, self-explanatory. It is inadvisable to try to lay down rules that will cover every case, but the following points should be borne in mind.

A thickened line (▬) denotes a bond projecting from the plane of the paper toward an observer, a broken line (- - -) denotes a bond projecting away from an observer, and, when this convention is used, a full line of normal thickness (—) denotes a bond lying in the plane of the paper. A wavy line (〰) may be used to denote a bond whose direction cannot be specified or, if it is explained in the text, a bond whose direction it is not desired to specify in the formula. Dotted lines (· · · · · ·) should preferably not be used to denote stereochemistry, and never when they are used in the same paper to denote mesomerism, intermediate states, etc. Wedges should not be used as complement to broken lines (but see below). Single large dots have sometimes been used to denote atoms or groups attached at bridgehead positions and lying above the plane of the paper, with open circles to denote them lying below the plane of the paper, but this practice is strongly deprecated.

Hydrogen or other atoms or groups attached at sterically designated positions should never be omitted.

In chemical formulas, rings are usually drawn with lines of normal thickness, that is, as if they lay wholly in the plane of the paper even though this may be known not to be the case. In a formula such as (I) it is then clear that the H atoms attached at the A/B ring junction lie further from the observer than these bridgehead atoms, that the H atoms attached at the B/C ring junction lie nearer to the observer than those bridgehead atoms, and that X lies nearer to the observer than the neighboring atom of ring C.

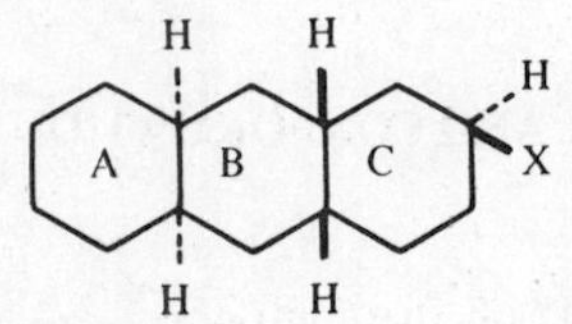

(I) 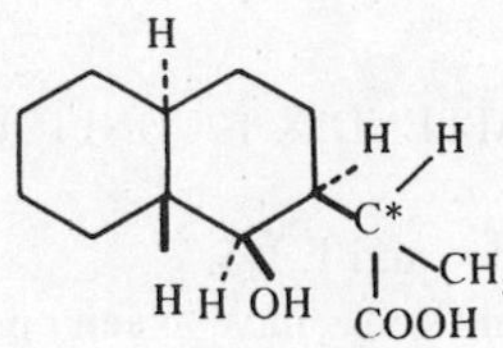 (II)

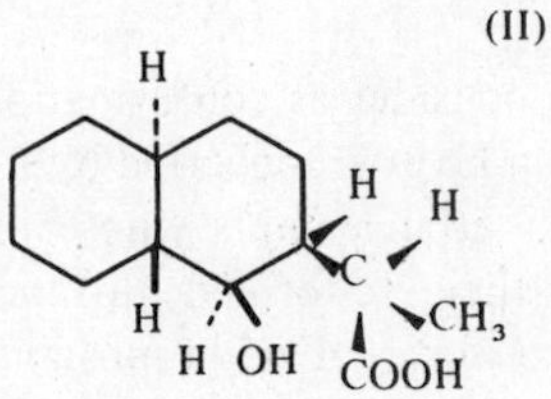

(III)

However, ambiguity can then sometimes arise, particularly when it is necessary to show stereochemistry within a group such as X attached to the rings that are drawn planar. For instance, in formula (II), the atoms O and C*, lying above the plane of the paper, are attached to ring B by thick bonds, but then, when showing the stereochemistry at C*, one finds that the bond *from* C* *to* ring B projects away from the observer and so should be a broken line. Such difficulties can be overcome by using wedges in place of lines, the broader end of the wedge being considered nearer to the observer, as in (III).

In some fields, notably for carbohydrates, rings are conveniently drawn as if they lay perpendicular to the plane of the paper, as represented in (IV); however, conventional formulas such as (V), with the lower bonds considered as the nearer to the observer, are so well established that is is rarely necessary to elaborate this to form (IV).

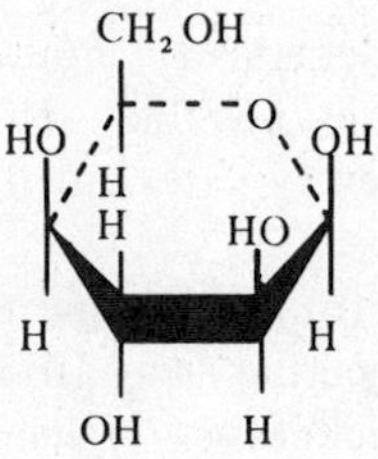

(IV) 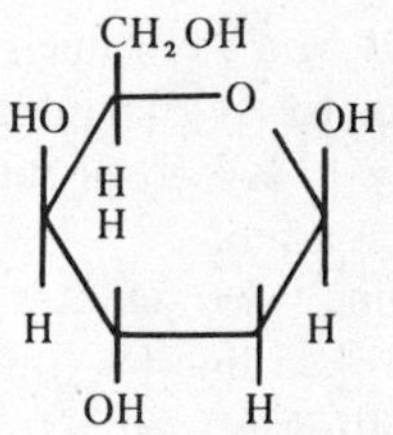 (V)

By a similar convention, in drawings such as (VI) and (VII), the lower sets of bonds are considered to be nearer than the upper to the observer. In (VII), note the gaps in the rear lines to indicate that the bonds crossing them pass in front (and thus obscure sections of the rear bonds). In some cases, when atoms have to be shown as lying in several planes, the various conventions may be combined, as in (VIII). In all cases the overriding aim should be clarity.

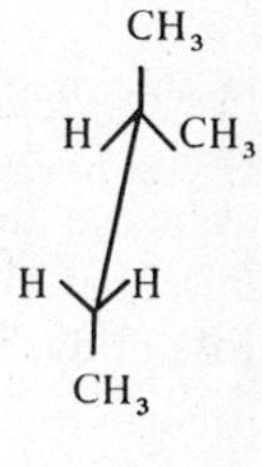

(VI) 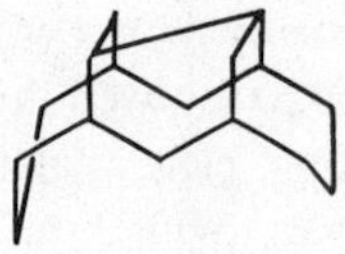 (VII) 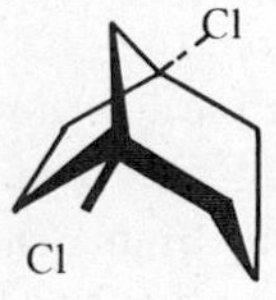 (VIII)

APPENDIX 1. CONFIGURATION AND CONFORMATION

See Rules E-1.4 and E-1.5.

Various definitions have been propounded to differentiate configurations from conformations.

The original usage was to consider as conformations those arrangements of the atoms of a molecule in space that can be interconverted by rotation(s) around a single bond, and as configurations those other arrangements whose interconversion by rotation requires bonds to be broken and then re-formed differently. Interconversion of different configurations will then be associated with substantial energies of activation, and the various species will be separable, but interconversion of different conformations will normally be associated with less activation energy, and the various species, if separable, will normally be more readily interconvertible. These differences in activation energy and stability are often large.

Nevertheless, rigid differentiation on such grounds meets formidable difficulties. Differentiation by energy criteria would require an arbitrary cut in a continuous series of values. Differentiation by stability of isolated species requires arbitrary assumptions about conditions and halflives. Differentiation on the basis of rotation around single bonds meets difficulties connected both with the concept of rotation and with the selection of single bonds as requisites, and these need more detailed discussion here.

Enantiomeric biaryls are nowadays usually considered to differ in conformation, any difficulty in rotation about the 1,1′ bond due to steric hindrance between the neighboring groups being considered to be overcome by bond bending and/or bond stretching, even though the movements required must closely approach bond breaking if these substituents are very large. Similar doubts about the possibility of rotation occur with a molecule such as (A), where rotation of the benzene ring around the oxygen-to-ring single bonds affords easy interconversion if x is large but appears to be physically impossible if x is small; and no critical size of x can be reasonably established. For reasons such as this, Rules E-1.4 and E-1.5 are so worded as to be independent of whether rotation appears physically feasible or not (see Note 2 to those Rules).

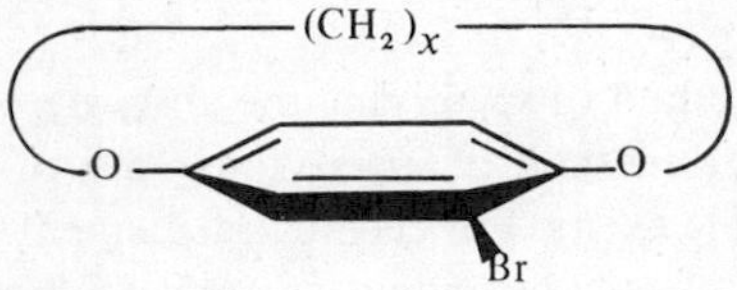

(A) 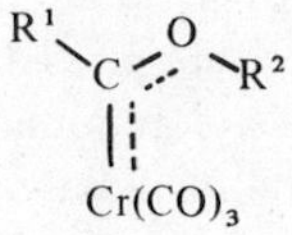 (B)

The second difficulty arises in the many cases where rotation is around a bond of fractional order between one and two, as in the helicenes, crowded aromatic molecules, metallocenes, amides, thioamides, and carbene-metal coordination compounds (such as B). The term conformation is customarily used in these cases and that appears a reasonable extension of the original conception, though it will be wise to specify the usage if the reader might be in doubt.

When interpreted in these ways, Rules E-1.4 and E-1.5 reflect the most frequent usage of the present day and provide clear distinctions in most situations. Nevertheless, difficulties remain and a number of other usages have been introduced.

It appears to some workers that once it is admitted that change of conformation may involve rotation about bonds of fractional order between one and two, it is then illogical to exclude rotation about classical double bonds because interconversion of open-chain *cis-trans* isomers depends on no fundamentally new principle and is often relatively easy, as for certain alkene derivatives such as stilbenes and for *azo* compounds, by irradiation. This extension is indeed not excluded by Rules E-1.4 and E-1.5, but if it is applied that fact should be explicitly stated.

A further interpretation is to regard a stereoisomer possessing some degree of stability (that is, one associated with an energy hollow, however shallow) as a configurational isomer, the other arrangements in space being termed conformational isomers; the term conformer (Rule E-6.1) is then superfluous. This definition, however, requires a knowledge of stability (energy relations) that is not always available.

In another view, a configurational isomer is any stereoisomer that can be isolated or (for some workers) whose existence can be established (for example, by physical methods); all other arrangements then represent conformational isomers; but it is then impossible to differentiate configuration from conformation without involving experimental efficiency or conditions of observation.

Yet another definition is to regard a conformation as a precise description of a configuration in terms of bond distances, bond angles, and dihedral angles.

In none of the above views except the last is attention paid to extension or contraction of the bond to an atom that is attached to only one other atom, such as –H or =O. Yet such changes in interatomic distance due to nonbonded interactions may be important, for instance, in hydrogen bonding, in differences due to crystal form, in association in solution, and in transition states. This area may repay further consideration.

Owing to the circumstances outlined above, the Rules E-1.4 and E-1.5 have been deliberately made imprecise, so as to permit some alternative interpretations, but they are not compatible with all the definitions mentioned above. The time does not seem ripe to legislate for other than the commoner usages or to choose finally between these. It is, however, encouraging that no definition in this field has (yet) involved atomic vibrations for which, in all cases, only time-average positions are considered.

Finally it should be noted that an important school of thought uses conformation with the connotation of "a particular geometry of the molecule, i.e., a description of atoms in space in terms of bond distances, bond angles, and dihedral angles," a definition much wider than any discussed above.

APPENDIX 2. OUTLINE OF THE SEQUENCE-RULE PROCEDURE

The sequence-rule procedure is a method of specifying the absolute molecular chirality (handedness) of a compound, that is, a method of specifying which of two enantiomeric forms each chiral element of a molecule exists. For each chiral element in the molecule it provides a symbol, usually *R* or *S*, which is independent of nomenclature and numbering. These symbols define the chirality of the specific compound considered; they may not be the same for a compound and some of its derivatives; and they are not necessarily constant for chemically similar situations within a chemical or a biogenetic class. The procedure is applied directly to a three-dimensional model of the structure, and not to any two-dimensional projection thereof.

The method has been developed to cover all compounds with ligancy up to four and with ligancy six,* and for all configurations and conformations of such compounds. The following is an outline confined to the most common situations; it is essential to study the original papers, especially the 1966 paper,† before using the sequence rule for other than fairly simple cases.

General Basis

The sequence rule itself is a method of arranging atoms or groups (including chains and rings) in an order of precedence, often referred to as an order of preference; for discussion this order can conveniently be generalized as a > b > c > d, where > denotes "is preferred to."

The first step, however, in considering a model is to identify the nature and position of each chiral element that it contains. There are three types of the chiral element, namely, the chiral center, the chiral axis, and the chiral plane. The chiral center, which is very much the most commonly met, is exemplified by an asymmetric carbon atom with the tetrahedral arrangement of ligands, as in (1). A chiral axis is present in, for instance, the chiral allenes such as (2) or the chiral biaryl derivatives. A chiral plane is exemplified by the plane containing the benzene ring and the bromine and oxygen atoms in the chiral compound (3), or by the underlined atoms in the cycloalkene (4). Clearly, more than one

*Ligancy refers to the number of bonds from an atom, independently of the nature of the bonds.
†Cahn, R. S., Ingold, C., and Prelog, V., *Angew. Chem. Int. Ed.*, 5, 385 (1966); errata, 5, 511 (1966); *Angew. Chem.*, 78, 413 (1966). Earlier papers: Cahn, R. S. and Ingold, C. K., *J. Chem. Soc.* (Lond.), 612 (1951); Cahn, R. S., Ingold, C., and Prelog, V., *Experientia*, 12, 81 (1956). For a partial, simplified account see Cahn, R. S., *J. Chem. Educ.*, 41, 116 (1964); errata, 41, 503 (1964).

type of chiral element may be present in one compound; for instance, group "a" in (2) migh be a *sec*-butyl group which contains a chiral center.

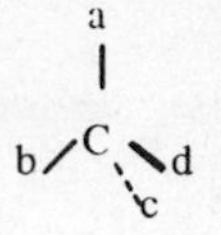

(1)

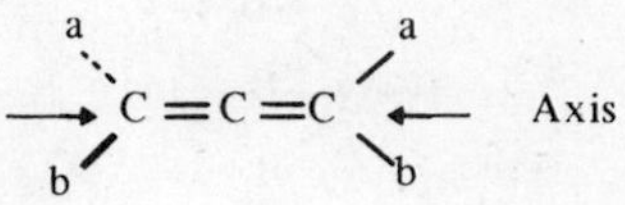

Axis

(2)

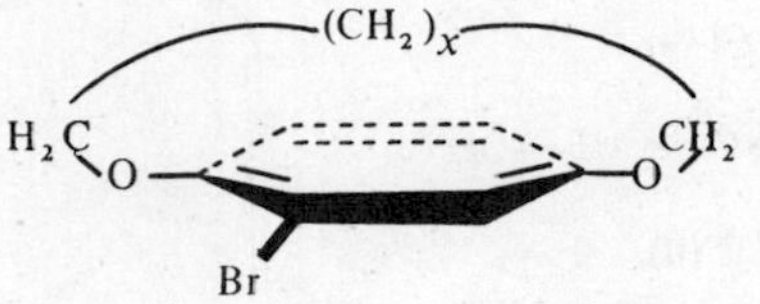

(3)

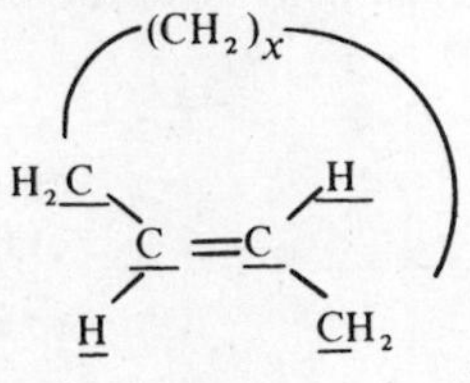

(4)

The Chiral Center

Let us consider first the simplest case, namely, a chiral center (such as carbon) with four ligands, a, b, c, and d, which are all different atoms tetrahedrally arranged as in CHFClBr. The four ligands are arranged in order of preference by means of the sequence rule; this contains five subrules, which are applied in succession so far as necessary to obtain a decision. The first subrule is all that is required in a great majority of actual cases; it states that ligands are arranged in order of decreasing atomic number, in the above case (a) Br > (b) Cl > (c) F > (d) H. There would be two (enantiomeric) forms of the compound and we can write these as (5) and (6). In the sequence-rule procedure the model is viewed from the side remote from the least-preferred ligand (d), as illustrated. Then, tracing a path from a to b to c in (5) gives a clockwise course, which is symbolized by (*R*) (Latin *rectus*, right; for right hand); in (6) it gives an anticlockwise course, symbolized as (*S*) (Latin *sinister*, left). Thus (5) would be named (*R*)-bromochlorofluoromethane, and (6) would be named (*S*)-bromochlorofluoromethane. Here already it may be noted that converting one enantiomer into another changes each *R* to *S*, and each *S* to *R*, always. It will be seen also that the chirality prefix is the same whether the alphabetical order is used, as above, for naming the substituents or whether this is done by the order of complexity (giving fluorochlorobromomethane).

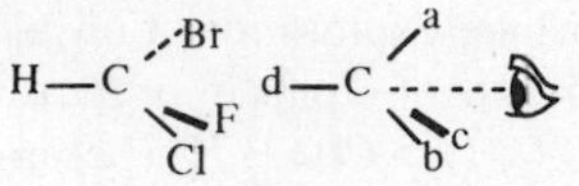

(5), (*R*)

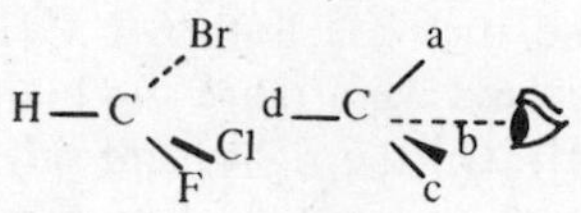

(6), (*S*)

Next, suppose we have H_3C–CHClF. We deal first with the atoms directly attached to the chiral center; so the four ligands to be considered are Cl > F > C (of CH_3) > H. Here the H's of the CH_3 are not concerned, because we do not need them in order to assign our symbol.

However, atoms directly attached to a center are often identical, as, for example, the underlined C's in $H_3\underline{C}$–CHCl–$\underline{C}H_2OH$. For such a compound we at once establish a preference (a) Cl > (b, c) $\underline{C},\underline{C}$ > (d) H. Then to decide between the two $\underline{C}$'s we work outward, to the atoms to which they in turn are directly attached and we then find which we can conveniently write as C(H,H,H) and C(O,H,H). We have to compare H,H,H with O,H,H, and since oxygen has a higher atomic number than hydrogen we have O > H and thence the complete order Cl > C (of CH_2OH) > C (of CH_3) > H, so that the chirality symbol can then be determined from the three-dimensional model.

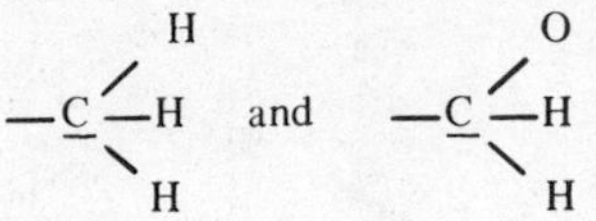

$$-\underline{C}\begin{matrix} H \\ -H \\ H \end{matrix} \quad \text{and} \quad -\underline{C}\begin{matrix} O \\ -H \\ H \end{matrix}$$

We must next meet the first complication. Suppose that we have a molecule (7).

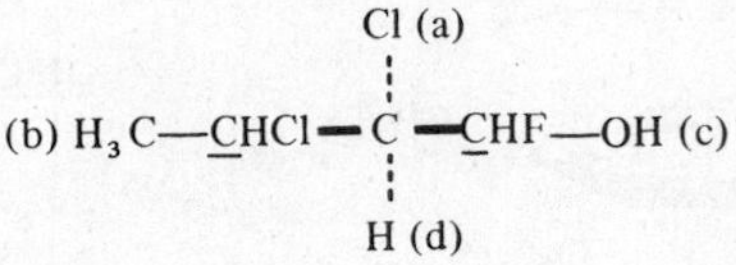

$$\begin{matrix} & \text{Cl (a)} & \\ \text{(b) } H_3C\text{—}\underline{C}HCl\text{—} & C & \text{—}\underline{C}HF\text{—OH (c)} \\ & \text{H (d)} & \end{matrix}$$

(7) (*S*)

To decide between the two C's we first arrange the atoms attached to them in *their* order of preference, which gives C(Cl,C,H) on the left and C(F,O,H) on the right. Then we compare the preferred atom of one set (namely, Cl) with the preferred atom (F) of the other set, and as Cl > F we arrive at the preferences a > b > c > d shown in (7) and chirality (S). If, however, we had a compound (8) we should have met C(Cl,C,H) and C(Cl,O,H) and, since the atoms of first preference are identical (Cl), we should have had to make the comparisons with the atoms of second preference, namely, O > C, which to the different chirality (R) as shown in (8).

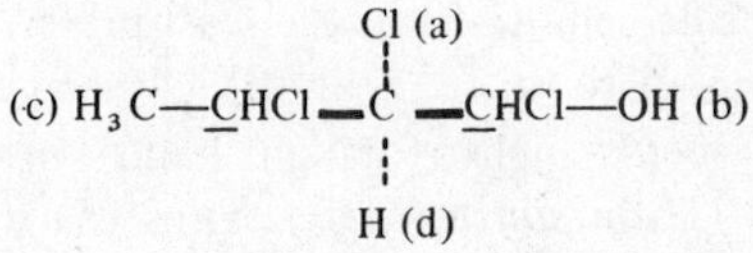

$$\begin{matrix} & \text{Cl (a)} & \\ \text{(c) } H_3C\text{—}\underline{C}HCl\text{—} & C & \text{—}\underline{C}HCl\text{—OH (b)} \\ & \text{H (d)} & \end{matrix}$$

(8) (*R*)

Branched ligands are treated similarly. Setting them out in full gives a picture that at first sight looks complex but the treatment is in fact simple. For instance, in compound (9) a first quick glance again shows (a) Cl > (b, c) C,C > (d) H: When we expand the two C's we find they are both C(C,C,H), so we continue exploration. Considering first the left-hand ligand we arrange the branches and their sets of atoms in order thus: C(Cl,H,H) > C(H,H,H). On the right-hand side we have C(O,C,H) > C(O,H,H) (because C > H). We compare first the preferred of these branches from each side and we find C(Cl,H,H) > C(O,C,H) because Cl > O, and that gives the left-hand branch preference over the right-hand branch. That is all we need to do to establish chirality (S) for this highly branched compound (9). Note that it is immaterial here that, for the lower branches, the right-hand C(O,H,H) would have been preferred to the left-hand C(H,H,H); we did not need to reach that point in our comparisons and so we are not concerned with it; but we should have reached it if the two top (preferred) branches had both been the same CH_2Cl.

Rings, when we met during outward exploration, are treated in the same way as branched chains.

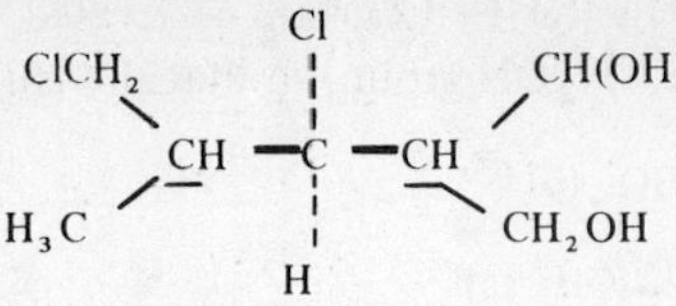

(9)

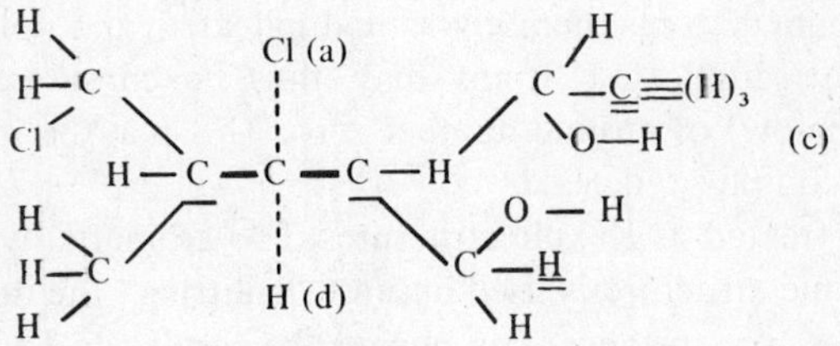

(9) (*S*)

With these simple procedures alone, quite complex structures can be handled; for instance, the analysis alongside Formula (10) for natural morphine explains why the specification is as shown. The reason for considering C-12 as C(C,C,C) is set out in the next paragraphs.

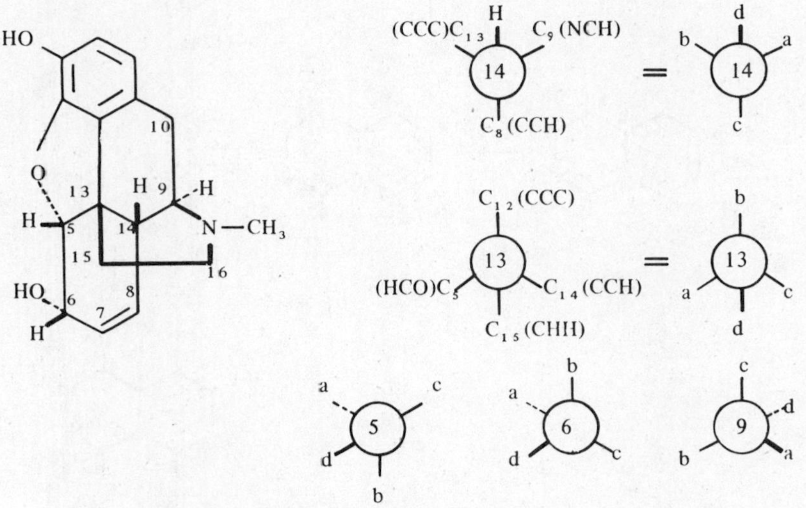

(10) (5R̲, 6S̲, 9R̲, 13S̲, 14R̲,)-Morphine

Now, using the sequence rule depends on exploring along bonds. To avoid theoretical arguments about the nature of bonds, simple classical forms are used. Double and triple bonds are split into two and three bonds, respectively. A > C=O group is treated as (i) (below) where the (O) and the (C) are duplicate representations of the atoms at the other end of the double bond. –C≡CH is treated as (ii) and –C≡N is treated as (iii).

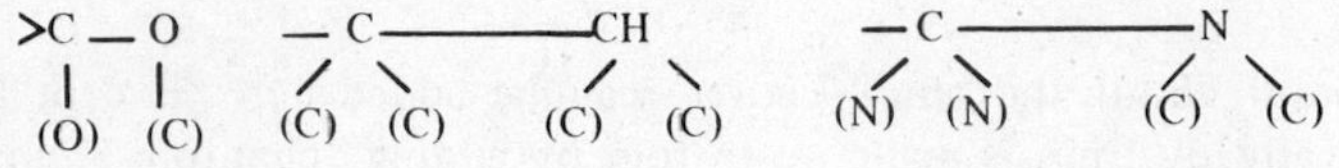

(i) (ii) (iii)

Thus in D-glyceraldehyde (11) the CHO group is treated as C(O,(O),H) and is thus preferred to the C(O,H,H) of the CH_2OH group, so that the chirality symbol is (*R*).

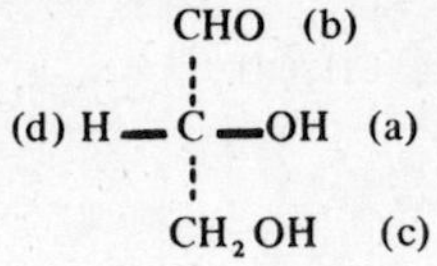

CHO (b)
(d) H—C—OH (a)
CH_2OH (c)

D-Glyceraldehyde
(11) (*R*)

Only the doubly bonded atoms themselves are duplicated, and not the atoms or groups attached to them; the duplicated atoms may thus be considered as carrying three phantom atoms (see below) of atomic number zero. This may be important in deciding preferences in certain complicated cases.

Aromatic rings are treated as Kekulé structures. For aromatic hydrocarbon rings it is immaterial which Kekulé structure is used because "splitting" the double bonds gives the same result in all cases; for instance, for phenyl the result can be represented as (12a) where "(6)" denotes the atomic number of the duplicate representations of carbon.

(12) (6) (6) (6) (6) (6) (6)

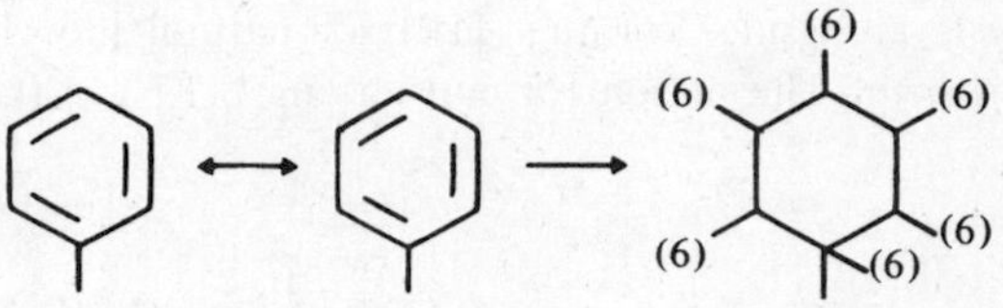

(12a)

(13) 5 4 4a 6 3 7 N N^2 8 9 1

(6) $(6\frac{1}{3})$ (6) (6) $(6\frac{1}{2})$ N N (6) (7) $(6\frac{1}{2})$

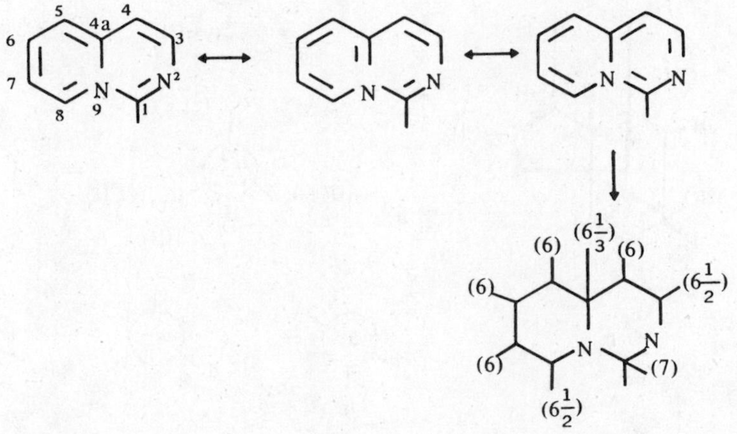

For aromatic hetero rings, each duplicate is given an atomic number that is the mean of what it would have if the double bond were located at each of the possible positions. A complex case is illustrated in (13). Here C-1 is doubly bonded to one or other of the nitrogen atoms (atomic number 7) and never to carbon, so its added duplicate has atomic number 7; C-3 is doubly bonded either to C-4 (atomic number 6) or to N-2 (atomic number 7), so its added duplicate has atomic number 6½; so has that of C-8; but C-4a may be doubly bonded to C-4, C-5, or N-9, so its added duplicate has atomic number 6.33.

One last point about the chiral center may be added here. Except for hydrogen, ligancy, if not already four, is made up to four by adding "phantom atoms" which have atomic number zero and are thus always last in order of preference. This has various uses but perhaps the most interesting is where nitrogen occurs in a rigid skeleton, as, for example, in α-isosparteine (14). Here the phantom atom can be placed where the nitrogen

SOME COMMON GROUPS IN ORDER OF SEQUENCE-RULE PREFERENCE[a]

A. Alphabetical Order (Higher Number Denotes Greater Preference)

64 Acetoxy
36 Acetyl
48 Acetylamino
21 Acetylenyl
10 Allyl
43 Amino
44 Ammonio $^{+}H_3N$–
37 Benzoyl
49 Benzoylamino
65 Benzoyloxy
50 Benzyloxycarbonylamino
13 Benzyl
60 Benzyloxy
41 Benzyloxycarbonyl
75 Bromo
42 *ter*-Butoxycarbonyl
5 *n*-Butyl
16 *sec*-Butyl
19 *tert*-Butyl
38 Carboxyl
74 Chloro
17 Cyclohexyl
52 Diethylamino
51 Dimethylamino
34 2,4-Dinitrophenyl
28 3,5-Dinitrophenyl
59 Ethoxy
40 Ethoxycarbonyl
3 Ethyl
46 Ethylamino
68 Fluoro
35 Formyl
63 Formyloxy
62 Glycosyloxy
7 *n*-Hexyl
1 Hydrogen
57 Hydroxy
76 Iodo
9 Isobutyl
8 Isopentyl
20 Isopropenyl
14 Isopropyl
69 Mercapto
58 Methoxy
39 Methoxycarbonyl
2 Methyl
45 Methylamino
71 Methylsulfinyl
66 Methylsulfinyloxy
72 Methylsulfonyl
67 Methylsulfonyloxy
70 Methylthio
11 Neopentyl
56 Nitro
27 *m*-Nitrophenyl
33 *o*-Nitrophenyl
24 *p*-Nitrophenyl
55 Nitroso
6 *n*-Pentyl
61 Phenoxy
22 Phenyl
47 Phenylamino
54 Phenylazo
18 Propenyl
4 *n*-Propyl
29 1-Propynyl
12 2-Propynyl
73 Sulfo
25 *m*-Tolyl
30 *o*-Tolyl
23 *p*-Tolyl
53 Trimethylammonio
32 Trityl
15 Vinyl
31 2,6-Xylyl
26 3,5-Xylyl

B. Increasing Order of Sequence Rule Preference

1 Hydrogen
2 Methyl
3 Ethyl
4 *n*-Propyl
5 *n*-Butyl
6 *n*-Pentyl
7 *n*-Hexyl
8 Isopentyl
9 Isobutyl
10 Allyl
11 Neopentyl
12 2-Propynyl
13 Benzyl
14 Isopropyl
15 Vinyl
16 *sec*-Butyl
17 Cyclohexyl
18 1-Propenyl
19 *tert*-Butyl
20 Isopropenyl
21 Acetylenyl
22 Phenyl
23 *p*-Tolyl
24 *p*-Nitrophenyl
25 *m*-Tolyl
26 3,5-Xylyl
27 *m*-Nitrophenyl
28 3,5-Dinitrophenyl
29 1-Propynyl
30 *o*-Tolyl
31 2,6-Xylyl
32 Trityl
33 *o*-Nitrophenyl
34 2,4-Dinitrophenyl
35 Formyl
36 Acetyl
37 Benzoyl
38 Carboxyl
39 Methoxycarbonyl[b]
40 Ethoxycarbonyl[b]
41 Benzyloxycarbonyl[b]
42 *tert*-Butoxycarbonyl[b]
43 Amino
44 Ammonio $^{+}H_3N$–
45 Methylamino
46 Ethylamino
47 Phenylamino
48 Acetylamino
49 Benzoylamino
50 Benzyloxycarbonylamino
51 Dimethylamino
52 Diethylamino
53 Trimethylammonio
54 Phenylazo
55 Nitroso
56 Nitro
57 Hydroxy
58 Methoxy
59 Ethoxy
60 Benzyloxy
61 Phenoxy
62 Glycosyloxy
63 Formyloxy
64 Acetoxy
65 Benzoyloxy
66 Methylsulfinyloxy
67 Methylsulfonyloxy
68 Fluoro
69 Mercapto HS–
70 Methylthio CH_3S–
71 Methylsulfinyl
72 Methylsulfonyl
73 Sulfo HO_3S–
74 Chloro
75 Bromo
76 Iodo

[a]ANY alteration to structure, or substitution, etc., may alter the order of preference.
[b]These groups are ROC(=O)–

lone pair of electrons is; then N-1 appears as shown alongside the formula; and the chirality (*R*) is the consequence. The same applies to N-16. Phantom atoms are similarly used when assigning chirality symbols to chiral sulfoxides (see example to Rule E-5.9).

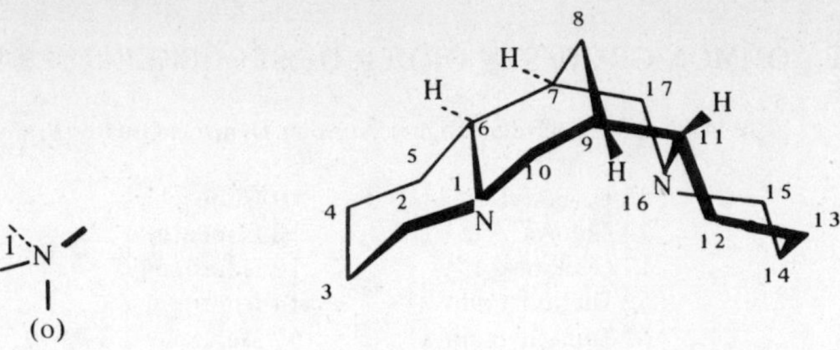

(14) (1*R*, 6*R*, 7*S*, 9*S*, 11*R*, 16*R*)-Sparteine

Symbolism

In names of compounds, the *R* and *S* symbols, together with their locants, are placed in parentheses, normally in front of the name, as shown for morphine (10) and sparteine (14), but this may be varied in indexes or in languages other than English. Positions within names are required, however, when more than a single series of numerals is used, as for esters and amines. When relative stereochemistry is more important than absolute stereochemistry, as for steroids or carbohydrates, a local system of stereochemical designation may be more useful and sequence-rule symbols need then be used only for any situations where the local system is insufficient.

Racemates containing a single center are labeled (*RS*). If there is more than one center the first is labeled (*RS*) and the others are (*RS*) or (*SR*) according to whether they are *R* or *S* when the first is *R*. For instance, the 2,4-pentanediols CH_3–CH(OH)–CH_2–CH(OH)–CH_3 are differentiated as

one chiral form (2*R*,4*R*)–
other chiral form (2*S*,4*S*)–
meso compound (2*R*,4*S*)–
racemic compound (2*RS*,4*RS*)–

Finally the principles by which some of the least rare of other situations are treated will be very briefly summarized.

Pseudoasymmetric Atoms

A subrule decrees that *R* groups have preference over *S* groups and this permits pseudoasymmetric atoms, as in abC(c-*R*)(c-*S*) to be treated in the same way as chiral centers, but as such a molecule is achiral (not optically active) it is given the lower case symbol *r* or *s*.

Chiral Axis

The structure is regarded as an elongated tetrahedron and viewed along the axis – it is immaterial from which end it is viewed; the nearer pair of ligands receives the first two positions in the order of preference, as shown in (15) and (16).

H F Br F = X H F F Br Y = b a d c (*S*) viewed from X or d c b a (*S*) viewed from Y

(15)

(16) (R) (R)

Chiral Plane

The sequence-rule-preferred atom directly attached to the plane is chosen as "pilot atom." In compound (3) this is the C of the left-hand CH_2 group. Now this is attached to the left-hand oxygen atom in the plane. The sequence-rule-preferred path from this oxygen atom is then explored in the plane until a rotation is traced which is clockwise (*R*) or anticlockwise (*S*) when viewed from the pilot atom. In (3) this path is O → C → C(Br) and it is clockwise (*R*).

Other Subrules

Other subrules cater for new chirality created by isotopic labeling (higher mass number preferred to lower) and for steric differences in the ligands. Isotopic labeling rarely changes symbols allotted to other centers.

Octahedral Structures

Extensions of the sequence rule enable ligands arranged octahedrally to be placed in an order of preference, including polydentate ligands, so that a chiral structure can then always be represented as one of the enantiomeric forms (17) and (18). The face 1–2–3 is observed from the side remote from the face 4–5–6 (as marked by arrows), and the path 1 → 2 → 3 is observed; in (17) this path is clockwise (R), and in (18) it is anticlockwise (S).

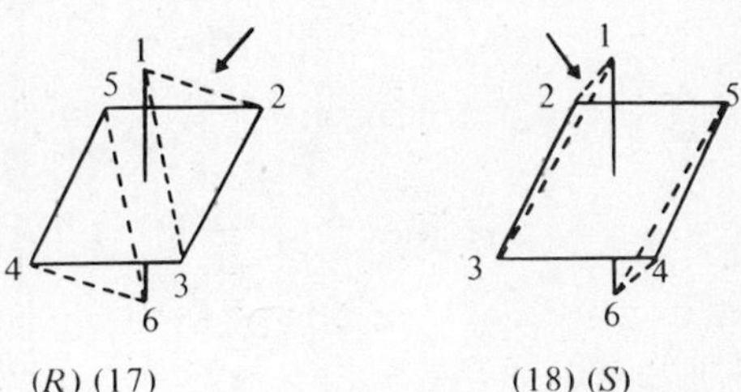

(*R*) (17) (18) (*S*)

Conformations

The torsion angle between selected bonds from two singly bonded atoms is considered. The selected bond from each of these two atoms is that to a unique ligand, or otherwise to the ligand preferred by the sequence rule. The smaller rotation needed to make the front ligand eclipsed with the rear one is noted (this is the rotatory characteristic of a helix); if this rotation is right-handed it leads to a symbol *P* (plus); if left-handed to *M* (minus). Examples are

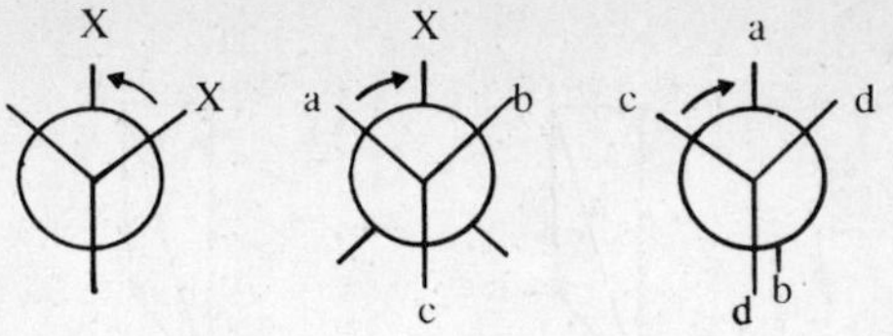

X
X
(M)
X a b c (P)
a c d d b (P)

Details and Complications

For details and complicating factors the original papers should be consulted. They include treatment of compounds with high symmetry or containing repeating units (e.g., cyclitols), also π bonding (metallocenes, etc.), mesomeric compounds and mesomeric radicals, and helical and other secondary structures.

ABBREVIATIONS AND SYMBOLS FOR THE DESCRIPTION OF THE CONFORMATION OF POLYPEPTIDE CHAINS TENTATIVE RULES (1969)*

IUPAC-IUB Commission on Biochemical Nomenclature

Preamble

These Rules are based on "A Proposal of Standard Conventions and Nomenclature for the Description of Polypeptide Conformation" (Edsall et al.),[8] and have been prepared by a subcommission set up by the IUPAC-IUB Commission on Biochemical Nomenclature in 1966.† The original proposals have been modified so as to bring them as far as possible into line with the system of nomenclature current in the fields of organic and polymer chemistry.

Two Recommendations are appended to the Rules, the first dealing with the terms configuration and conformation, and the second with primary, secondary, and tertiary structure. These are formulated as recommendations rather than rules because there is at present no general agreement about their definition.

Note: Two alternative notations are recommended throughout. That with superscripts and subscripts may be used when it is unlikely to cause confusion, e.g., in printed or manuscript material; that without is to be used where superscripts or subscripts may cause confusion, or are technically difficult or impossible, e.g., in computer outputs. In the latter connection the following Roman equivalents of Greek letters are recommended:

α	A	ρ	R
β	B	τ	T
γ	G	ν	U
δ	D	ϕ	F
ϵ	E	χ	X
ξ	Z	ψ	Q
η	H	ω	W

Rule 1. General Principles of Notation

1.1. *Designation of Atoms.* The atoms of the main chain are denoted thus:

$$—\mathrm{NH}—\underset{|}{\mathrm{C}^{\alpha}\mathrm{H}^{\alpha}}—\mathrm{CO}—$$

Where confusion might arise the following additional symbolism may be used:

$$—\mathrm{N'H'}—\underset{|}{\mathrm{C}^{\alpha}\mathrm{H}^{\alpha}}—\mathrm{C'O'}$$

1.2. Amino-acid residues, –NH-CHR-CO–, are numbered sequentially from the amino-terminal to the carboxyl-terminal end of the chain, the residue number being denoted *i*.

Example:

C^{α} of the *i*th residue is written C_i^{α} or $\mathrm{C}\alpha(i)$

†The members of the subcommission were J. C. Kendrew (Chairman), W. Klyne, S. Lifson, T. Miyazawa, G. Némethy, D. C. Phillips, G. N. Ramachandran, and H. A. Scheraga. In addition, the following assisted in the work of the subcommission: R. S. Cahn, R. Diamond, J. T. Edsall, P. J. Flory, C. K. Ingold, A. Liquori, V. Prelog, and J. A. Schellman.

*From IUPAC-IUB Commission on Biochemical Nomenclature, *Pure Appl. Chem.*, 40(3), in press. With permission.

1.3. For some purposes it is more convenient to group together the atoms –CHR-CO-NH–. These groups are described as "peptide units," and the peptide unit number, like the residue number, is denoted i. It will be noted that the two numbers are identical for all atoms except NH; generally there will be no confusion, because a single document will use either "residues" alone, or "peptide units" alone, but in the latter case explicit reference must be made to this usage at the beginning. If confusion might arise, the symbols N_i^* and H_i^* are to be used for these atoms in the ith peptide unit, N_i and H_i in the ith residue (so that $N_i^* = N_{i+1}$).

Example:

Peptide Unit No. 2

NH_2 — CHR_1 — CO ┼ NH ┼ CHR_2 — CO ┼ NH ┼ CHR_3

Residue No. 2

Residue notation	N_2	C_2^α	C_2	N_3
Peptide unit notation	N_1^*	C_2^α	C_2	N_2^*

Notes: (i) Residue notation is used throughout these Rules.

(ii) Whether "residues" or "peptide units" are being used, ϕ_i and ψ_i always refer to torsion angles about the same C_i^α.

1.4. *Bond Lengths*. If a bond A–B be denoted A_i–B_j or A_i (see Rules 3.1, 4.5), the bond length is written $b(A_i, B_j)$ [or $b(Ai, Bj)$, or b_i^A (or $bA(i)$)]. An abbreviated notation for use in side chains is indicated in Rule 4.5.

Note: The symbol previously recommended for bond length was l. This symbol is no longer recommended, partly because it is easily confused with 1 in many type fonts, and partly because it is also used for vibration amplitude in electron diffraction and spectroscopy.

1.5. *Bond Angles*. The bond angle included between three atoms A_i—B_j—C_k (with B_j at the apex) is written $\tau(A_i, B_j, C_k)$, which may be abbreviated, if there is no ambiguity, to $\tau(B_j)$ or τ_j^B or $\tau B(j)$.

1.6. *Torsion Angles*. If a system of four atoms A—B—C—D (A attached to B, D attached to C) is projected onto a plane normal to bond B–C, the angle between the projection of A–B and the projection of C–D is described as the *torsion angle** of A and D about bond B–C; this angle may also be described as the angle between the plane containing A, B, and C, and the plane containing B, C, and D. The torsion angle is written in full as $\theta(A_i, B_j, C_k, D_l)$ which may be abbreviated, if there is no ambiguity, to $\theta(B_j, C_k)$, $\theta(B_j)$, or θ_j^B, etc. In the eclipsed conformation in which the projections of A–B and C–D coincide, θ is given the value 0° (synplanar conformation). A torsion angle is considered positive ($+\theta$) or negative ($-\theta$) according as, when the system is viewed along the central bond in the direction B → C (or C → B), the bond to the front atom A (or D) must be rotated to the right or to the left, respectively, in order that it may eclipse the bond to the rear atom D (or A); note that it is immaterial whether the system be viewed from one end or the other. These relationships are illustrated in Figure 1.

*The terms *dihedral angle* and *internal rotation angle* are also used to describe this angle, and may be regarded as alternatives to *torsion angle* though the latter has been used throughout these Rules.

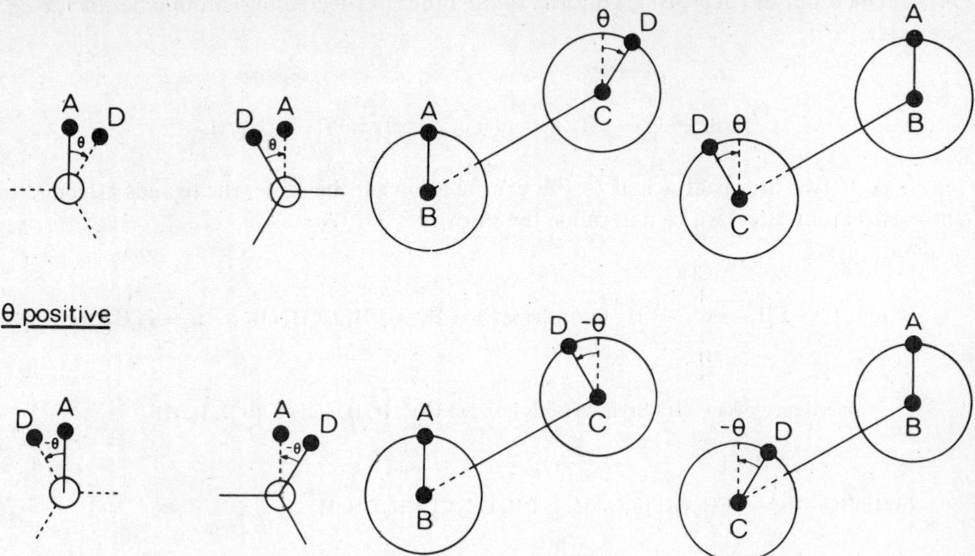

FIGURE 1. Newman and perspective projections illustrating positive and negative torsion angles. Note that a right-handed turn of the bond to the front atom about the central bond gives a positive value of θ from whichever end the system is viewed.

Notes: (i) Angles are measured in the range $-180° < \theta \leqslant +180°$, rather than from 0° to 360°, so that the relationship between enantiomeric configurations or conformations can be readily appreciated.

(ii) The symbols actually used to describe the various torsion angles important in polypeptides are ϕ, ψ, ω, ν, and χ (see Rules 3.2, 4.5.2). In the above, θ is used simply as an illustrative generic symbol covering all these.

Rule 2. The Sequence Rule, and Choice of Torsion Angle

2.1. The Rules here enunciated for use in the field of synthetic polypeptides and proteins are in general harmony with the Sequence Rule of Cahn, Ingold, and Prelog,* with the exceptions of Rules 2.1.1 and 2.2.2 (cases II and III), and later Rules dependent upon these. The Sequence Rule was formulated as a universal and unambiguous means of designating the "handedness" or chirality of an element of asymmetry. It includes Subrules for the purpose of arranging atoms or groups in an order of precedence or preference, and this system may conveniently be used in the description of steric relationships across single bonds (see Klyne and Prelog).[11] Here its function is to determine the priority of precedence of different atoms or groups attached to the same atom. However, Rule 2.1.1 below overrides the precedences of the Sequence Subrules, providing a new "local" (specialist) system for use with the general Sequence Rule.† After application of Rule 2.1.1, the normal procedure of the Sequence Rule is applied, but modified by Rule 2.2.2; in this connection the only parts of the Sequence Rule required are given in Rules 2.1.2 to 2.1.5.

2.1.1. The main chain is given formal priority over branches, notwithstanding any conflict with the following rules. Thus the main chain has precedence at C^{α} over the side chain, and at C′ over O′.

Note: This rule has not yet been formally accepted except in the present context.

*See Cahn, Ingold, and Prelog[7] IUPAC Tentative Rules for the Nomenclature of Organic Chemistry, Section E, in IUPAC Information Bulletin No. 35, pp. 36–80.[10] Earlier papers: Cahn and Ingold[5] Cahn, Ingold and Prelog.[6] For a partial, simplified account see Cahn[4] Eliel.[9]

†Other local systems are available analogously for steroids, carbohydrates and cyclitols, where the Sequence Rule is applied when the local system does not suffice.

2.1.2. The order of (decreasing) priority is the order of (decreasing) atomic number. Example:

In $\mathrm{Br—\underset{H}{\overset{Cl}{C}}—CH_3}$ the order of priority is Br, Cl, CH_3, H.

2.1.3. If two atoms attached to the central atom are the same, the ligands attached to these two atoms are used to determine the priority:

Examples:

(i) In $\mathrm{CH_3—CH_2—\underset{H}{\overset{Cl}{C}}—CH_3}$ the order is Cl, ($CH_3—CH_2$), CH_3, H. ($C^xH_2—CH_3$ takes precedence over C^yH_3 because C^x is bonded to C, H, H, and C^y to H, H, H).

(ii) In $\mathrm{HO—\underset{H}{\overset{CH_2Cl}{C}}—CH_2OH}$ the order is OH, CH_2Cl, CH_2OH, H.

(iii) In $\mathrm{CH_3—CH_2—\underset{H}{\overset{OH}{C}}—CH(CH_3)_2}$ the order is OH, $CH(CH_3)_2$, $CH_2—CH_3$, H.

2.1.4. A double bond is formally treated as though it were split. Thus $> C = O$ is treated as $>\underset{(O)}{C}-\underset{(C)}{O}$.

Example:

In $CH_3—CO—OH$ the order is $=O$, $—OH$, CH_3.

2.1.5. If two ligands are distinguished only by having different masses (e.g., deuterium and hydrogen), the heavier takes precedence.

Example:

In $\mathrm{Br—\underset{H}{\overset{D}{C}}—CH_3}$ the order is Br, CH_3, D, H.

Note: This rule is to be used only if the two previous rules do not give a decision.

2.2 Choice of Torsion Angle and Numbering of Branches (Tetrahedral Configurations)

2.2.1. If, in a compound P–B–C–E (with A and Q attached to B, and D and F attached to C), the Sequence Rule gives the priorities A > P, Q and D > E > F, then the Principal Torsion Angle, θ, is that measured by reference to the atoms A–B–C–D as in Rule 1.6 above.

The branches beginning at C are numbered $C\overline{1}D$, $C\overline{2}E$ and $C\overline{3}F$.

2.2.2. If two branches are identical, and the third is different (or nonexistent), they are numbered in a clockwise sense when viewed in the direction B → C, as follows (see Figure 2).

2.2.3. If all three branches are identical, that giving the smallest positive or negative

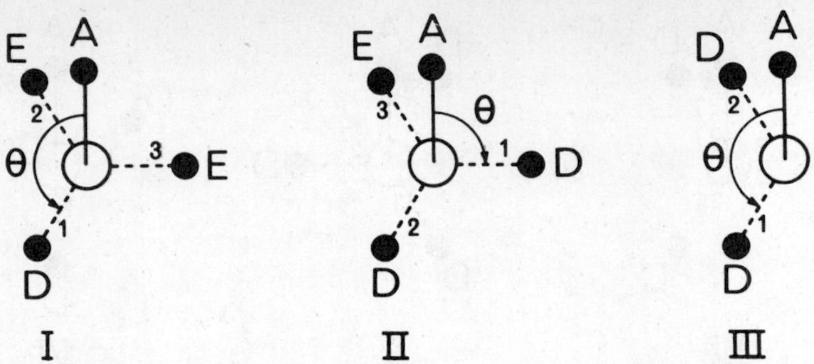

FIGURE 2. Tetrahedral configurations. Case I: D > E = E. D has the highest priority and is given the smallest number (1). Case II: D = D > E. E has the lowest priority and is given the largest number (3). Case III: D = D, numbered 1 and 2 (E is nonexistant). In each case the Principal Torsion Angle is measured between A–B and Branch 1.

Notes: (i) The rule given in Case II differs from Conformational Selection Rule (b) of the Sequence Rule (see Cahn, Ingold, and Prelog, p. 406)[7] according to which if an identity among the groups of a set leaves one group unique, the unique group is fiducial. The reason for the difference is that the Sequence Rule would define Principal Torsion Angle in terms of a hydrogen atom whenever a single such atom formed part of the set; in the X-ray technique, nearly always used to establish structures of the type under discussion, hydrogen atoms are usually unobservable, and even at best not accurately locatable, so that the position of one used to define a Principal Torsion Angle could only be established by calculation based on (perhaps unjustified) assumptions about the bond angles concerned. These considerations apply with even more force to Case III, where one branch is nonexistent; The "phantom atom" of zero atomic number would be given highest priority because it is unique.
(ii) In Case III the clockwise passage from CD^1 to CD^2 shall be by the shorter of the two possible routes.

value of the Principal Torsion Angle is normally* assigned the highest priority and the lowest number (1) (see Figure 3, IV, V); if two branches have torsion angles respectively +60° and −60°, the former is chosen (see Figure 3, VI). The others are numbered in a clockwise sense when viewed in the direction B → C.

Note: Rule 2.2.3 introduces a new principle, not invoked in 2.2.1 or 2.2.2, that the precedence depends on the conformation. This must necessarily be done since in this case the branches are distinguishable only in this respect. (The same applies to Rule 2.3.2 below).

2.3 *Choice of Torsion Angle and Numbering of Branches (Planar Trigonal Configurations)*

```
      A     D
       \   /
     P–B–C
       /   \
      Q     E
```

2.3.1. If, in a compound P–B–C (with A, Q on B and D, E on C as shown) such that B, C, D, and E are coplanar, or nearly so, the Sequence Rule gives the priorities A > P, Q and D > E, then the Principal Torsion Angle is that measured by reference to atoms A–B–C–D as in Rule 1.6 above.

The branches beginning at C are numbered $C_{\overline{1}}D$, $C_{\overline{2}}E$.

2.3.2. If the branches are identical, that giving the smallest positive or negative value of

*The qualification "normally" is added to avoid the need to renumber the branches, if by chance the rule would demand this in consequence of a movement during refinement of a structure. In this or similar cases, the symbolism should remain unchanged.

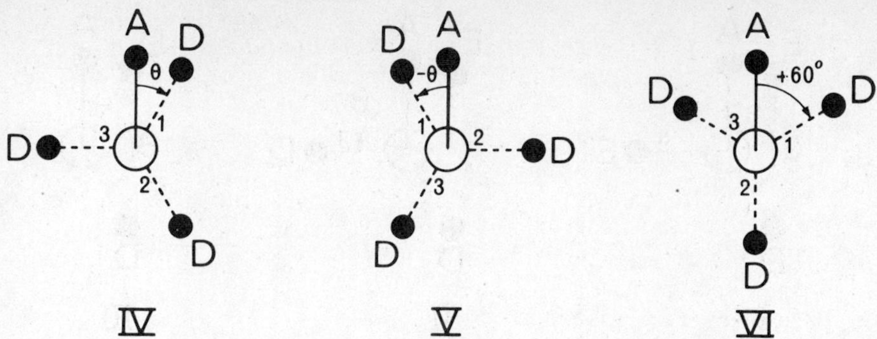

FIGURE 3. Tetrahedral configurations. Three identical branches: IV – general case, θ positive; V – general case, θ negative; VI – θ = +60°.

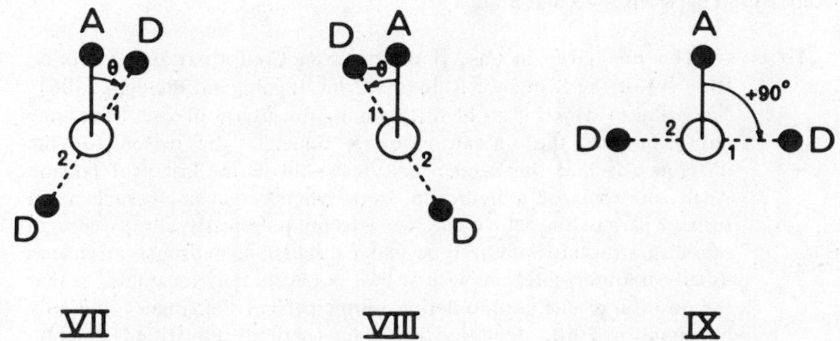

FIGURE 4. Planar trigonal configurations. Identical branches: VII – θ positive; VIII – θ negative; IX – θ = +90°.

the Principal Torsion Angle is normally assigned the highest priority and the lowest number (1); if the two branches have torsion angles respectively +90° and −90°, the former is chosen (see Figure 4).

Rule 3. The Main Chain (or Polypeptide Backbone)

3.1. *Designation of Bonds*. Bonds between main-chain atoms are denoted by the symbols of the two atoms terminating them, e.g., N_i–C_i^α, C_i^α–C_i, C_i–N_{i+1}, C_i–O_i, N_i–H_i. Abbreviated symbols should not be used. Bond lengths are written $b(C_i, N_{i+1})$, etc.

3.2. *Torsion Angles*

3.2.1. The Principal Torsion Angle describing rotation about N–C$^\alpha$ is denoted by ϕ, that describing rotation about C$^\alpha$–C is denoted by ψ, and that describing rotation about C–N is denoted by ω. The symbols ϕ_i, ψ_i, and ω_i are used to denote torsion angles of bonds within the ith residue in the case of ϕ and ψ, and between the ith and $(i+1)$th residues in the case of ω; specifically, ϕ_i refers to the torsion angle of the sequence of atoms C_{i-1}, N_i, C_i^α, C_i; ψ_i to the sequence N_i, C_i^α, C_i, N_{i+1}; and ω_i to the sequence C_i^α, C_i, N_{i+1}, C_{i+1}^α (see Figure 5). In accordance with Rules 1.6 and 2.1.1, these torsion angles are ascribed zero values for eclipsed conformation of the main-chain atoms N, C$^\alpha$, and C, that is, for the so-called *cis*-conformations (see Table 1).

Notes: (i) This convention differs from that proposed by Edsall et al.[8] The new designation of angles may be derived from the old by adding 180° to, or subtracting 180°

Table 1
MAIN-CHAIN TORSION ANGLES FOR VARIOUS CONFORMATIONS IN PEPTIDES OF L-AMINO ACIDS

ϕ	Rotation about N–C$^\alpha$		ψ	Rotation about C$^\alpha$–C	
0°	C$^\alpha$–C *trans*	to N–H	0°	C$^\alpha$–N *trans*	to C–O
+60°	C$^\alpha$–H *cis*		+60°	C$^\alpha$–R *cis*	
+120°	C$^\alpha$–R *trans*		+120°	C$^\alpha$–H *trans*	
+180°	C$^\alpha$–C *cis*		+180°	C$^\alpha$–N *cis*	
–120°	C$^\alpha$–H *trans*		–120°	C$^\alpha$–R *trans*	
–60°	C$^\alpha$–R *cis*		–60°	C$^\alpha$–H *cis*	

Notes: (i) *trans* to N_i–H_i is the same as *cis* to N_i–$C_i'^{-1}$; *trans* to C_i–O_i is the same as *cis* to C_i–N_{i+1} (see Figure 5).
(ii) For the description of D-amino acids, interchange C$^\alpha$–H and C$^\alpha$–R in the Table.

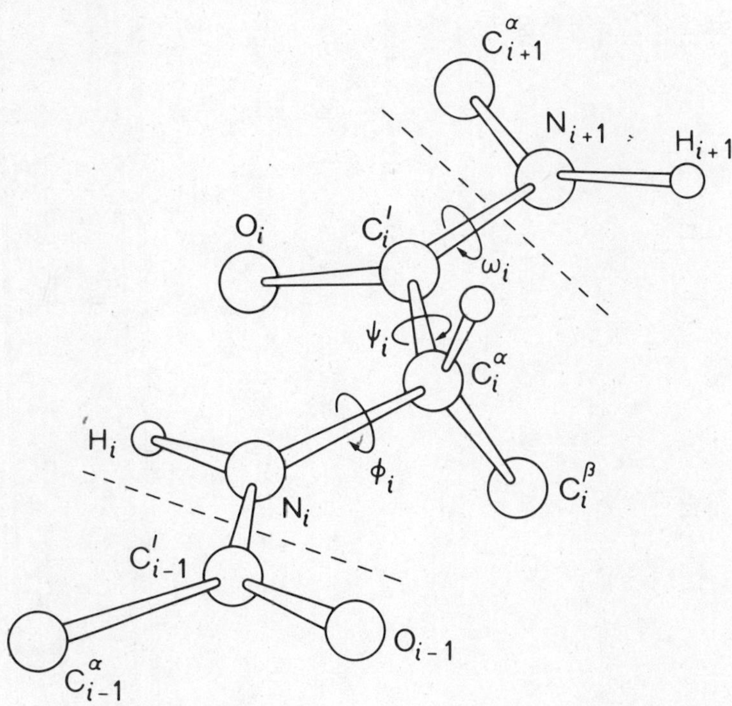

FIGURE 5. Perspective drawing of a section of polypeptide chain representing two peptide units. The limits of a *residue* are indicated by dashed lines, and recommended notations for atoms and torsion angles are indicated. The chain is shown in a fully extended conformation ($\phi_i = \psi_i = \omega_i = +180°$), and the residue illustrated is L-.

from, the latter. (This statement is precisely correct only if the peptide bond is exactly planar, which is not generally the case in experimentally determined structures.)

(ii) Owing to the partial double-bond character of CO=NH, it is normally possible for ω to assume values only in the neighborhood of 0° or 180°. $\omega \sim 180°$ is the value which is generally found (i.e., the *trans*-conformation).

(iii) A "fully-extended" polypeptide chain is characterized by $\phi = \psi = \omega = +180°$. The case of $\phi = \psi = 0°$ would involve the relations indicated in Table 1.

(iv) Table 2 gives values of ϕ and ψ for various well-known regular structures. It is noteworthy that a right-handed α-helix has negative torsion angles.

Table 2
APPROXIMATE TORSION ANGLES FOR SOME REGULAR STRUCTURES

(*Note*: For a fully extended chain $\phi = \psi = \omega = +180°$)

	ϕ	ψ	ω	Reference
Right-handed α-helix (α-poly-L-alanine)	−57°	−47°	+180°	1
Left-handed α-helix	+57°	+47°	+180°	1
Parallel-chain pleated sheet	−119°	+113°	+180°	16
Antiparallel-chain pleated sheet (β-poly-L-alanine)	−139°	+135°	−178°	3
Polyglycine II	−80°	+150°	+180°	15
Collagen	−51°, −76°, −45°	153°, 127°, 148°	+180°	19
Poly-L-proline I	−83°	+158°	0°	14, 17
Poly-L-proline II	−80°	+85°	+180°	2

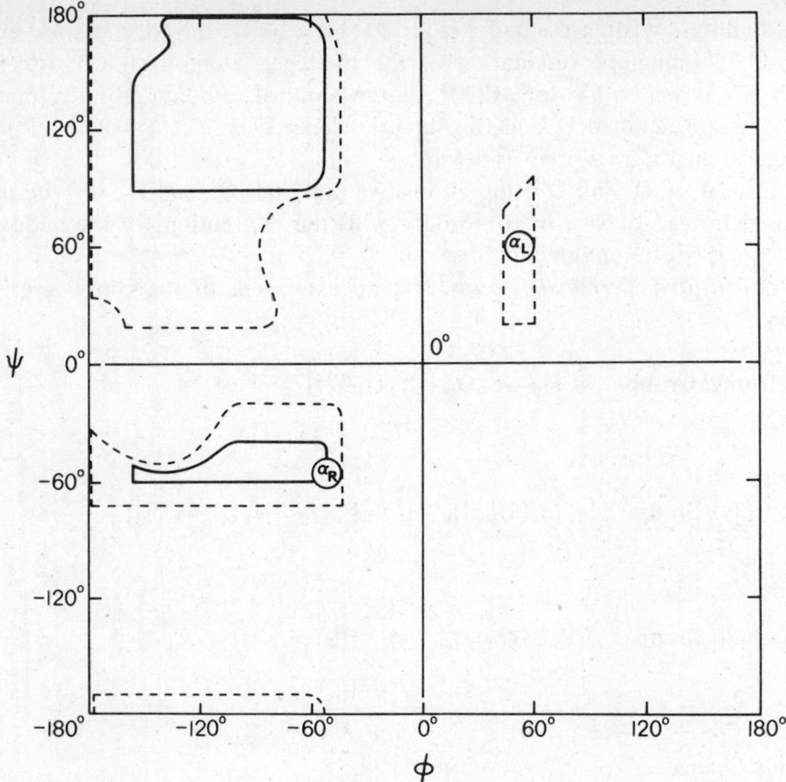

FIGURE 6. Typical conformational map (Ramachandran, Ramakrishnan, and Sasisekharan[13]) transposed into the standard conventions.

Note: This diagram is identical to that of Edsall et al.[8] except that the origin is now at the center, instead of at the lower left-hand corner. The solid lines enclose the freely allowed values of ϕ and ψ for an alanyl residue in a polypeptide; the dotted lines enclose "outer limit" values based on the shortest known van der Waals' radii in related structures. Analogous diagrams for other residues, and for slightly different assumptions, are given by Ramachandran and Sasisekharan;[14] note that these authors used the earlier convention with the origin at the corner.

(v) Figure 6 is a typical conformational map [(ϕ–ψ) plot] using the Rules enunciated above.

3.2.2. There may occasionally be a need to consider torsion angles differing from zero for the sequences of atoms O=C–N–C$^{\alpha}$ and C$^{\alpha}$–C–N–H, in cases where C=O or N–H lie out of the peptide plane. These angles may be represented ν^{O} and ν^{H} (Greek *upsilon*).

3.3. *Chain Terminations*

3.3.1. If the terminal amino group of the chain is protonated the three hydrogen atoms are denoted H_1^1, H_1^2, and H_1^3; the hydrogen atom giving the smallest (positive or negative) value of the Principal Torsion Angle H–N–C$^{\alpha}$–C is denoted H_1^1, and the others are numbered in a clockwise sense when viewed in the direction C$^{\alpha}$ → N. The corresponding torsion angles are denoted ϕ_1^1, ϕ_1^2, and ϕ_1^3. If the terminal amino group is not protonated, the hydrogen atoms are denoted H_1^1 and H_1^2 in accordance with Rule 2.2.2 and the corresponding torsion angles ϕ_1^1 and ϕ_1^2.

3.3.2. At the carboxyl-terminus of the chain (i = T) the double bonded oxygen is written as O′ and the other oxygen as O″, thus C$^{\alpha}$–C(=O′)–O″–H″. The torsion angles about the C^α–C bond are written ψ_T^1 and ψ_T^2 [or $\psi 1(T)$, $\psi 2(T)$]; the torsion angle about the C–O″ bond, defining the orientation of the hydrogen atom of the hydroxyl group relative to C^α, is writen θ_T^C [or $\theta C(T)$]. If the terminal carboxyl group is ionized, the oxygen atoms are denoted O′ and O″, the precedence being determined by Rule 2.3.2, and the torsion angles are written as before.

Note: Instead of O′ and O″, the alternative notations O^1 and O^2 may be used. ψ_T may be used instead of ψ_T^1, in conformity with the convention for the middle of the chain, so long as confusion does not arise.

3.3.3. *Substituted Terminal Groups.* Natural extensions of the above rules may be devised, e.g.,

i. *N*-formyl Group $\quad H_0—C_0O_0—N_1H_1C_1^\alpha H_1^\alpha— \cdots$

ii. *N*-acetyl Group $\quad C_0(H_0^1, H_0^2, H_0^3)—C_0O_0—N_1H_1—C_1^\alpha H_1^\alpha— \cdots$

iii. C-amido Group $\quad C_T^\alpha H_T^\alpha—C_T(=O'_T)—N_{T+1}(H_{T+1}^1)(H_{T+3}^2)$

Rule 4. Side Chains

4.1. Atoms are lettered or lettered and numbered from C^α, and bonds are numbered from C^α, working outwards away from the main chain.

4.2. *Designation of Atoms other than Hydrogen.* Atoms other than hydrogen are designated in the usual way by Greek letters, β, γ, δ, etc., e.g., C_i^β [or $C\beta(i)$], N_i^ζ [or $N\zeta(i)$].

Note: The notations for the amino acids normally occurring in proteins are given in Table 3.

4.3. *Designation of Branches.* If a side chain is branched, the branches are numbered 1 and 2, the order being determined.

i. in cases where the branches are different, by application of Rule 2.2.1 or 2.3.1,
ii. in cases where two branches are identical (e.g., in valine, phenylalanine), by the application of Rule 2.2.2 (valine) or 2.3.2 (phenylalanine).

Nonhydrogen atoms in different branches are designated by the Greek letter indicating their degree of remoteness from C^α and by the number of their branch (see Rule 2.2 and 2.3); e.g., in valine $C_i^{\gamma 1}$ and $C_i^{\gamma 2}$ [or $C\gamma 1(i)$, $C\gamma 2(i)$]. The branch number need not be indicated where no ambiguity results, e.g., in threonine O^γ and C^γ instead of $O^{\gamma 1}$ and $C^{\gamma 2}$, in hydroxyproline O^δ, C^δ instead of $O^{\delta 1}$, $C^{\delta 2}$, and in histidine C^δ, N^ϵ etc. instead of $C^{\delta 2}$, $N^{\epsilon 2}$. For asparagine or glutamine, in cases where nitrogen and oxygen in the amide group have not yet been distinguished, these atoms may be written $(NO)^{\delta 1}$, $(NO)^{\delta 2}$, or $(NO)^{\epsilon 1}$, $(NO)^{\epsilon 2}$, the indices 1 and 2 being determined by Rule 2.3.2.

4.4. *Designation of Hydrogen Atoms.* Hydrogen atoms are designated by the Greek letter and/or number of the atom to which they are attached, e.g., in valine H_i^β [or $H\beta(i)$]. Where three hydrogen atoms are attached to a single nonhydrogen atom, they are designated 1, 2, and 3; in the situation A–B–C(H)(H)–H, the hydrogen atom giving the smallest

Table 3
SYMBOLS FOR ATOMS AND BONDS IN THE SIDE CHAINS OF THE COMMONLY OCCURRING L-AMINO ACIDS

(a) Unbranched Side Chains

Alanine	$C^{\alpha} \underset{1}{—} C^{\beta}$
Serine	$C^{\alpha} \underset{1}{—} C^{\beta} \underset{2}{—} O^{\gamma}$
Cysteine	$C^{\alpha} \underset{1}{—} C^{\beta} \underset{2}{—} S^{\gamma}$
Cystine	$C^{\alpha}_{i} \underset{1i}{—} C^{\beta}_{i} \underset{2i}{—} S^{\gamma}_{i} \underset{3i}{—} S^{\gamma}_{k} \underset{2k}{—} C^{\beta}_{k} \underset{1k}{—} C^{\alpha}_{k}$
Methionine	$C^{\alpha} \underset{1}{—} C^{\beta} \underset{2}{—} C^{\gamma} \underset{3}{—} S^{\delta} \underset{4}{—} C^{\epsilon}$
Lysine	$C^{\alpha} \underset{1}{—} C^{\beta} \underset{2}{—} C^{\gamma} \underset{3}{—} C^{\delta} \underset{4}{—} C^{\epsilon} \underset{5}{—} N^{\zeta}$

(b) Branched Side Chains

Valine	$C^{\alpha} \underset{1}{—} C^{\beta}$; $C^{\beta} \underset{2,1}{—} C^{\gamma 1}$; $C^{\beta} \underset{2,2}{—} C^{\gamma 2}$
Threonine	$C^{\alpha} \underset{1}{—} C^{\beta}$; $C^{\beta} \underset{2,1}{—} O^{\gamma 1}$; $C^{\beta} \underset{2,2}{—} C^{\gamma 2}$
Isoleucine	$C^{\alpha} \underset{1}{—} C^{\beta}$; $C^{\beta} \underset{2,1}{—} C^{\gamma 1} \underset{3,1}{—} C^{\delta 1}$; $C^{\beta} \underset{2,2}{—} C^{\gamma 2}$
Leucine	$C^{\alpha} \underset{1}{—} C^{\beta} \underset{2}{—} C^{\gamma}$; $C^{\gamma} \underset{3,1}{—} C^{\delta 1}$; $C^{\gamma} \underset{3,2}{—} C^{\delta 2}$
Aspartic acid	$C^{\alpha} \underset{1}{—} C^{\beta} \underset{2}{—} C^{\gamma}$; $C^{\gamma} \underset{3,1}{—} O^{\delta 1}$; $C^{\gamma} \underset{3,2}{—} O^{\delta 2}$ or $— C^{\gamma}$; $C^{\gamma} \underset{3,1}{=} O^{\delta 1}$; $C^{\gamma} \underset{3,2}{—} O^{\delta 2} — H$
Asparagine	$C^{\alpha} \underset{1}{—} C^{\beta} \underset{2}{—} C^{\gamma}$; $C^{\gamma} \underset{3,1}{—} O^{\delta 1}$; $C^{\gamma} \underset{3,2}{—} N^{\delta 2}$

Table 3 (continued)
SYMBOLS FOR ATOMS AND BONDS IN THE SIDE CHAINS OF THE COMMONLY OCCURRING L-AMINO ACIDS

(b) Branched Side Chains (continued)

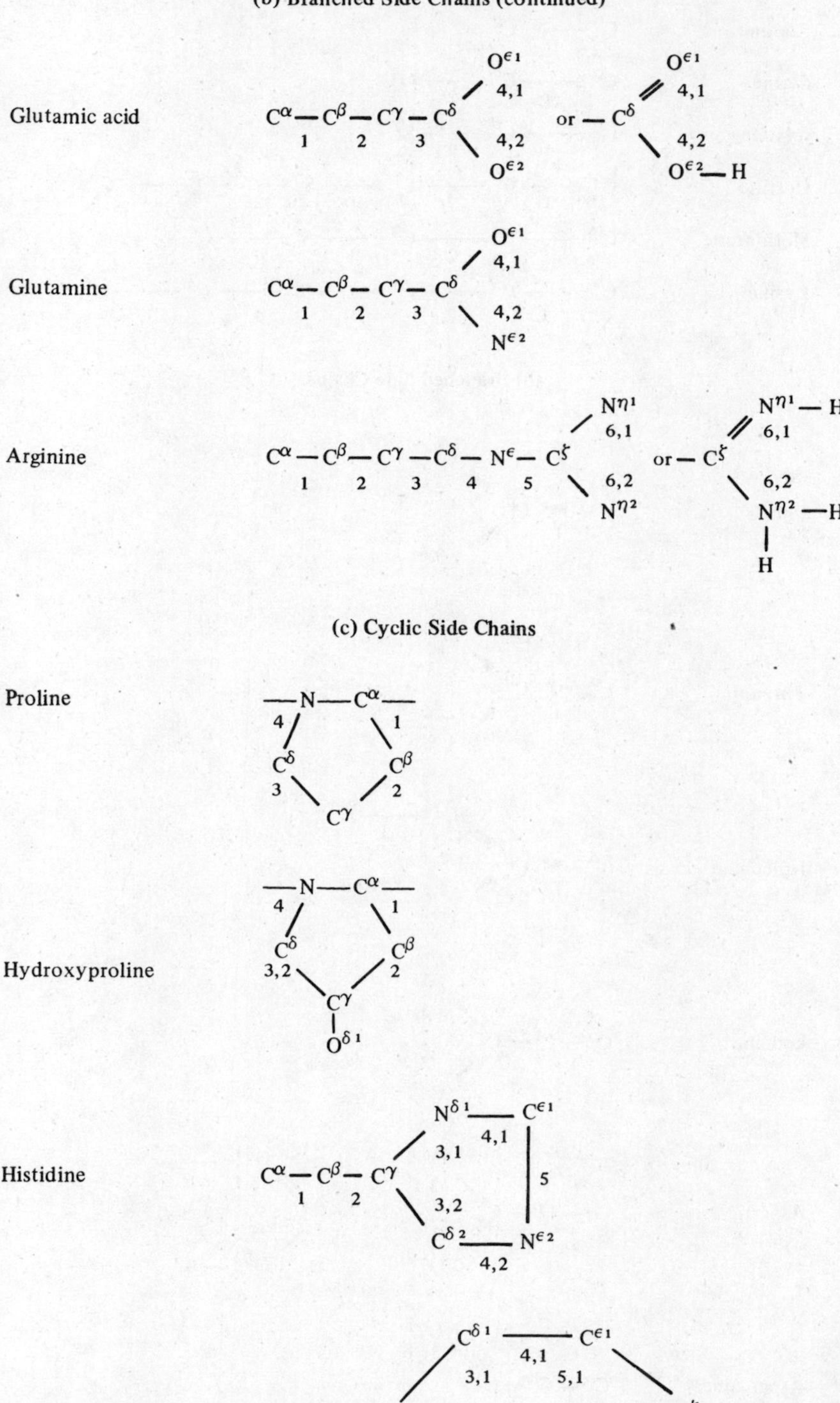

Table 3 (continued)
SYMBOLS FOR ATOMS AND BONDS IN THE SIDE CHAINS OF THE COMMONLY OCCURRING L-AMINO ACIDS

(c) Cyclic Side Chains (continued)

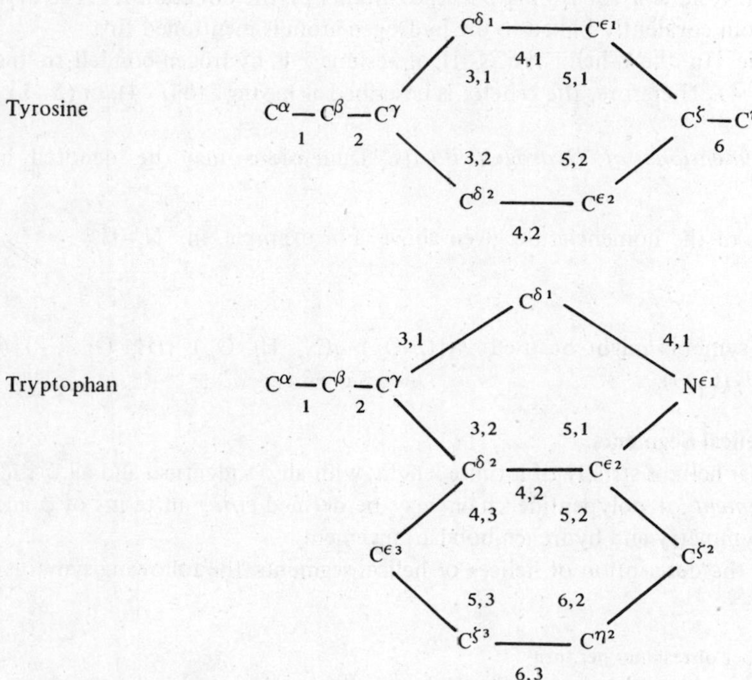

(positive or negative) value of the Principal Torsion Angle is designated 1, and the others are numbered in a clockwise sense when viewed in the direction B → C (see Rule 2.2.3, which also covers the case where $\theta = \pm 60°$); e.g., in valine $H_i^{\gamma 11}$, $H_i^{\gamma 12}$, $H_i^{\gamma 13}$ and $H_i^{\gamma 21}$, $H_i^{\gamma 22}$, $H_i^{\gamma 23}$ [or Hγ11(*i*), etc.]. Where only two hydrogen atoms are present, they are designated in accordance with Rule 2.2.2, Case I for $-CH_2-R$ and Case III for $-NH_2$.

4.5. *Designation of Bonds and Torsion Angles*

4.5.1. Bonds are designated by means of the two atoms terminating them, e.g., $C_i^{\alpha}-C_i^{\beta}$, $N_i^{\zeta}-H_i^{\zeta 2}$, or, if no ambiguity results, by the symbol of the first atom of the bond, e.g., C_i^{α}, $C_i^{\gamma 1}$. In superscripts the bond may be denoted either by α; β; γ1; γ2 etc. or by 1; 2; 3,1; 3,2 etc. Bond lengths are denoted $b(C_i^{\alpha}, C_i^{\beta})$, bC_i^{α}, b_1^1, $b_i^{3,1}$ etc.

4.5.2. Torsion angles are denoted by χ, and are specified by two (or three) superscripts, the first one (or two) (in the situation A–B–C–E) indicating the bond B–C about which the angle is measured, and the last indicating whether the angle is measured relative to D, E, or F. The Principal Torsion Angle is defined by Rule 2.2.1, and if there is no ambiguity the last superscript may be omitted in referring to it.

$$\begin{array}{l} \quad\;\; D \\ \quad\; / \\ A-B-C-E \\ \quad\; \backslash \\ \quad\;\; F \end{array}$$

Thus, in valine, $\chi_i^{2,1}$ and $\chi_i^{2,2}$ refer to the torsion angles specifying atoms $C_i^{\gamma 1}$ and $C_i^{\gamma 2}$; in leucine, $\chi_i^{3,1,1}$, $\chi_i^{3,1,2}$, and $\chi_i^{3,1,3}$ refer to the torsion angles specifying the three hydrogen atoms attached to $C^{\delta 1}$. If there is no ambiguity, the Principal Torsion Angles may be referred to, in valine and leucine, as χ_i^2 and $\chi_i^{3,1}$, respectively. Corresponding notations without superscripts are χ2,1(*i*), χ2(*i*); χ3,1,1(*i*), χ3,1(*i*).

Note: By the Sequence Rule, when $\chi_1 = 0$, C^γ (or $C^{\gamma 1}$) is in the eclipsed position relative to N.

Rule 5.

5.1. *Polarity of Hydrogen Bonds*. In specifying a hydrogen bond as directed from residue i to residue k (or from atom X_i to atom Y_k), the direction X–H to :Y is implied; i.e., the atom covalently linked to the hydrogen atom is mentioned first.

Example: In the α-helix the N–H of residue i is hydrogen-bonded to the C=O of residue (i–4). Therefore, the α-helix is described as having i to (i–4), or (5–1), hydrogen bonding.

5.2. *Dimensions of Hydrogen Bonds*. Dimensions may be denoted by natural extensions of the nomenclature given above. For example, in $N_i–H_i \cdots O_k = C_k$, the following symbols might be used: $b(H_i, O_k)$, $\tau(N_i, H_i, O_k)$, $\tau(H_i, O_k, C_k)$, $\theta(H_i, O_k)$, $\theta_i(N, H)$, $\theta_k(C, O)$.

Rule 6. Helical Segments

A regular helix is strictly of infinite length, with all ϕ's identical and all ψ's identical. A helical *segment* of polypeptide chain may be defined *either* in terms of ϕ and ψ, *or* in terms of symmetry and hydrogen-bond arrangement.

6.1. In the description of helices or helical segments, the following symbols should be used:

n = number of residues per turn,
h = unit height (translation per residue along the helix axis),
t = $360°/n$ = unit twist (angle of rotation per residue about the helix axis).

6.2. *Definition in Terms of ϕ and ψ*. Under this definition, a helical segment is referred to as (ϕ, ψ) helix; thus a right-handed α-helix would be a (−57°, −47°) helix. The *first* and *last* residues of the helical segment are taken to be the first and last residues which have ϕ and ψ values equal to those defining the helix, within limits which should be defined in the context. No account is taken of hydrogen-bonding arrangements.

6.3. *Definition in Terms of Symmetry and Hydrogen-bond Arrangement*. A helix is referred to as an n_r helix, where

n = number of residues per turn,
r = number of atoms in ring formed by a hydrogen bond and the segment of main chain connecting its extremities.

Thus an α-helix would be 3.6_{13}. The *first* helical residue is taken as the first whose CO group is *regularly* bonded to NH along the helix (in the case of an α-helix, to the NH of the fifth residue); the *last* helical residue is the last whose NH is *regularly* hydrogen-bonded to CO along the helix (in the case of an α-helix, to the CO of the residue last but four). Irregular hydrogen-bonding arrangements are not considered to form part of the helix.

Notes: (i) A helical segment defined by Rule 6.2 may, but need not necessarily, be two residues shorter than the same segment defined by Rule 6.3.

(ii) These rules prescribe no definitions for irregular helical segments.

APPENDIX

Recommendation A. Conformation and Configuration

There is at present no agreed definition of these two terms for general stereochemical usage.

In polypeptide chemistry, the term "conformation" should be used, in conformity with current usage, to describe different spatial arrangements of atoms produced by rotation about covalent bonds; a change in conformation does not involve the breaking of chemical bonds (except hydrogen bonds) or changes in chirality (see Cahn, Ingold, and Prelog.[7])

On the other hand, in polypeptide chemistry the term "configuration" is currently used to describe spatial arrangements of atoms whose interconversion requires the formal breaking and making of covalent bonds (*note*: this usage takes no account of the breaking or making of hydrogen bonds). [Cf. a more extensive discussion in IUPAC Tentative Rules for the Nomenclature of Organic Chemistry, Section E, Fundamental Stereochemistry, IUPAC Information Bulletin No. 35, pp. 71–80 (*Biochim. Biophys. Acta,* 208, 1 (1970)).]

Recommendation B. Definitions of Primary, Secondary, Tertiary, and Quaternary Structure

These concepts, originally introduced by Linderstrøm-Lang,[12]* cannot be defined with precision, but the definitions given below may be helpful.

B.1. The *primary structure* of a segment of polypeptide chain or of a protein is the amino-acid sequence of the polypeptide chain(s), without regard to spatial arrangement (apart from configuration at the α-carbon atom).

Note: This definition does not include the positions of disulphide bonds, and is therefore not identical with "covalent structure."

B.2. The *secondary structure* of a segment of polypeptide chain is the local spatial arrangement of its main-chain atoms without regard to the conformation of its side chains or to its relationship with other segments.

B.3. The *tertiary structure* of a protein molecule, or of a subunit of a protein molecule, is the arrangement of all its atoms in space, without regard to its relationship with neighboring molecules or subunits.

B.4. The *quaternary structure* of a protein molecule is the arrangement of its subunits in space and the ensemble of its inter-subunit contacts and interactions, without regard to the internal geometry of the subunits.

Note: A protein molecule not made up of at least potentially separable subunits (not connected by covalent bonds) possesses no quaternary structure. Examples of proteins without quaternary structure are ribonuclease (1 chain) and chymotrypsin (3 chains).

REFERENCES

1. **Arnott and Dover,** *J. Mol. Biol.,* 30, 209 (1967).
2. **Arnott and Dover,** *Acta Cryst.,* B24, 599 (1968).
3. **Arnott, Dover, and Elliott,** *J. Mol. Biol.,* 30, 201 (1967).
4. **Cahn,** *J. Chem. Educ.,* 41, 116 (1964).
5. **Cahn and Ingold,** *J. Chem. Soc.,* 612 (1951).
6. **Cahn, Ingold, and Prelog,** *Experientia,* 12, 81 (1956).
7. **Cahn, Ingold, and Prelog,** *Angew. Chem. Int. Ed.,* 5, 385, 511 (1966); *Angew. Chem.,* 78, 413 (1966).

*The use of the terms "primary, secondary, tertiary, and quaternary structure" has been criticized as being imprecise by Wetlaufer.[18] He has proposed an alternative terminology.

8. **Edsall, Flory, Kendrew, Liquori, Némethy, Ramachandran, and Scheraga,** *J. Biol. Chem.,* 241, 1004 (1966); *Biopolymers,* 4, 130 (1966); *J. Mol. Biol.,* 15, 339 (1966).
9. **Eliel,** *Stereochemistry of Carbon Compounds,* McGraw Hill, New York, 1962, 92.
10. IUPAC Information Bulletin No. 35, 36 (1969).
11. **Klyne and Prelog,** *Experientia,* 16, 521 (1960).
12. **Linderstrøm-Lang,** *Proteins and Enzymes, Lane Memorial Lectures,* Stanford University Press, 1952, 54.
13. **Ramachandran, Ramakrishnan, and Sasisekharan,** *J. Mol. Biol.,* 7, 95 (1963).
14. **Ramachandran and Sasisekharan,** *Advan. Prot. Chem.,* 23, 283 (1968).
15. **Ramachandran, Sasisekharan, and Ramakrishnan,** *Biochim. Biophys. Acta,* 112, 168 (1966).
16. **Schellman and Schellman,** in *The Proteins,* Vol. 2, Neurath, Ed., Academic Press, New York, 1964, 1.
17. **Traub and Schmueli,** in *Aspects of Protein Structure,* Ramachandran, Ed., Academic Press, London, 1963, 81.
18. **Wetlaufer,** *Nature,* 190, 1113 (1961).
19. **Yonath and Traub,** *J. Mol. Biol.,* 43, 461 (1969).

A ONE-LETTER NOTATION FOR AMINO ACID SEQUENCES*† TENTATIVE RULES

IUPAC-IUB Commission on Biochemical Nomenclature (CBN)

INTRODUCTION

1. General Considerations

Various difficulties are encountered when presenting the formulas of long protein sequences in the usual three-letter symbols.[1] Space is often at a premium. A one-letter code minimizes this difficulty and has other distinct advantages. In summarizing large amounts of data or in the alignment of homologous protein sequences, it is important that the patterns in the sequences be condensed and simplified as much as possible. Computer techniques are increasingly applied for the storage of sequences of hundreds of amino acid residues and for their evaluation. For this purpose, a one-letter code is the best solution. Finally, a one-letter code is useful in the labeling of individual amino acid side-chains in three-dimensional pictures of protein molecules.

The possibility of using one-letter symbols was mentioned by Gamow and Ycas[2] in 1958. The idea was systematized by Šorm et al.[3] in 1961. It was used by this group[4–10] and also by Fitch[11] in several papers on the structure of proteins. In extensive compilations of protein structures, Eck and Dayhoff[12–14] systematically used one-letter symbols derived partly from the code of Šorm and Keil. Independent proposals were made by Wiswesser[15] and by Braunstein.[16]

In view of the increasing number of different notations and the attending problems, the IUPAC-IUB Commission on Biochemical Nomenclature (CBN) has undertaken the task of drafting a single notation for one-letter symbols. The present proposal was evolved by a CBN subcommission (composed of B. Keil, R. V. Eck, M. O. Dayhoff, and W. E. Cohn); it is based principally on the most recent summary published by Dayhoff and Eck.[14]

2. Limits of Application

In publications, CBN recommends that one-letter symbols be used only in comparisons of long sequences in tables, lists, or figures, and for such special use as tagging three-dimensional models of proteins. They should not be used in simple text nor for original reports of experimental details of sequences. This system is not suitable for reporting the details of peptide synthesis, for example, where a fuller description of substituents is needed and where uncommon amino acids may occur. It should not be used in papers where the single-letter system for nucleoside sequences is employed (Reference 1a, sections 5.4 and 5.5), as in representing codons, etc.

RULES

3. Principles of the One-letter Code

3.1. The letter written at the left-hand end is that of the amino acid residue carrying the free amino group and the letter written at the right-hand end is that of the amino acid residue carrying the free carboxyl group. The absence of punctuation beyond either end

*Document of the IUPAC-IUB Commission on Biochemical Nomenclature (CBN), approved by IUPAC and IUB in March 1968 and published by permission of the International Union of Pure and Applied Chemistry, the International Union of Biochemistry, and the official publishers to the International Union of Pure and Applied Chemistry, Messrs. Butterworths Scientific Publications.

†From *Pure Appl. Chem.*, 31(4), 639–645 (1972). With permission.

of a sequence implies that it is known to be the amino or carboxyl end of the protein. A fragmentary sequence is to be preceded or followed by a slash (/) to indicate that it is not known to be the end of the complete protein (see comment in section 8.2).

3.2. Initial letters are used where there is no ambiguity. There are six such cases – cysteine, histidine, isoleucine, methionine, serine and valine. All the other amino acids share the initial letters A, G, L, P, or T and assignments of them must therefore be somewhat arbitrary. These letters are assigned to the most frequently occurring and structurally most simple amino acids. On the basis, the letters A, G, L, P, and T are assigned to alanine, glycine, leucine, proline, and threonine, respectively.

3.3. The assignment of the other abbreviations is more arbitrary. However, certain clues are helpful. Two are phonetically suggestive, F for *ph*enylalanine, and R for *ar*ginine. For tryptophan, the double ring in the molecule is associated with bulky letter W. The letters N and Q are assigned to asparagine and glutamine, respectively; D and E are assigned to aspartic acid and glutamic acid, respectively. This leaves lysine and tyrosine, to which K and Y are assigned. These are chosen rather than any of the few other remaining letters because they are alphabetically nearest the initial letters L and T. U and O are avoided because U is easily confused with V in handwritten work and O is confused with G, Q, C, and D in imperfect computer print-outs and also with zero. J is avoided for linguistic reasons.

3.4. Two other abbreviations are necessary in order to avoid ambiguity. B is assigned to aspartic acid or asparagine when this distinction has not been determined. Z is assigned when glutamic acid and glutamine have not been distinguished. X means that the identity of an amino acid is undetermined, or the amino acid is atypical.

4. Abbreviations in Alphabetical Order

These are listed in Tables 1 and 2.

5. Spacing

A very important use of the one-letter notation is in presenting alignments of many homologous sequences. In printing, it often happens that the alignment is not perfectly maintained because of the variable size of the letters and the variable amount of punctuation. This effect can be very troublesome in extensive comparisons. Therefore, a single typewriter space is left between letters, either as a blank or occupied by punctuation (see Sections 6, 7, 8). The alignment is preserved by allowing exactly the same spacing for each letter, each blank, and each punctuation mark, as in typewritten material or, if printed, as in "typewriter type font."

6. Known and Unknown Sequences*

A blank between letters indicates that the sequence is known. (See also comment in Section 8.2.) As in the three-letter notation, parentheses and commas are used to indicate regions in which the sequence is unknown or undetermined.

Example. The β-corticotropin releasing factor, using three-letter symbols:

Ser-Tyr-Cys-Phe-His(Asn, Gln)Cys(Pro, Val)Lys-Gly

or one-letter symbols:

S Y C F H(N,Q)C(P,V)K G

*The sequence quoted for β-corticotropin releasing factor has been withdrawn from the 1969 *Atlas* (see References 12–14). Hence the sequence used in Section 6 should be regarded only as a hypothetical example for purposes of illustration.

7. Juxtaposition of Unknown Sequences Known to be Connected

Consider the two sequences, one completely known, the other containing peptides of unknown internal sequence.

a) Ala-Cys-Asp-Glu-Phe-Gly-His-Ile-Lys-Leu-Met-Asn- Pro-Gln
b) (Ala,Cys,Asp)(Arg,Ser)(Gly,His,Ile)Lys-Leu-Met-Asn- Pro-Gln

In one-letter notation, these become:

a) A C D E F G H I K L M N P Q
b) (A , C , D) (R , S) (G , H , I) K L M N P Q
↑ ↑

In the second illustration, two punctuation marks have been crowded into each of two single spaces (indicated by the arrows). In a computer print-out, this would not be possible. A single one-space symbol must be used. Here = is used for)(to indicate the end of one unknown sequence and the beginning of another, as shown below.

a) A C D E F G H I K L M N P Q
b) (A , C , D = R , S = G , H , I) K L M N P Q
↑ ↑

8. Juxtaposition of Residues Inferred, but not Known, to be Connected

Consider the following case in which peptides from a second sequence (d) can be aligned with a known, related sequence (c).

c) A C D E F G H I K L M N P Q
d) (A . C . D = R , S = G . H . I) K L / M N / P Q /

8.1. In this illustration, the sequences of two of the fragments (A.C.D and G.H.I in d), while not determined, are inferred with good confidence, which is indicated by dots instead of commas between their residues. Where such inferences cannot be made with confidence, commas, which retain their original connotation of "unknown sequence" (Section 6), should be used, as in the R,S dipeptide.

8.2. The two internal slashes (/) separate adjacent amino acids that come from different peptides not proven experimentally to be connected. The third (end) slash indicates that Q is not experimentally proven to be at the carboxyl end of the protein, although it is at the carboxyl end of the P–Q dipeptidyl residue.

Comment. The absence of punctuation at the beginning or end of a complete polypeptide or protein sequence indicates the known amino or carboxyl terminal, respectively (see Section 3.1).

8.3. Depending on the experimental details and the nature of the inferences to be represented, even more elaborate punctuation may sometimes be required. It is essential, however, that only one character (or a blank space of similar size) appear between the single letters to preserve the spacing that is essential for comparisons (see Section 5).

Table 1

Amino acid	One-letter symbol	
*A*lanine	A	
*Ar*ginine	R	
Asparagine	N	B[a]
Aspartic acid	D	
*C*ysteine	C	
Glutamine	Q	Z[b]
Glutamic acid	E	
*G*lycine	G	
*H*istidine	H	
*I*soleucine	I	
*L*eucine	L	
Lysine	K	
*M*ethionine	M	
*Ph*enylalanine	F	
*P*roline	P	
*S*erine	S	
*T*hreonine	T	
Tryptophan	W	
T*y*rosine	Y	
*V*aline	V	
Unknown or "other".	X	

[a]For aspartic acid *or* asparagine.
[b]For glutamic acid *or* glutamine.

Table 2

One-letter symbol	Three-letter symbol	Amino acid
A	Ala	*A*lanine
B	Asx	Aspartic acid *or* asparagine
C	Cys	*C*ysteine
D	Asp	Aspartic acid
E	Glu	Glutamic acid
F	Phe	*Ph*enylalanine
G	Gly	*G*lycine
H	His	*H*istidine
I	Ile	*I*soleucine
K	Lys	Lysine
L	Leu	*L*eucine
M	Met	*M*ethionine
N	Asn	Asparagine
P	Pro	*P*roline
Q	Gln	Glutamine
R	Arg	*Ar*ginine
S	Ser	*S*erine
T	Thr	*T*hreonine
V	Val	*V*aline
W	Trp	Tryptophan
X	–	Unknown or "other"
Y	Tyr	T*y*rosine
Z	Glx	Glutamic acid *or* glutamine

REFERENCES

1a. IUPAC-IUB Tentative Rules, *J. Biol. Chem.*, 241, 527 (1966); see also *Eur. J. Biochem.*, 1, 259, (1967).

1b. IUPAC-IUB Tentative Rules, *J. Biol. Chem.*, 241, 2491 (1966); see also *Eur. J. Biochem.*, 1, 375 (1967). (Replaced in 1971. See "Symbols for Amino-acid Derivatives and Peptides.")

2. **Gamow and Yčas,** *Symp. on Information Theory in Biology,* Pergamon Press, New York, 1958.

3. **Šorm, Keil, Vaněček, Tomášek, Mikeš, Meloun, Kostka, and Holeyšovský,** *Collect. Czech. Chem. Commun.*, 26, 531 (1961).

4. **Mikeš, Holeyšovský, Tomášek, Keil, and Šorm,** *Collect. Czech. Chem. Commun.*, 27, 1964 (1962).

5. **Holeyšovský, Alexijev, Tomášek, Mikeš, and Šorm,** *Collect. Czech. Chem. Commun.*, 27, 2662 (1962).

6. **Šorm and Keil,** *Adv. Protein Chem.*, 17, 1967 (1962).

7. **Mikeš, Holeyšovský, Tomášek, and Šorm,** *6th Int. Congr. Biochem.*, New York, 1964, Abstr. II-136, p. 169.

8. **Keil, Prosík, and Šorm,** *Biochim. Biophys. Acta,* 78, 559 (1963).

9. **Mikeš, Prusik, and Svoboda,** *Collect. Czech. Chem. Commun.*, 29, 1193 (1964).

10. **Šorm, Holeyšovský, Mikeš, and Tomášek,** *Collect. Czech. Chem. Commun.*, 30, 2103 (1965).

11. **Fitch,** *J. Mol. Biol.*, 16, 1, 9, 17 (1966).

12. **Dayhoff, Eck, Chang, and Sochard,** *Atlas of Protein Sequence and Structure,* National

13. **Eck and Dayhoff,** *Atlas of Protein Sequence and Structure,* National Biomedical Research Foundation, Washington, D.C., 1968.

14. **Dayhoff and Eck,** *Atlas of Protein Sequence and Structure,* National Biomedical Research Foundation, Washington, D.C., 1968.

15. **Wiswesser,** *Chem. Eng. News,* 42, 4 (1964).

16. **Braunstein,** personal proposal to CBN.

17. **Schally and Bowers,** *Metabolism,* 13, 1190 (1964).

SYMBOLS FOR AMINO-ACID DERIVATIVES AND PEPTIDES* † RECOMMENDATIONS (1971)

IUPAC-IUB Commission on Biochemical Nomenclature

The revised "Tentative Rules" published by CBN in 1966[1] were an attempt to achieve a broad systematization of various types of abbreviated notation already in use [e.g., Brand and Edsall, *Annu. Rev. Biochem.,* 16, 224 (1947); Report of the Committee on Abbreviations of the American Society of Biological Chemists, December 18, 1959; Report of the Committee on Nomenclature of the European Peptide Symposium, Pergamon Press, 1963, pp. 261–269; "Tentative Rules for Abbreviations and Symbols of Chemical Names of Special Interest in Biological Chemistry"[2]]. They sought to reconcile the needs of the protein chemist, i.e., indication of amino-acid sequences, with those of persons concerned more with the chemical reactions of proteins and the synthesis of polypeptides, i.e., the need for conveying more detailed chemical information in abbreviated form.

Recent progress in the field of peptide synthesis and in the chemical modification of proteins has made necessary a revision of these "Tentative Rules." This revision has been aided by the work of an expert group consisting of J. S. Fruton, B. S. Hartley, R. R. Porter, J. Rudinger, R. Schwyzer, and G. T. Young. They are greatly indebted to many colleagues, notably W. H. Stein, for helpful suggestions.

1. General Considerations

1.1. The symbols chosen are derived from the trivial names or chemical names of the amino acids and of chemicals reacting with amino acids and polypeptides. For the sake of clarity, brevity, and listing in tables, the symbols for amino-acid residues have been, wherever possible, restricted to three letters, usually the first letters of the trivial names.

1.2. The symbols represent not only the names of the compounds but also their structural formulas.

1.3. The amino-acid symbols by themselves represent the amino acids. The use of the symbols to represent the free amino acids is *not* recommended in textual material, but such use may occasionally be desirable in tables, diagrams, or figures. Residues of amino acids are represented by addition of hyphens in specific positions as indicated in Section 3.

1.4. Heteroatoms of amino-acid residues (e.g., O^3 and S^3 of serine and cysteine, respectively, N^6 of lysine, N^2 of glycine, etc.) do not explicitly appear in the symbol; such features are understood to be encompassed by the abbreviation.

1.5. Amino-acid symbols denote the L configuration unless otherwise indicated by D or DL appearing before the symbol and separated from it by a hyphen. When it is desired to make the number of amino-acid residues appear more clearly, the hyphen between the configurational prefix and the symbol may be omitted (see 6.3.1.1 et seq.). (Note: The designation of an amino-acid residue as DL is inappropriate for compounds having another amino-acid residue with an asymmetrical center.)

1.6. Structural formulas of complicated features may be used along with the abbreviated notation wherever necessary for clarity.

1.7. All symbols listed below are to be printed or typed as one capital letter followed by two lower-case letters, e.g., Gln, not GLN or gln or GlN or glN, regardless of position

*Document of the IUPAC-IUB Commission of Biochemical Nomenclature (CBN), approved by CBN in May 1971, and published by permission of the International Union of Pure and Applied Chemistry and the International Union of Biochemistry.

†From *Pure Appl. Chem.,* 40(3), 315–331 (1974). With permission.

in a sentence or structure. However, when used for purposes other than to represent an amino-acid residue (e.g., to designate a genetic factor), three lower-case italic letters (i.e., *gln*) should be used.

2. Symbols for Amino Acids

2.1. *Common Amino Acids*

Alanine	Ala	Leucine	Leu
Arginine	Arg	Lysine	Lys
Asparagine	Asn*	Methionine	Met
Aspartic acid	Asp	Phenylalanine	Phe
Cysteine	Cys	Proline	Pro
Glutamic acid	Glu	Serine	Ser
Glutamine	Gln*	Threonine	Thr
Glycine	Gly	Tryptophan	Trp (not Try)
Histidine	His	Tyrosine	Tyr
Isoleucine	Ile	Valine	Val

2.2. *Less-common Amino Acids* – Symbols for less-common amino acids should be defined in each publication in which they appear. The following principles and notations are recommended.

2.2.1. *Hydroxyamino Acids*

		Preferred alternatives		
5-Hydroxylysine	5Hyl	Lys(5OH)	or	Lys 5 | OH
3-Hydroxyproline	3Hyp	Pro(3OH)	or	OH 3 | Pro
4-Hydroxyproline	4Hyp	Pro(4OH)	or	Pro 4 | OH

2.2.2 *Allo-amino Acids*

Alloi-Isoleucine	aIle			
Alloh-Hydroxylysine	aHyl	aLys(5OH)	or	OH 5 | aLys

2.2.3.*"Nor" and "Homo" Amino Acids* – "Nor" (e.g., in norvaline) is not used in its accepted sense (denoting a lower homologue) but to change the trivial name of a branched-chain compound into that of a straight-chain compound (compare with "iso," paragraph 2.1). "Nor" should therefore be treated as part of the trivial name without special emphasis. "Homo," used in the sense of a higher homologue, may also be incorporated into the trivial name.

Norvaline	Nva	Homoserine	Hse
Norleucine	Nle	Homocysteine	Hcy

*Asparagine and glutamine may also be denoted as Asp(NH_2) or Asp
|
NH_2, and Glu(NH_2) or Glu
|
NH_2, respectively, if necessary (as when the NH_2 is substituted, or its removal or modification is under discussion). See 4.2.

Glx may be used when the residue denoted could be "glutamic acid or glutamine;" similarly, Asx for "aspartic acid or asparagine."

2.2.4. *Higher Unbranched Amino Acids* – The functional prefix "amino" is included in the symbol as the letter "A," diamino as "A_2."* The trivial name of the parent acid is abbreviated to two letters. The word "acid" ("-säure," etc.) is omitted from the symbol as carrying no significant information. Unless otherwise indicated, single groups are in the 2 position, two amino groups in the 2 and terminal positions (monocarboxylic acids) or 2 and 2′ positions (dicarboxylic acids). The location of amino groups in positions other than these is shown by appropriate prefixes.

Examples:

2-Aminobutyric acid	Abu
2-Aminoadipic acid	Aad
2-Aminopimelic acid	Apm
2,4-Diaminobutyric acid	A_2bu*
2,2-Diaminopimelic acid	A_2pm*
2,3-Diaminopropionic acid	A_2pr*

or Ala($3NH_2$) or $\overset{\overset{\displaystyle NH_2}{3|}}{\text{Ala}}$ (see 4.3)

β-Alanine	βAla
Ornithine (2,4-diaminovaleric acid)	Orn
6-Aminohexanoic acid	εAhx**
3-Aminoadipic acid	βAad

2.2.5. *N^2-Alkylated Amino Acids* – N^2-Alkylamino acids are becoming more and more common (e.g., in the large group of depsipeptides). This justifies special symbols.

Examples:

N-Methylglycine (sarcosine) (see 6.2)	MeGly or Sar
N-Methylisoleucine	MeIle
N-Methylvaline, etc.	MeVal, etc.
N-Ethylglycine, etc.	EtGly, etc.

2.3. *Nonamino-acid Residues Linked to Peptides* – For residues of muramic, sialic, neuraminic, etc., acids linked to amino-acid residues, as in bacterial-cell-wall components, the symbols Mur, Sia, Neu, etc. (preceded by Ac if *N*-acetylated) are recommended. The symbols for sugar residues (Glc, Gal, etc.)[2] and nucleosides (Ado, Cyd, etc.)[3] may also be used.

3. Amino-acid Residues

The links between residues have frequently been shown by peptide chemists as full points (periods, dots: ·) and by carbohydrate chemists (generally) as short strokes (dashes, hyphens: -). At times, special symbols have been used (> or →) to show the direction of what is in all cases an unsymmetrical link (peptide or glycoside).

For consistency and ease of typing as well as economy in printing, the hyphen, representing the peptide bond, should be the standard connecting symbol.[2]

*The symbols for diamino compounds previously (1) utilized the letter "D" for "diamino". However, the overuse of D as the initial letter for many compounds beginning with "di" (and of "T" for "tri" and "tetra"), in addition to the fact that standard chemical symbolism utilizes subscript numerals for multipliers, leads to the proposal that diamino should be represented by A_2. This eliminates the ambiguity attached to "D" and makes more clear the chemical relationship between the diamino and monoamino derivatives. It is in keeping with the increasing use of Me_2SO instead of DMSO and of Me_3Si- in place of TMS- and with the earlier proposal of H_4 for tetrahydro.[4]

**Recommended in place of the previous[1] εAcp, in which "cp" for caproic may be confused with capric and caprylic.

The simple usage by which Gly-Gly-Gly stands for glycylglycylglycine appears to involve the employment of the same three letters (Gly) for three different residues or radicals (b), (c), (d) below. However, if the dashes or hyphens are considered as part of each symbol, we have four distinct forms, for the free amino acid and the three residues, viz:

(a)	Gly	=	$NH_2-CH_2-CO_2H$	the free amino acid
(b)	Gly-	=	NH_2-CH_2-CO-	the left-hand unit
(c)	-Gly-	=	$-NH-CH_2-CO-$	the middle unit
(d)	-Gly	=	$-NH-CH_2-CO_2H$	the right-hand unit

For peptides, a distinction may be made between the *peptide,* e.g., Gly-Glu (shown *without* dashes at the ends of the symbols), and the *sequence,* e.g., -Gly-Glu- (shown *with* dashes at the ends of the symbols).

3.1. *Lack of Hydrogen on the 2-Amino Group* – The 2-amino group is understood to be at the left-hand side of the symbol when hyphens are used, and – in special cases – at the point of the arrow when arrows are used to indicate the direction of the peptide bond ($-CO \rightarrow NH-$, $-NH \leftarrow CO-$). (For substitution for 2-amino hydrogen, see 4.1.)

Examples:

-Gly: $-HNCH_2COOH$

>Gly or $^{\perp}$Gly: $>NCH_2COOH$

-Ala:
$$\begin{array}{c} CH_3 \\ | \\ -HNCHCOOH \end{array}$$

>Ala or $^{\perp}$Ala:
$$\begin{array}{c} CH_3 \\ | \\ >NCHCOOH \end{array}$$

3.2. *Lack of Hydroxyl on the 1-Carboxyl Group* – The 1-carboxyl group is understood to be on the right-hand side of the symbol when hyphens are employed and – in such special cases as 6.3.1.3 – at the tail of the arrow when arrows are used to indicate the direction of the peptide bond ($-CO \rightarrow NH-$, $-NH \leftarrow CO-$).

Example:

Gly-: H_2NCH_2CO-

It is generally convenient to use the same abbreviated formula for a polypeptide no matter what its state of ionization. To show that a peptide is acting as a cation or anion, the amino-terminal and carboxyl-terminal ends of the peptide are amplified with H and OH, respectively (I); these may be modified to show the appropriate state of ionization (II or III).

H-Gly-Val-Thr-OH	or	Gly-Val-Thr	(I)
$^+H_2$-Gly-Val-Thr-OH	or	$^+$HGly-Val-Thr	(II)
H-Gly-Val-Thr-O^-	or	Gly-Val-ThrO^-	(III)

3.3. *Lack of Hydrogen on Amino, Imino, Guanidino, Hydroxyl, and Thiol Functions in the Side Chain* (for substitution in such positions, see 4.2)

Lys or Lys:
$$\begin{array}{c} H_2NCHCOOH \\ | \\ (CH_2)_4 \\ | \\ NH \\ | \end{array}$$

His or His:* $H_2NCHCOOH$ (CH_2, imidazole, πN–) or $H_2NCHCOOH$ (CH_2, imidazole, $N\tau$–)

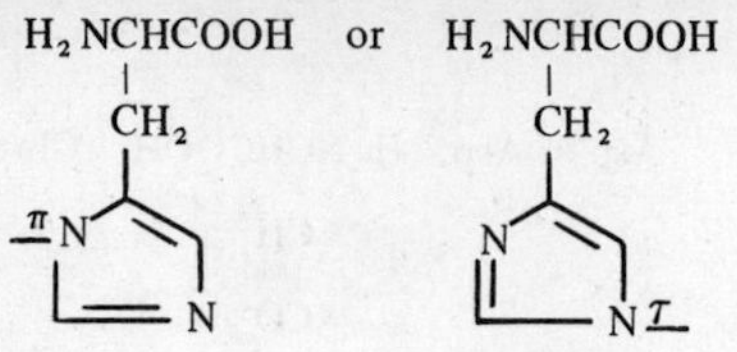

Trp or Trp: $H_2NCHCOOH$ (CH_2, indole N–)

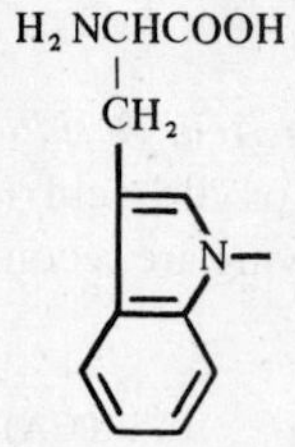

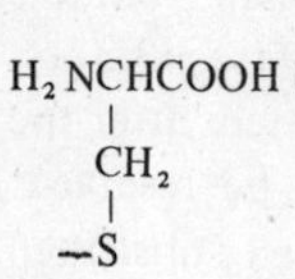

Arg or Arg: $H_2NCHCOOH$ — $(CH_2)_3$ — NH — C(H_2N)=N–

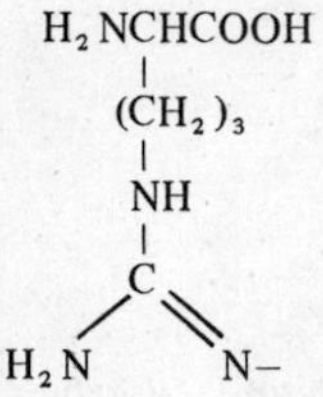

Ser or Ser: $H_2NCHCOOH$ — CH_2 — O —

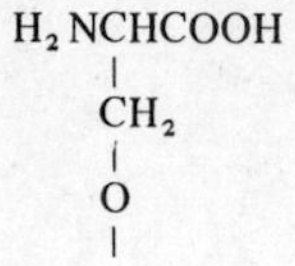

Tyr or Tyr: $H_2NCHCOOH$ — CH_2 — C_6H_4 — O —

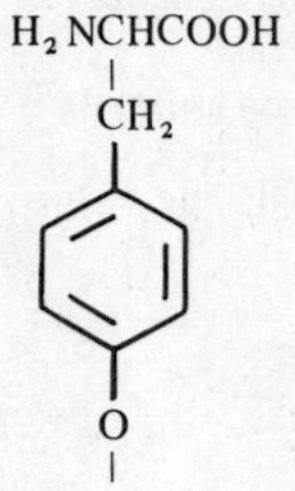

Cys or Cys: $H_2NCHCOOH$ ("half-cystine") — CH_2 — S—

(Cystine would be: Cys | Cys, Cys Cys, or Cys Cys, *not* Cys–Cys).

*The prolonged and well-entrenched ambiguity in the nomenclature of the *N*-methylhistidines (the chemist's *N*-1 being the biochemist's *N*-3 and vice versa) leads to the proposal that a new trivial system for designating these substances is necessary. It is therefore proposed that the imidazole N *nearer* the alanine residue be designated *pros* (symbol π) and the one *far*ther *tele* (symbol τ), to give the following names and symbols:

*pros*methylhistidine or N^{π}-methylhistidine, His(πMe);
*tele*methylhistidine or N^{τ}-methylhistidine, His(τMe).

3.4. *Lack of Hydroxyl on Carboxyl Groups in the Side Chain*

```
|                                 |
Asp or Asp:  H2NCHCOOH    Glu or Glu:  H2NCHCOOH
  |            |                    |       |
               CH2                          CH2
               |                            |
               CO                           CH2
               |                            |
                                            CO
                                            |
```

3.5. *Cyclic Derivatives of Amino Acid Residues* – For the special cases of the residues derived from pyrrolid-2-one-5-carboxylic acid (also known as pyroglutamic acid) and from homoserine lactone, the following are recommended:

```
┌─┐                                  OC–(CH2)2
└Glu-  or    &lt;Glu-    (not PCA)      |    |
                                     HN–CH–CO–

┌──┐                                 (H2C)2–O
-Hse┘   or   -Hse&gt;                      |     |
                                     –NH–CH–CO
```

4. Substituted Amino Acids

4.1. *Substitution in the 2-Amino and 1-Carboxyl Groups* – This follows logically from 3.1 and 3.2. The following examples will make the usage clear. (See also 6.2.)

N-Acetylglycine	Ac-Gly
Glycine ethyl ester	Gly-OEt
N^2-Acetyllysine	Ac-Lys
Serine methyl ester	Ser-OMe
O^1-Ethyl *N*-acetylglutamate	Ac-Glu-OEt
Isoglutamine	Glu-NH_2
O^1-Methyl hydrogen aspartate	Asp-OMe
N-Ethyl-*N*-methylglycine	Et-MeGly, $\genfrac{}{}{0pt}{}{Et}{Me}$ > Gly, $\genfrac{}{}{0pt}{}{Et}{\text{Me—Gly}}$ (Et above, vertical stroke)

4.2. *Substitution in the Side Chain* – Side-chain substituents may be portrayed above or below the amino-acid symbol (see 3.3 or 3.4), or by placing the symbol for the substituent in parentheses immediately after the amino-acid symbol.

The use of parentheses should be reserved for a *single* symbol denoting a side-chain substituent. Where a more complex substituent is involved, it is recommended that the vertical stroke and the two-line abbreviation be used.[5] In general, the one-line abbreviation should be used only when the structure of a substituted peptide is given in textual material.

O^4-Methyl hydrogen aspartate	OMe above Asp (vertical stroke) or Asp with OMe below (vertical stroke) or Asp(OMe)
O^5-Ethyl hydrogen *N*-acetyl-glutamate	Ac-Glu(OEt)

N^6-Acetyllysine	Ac | Lys or Lys or Lys(Ac) | Ac
O^3-Acetylserine	Ac | Ser or Ser or Ser(Ac) | Ac
O^4-Methyltyrosine	Me | Tyr or Tyr or Tyr(Me) | Me
S-Ethylcysteine	Et | Cys, or Cys or Cys(Et) | Et
S-Sulfocysteine (*S*-Cysteine-sulfonic acid)	SO_3H Cys | or | or Cys(SO_3H) Cys SO_3H
Cysteinesulfenic acid	OH Cys | or | or Cys(OH) Cys OH
Cysteinesulfinic acid	O_2H Cys | or | or Cys(O_2H) Cys O_2H
Cysteic acid (3-Sulfoalanine)	O_3H Cys | or | or Cys(O_3H) Cys O_3H
S-Cyanocysteine	CN Cys | or | or Cys(CN) Cys CN
Methionine sulfoxide	O | Met or Met or Met(O) | O
Methionine sulfone	O_2 | Met or Met or Met(O_2) | O_2
O^3-Phosphonoserine (Phosphoserine)	P | Ser or Ser or Ser(P) | P

	Me τ |
N^{τ}-Methylhistidine* (see 3.3) (*tele*methylhistidine)	His or His or His(τMe) τ | Me

similarly for N^{π} substitution (*pros*methylhistidine)

4.3. *Substitution on Carbon Side Chain* – This may use the same convention as in 4.2, with the addition of locant numerals where necessary, e.g.,

	NO_2 3 |
3-Nitrotyrosine	Tyr or Tyr or Tyr($3NO_2$) 3 | NO_2
	NH_2 3 |
2,3-Diaminopropionic acid (see 2.2.4) (3-aminoalanine)	Ala or Ala or Ala($3NH_2$) 3 | NH_2
Diiodotyrosine	Tyr(I_2)

5. Symbols for Substituents

Groups substituted for hydrogen or for hydroxyl may be indicated either by their structural formulas or by symbols or by combinations of both, e.g.,

Benzoylglycine (hippuric acid)	Ph-CO-Gly or C_6H_5CO-Gly or **Bz-Gly
Glycine methyl ester	Gly-OCH_3 or Gly-OMe
Trifluoroacetylglycine	CF_3CO-Gly

Suggestions for symbols designating substituent (or protecting) groups common in polypeptide and protein chemistry follow.

5.1. *N-Substituents (Protecting Groups) of the Urethane Type*

Benzyloxycarbonyl-	Z- or Cbz-
p-Nitrobenzyloxycarbonyl-	Z(NO_2)-
p-Bromobenzyloxycarbonyl-	Z(Br)-
p-Methoxybenzyloxycarbonyl	Z(OMe)-
p-Methoxyphenylazobenzyloxycarbonyl-	Mz-
p-Phenylazobenzyloxycarbonyl-	Pz-
t-Butoxycarbonyl-	Boc- or Bu^tOCO-
Cyclopentyloxycarbonyl-	Poc- or *c*PeOCO-

*The prolonged and well-entrenched ambiguity in the nomenclature of the *N*-methylhistidines (the chemist's *N*-1 being the biochemist's *N*-3 and vice versa) leads to the proposal that a new trivial system for designating these substances is necessary. It is therefore proposed that the imidazole N *nearer* the alanine residue be designated *pros* (symbol π) and the one *far*ther *tele* (symbol τ), to give the following names and symbols:

*pros*methylhistidine or N^{π}-methylhistidine, His(πMe);
*tele*methylhistidine or N^{τ}-methylhistidine, His(τMe).

**Bz- is the symbol generally used for *benzoyl* in organic chemistry. It should not be used for *benzyl* ($C_6H_5CH_2$ – or $PhCH_2$ –), for which the symbol is Bzl-. However, $PhCH_2$ – is unambiguous.

5.2. *Other N-Substituents*

Acetyl-	Ac-
Benzoyl- (C_6H_5CO-)	PhCO- or Bz-
Benzyl- ($C_6H_5CH_2$-)	$PhCH_2$- or * Bzl
Benzylthiomethyl-	$PhSCH_2$- or Btm-
Carbamoyl-	NH_2CO- (preferred to Cbm)
1-Carboxy-2-nitrophenyl-5-thio-	**Nbs-
3-Carboxypropionyl- (HOOC–CH_2–CH_2–CO–)†	Suc-
Dansyl- (5-dimethylaminonaphthalene-1-sulfonyl)	Dns-
Dinitrophenyl-	††N_2ph- or Dnp
Formyl-	HCO- or CHO-
p-Iodophenylsulfonyl (pipsyl)	Ips
Maleoyl- (–OC–CH=CH–CO–)	–Mal- or Mal<
Maleyl- (HOOC–CH=CH–CO–)	Mal-
Methylthiocarbamoyl-‡	MeNHCS- or ‡Mtc-
o-Nitrophenylthio-	Nps-
Phenylthiocarbamoyl-‡	PhNHCS- or ‡Ptc-
Phthaloyl-	–Pht- or Pht<
Phthalyl-	Pht-
Succinyl- (–OC–CH_2–CH_2CO–)‡‡	–Suc- or Suc<
Tetrahydropyranyl-	H_4pyran- (preferred to Thp††)
Tosyl- (*p*-tolylsulfonyl)	Tos-
Trifluoroacetyl-	††CF_3CO-
Trityl- (triphenylmethyl)-	Ph_3C- or Trt-

5.3. *Substituents at Carboxyl Group*

Benzyloxy- (benzyl ester)	–OCH_2Ph or –OBzl
Cyanomethoxy-	–OCH_2CN
Diphenylmethoxy- (benzhydryl ester)	–$OCHPh_2$ or –OBzh
Ethoxy- (ethyl ester)	–OEt
Methoxy- (methyl ester)	–OMe
p-Nitrophenoxy- (*p*-nitrophenyl ester)	–ONp
p-Nitrophenylthio-	–SNp
Phenylthio- (phenylthiolester)	–SPh
1-Piperidino-oxy-	–OPip
8-Quinolyloxy-	–OQu
Succinimido-oxy-	–ONSu
Tertiary butoxy- (*t*-butyl ester)	–OBu^t

*Bz is the symbol generally used for *benzoyl* in organic chemistry. It should not be used for *benzyl* ($C_6H_5CH_2$– or $PhCH_2$–), for which the symbol is Bzl-. However, $PhCH_2$– is unambiguous.
**See Comment following 5.3.
†Not succinyl, although it is the monovalent radical of succinic acid. See succinyl and Footnote ‡‡.
††The use of D for "di" and T for "tri" or "tetra" (and DH and TH for "dihydro" and "tetrahydro," respectively) is discouraged. Recognized symbols and subscripts are recommended.
‡The symbol Pth has been used to denote a phenylthiohydantoin (e.g., Pth-Leu). Since this incorrectly implies the substitution of an amino acid by a "phenylthiohydantoyl" group, it is suggested that the abbreviated symbol for such compounds be of the type CS-Leu-NPh or PhNCS-Leu (or Leu> PhNCS in textual material).
‡‡Not succinoyl.[6]

Comment

Many reagents used in peptide and protein chemistry for the modification (protection) of amino, carboxyl and side-chain groups in amino-acid residues have been designated by a variety of acronymic abbreviations, too numerous to be listed here. Extensive and indiscriminate use of such abbreviations is discouraged, especially where the accepted trivial name of a reagent is short enough, e.g., tosyl chloride, bromosuccinimide, trityl chloride, dansyl chloride, etc., or may be formulated in terms of the group transferred, e.g. N_2ph-F instead of FDNB for 1-fluoro-2,4-dinitrobenzene, Dns-Cl or dansyl-Cl in place of DNS, Nbs_2 in place of DTNB for 5,5′-dithio-bis(2-nitrobenzoic acid) (Ellman's reagent), $(Pr^iO)_2$PO-F, Pr^i_2*P*-F, iPr_2*P*-F, or Dip-F instead of DFP for diisopropylfluorophosphate. Other commonly used substances that may be expressed more clearly in terms of symbols are MalNEt (instead of NEM) for *N*-ethylmaleimide, Tos-PheCH_2Cl (instead of TPCK) for L-1-tosylamido-2-phenylethyl chloromethyl ketone, Tos-Arg-OMe (instead of TAME) for tosyl-L-arginine methyl ester, Me_3Si- (instead of TMS-) for trimethylsilyl, CF_3CO- (instead of TFA) for trifluoroacetyl (see 5.2), H_4furan (instead of THF), etc.

Some additional symbolic terms for substituents (and reagents), as examples, are

2-Aminoethyl-	$-(CH_2)_2NH_2$ (preferred to Aet)
Carbamoylmethyl-	$-CH_2CONH_2$ (preferred to Cam)
Carboxymethyl-	$-CH_2CO_2H$ (preferred to Cm)
Chloroethylamine	$Cl(CH_2)_2NH_2$
Ethyleneimine	$(CH_2)_2NH$
Chloroacetamide	$ClCH_2CONH_2$
Chloroacetic acid	$ClCH_2CO_2H$
p-Carboxyphenylmercuri-	–HgBzOH
p-Chloromercuribenzoate	*p*Cl-HgBzO–
Diazoacetyl-	N_2CHCO-
Hydroxyethyl-	$-(CH_2)_2OH$
Ethylene oxide	$(CH_2)_2O$

6. Polypeptides

6.1. *Polypeptide Chains*[5] – Polypeptides may be dealt with in the same manner as substituted amino acids, e.g.,

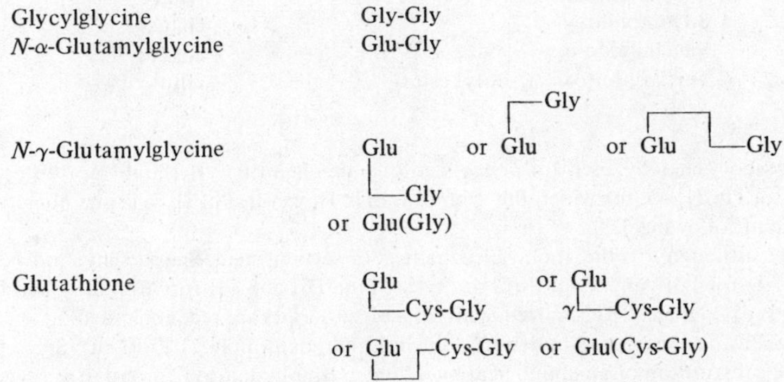

(Note that Glu over Cys-Gly (joined by a vertical bond) would represent the corresponding thiolester with a bond between the γ-carboxyl of glutamic acid and the thiol group of cysteine).

N^2-α-Glutamyllysine	Glu-Lys
N^6-α-Glutamyllysine	Glu—Lys (Lys below Glu) or Glu—Lys
N^2-γ-Glutamyllysine	Glu—Lys (Lys below Glu) or Glu—Lys
N^6-γ-Glutamyllysine	Glu over Lys or Glu Lys or Glu Lys

The presence of free, substituted, or ionized functional groups can be represented (or stressed) as follows:

Glycyllysylglycine	H-Gly-Lys(H)-Gly-OH
Its dihydrochloride	$^+H_2$-Gly-Lys($^+H_2$)-Gly-OH·2Cl$^-$
Its sodium salt	Gly-Lys-Gly-O$^-$ Na$^+$
Its N^6-formyl derivative	Gly-Lys(CHO)-Gly (CHO below Lys) or Gly-Lys(CHO)-Gly

etc.

6.2. *Peptides Substituted at N^2* (see 4.1)

Glycylnitrosoglycine	Gly — Gly (NO below) or Gly-(NO)Gly
Glycylsarcosine (see 2.2.5)	Gly — Gly (Me below) or Gly-MeGly or Gly-Sar
N-Glycyl-*N*-acetylglycine	Gly — Gly (Ac above) or Gly-(Ac)Gly
N,N-bisglycylglycine	Glu / Gly — Gly or $(Gly)_2$>Gly

6.3. *Cyclic Polypeptides*

6.3.1. *Homodetic Cyclic Polypeptides* – (The ring consists of amino-acid residues in peptide linkage only). Three representations are possible.

6.3.1.1. The sequence is formulated in the usual manner but placed in parentheses and preceded by (an italic) *cyclo*.

Example: Gramicidin S =

cyclo(-Val-Orn-Leu-D-Phe-Pro-Val-Orn-Leu-D-Phe-Pro-)

or (see 1.5, sentence 2).

cyclo(-Val-Orn-Leu-DPhe-Pro-Val-Orn-Leu-DPhe-Pro-)

6.3.1.2. The terminal residues may be written on one line, as in 6.3.1.1, but joined by a lengthened bond. Using the same example in the two forms (see 1.5):

Val-Orn-Leu-D-Phe-Pro-Val-Orn-Leu-D-Phe-Pro

or

Val-Orn-Leu-DPhe-Pro-Val-Orn-Leu-DPhe-Pro

6.3.1.3. The residues are written on more than one line, in which case the CO → NH direction must be indicated by arrows, thus (in the optional manner of 1.5):

→ Val → Orn → Leu → DPhe → Pro
Pro ← DPhe ← Leu ← Orn ← Val ←

6.3.2. *Heterodetic-cyclic Polypeptides* – (The ring consists of other residues in addition to amino-acid residues in peptide linkage): These follow logically from the formulation of substituted amino acids.

Example:

Oxytocin

Cys-Tyr-Ile-Gln-Asn-Cys-Pro-Leu-Gly-NH_2

Cyclic ester of threonylglycylglycylglycine

Thr-Gly-Gly-Gly or H-Thr-Gly-Gly-Gly

REFERENCES

1. Abbreviated Designation of Amino-Acid Derivatives and Peptides, *J. Biol. Chem.*, 241, 2491 (1966); *Biochemistry*, 5, 2485 (1966); *Biochim. Biophys. Acta*, 121, 1 (1966); *Biochem. J.*, 102, 23 (1967); *Arch. Biochem. Biophys.*, 121, 1 (1967); *Eur. J. Biochem.*, 1, 375 (1967); *Hoppe-Seyler's Z. Physiol. Chem.*, 348, 256 (1967); *Bull. Soc. Chim. Biol.*, 49, 121 (1967); *Molek. Biol.*, 2, 282 (1968); *Molek. Biol.*, 5, 492 (1971).
2. Abbreviations and Symbols for Chemical Names of Special Interest in Biological Chemistry (Section 5 revised by Reference 3 below), *J. Biol. Chem.*, 241, 527 (1966); *Biochemistry*, 5, 1445 (1966); *Biochem. J.*, 101, 1 (1966); *Virology*, 29, 480 (1966); *Arch. Biochem. Biophys.*, 115, 1 (1966); *Eur. J. Biochem.*, 1, 259 (1967); *Hoppe-Seyler's Z. Physiol. Chem.*, 348, 245 (1967); *Bull. Soc. Chim. Biol.*, 50, 3 (1968); *Molek. Biol.*, 1, 872 (1967). (Nom. I preceding).
3. Abbreviations and Symbols for Nucleic Acids, Polynucleotides and Their Constitutents, *Biochem. J.*, 120, 449 (1970); *Biochemistry*, 9, 4022 (1970); *Eur. J. Biochem.*, 15, 203 (1970); *J. Biol. Chem.*, 245, 5171 (1970); *Hoppe-Seyler's Z. Physiol. Chem.*, 351, 1055 (1970); *J. Mol. Biol.*, 55, 299 (1971); and elsewhere, including Nucleic Acids, Vol. I. of this *Handbook*.
4. Nomenclature of Vitamins, Coenzymes, and Related Compounds: Trivial Names of Miscellaneous Compounds of Importance in Biochemistry, Nomenclature of Quinones with Isoprenoid Side Chains, Nomenclature and Symbols for Folic Acid and Related Compounds, Nomenclature of Corrinoids, *Arch. Biochem. Biophys.*, 118, 505 (1967); *Biochem. J.*, 102, 15 (1967); *Biochim. Biophys. Acta*, 107, 1 (1965); *Biochim. Biophys. Acta*, 117, 285 (1966); *Eur. J. Biochem.*, 2, 1 (1967); *J. Biol. Chem.*, 241, 2987 (1966); *Bull. Soc. Chim. Biol.*, 49, 331 (1967); *Hoppe-Seyler's Z. Physiol. Chem.*, 348, 266 (1967).
5. Abbreviated Nomenclature of Synthetic Polypeptides (Polymerized Amino Acids), *Biopolymers*, 8, 161 (1969); *Arch. Biochem. Biophys.*, 123, 633 (1968); *Biochem. J.*, 106, 577 (1968); *Biochemistry*, 7, 483 (1968); *Biochem. Biophys. Acta*, 168, 1 (1968); *Eur. J. Biochem.*, 3, 129 (1968); *J. Biol. Chem.*, 243, 2451 (1968); *Bull. Soc. Chim. Biol.*, 51, 205 (1969); *Hoppe-Seyler's Z. Physiol. Chem.*, 349, 1013 (1969). Revised 1971 (Nom. V following).
6. IUPAC, Nomenclature of Organic Chemistry (Definitive Rules for Section C), Rule C-404 and Table VI, *Pure Appl. Chem.*, 11, Nos. 1–2 (1965).

ABBREVIATED NOMENCLATURE OF SYNTHETIC POLYPEPTIDES (POLYMERIZED AMINO ACIDS)*† REVISED RECOMMENDATIONS (1971)**

IUPAC-IUB Commission on Biochemical Nomenclature (CBN)

The numerous studies on the physical, chemical, and biological properties of synthetic polypeptides have brought with them different ways of describing, in abbreviated form, these products whose structures are often incompletely known. The use of a variety of nomenclatures complicates the literature; hence, a consistent and clearly defined system for naming such polypeptides is desirable. The proposals set forth here, which represent the consensus of many discussions and suggestions, should aid in systematizing the nomenclature of a wide variety of synthetic polypeptides.

These proposals are based in large part on the abbreviated nomenclature devised by Gill[2] and by Sela[3] and others. They utilize the symbols and conventions set forth in Section 2 of "Revised Tentative Rules for Abbreviations and Symbols of Chemical Names of Special Interest in Biological Chemistry"[4] and in "Abbreviated Designation of Amino Acid Derivatives and Peptides,"[5] and they add only those terms or conventions needed for the specification of polymers but not encompassed by these schemes.

The symbols and conventions of the previous "Tentative Rules"[4,5] used in this nomenclature system are summarized as follows. The symbols of the amino acid residues and their derivatives or modifications are those indicated in the "Tentative Rules"[4,5] or formulated according to the principles set out in them. Hyphens or commas between the symbols for residues or groups of residues indicate known or unknown sequence, respectively, and involve only the α-NH_2 and α-COOH groups (the peptide link). Commas may be omitted when other symbols (e.g., subscripts or superscripts) separate symbols in unknown sequences. Vertical strokes indicate covalent bonds involving functional groups or the remaining H-atom of the peptide bond, depending upon their placement.[5] L-Amino acids and α-peptide links, read from left (NH_2 terminus) to right (COOH terminus), are assumed unless indicated otherwise.[4,5]

Definitions

△ 1. **Linear Polymer** – All amino acid residues (constitutional units) are linked in an unbranched chain.

2. **Block** – A polymer that forms a distinct part of a larger polymer (e.g., a block or graft polymer may contain several blocks).

3. **Graft Polymer** – One or more blocks are linked to the functional groups of a linear polymer, thus creating a branch or branches. (Functional groups include ϵ-NH_2, β- or γ-COOH, etc., and the remaining H-atom of an α-peptide link.)

4. **Block Polymer** – Two or more species of block are linked to form a larger linear polymer.

*From IUPAC-IUB Commission on Biochemical Nomenclature, *Eur. J. Biochem.*, 26, 301 (1972). With permission.

†Document of the IUPAC-IUB Commission on Biochemical Nomenclature (CBN), originally approved and published by IUPAC and IUB in October 1967,[1] and again, in the present revised form, in 1971, approved and published by permission of the International Union of Pure and Applied Chemistry and the International Union of Biochemistry.

**This revision differs from the original[1] essentially in the addition of comments after Recommendation 1 (marked with ▲ in the margin) and in the relevant "Examples." These revisions were made to conform with the practices in polymer chemistry and were recommended to CBN by the IUPAC Commission on Macromolecular Nomenclature (*IUPAC Information Bulletin*, Appendices on Nomenclature, etc., No. 13, February 1971) and the Nomenclature Committee of the Division of Polymer Chemistry of the American Chemical Society [see *Macromolecules*, 1, 193 (1968)]. Minor revisions are indicated by △.

Recommendations

1. **Designation of Blocks or Linear Polymers** – The prefix "poly" or the subscript *n* indicates "polymer of." It is attached to each main chain and is repeated in each block within a larger polymer unless there is sufficient indication of size and of structure to make this repetition unnecessary. For example, poly(Glu) or $(Glu)_n$ represent poly(glutamic acid), and $(Glu)_{10}$, a decapeptide of glutamic acid. "Oligo" may replace "poly" for short chains.

Comments

(a) n replaces the p as originally, but no longer, used in the polymer nomenclature scheme devised by the IUPAC Subcommission on the Nomenclature of Macromolecules.[6] It is used in designating polynucleotides,[7] and it is chosen in place of p in order to avoid confusion with the "p" used for a terminal phosphoric residue in the latter scheme. The n may be replaced by a definite number (e.g., 10 above), an average (e.g., $\overline{10}$), or a range (e.g., 8–12), as appropriate. However, two n's should not appear in the same formula unless equal length is implied. When equal length is not the case, different letters should ▲ be used, such as m, j, k.

(b) If "poly" is used rather than the subscript n, the symbol(s) following "poly" should ▲ be enclosed in parentheses with no intervening space, e.g., poly(Lys). If "poly" is followed by a single, simple word, the whole is written as one word, e.g., polylysine. If what follows "poly" is complex, it should be enclosed in parentheses (again without following space), e.g., poly(amino acid), not polyamino acid or polyaminoacid; poly(glutamic acid) or polyglutamate, but not polyglutamic acid; poly(DL-alanine, L-lysine) for the substance shown in Example 2; and poly(DL-alanine-L-lysine) for the substance shown in Example 3. The format poly(L-lysine) is preferred to poly-L-lysine, i.e., L-lysine is regarded as a complex term. Similarly, poly(hydroxyproline), not polyhydroxyproline.

2. **Designation of Branches and Branch Points** – Branches (side chains) connected to the main chain can be designated in one of three ways: by a vertical line joining the main chain and the branch (side chain); by an extended bond joining the appropriate residues with the main chain wrtten first; or by a horizontal double dash (not preferred).

The branch points are indicated by the origin and terminus of the vertical line. If the origin is unknown, the line originates at the "p" in "poly," if "poly" is used, or at the first parenthesis (bracket), if the subscript n is used (see Recommendation 1). If the origin is known, the line originates: (a) vertically at the initial letter of the appropriate symbol, if functional groups other than α-NH_2 or α-COOH residues are involved; (b) vertically at the position of the appropriate link, if substitution for the remaining H-atom of a peptide link is involved; or (c) horizontally to the left or right of the symbol, respectively, if α-NH_2 or α-COOH groups are involved. The same rules apply to the termination of the line. Thus, the linkage between a side chain functional group and an α-NH_2 or α-COOH group in the main chain is indicated by two perpendicular lines with the vertical line originating in the functional group and the horizontal line in the α-NH_2 or α-COOH group. A number in parentheses lying beside the line indicates the number of such links per 100 residues of polymer, if known.

Comment

A limitation of the double dash as a connecting link lies in its inability to originate or to terminate definitively in a specific residue. Either the arrangement of the symbols must be such that connected ones are adjacent, or the information must be given independently.

Δ 3. **Block size** – A superscript outside the parentheses enclosing a block indicates the number of repeating sequences per 100 residues of polymer, and it is given to the first decimal place.

4. **The Molar Percentage** of a single type of amino-acid, residue within a copolymer, derived from the amino-acid analysis and assuming copolymerization, is indicated by a superscript attached to the symbol of the residue. The molar percentages are given in whole numbers and should total 99 to 101%.

Examples:

1. Simple homopolymer:

$$\text{poly (Ala) or (Ala)}_n$$

2. Linear copolymer, unknown sequence, composition not specified:

$$\text{poly (DLAla,Lys) or (DLAla,Lys)}_n$$

3. Linear copolymer, regular alternating sequence.

$$\text{poly(DLAla-Lys) or (DLAla-Lys)}_n$$

4. Linear sequence of unknown order [Composition: 56% Glu, 38% Lys, and 6% Tyr (Σ = 100%)] :

(a) poly($Glu^{56}Lys^{38}Tyr^{6}$) or $(Glu^{56}Lys^{38}Tyr^{6})_n$ (all L)
(b) poly ($DGlu^{56}DLys^{38}Tyr^{6}$) (only Tyr is L)
(c) poly ($DLGlu^{56}Lys^{38}DTyr^{6}$) (Glu is DL, Tyr is D)

5. Block polymer of poly(Glu) combined through the α-COOH terminus to the α-NH_2 terminus of poly(Lys) [Composition: 56% Glu, 44% Lys (Σ = 100%)] :

$$\text{poly}(Glu^{56})\text{-poly}(Lys^{44}) \text{ or } (Glu^{56})_n\text{-}(Lys^{44})_m$$

6. (a) Known, repeating sequence (a polymer of Glu-Lys-Lys-Tyr):

$$\text{poly(Glu-Lys}_2\text{-Tyr) or (Glu-Lys}_2\text{-Tyr)}_n$$

(b) Known, repeating sequences within each of two constituent blocks of a linear polymer [Composition: 37.5% Glu, 25% Lys, 25% Tyr, 12.5% Ala (Σ = 100%)] :

$$\text{poly(Glu-Lys)}^{25}\text{-poly(Ala-Tyr}_2\text{-Glu)}^{12.5}$$

or

$$\text{(Glu-Lys)}_n^{25}\text{-(Ala-Tyr}_2\text{-Glu)}_m^{12.5}$$

[The connection between the polymeric segments shown here is from the α-COOH of Lys to the α-NH_2 of Ala. Origin or termination in any other residue or functional group can be shown by rearranging the order of residues and by the orientation of the connecting line at its origin and terminus (see Examples 7, 8, and 9).]

(c) Known, repeating sequence in the main chain connected by the ϵ-NH_2 of a lysine (which of the two is not known) to an unknown point in an unknown sequence in the side chain [Composition: 30% Asp, 55% Glu, 10% Lys, 5% Tyr (Σ = 100%)]:

$$\begin{array}{c} \text{poly}(\text{Asp}^{30}\text{Glu}^{50}) \\ | \\ \text{poly}(\text{Glu-Lys}_2\text{-Tyr})^5 \end{array} \quad \text{or} \quad \begin{array}{c} (\text{Asp}^{30}\text{Glu}^{50})_m \\ | \\ (\text{Glu-Lys}_2\text{-Tyr})^5_n \end{array}$$

or poly(Glu-Lys_2-Tyr)^5 ⌐———⌐ poly($Asp^{30}Glu^{50}$)

or $(Glu\text{-}Lys_2\text{-}Tyr)^5_n$ ⌐———⌐ $(Asp^{30}Glu^{50})_m$

(*Note:* The double hyphen system is not applicable here.) If it is known which lysine residue is connected to the side chain, the main chain would be written in the form,

poly(Glu-Lys-Lys-Tyr)^5 or poly(Glu-Lys-Lys-Tyr)^5,

as appropriate.

7. Graft polymer with the main chain of DL-alanine and L-lysine connected through the ϵ-NH_2 group of lysine to the α-COOH group of L-tyrosine in the side chain, which consists of a block polymer of L-tyrosine and L-alanine (no analytical data for the main chain):

$$\begin{array}{l} \text{poly(Ala)-poly(Tyr)}\text{—⌐} \\ \qquad\qquad \text{poly(DLAla,Lys)} \end{array} \quad \text{or} \quad \begin{array}{l} (\text{Ala})_n\text{-}(\text{Tyr})_m\text{—⌐} \\ \qquad\qquad (\text{DLAla,Lys})_i \end{array}$$

or poly(DLAla,Lys) ⌐———⌐ poly(Ala)-poly(Tyr)—┘

or poly(DLAla,Lys)– –poly(Ala)-poly(Tyr)

(*Note:* The points of attachment of Lys and Tyr cannot be specified in the last example.)

8. Graft polymer with an unknown sequence in the main chain and in the side chain [Composition: 16% Lys, 20% Ala, 35% Glu, 29% Tyr (Σ = 100%)]:

(a) Number and position of the points of attachment in the main chain unknown, but terminating in the lysine residues of the side chain:

$$\begin{array}{l} \text{poly}(\text{Lys}^{16}\text{Ala}^{20}) \\ \;| \\ \text{poly}(\text{Glu}^{35}\text{Tyr}^{29}) \end{array} \quad \text{or} \quad \begin{array}{l} (\text{Lys}^{16}\text{Ala}^{20})_m \\ \;| \\ (\text{Glu}^{35}\text{Tyr}^{29})_n \end{array}$$

or poly($Glu^{35}Tyr^{29}$) ⌐———⌐ poly($Lys^{16}Ala^{20}$)

or $(Glu^{35}Tyr^{29})_n$ └———┘ $(Lys^{16}Ala^{20})_m$

(b) Same, but attachments are 3 in number and connect the ϵ-NH_2 groups of the lysine residues in the side chain and the γ-COOH groups of the glutamic acid residues in the main chain:

$$\begin{array}{l} \text{poly}(\text{Lys}^{16}\text{Ala}^{20}) \\ (3)| \\ \text{poly}(\text{Glu}^{35}\text{Tyr}^{29}) \end{array} \quad \text{or} \quad \begin{array}{l} (\text{Lys}^{16}\text{Ala}^{20})_m \\ (3)| \\ (\text{Glu}^{35}\text{Tyr}^{29})_n \end{array}$$

or poly($Glu^{35}Tyr^{29}$) └——(3)——┘ poly($Lys^{16}Ala^{20}$)

or $\text{poly}(\text{Tyr}^{29}\text{Glu}^{35})\underset{(3)}{\overset{\gamma\epsilon}{-}}\text{poly}(\text{Lys}^{16}\text{Ala}^{20})$

9. Graft polymer with a block polymer and an unknown sequence in the side chain (upper) attached to an unknown sequence in the main chain (lower); the points of attachment are between the γ-COOH groups of glutamic acid in the side chain and the ε-NH_2 groups of lysine in the main chain [Composition: 12% Glu, 21% Lys, 24% Tyr, 24% Leu, 20% Ala (Σ = 101%)].

$$\begin{array}{ll} & \text{poly(Glu}^{12}\text{)-poly(Lys}^{6}\text{)-poly(Tyr}^{24}\text{Leu}^{24}\text{)} \\ & \quad | \\ & \text{poly(Lys}^{15}\text{Ala}^{20}\text{)} \\ \text{or} & \text{(Glu}^{12}\text{)}_m\text{-(Lys}^{6}\text{)}_j\text{-(Tyr}^{24}\text{Leu}^{24}\text{)}_k \\ & \quad | \\ & \text{(Lys}^{15}\text{Ala}^{20}\text{)}_n \end{array}$$

$$\text{or } \underbrace{\text{poly(Lys}^{15}\text{Ala}^{20}\text{)} \quad \text{poly(Glu}^{12}}\text{)-poly(Lys}^{6}\text{)-poly(Tyr}^{24}\text{Leu}^{24}\text{)}$$

$$\text{or } \underbrace{\text{(Lys}^{15}\text{Ala}^{20}\text{)}_n \quad \text{(Glu}^{12}}\text{)}_m\text{-(Lys}^{6}\text{)}_j\text{-(Tyr}^{24}\text{Leu}^{24}\text{)}_k.$$

$$\text{or poly(Ala}^{20}\text{Lys}^{15}\text{)} \overset{\epsilon\gamma}{-\!-} \text{poly(Glu}^{12}\text{)-poly(Lys}^{6}\text{)-poly(Tyr}^{24}\text{Leu}^{24}\text{)}$$

REFERENCES

1. *Biopolymers,* 8, 161 (1969); *Arch. Biochem. Biophys.,* 123, 633 (1968); *Biochem. J.,* 106, 577 (1968); *Biochemistry,* 7, 483 (1968); *Biochim. Biophys. Acta,* 168, 1 (1968); *Eur. J. Biochem.,* 3, 129 (1968); *J. Biol. Chem.,* 243, 2451 (1968); *Bull. Soc. Chim. Biol.,* 51, 205 (1969); *Z. Physiol. Chem.,* 349, 1013 (1969); *J. Mol. Biol.,* 5, 492 (1971).
2. **Gill,** *Biopolymers,* 2, 283 (1964); also *J. Biol. Chem.,* 240, 3227 (1965); *Biochim. Biophys. Acta,* 124, 374 (1966)
3. **Sela,** *Adv. Immunol.,* 5, 30 (1966).
4. *J. Biol. Chem.,* 241, 527 (1966); *Biochemistry,* 5, 1445 (1966); *Biochem. J.,* 101, 1 (1966); *Virology,* 29, 480 (1966); *Arch. Biochem. Biophys.,* 115, 1 (1966); *Eur. J. Biochem.,* 1, 259 (1967); *Z. Physiol. Chem.,* 348, 245 (1967). (See also Reference 7).
5. *J. Biol. Chem.,* 241, 2491 (1966); *Biochemistry,* 5, 2485 (1966); *Biochim. Biophys. Acta,* 121, 1 (1966); *Biochem. J.,* 102, 23 (1967); *Arch. Biochem. Biophys.,* 121, 1 (1967); *Eur. J. Biochem.,* 1, 375 (1967); *Z. Physiol. Chem.,* 348, 256 (1967); *Bull. Soc. Chim. Biol.,* 49, 121 (1967). (Revision 1971). (See "Symbols for Amino-acid Derivatives and Peptides.")
6. *J. Polymer Sci.,* 8, 257 (1952); revised in 1967 (unpublished; see references in the 2nd footnote of this article).
7. *Biochem. J.,* 120, 449 (1970); *Biochemistry,* 9, 4022 (1970); *Eur. J. Biochem.,* 15, 203 (1970); *J. Biol. Chem.,* 245, 5171 (1970); *Z. Physiol. Chem.,* 351, 1055 (1970); *J. Mol. Biol.,* 55, 299 (1971); and elsewhere.

STRUCTURES AND SYMBOLS FOR SYNTHETIC AMINO ACIDS INCORPORATED INTO SYNTHETIC POLYPEPTIDES

M. C. Khosla and W. E. Cohn

The amino acids included in this list are those that have been incorporated into biologically active peptides, e.g., angiotensin II,[a] to study structure-activity relationships. Most of these amino acids are synthetic and are either available commercially or have been synthesized by various investigators as structural variants of naturally occurring amino acids. However, a few of these are also naturally occurring.[b] The selection here is of those most widely used and whose representation by symbols in peptide sequences has caused problems for authors and editors. The symbols listed are those considered most in keeping with the system originated by the IUPAC-IUB Commission on Biochemical Nomenclature[c,d] and have been chosen with an eye to internal consistency, ability to evoke the proper name, and suitability for use in sequences. Only one new one has been invented: ▲ for -yn- (triple bond), by analogy with Δ for -en- (double bond).

The list may also be useful in selecting suitable isosters of natural amino acids. The bibliography may be helpful in synthesis, resolution, or studies of the effects of these substances on the biological activity of various peptides.

The following trivial names are listed under other names (given by the number of the entry): *N*-amidinoglycine (67), 6-aminocaproic acid (18), 2-aminoethanesulfonic acid (112), β,β-bis(trifluoromethyl)alanine (70), carbamoylglycine (73), 2-(2-carboxyhydrazino)propane (83), cycloleucine (15), diethylalanine (17), dihydrophenylalanine (46), dopa (57), glycocyamine (67), isolysine (52), β-lysine (52), mercaptovaline (60), α-methylalanine (21), penicillamine (60), 5-pyrrolidone-2-carboxylic acid (109), surinamine (94), tetrahydrophenylalanine (47), trimethylammoniocaproic acid (116).

[a]For a review of structure-activity relationships and a listing of various analogs in which a number of these amino acids have been incorporated in angiotensin molecule, see Khosla, Smeby, and Bumpus.[26]

[b]IUPAC Commission on the Nomenclature of Organic Chemistry and IUPAC-IUB Commission on Biochemical Nomenclature, Nomenclature of α-Amino Acids (Recommendations 1974), *Biochemistry*, 14, 449 (1975).

[c]IUPAC-IUB Commission on Biochemical Nomenclature, Symbols for Amino-acid Derivatives and Peptides (Recommendations 1971), *Biochemistry*, 11, 1726 (1972). See Nomenclature section.

[d]IUPAC Commission on Biochemical Nomenclature, Abbreviations and Symbols for Nucleic Acids, Polynucleotides, and Their Constituents (Recommendations 1970), *J. Biol. Chem.*, 245, 5171 (1970). See Nomenclature Section.

STRUCTURE AND SYMBOLS FOR SYNTHETIC AMINO ACIDS INCORPORATED INTO SYNTHETIC POLYPEPTIDES

No.	Structure	Name/reference	Symbol
1	$H_2C{=}C(NHCOCH_3)COOH$	2-Acetamidoacrylic acid	AcAacr
2	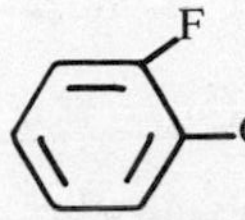 F; $-CH_2CH(NHCOCH_3)COOH$	N^{α}-Acetyl-2-fluorophenylalanine	AcPhe(2F)
3	$H_2N(CH_2)_4CH(NHCOCH_3)CONHCH_3$	N^{α}-Acetyllysine-*N*-methylamide	Ac-Lys-NHMe
4	$H_2NCH_2CH_2COOH$	β-Alanine[b]	βAla[c]
5	$CH_3CH_2CH(CH_3)CH(NH_2)COOH$	Alloisoleucine[b]	aIle[c]
6	$HOOC(CH_3)CH(NH_2)COOH$	2-Aminoadipic acid	Aad[c]
7	$HOOCCH_2CH_2CH(NH_2)CH_2COOH$	3-Aminoadipic acid	βAad[c]
8	$H_2NCH_2CH(CH_2C_6H_5)COOH$	3-Amino-2-benzylpropionic acid	βApr(αBzl)
9	$H_2NCH(CH_2C_6H_5)CH_2COOH$	3-Amino-3-benzylpropionic acid	βApr(βBzl)
10	$CH_3CH_2CH(NH_2)COOH$	2-Aminobutyric acid	Abu[c]
11	$CH_3CH(NH_2)CH_2COOH$	3-Aminobutyric acid	βAbu[c]
12	$H_2N(CH_2)_3COOH$	4-Aminobutyric acid	γAbu
13	$CH_3CH{=}C(NH_2)COOH$	2-Aminocrotonic acid	ACrt
14	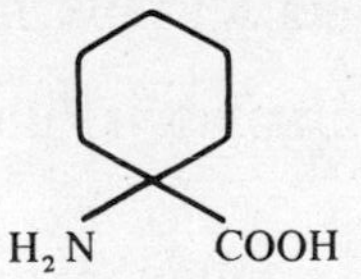 H_2N COOH	1-Aminocyclohexane-1-carboxylic acid (cyclonorleucine)	cHxA(αCx); cNle[b]
15	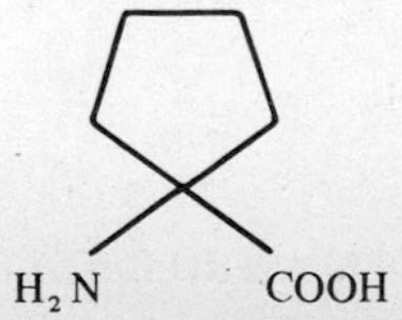 H_2N COOH	1-Aminocyclopentane-1-carboxylic acid (cycloleucine)	cPeA(αCx); cLeu[c]

STRUCTURE AND SYMBOLS FOR SYNTHETIC AMINO ACIDS INCORPORATED INTO SYNTHETIC POLYPEPTIDES (continued)

No.	Structure	Name/reference	Symbol
16	$(CH_3)_2NCH_2C{\equiv}CCH_2CH(NH_2)COOH$	2-Amino-6-dimethylamino-4-hexynoic acid (1)	$\alpha\epsilon A_2$ hx($\blacktriangle^{\gamma}$,$N^{\epsilon}Me_2$)
17	$CH_3CH_2CH(CH_2CH_3)CH(NH_2)COOH$	2-Amino-3-ethylvaleric acid (diethylalanine)	Ala(βEt_2)
18	$H_2N(CH_2)_5COOH$	6-Aminohexanoic acid (6-aminocaproic acid)	ϵAhx[c]
19	$(CH_3)_2CHCH_2CH(NH_2)CH(OH)CH_2COOH$	4-Amino-3-hydroxy-6-methylheptanoic acid (2,3)	γAhp(βOH, ϵMe)
20	(2-imidazolyl ring: N, N–H)–$CH_2CH(NH_2)COOH$	2-Amino-3-(2-imidazolyl)propionic acid	Apr(βIm-2)
21	$(CH_3)_2C(NH_2)COOH$	2-Aminoisobutyric acid (α-methylalanine)	Ala(αMe)
22	$H_2NCH_2SO_3H$	Aminomethanesulfonic acid	Ams
23	H_2NCH_2–(benzene ring)–COOH	4-Aminomethylbenzoic acid	Bz(4AMe); Bz(4CH_2NH_2)
24	$CH_3CH_2CH(CH_3)CH_2CH(NH_2)COOH$	2-Amino-4-methyl-hexanoic acid (4)	Ahx(γMe)
25	$CH_3CH{=}C(CH_3)CH_2CH(NH_2)COOH$	2-Amino-4-methyl-4-hexenoic acid (4)	Ahx(Δ^{γ}, γMe)
26	$CH_2{=}C(CH_3)CH_2CH_2CH(NH_2)COOH$	2-Amino-5-methyl-5-hexenoic acid (4)	Ahx(Δ^{δ}, δMe)
27	$H_2N(CH_2)_7COOH$	8-Aminooctanoic acid	ωAoc
28	H_2N–(benzene ring)–$CH_2CH(NH_2)COOH$	(4-Amino)phenylalanine (5)	Phe($4NH_2$)

STRUCTURE AND SYMBOLS FOR SYNTHETIC AMINO ACIDS INCORPORATED INTO SYNTHETIC POLYPEPTIDES (continued)

No.	Structure	Name/reference	Symbol
29	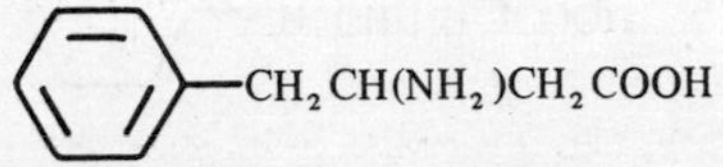 $-CH_2CH(NH_2)CH_2COOH$	3-Amino-4-phenylbutyric acid	βAbu(γPh)
30	$HOOCCH(NH_2)(CH_2)_4COOH$	2-Aminopimelic acid	αApm
31	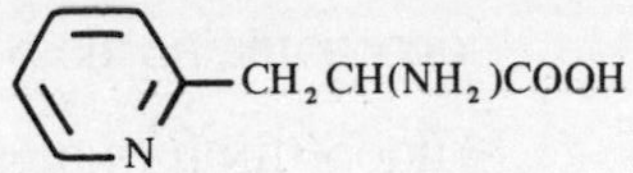 N $-CH_2CH(NH_2)COOH$	2-Amino-3-(2-pyridyl)propionic acid	Apr(βPrd-2)
32	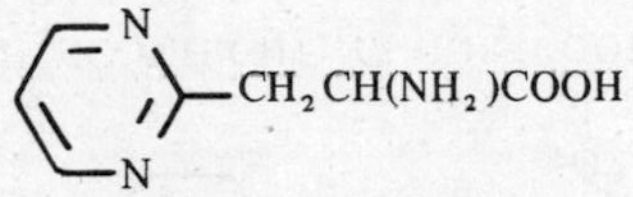 N N $-CH_2CH(NH_2)COOH$	2-Amino-3-(2-pyrimidyl)propionic acid	Apr(βPyr-2)[c,d]
33	$(CH_3)_3\overset{+}{N}(CH_2)_4CH(NH_2)COOH$	2-Amino-6-(trimethylammonio)hexanoic acid	$\alpha,\epsilon A_2hx(N^{\epsilon}Me_3)$
34	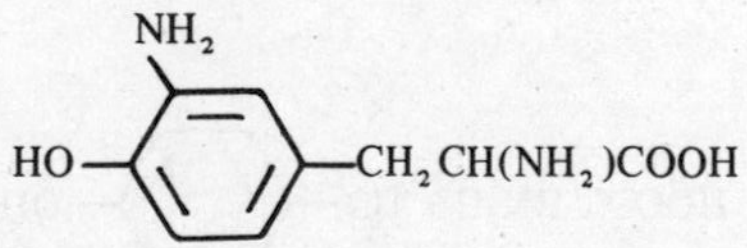 NH_2 HO— $-CH_2CH(NH_2)COOH$	3-Aminotyrosine	$Tyr(3NH_2)$
35	$HOOCCH_2CH(NH_2)CONHCH_3$	Aspartic α-methylamide	Asp-NHMe
36	$CH_3NHCOCH_2CH(NH_2)COOH$	Aspartic β-methylamide	Asn(Me); Asp(NHMe)
37	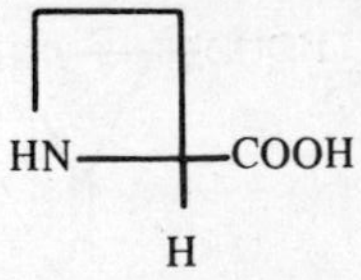 HN— —COOH H	Azetidine-2-carboxylic acid	Azt
38	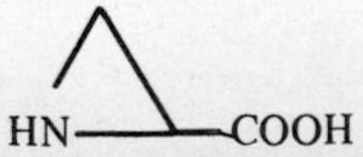 HN— —COOH	Aziridinecarboxylic acid	Azr

STRUCTURE AND SYMBOLS FOR SYNTHETIC AMINO ACIDS INCORPORATED INTO SYNTHETIC POLYPEPTIDES (continued)

No.	Structure	Name/reference	Symbol
39	O; HN—COOH	Aziridinonecarboxylic acid (6)	Azro
40	—$CH_2C(CH_2C_6H_5)(NH_2)COOH$	(α-Benzyl)phenylalanine (7)	Phe(αBzl)
41	$CH_2C_6H_5$; HO— —$CH_2CH(NH_2)COOH$	3-Benzyltyrosine (8)	Tyr(3Bzl)
42	Cl— —$CH_2CH(NH_2)COOH$	(4-Chloro)phenylalanine (5)	Phe(4Cl)
43	$H_2NCONH(CH_2)_3CH(NH_2)COOH$	Citrulline[b]	Ctr
44	$NCCH_2CH(NH_2)COOH$	β-Cyanoalanine	Ala(βCN)
45	$NCCH_2CH_2NHCH_2COOH$	*N*-(2-Cyanoethyl)glycine	(CNEt)Gly; CNEt-Gly
46	—$CH_2CH(NH_2)COOH$	β-(1,4-Cyclohexadienyl)alanine (9) (dihydrophenylalanine)	Ala(βcHx$\Delta^1\Delta^5$); Phe(H_2)
47	—$CH_2CH(NH_2)COOH$	β-(Cyclohexyl)alanine (10, 20) (hexahydrophenylalanine)	Ala(βcHx); Phe(H_6)
48	—$CH(NH_2)COOH$	α-(Cyclohexyl)glycine	Gly(cHx)

STRUCTURE AND SYMBOLS FOR SYNTHETIC AMINO ACIDS INCORPORATED INTO SYNTHETIC POLYPEPTIDES (continued)

No.	Structure	Name/reference	Symbol
49	$-CH_2CH(NH_2)COOH$	β-(Cyclopentyl)alanine	Ala(βcPe)
50	$-CH(NH_2)COOH$	α-(Cyclopentyl)glycine	Gly(cPe)
51	$H_2NCH_2CH_2CH(NH_2)COOH$	2,4-Diaminobutyric acid	A_2bu [c]
52	$H_2N(CH_2)_3CH(NH_2)CH_2COOH$	3,6-Diaminohexanoic acid (isolysine;[b] β-lysine[b])	$\beta\epsilon A_2hx$; βLys
53	$H_2NCH_2C{\equiv}CCH_2CH(NH_2)COOH$	2,6-Diamino-4-hexynoic acid (11)	$\alpha\epsilon A_2hx(\blacktriangle^{\gamma})$
54	$HOOCCH(NH_2)(CH_2)_3CH(NH_2)COOH$	2,2′-Diaminopimelic acid	A_2pm [c]
55	$H_2NCH_2CH(NH_2)COOH$	2,3-Diaminopropionic acid	A_2pr [c]
56	HO, HO, $-CH_2C(CH_3)(NH_2)COOH$	3,4-Dihydroxy-(α-methyl)phenylalanine .[β-(3,4-Dihydroxyphenyl)-α-methylalanine	Dopa(αMe)
57	HO, HO, $-CH_2CH(NH_2)COOH$	3,4-Dihydroxyphenylalanine[b]	Dopa [b]
58	OH, HO, $-CH(OH)CH(NH_2)COOH$	(3,4-Dihydroxyphenyl)serine	Dopa(βHO)

STRUCTURE AND SYMBOLS FOR SYNTHETIC AMINO ACIDS INCORPORATED INTO SYNTHETIC POLYPEPTIDES (continued)

No.	Structure	Name/reference	Symbol
59	$HOOCCH_2CH(Me_2N \rightarrow O)COOH$	*N,N*-Dimethylaspartic *N*-oxide (12)	(O,Me_2)Asp; Me_2(O)Asp
60	$(CH_3)_2C(SH)CH(NH_2)COOH$	β,β-Dimethylcysteine (β-mercaptovaline; penicillamine[b])	Val(βSH); Cys(βMe_2)
61	$(CH_3)_2CHCH(CH_3)CH(NHCH_3)COOH$	*threo-N*, β-Dimethylleucine (13)	MeLeu(βMe)
62	HO–$C_6H_3(CH_3)$–$CH_2C(CH_3)(NH_2)COOH$ (3-methyl-4-hydroxyphenyl)	α,3-Dimethyltyrosine	Tyr(α,3-Me_2)
63	$(C_6H_5)_2C(NH_2)COOH$	α,α-Diphenylglycine	Gly(Ph_2)
64	$CH_3CH_2SCH_2CH_2CH(NH_2)COOH$	Ethionine[b]	Eth
65	$CH_3CH_2NHCH_2COOH$	*N*-Ethylglycine	EtGly
66	imidazole ring (N, N–Et) with $CH_2CH(NH_2)COOH$	*tele*Ethylhistidine;[b,c] "1-Ethylhistidine" (14) (cf. 88, 89)	His(τEt)[b,c]
67	$H_2NC(=NH)NHCH_2CO$—	Guanidinoacetyl (*N*-amidinoglycyl; glycocyamine)	GdnAc-; AmdGly—
68	$H_2NC(=NH)NHCH_2CH(NH_2)COOH$	β-Guanidinoalanine	Ala(βGdn)
69	$H_2NC(=NH)NH(CH_2)_4COOH$	5-Guanidinovaleric acid	Vlr(δGdn)
70	$(CF_3)_2CHCH(NH_2)COOH$	γ_6-Hexafluorovaline [β,β-bis(trifluoromethyl)alanine]	Val(γF_6)
71	$H_2NC(=HN)NH(CH_2)_4CH(NH_2)COOH$	Homoarginine[b]	Har[c]

STRUCTURE AND SYMBOLS FOR SYNTHETIC AMINO ACIDS INCORPORATED INTO SYNTHETIC POLYPEPTIDES (continued)

No.	Structure	Name/reference	Symbol
72	$H_2N(CH_2)_5CH(NH_2)COOH$	Homolysine[b] (15)	Hly[c]
73	$H_2NCONHCH_2COOH$	Hydantoic acid; (carbamoylglycine)	CbmGly
74	(hydantoin ring: O, HN, NH, O) $—CH_2CO—$	5-Hydantoinacetyl	HydAc-
75	$CH_3CH(OH)CH_2CH_2CH(NH_2)COOH$	ε-Hydroxynorleucine (16)	Nle(εOH)
76	(piperidine ring) N(—OH)—COOH	1-Hydroxypipecolic acid (17)	Pip(1HO)
77	(pyrrolidine ring) HO — N —COOH	1-Hydroxyproline (17)	Pro(1HO)[c]
78	(pyrrolidine ring, OH) HN—COOH	3-Hydroxyproline	Pro(3HO)[c]
79	(pyrrolidine ring, OH) HN—COOH	4-Hydroxyproline	Pro(4HO)[c]
80	$CH_3N(OH)CH_2COOH$	*N*-Hydroxysarcosine (17)	Sar(*N*-HO)

STRUCTURE AND SYMBOLS FOR SYNTHETIC AMINO ACIDS INCORPORATED INTO SYNTHETIC POLYPEPTIDES (continued)

No.	Structure	Name/reference	Symbol
81	$HOOCCH_2CH(NH_2)CONH_2$	Isoasparagine[b]	$Asp\text{-}NH_2$ [c]
82	$HOOCCH_2CH_2CH(NH_2)CONH_2$	Isoglutamine[b]	$Glu\text{-}NH_2$ [c]
83	$H_2NN(CHMe_2)COOH$	2-Isopropylcarbazic acid [2-(1-carboxyhydrazino)propane]	Hdz(iPr)
84	$(CH_3)_2CHNH(CH_2)_3CH(NH_2)COOH$	N^{δ}-Isopropylornithine	Orn(δiPr)
85	$CH_3C(SH)(NH_2)COOH$	α-Mercaptoalanine (18)	Ala(αSH)
86	$CH_3CH(NHCH_3)COOH$	*N*-Methylalanine (19)	MeAla
87	$CH_3CH_2CH(CH_3)CH(NHCH_3)COOH$	*N*-Methylalloisoleucine (20)	(Me)aIle
88	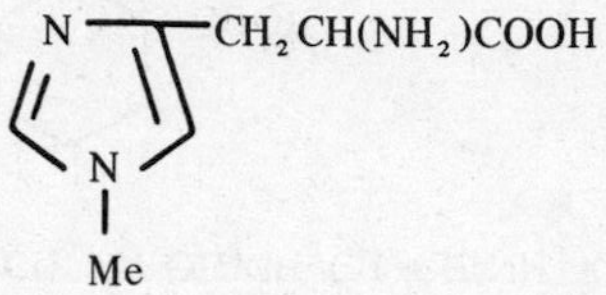 N, N, Me — $CH_2CH(NH_2)COOH$	*tele*Methylhistidine;[b,c] "1-Methylhistidine" (14, 21) (cf. 89)	His(τMe)[b,c]
89	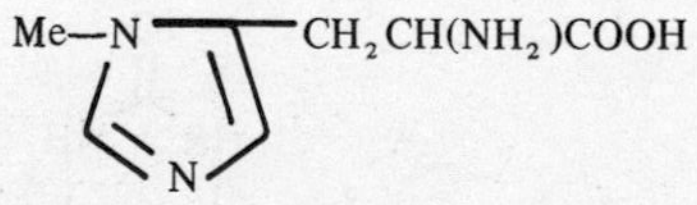 Me—N, N — $CH_2CH(NH_2)COOH$	*pros*Methylhistidine;[b,c] "3-Methylhistidine" (21) (cf. 88)	His(πMe)[b,c]
90	$HOOCCH_2CH(NHCH_3)CONH_2$	*N*-Methylisoasparagine (22)	$MeAsp\text{-}NH_2$
91	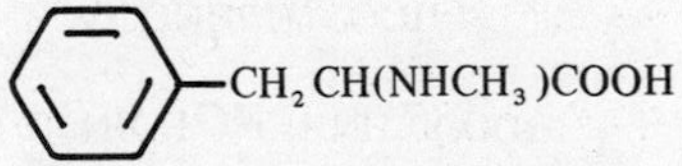 C_6H_5—$CH_2CH(NHCH_3)COOH$	(*N*-Methyl)phenylalanine (23)	MePhe
92	$CH_3OCH_2CH(NH_2)COOH$	*O*-Methylserine	Ser(Me)[c]
93	$CH_3CH(OCH_3)CH(NH_2)COOH$	*O*-Methylthreonine	Thr(Me)[c]

STRUCTURE AND SYMBOLS FOR SYNTHETIC AMINO ACIDS INCORPORATED INTO SYNTHETIC POLYPEPTIDES (continued)

No.	Structure	Name/reference	Symbol
94	HO–C_6H_4–$CH_2CH(NHCH_3)COOH$	*N*-Methyltyrosine (22) (surinamine)[b]	MeTyr
95	HO–C_6H_4–$CH_2C(CH_3)(NH_2)COOH$	α-Methyltyrosine	Tyr(αMe)
96	CH_3O–C_6H_4–$CH_2CH(NH_2)COOH$	*O*-Methyltyrosine	Tyr(O^4Me); Phe(4-OMe)
97	1-Naphthyl–$CH_2CH(NH_2)COOH$	β-(1-Naphthyl)alanine	Ala(βNap-1)
98	2-Naphthyl–$CH_2CH(NH_2)COOH$	β-(2-Naphthyl)alanine	Ala(βNap-2)
99	$O_2NHNC(=NH)NHCH_2CO$—	Nitroguanidinoacetyl	NGdnAc-
100	$CH_3(CH_2)_3CH(NH_2)COOH$	Norleucine[b] (2-aminohexanoic acid)	Nle[c]
101	$CH_3CH_2CH_2CH(NH_2)COOH$	Norvaline (2-aminovaleric acid)	Nva[c]
102	C_6F_5–$CH_2CH(NH_2)COOH$	(Pentafluorophenyl)alanine	Ala(βPhF_5)

STRUCTURE AND SYMBOLS FOR SYNTHETIC AMINO ACIDS INCORPORATED INTO SYNTHETIC POLYPEPTIDES (continued)

No.	Structure	Name/reference	Symbol
103	Phenyl–$CH(NH_2)COOH$	Phenylglycine	Gly(Ph)
104	Piperidine ring (N–H) with –COOH	Pipecolic acid (piperidine-2-carboxylic acid)	Pip
105	Pyrazole ring N–$CH_2CH(NH_2)COOH$	β-(1-Pyrazolyl)alanine	Ala(βPz1)
106	Pyrazole ring (N–H) with $CH_2CH(NH_2)COOH$	β-(3-Pyrazolyl)alanine (24, 25)	Ala(βPz3)
107	Pyrazole ring (N–H) with $CH_2CH(NH_2)COOH$	β-(4-Pyrazolyl)alanine (25)	Ala(βPz4)
108	O= (ring) HN — CH— COOH	Pyro-2-aminoadipic acid	pAad; <Aad
109	O= (ring) HN — COOH	Pyroglutamic acid[b] 5-pyrrolidone-2-carboxylic acid	pGlu; <Glu[c]

STRUCTURE AND SYMBOLS FOR SYNTHETIC AMINO ACIDS INCORPORATED INTO SYNTHETIC POLYPEPTIDES (continued)

No.	Structure	Name/reference	Symbol
110	CH_3HNCH_2COOH	Sarcosine[b]; (*N*-methylglycine)	Sar[c]; MeGly
111	$HOOCCH_2CH_2CONH_2$	Succinamic acid	Suc-NH_2
112	$H_2NCH_2CH_2SO_3H$	Taurine (2-aminoethanesulfonic acid)	Tau
113	S; HN—COOH	Thiazolidine-4-carboxylic acid	Tzl
114	S; —$CH_2CH(NH_2)COOH$	*β*-(2-Thienyl)alanine	Ala(*β*Thi2)
115	S; —$CH(OH)CH(NH_2)COOH$	*β*-(2-Thienyl)serine	Ser(*β*Thi2)
116	$(CH_3)_3\overset{+}{N}(CH_2)_5COOH$	*ε*-(Trimethylammonio)hexanoic acid [(*ε*-trimethylammonio)caproic acid]	*ε*Ahx($N^{\epsilon}Me_3$)
117	OH; —$CH_2CH(NH_2)COOH$	*o*-Tyrosine	Phe(2HO)
118	OH; —$CH_2CH(NH_2)COOH$	*m*-Tyrosine[b]	Phe(3HO)

Compiled by M. C. Khosla and W. E. Cohn.

REFERENCES

1. **Jansen, Weustink, Kerling, and Havinga,** *Rec. Trav. Chim. Pays-Bas,* 88, 819 (1969).
2. **Umezawa, Aoyagi, Morishima, Matsuzaki, Hamada, and Takeuchi,** *J. Antiobiot.* (Tokyo), 23, 259 (1970).
3. **Morishima, Takita, and Umezawa,** *J. Antibiot.* (Tokyo), 26, 115 (1973).
4. **Edelson, Skinner, Ravel, and Shive,** *J. Am. Chem. Soc.,* 81, 5150 (1959).
5. **Houghten and Rapoport,** *J. Med. Chem.,* 17, 556 (1974).
6. **Miyoshi,** *Bull. Chem. Soc. Jap.,* 46, 1489 (1973).
7. **Rydon,** *J. Chem. Soc. Perkin Trans. 1,* 2634 (1972).
8. **Erickson and Merrifield,** *J. Am. Chem. Soc.,* 95, 3750 (1973).
9. **Nagarajan, Diamond, and Ressler,** *J. Org. Chem.,* 38, 621 (1973).
10. **Khosla, Leese, Maloy, Ferreira, Smeby, and Bumpus,** *J. Med. Chem.,* 15, 792 (1972).
11. **Jansen, Kerling, and Havinga,** *Rec. Trav. Chim. Pays-Bas,* 89, 861 (1970).
12. **Ikutani,** *Bull. Chem. Soc. Jap.,* 43, 3602 (1970).
13. **Sheehan and Ledis,** *J. Am. Chem. Soc.,* 95, 875 (1973).
14. **Beyerman, Maat, and Van Zon,** *Rec. Trav. Chim. Pays-Bas,* 91, 246 (1972).
15. **Bodanszky and Lindeberg,** *J. Med. Chem.,* 14, 1197 (1971).
16. **Dreyfuss,** *J. Med. Chem.,* 17, 252 (1974).
17. **Nagasawa, Kohlhoff, Fraser, and Mikhail,** *J. Med. Chem.,* 15, 483 (1972).
18. **Patel, Currie, Jr., and Olsen,** *J. Org. Chem.,* 38, 126 (1973).
19. **Khosla, Hall, Smeby, and Bumpus,** *J. Med. Chem.,* 17, 431 (1974).
20. **Khosla, Hall, Smeby, and Bumpus,** *J. Med. Chem.,* 17, 1156 (1974).
21. **Needleman, Marshall, and Rivier,** *J. Med. Chem.,* 16, 968 (1973).
22. **Khosla, Smeby, and Bumpus,** Abstr. 169th Natl. Meet. Am. Chem. Soc. Philadelphia, April 1975, MEDI 57.
23. **Khosla, Smeby, and Bumpus,** *J. Am. Chem. Soc.,* 94, 4721 (1972).
24. **Hofmann and Bowers,** *J. Med. Chem.,* 13, 1099 (1970).
25. **Seeman, McGandy, and Rosenstein,** *J. Am. Chem. Soc.,* 94, 1717 (1972).
26. **Khosla, Smeby, and Bumpus,** in *Handbook of Experimental Pharmocology,* Vol. 37, Page and Bumpus, Eds., Springer-Verlag, Heidelberg, 1974, 126.

Amino Acids

DATA ON THE NATURALLY OCCURRING AMINO ACIDS

Elizabeth Dodd Mooz

The amino acids included in these tables are those for which reliable evidence exists for their occurrence in nature. These tables are intended as a guide to the primary literature in which the isolation and characterization of the amino acids are reported. Originally, it was planned to include more factual data on the chemical and physical properties of these compounds; however, the many different conditions employed by various authors in measuring these properties (i.e., chromatography and spectral data) made them difficult to arrange into useful tables. The rotation values are as given in the references cited; unfortunately, in some cases there is no information given on temperature, solvent, or concentration.

The investigator employing the data in these tables is urged to refer to the original articles in order to evaluate for himself the reliability of the information reported. These references are intended to be informative to the reader rather than to give credit to individual scientists who published the original reports. Thus not all published material is cited.

The compounds listed in Sections A to N are known to be of the L configuration. Section O contains some of the D amino acids which occur naturally. This last section is not intended to be complete since most properties of the D amino acids correspond to those of their L enantiomorphs. Therefore, emphasis was placed on including those D amino acids whose L isomers have not been found in nature. The reader will find additional information on the D amino acids in the review by Corrigan[263] and in the book by Meister.[1]

Compilation of data for these tables was completed in December 1974. Appreciation is expressed to Doctors L. Fowden, John F. Thompson, Peter Müller, and M. Bodanszky who were helpful in supplying recent references and to Dr. David Pruess who made review material available to me prior to its publication. A special word of thanks to Dr. Alton Meister who made available reprints of journal articles which I was not able to obtain.

DATA ON THE NATURALLY OCCURRING AMINO ACIDS (continued)

No.	Amino acid (synonym)	Source	Structure	Formula (mol wt)	Melting point °C[a]	$[\alpha]_D$[b]	pK_a	References: Isolation and purification	References: Chromatography	References: Chemistry	References: Spectral data
			A. L-Monoamino, Monocarboxylic Acids								
1	Alanine (α-aminopropionic acid)	Silk fibroin	$CH_3CH(NH_2)COOH$	$C_3H_7NO_2$ (89.09)	297°	$+1.8^{25}$ (c 2, H_2O) (1) $+14.6^{25}$ (c 2, 5 *N* HCl) (1)	2.34 9.69	2	3	4	4
2	β-Alanine (β-aminopropionic acid)	*Iris tingitana*	$NH_2CH_2CH_2COOH$	$C_3H_7NO_2$ (89.09)	196° (dec)	–	3.55 10.24	5	5	5	–
3	α-Aminobutyric acid	Yeast protein	$CH_3CH_2CH(NH_2)COOH$	$C_4H_9NO_2$ (103.12)	292° (dec)	$+20.5^{25}$ (c 1–2, 5 *N* HCl) (290) $+9.3^{25}$ (c 1–2, (H_2O) (290) $+42^{25}$ (c 1–2, gl acetic) (290)	2.29 9.83	6	7	6	–
4	γ-Aminobutyric acid (piperidinic acid)	Bacteria	$NH_2CH_2CH_2CH_2COOH$	$C_4H_9NO_2$ (103.12)	203° (dec)	–	4.03 10.56 (290)	8–10	9, 10	11	–
5	1-Aminocyclopropane-1-carboxylic acid	Pears and apples	H_2C—CH_2—C(–COOH)(NH_2) (cyclopropane ring)	$C_4H_7NO_2$ (101.11)	–	–	–	11	11	12	12
6	2-Amino-3-formyl-3-pentenoic acid	*Bankera fulgineoalba* (a mushroom)	$CH_3CH{=}C(CHO)CH(NH_2)COOH$	$C_6H_9NO_3$ (143.15)	–	–	–	13	13	13	13
7	α-Aminoheptanoic acid	*Claviceps purpurea*	$CH_3(CH_2)_4CH(NH_2)COOH$	$C_7H_{15}NO_2$ (145.21)	–	–	–	14	14	–	–

Note: Footnotes appear on page 166.

DATA ON THE NATURALLY OCCURRING AMINO ACIDS (continued)

No.	Amino acid (synonym)	Source	Structure	Formula (mol wt)	Melting point °C[a]	$[\alpha]_D$[b]	pK_a	References: Isolation and purification	References: Chromatography	References: Chemistry	References: Spectral data
7a	2-Amino-4,5-hexadienoic acid	*Amanita solitaria*	$H_2C{=}C{=}CHCH_2{-}CH(NH_2)COOH$	$C_6H_9NO_2$ (127.16)	200° (dec) (14a)	–	–	14a	–	–	14a
8	2-Amino-4-hexenoic acid	Ilamycin	$CH_3CH{=}CHCH_2CH(NH_2)COOH$	$C_6H_{11}NO_2$ (129.17)	–	–	–	15	15	–	–
8a	2-Amino-4-hydroxy-hept-6-ynoic acid	*Euphoria longan*	$HC{\equiv}C{-}CH_2CH(OH)CH_2CH(NH_2)COOH$	$C_7H_{11}NO_3$ (157.19)	–	-27^{20} (*c* 2, H_2O) -8^{20} (*c* 1, 5 *N* HCl) (15a)	–	15a	15a	–	15a
8b	2-Amino-6-hydroxy-4-methyl-4-hexenoic acid	*Aesculus Californica* seeds	$(HOH_2C)(H)C{=}C(H_3C)(CH_2CH(NH_2)COOH)$	$C_7H_{13}NO_3$ (159.21)	–	-31^{20} (*c* 2.2, H_2O) $+2^{20}$ (*c* 1.1, 5 *N* HCl) (23b)	–	23b	23b	–	15b
8c	2-Amino-4-hydroxy-5-methyl hexenoic acid	*Euphoria longan*	$(HC{\equiv}C)(HOH_2C)CHCH_2CH(NH_2)COOH$	$C_7H_{11}NO_3$ (157.19)	–	-27^{20} (*c* 2, H_2O) -13^{20} (*c* 1, 5 *N* HCl) (15a)	–	15a	–	15a	15a
8d	2-Amino-3-hydroxy-methyl-3-pentenoic acid	*Bankera fuligineoalba*	$H_3CCH{=}C(CH_2OH)CH(NH_2)COOH$	$C_6H_{11}NO_3$ (145.18)	160–161° (dec) (13)	$+182^{25°}$ (*c* 0.8, H_2O) $+201^{25}$ (*c* 0.8, 0.3 *N* HCl) (13)	–	13	13	13	13
9	α-Aminoisobutyric acid	*Iris tingitana*, muscle protein	$(CH_3)_2C(NH_2)COOH$	$C_4H_9NO_2$ (103.12)	200° (dec)	–	2.36 10.21 (290)	16	16	–	–
10	β-Aminoisobutyric acid	*Iris tingitana*	$NH_2CH_2CH(CH_3)COOH$	$C_4H_9NO_2$ (103.12)	179° (17)	-21^{26} (*c* 0.43, H_2O) (17)	–	17	17	17	17

DATA ON THE NATURALLY OCCURRING AMINO ACIDS (continued)

No.	Amino acid (synonym)	Source	Structure	Formula (mol wt)	Melting point °C[a]	$[\alpha]_D$[b]	pK_a	References: Isolation and purification	References: Chromatography	References: Chemistry	References: Spectral data
10a	2-Amino-4-methoxy-*trans*-3-butenoic acid	*Pseudomonas aeruginosa*	$H_3CO(H)C=C(H)CH(NH_2)COOH$	$C_5H_9NO_3$ (131.15)	–	–	–	17a	17b	–	17a, 17b
11	γ-Amino-α-methylene butyric acid	*Arachis hypogaea* (groundnut plants)	$NH_2CH_2CH_2C(=CH_2)COOH$	$C_4H_9NO_2$ (115.13)	152° (18)	–	–	18	18	18	–
12	2-Amino-4-methyl-hexanoic acid (homoisoleucine)	*Aesculus Californica* seeds	$(CH_3CH_2)(CH_3)CHCH_2CH(NH_2)COOH$	$C_7H_{15}NO_2$ (145.21)	–	-2^{20} (c 1, H_2O) (19) $+24^{20}$ (c 0.87, 5 N HCl) (19)	–	19	19	19	19
13	2-Amino-4-methyl-4-hexenoic acid	*Aesculus Californica* seeds	$H_3C(H)C=C(H_3C)CH_2CH(NH_2)COOH$	$C_7H_{13}NO_2$ (143.19)	–	-61^{20} (c 2.4, H_2O) (19) -36 (c 1.2, 6 N HCl) (19)	–	19	19	19	19
13a	2-Amino-4-methyl-5-hexenoic acid	*Streptomyces* species	$CH_2=CHCH(CH_3)CH_2CH(NH_2)COOH$	$C_7H_{13}NO_2$ (143.21)	260° (dec) (19a)	$-9.6°$ (c 1.78, H_2O) $+5.7°$ (c 0.7, 1 N HCl) (99a)	–	19a	19a	–	19a
14	2-Amino-5-methyl-4-hexenoic acid	*Leucocortinarius bulbiger*	$CH_3C(CH_3)=CHCH_2CH(NH_2)COOH$	$C_7H_{13}NO_2$ (143.19)	260–270° (dec) (22a)	-45.9^{23} (c 0.47, H_2O) -7^{23} (c 0.4, 1 N HCl) (22a)	–	22, 22a	22a	–	22a

DATA ON THE NATURALLY OCCURRING AMINO ACIDS (continued)

No.	Amino acid (synonym)	Source	Structure	Formula (mol wt)	Melting point °C[a]	$[\alpha]_D$[b]	pK_a	References: Isolation and purification	References: Chromatography	References: Chemistry	References: Spectral data
14a	2-Amino-4-methyl-5-hexenoic acid	*Euphoria longan*	$HC{\equiv}C(H_3C)CHCH_2CH(NH_2)COOH$	$C_7H_{11}NO_2$ (141.19)	–	-33^{20} (c 2, H_2O) -27^{20} (c 1, 5 N HCl) (15a)	–	15a	–	15a	15a
15	α-Amino-octanoic acid	*Aspergillus atypique*	$CH_3(CH_2)_5CH(NH_2)COOH$	$C_8H_{17}NO_2$ (159.23)	–	–	–	23	23	23	23
15a	2-Amino-4-pentynoic acid	*Streptomyces* sp. #8–4	$HC{\equiv}CCH_2CH(NH_2)COOH$	$C_5H_7NO_2$ (113.13)	241–242° (dec) (23a)	-31.1^{25} (c 1, H_2O) -5.5^{25} (c 1, 5 N HCl) (23a)	–	23a	23a	23a	23a
15a′	*cis*-α-(Carboxy-cyclopropyl) glycine	*Aesculus parviflora*	HOOCHC—CHCH (NH₂)COOH with bridging CH_2 (cyclopropane)	$C_6H_9NO_4$ (159.16)	–	$+25^{20}$ (c 1, H_2O) +58 (c 0.5, 5 N HCl) (23a′)	–	23a′	23a′	–	23a′
15b	*trans*-α-(Carboxy-cyclopropyl) glycine	*Blighia sapida*	HOOCHC—CHCH(NH₂)COOH with bridging CH_2 (cyclopropane)	$C_6H_9NO_4$ (159.16)	–	$+107^{20}$ (c 2, H_2O) $+146^{20}$ (c 1, 5 NHCl) (23a′)	–	23a′	23a′	–	23a′
15b′	*trans*-α-(2-Carboxy-methylcyclopropyl glycine	*Blighia unijugata*	HC(CH₂COOH)—CHCH(NH₂)COOH with bridging CH_2 (cyclopropane)	$C_7H_{11}NO_4$ (173.19)	–	$+12^{20}$ (c 1, H_2O) $+45^{20}$ (c 0.5, 5 N HCl) (99a)	–	99a	99a	–	99a
15c	γ-Glutamyl-2-amino-4-methyl-hex-4-enoic acid	*Aesculus Californica* seeds	$(H_3C)(H)C{=}C(H_3C)CH_2-CH(COOH)-NH-C(=O)-CH_2-CH_2-CH(NH_2)COOH$	$C_{12}H_{20}N_2O_5$ (272.34)	–	$+17^{20}$ (c 3, H_2O) (23b)	–	23b	23b	–	–

DATA ON THE NATURALLY OCCURRING AMINO ACIDS (continued)

No.	Amino acid (synonym)	Source	Structure	Formula (mol wt)	Melting point °C[a]	$[\alpha]_D$[b]	pK_a	References: Isolation and purification	References: Chromatography	References: Chemistry	References: Spectral data
			B. L-Monoamino, Dicarboxylic Acids								
16	Glycine (α-aminoacetic acid)	Gelatin hydrolyzate	$CH_2(NH_2)COOH$	$C_2H_6NO_2$ (75.07)	290° (dec) (1)	–	2.35 9.78 (290)	24	3	25	25
17	Hypoglycin A [α-amino-β-(2-methylene cyclopropyl)-propionic acid]	*Blighia sapida*	$H_2C{=}C$—CH_2 ring via CH; CH—CH_2—$NH_2CHCOOH$	$C_7H_{11}NO_2$ (141.18)	280–284° (26)	+9.2 (*c* 1, H_2O) (26)	–	26	26	27	27
18	Isoleucine (α-amino-β-methyl-valeric acid)	Sugar beet molasses	CH_3CH_2 / CH_3 >$CHCH(NH_2)COOH$	$C_6H_{13}NO_2$ (131.17)	284° (1)	$+39.5^{25}$ (*c* 1, 5 *N* HCl) (290) $+12.4^{25}$ (*c* 1, H_2O) (290)	2.36 9.68	3	9	29	1
19	Leucine (α-aminoisocaproic acid)	Muscle fiber, wool	$(CH_3)_2CHCH_2CH(NH_2)COOH$	$C_6H_{13}NO_2$ (131.17)	337° (1)	-11^{25} (*c* 2, H_2O) $+16^{25}$ *c* 2, 5 *N* HCl) (1)	2.36 960 (1)	30	3	31	31
19a	*N*-Methyl-γ-methyl-*allo*isoleucine	Etamycin	$CH_3CH(CH_3)CH(CH_3)CH(NHCH_3)COOH$	$C_8H_{17}NO_2$ (159.26)	–	–	–	31a	31a	–	–
19b	β-Methyl-β-(methylene-cyclopropyl) alanine	*Aesculus Californica* seeds	$H_2C{=}C$—$CH(CH_3)CH(NH_2)COOH$ (ring via CH_2)	$C_8H_{13}NO_2$ (155.22)	–	1.5^{20} (*c* 2, H_2O) $+45^{20}$ (*c* 1, 5 *N* HCl) (23b)	–	23b	23b	–	15b
20	α-(Methylene-cyclopropyl) glycine	*Litchi chinensis*	$H_2C{=}C$—$CHCH(NH_2)COOH$ (ring via CH_2)	$C_6H_9NO_2$ (127.15)	–	$+43^{22.5}$ (*c* 0.5, 5 *N* HCl) (32)	–	32	32	32	32

DATA ON THE NATURALLY OCCURRING AMINO ACIDS (continued)

No.	Amino acid (synonym)	Source	Structure	Formula (mol wt)	Melting point °C[a]	$[\alpha]_D$[b]	pK_a	References: Isolation and purification	References: Chromatography	References: Chemistry	References: Spectral data
21	β-(Methylene-cyclopropyl)-β-methylalanine	*Aesculus Californica*	$H_2C{=}C$(—CH_2—, ring)—CHCH(CH_3)CH(NH_2)COOH	$C_8H_{13}NO_2$ (155.19)	–	$+1.5^{20}$ (c 2, H_2O) $+45^{20}$ (c 1, 5 N HCl)	–	19	19	19	19, 21
21a	β-Methylenenor-leucine	*Amanita vaginata*	$H_3C—CH_2—CH_2—C(=CH_2)—CH(NH_2)COOH$	$C_7H_{13}NO_2$ (143.21)	171° (35a)	$+158^{20}$ (c 0.51, 1 N HCl) $+149^{20}$ (c 0.56, H_2O) (35a)	–	–	35a	35a	–
22	Valine (α-aminoisovaleric acid	Casein	$(CH_3)_2CHCH(NH_2)COOH$	$C_5H_{11}NO_2$ (117.15)	292–295° (1)	$+28.3^{25}$ (c 1, 2, 5 N HCl) $+5.63^{25}$ (c 1–2, H_2O) (290)	2.32 9.62 (1)	35	3	35, 36	36
23	α-Aminoadipic acid	*Pisum sativum*	$HOOC(CH_2)_3(CH(NH_2)COOH$	$C_6H_{11}NO_4$ (161.18)	195° (37)	$+3.2^{25}$ (c 2, H_2O (290) $+23^{22}$ (c 2, 6 N (HCl) (37)	2.14 4.21 9.77 (290)	37	37	37	–
24	3-Aminoglutaric acid	*Chondria armata*	$HOOCCH_2CH(NH_2)CH_2COOH$	$C_5H_9NO_4$ (162.13)	280–282° (38)	±0[c] (c 2, 5 N HCl) (38)	–	38	38	38	38
25	α-Aminopimelic acid	*Asplenium septentrionale*	$HOOC(CH_2)_4CH(NH_2)COOH$	$C_7H_{13}NO_4$ (175.19)	204° (39)	–	–	39	39	39	–
26	Asparagine (α-aminosuccinamic acid)	Asparagus	$H_2NCOCH_2CH(NH_2)COOH$	$C_4H_8N_2O_3$ (132.12)	236° (1)	-5.6^{25} (c 2, H_2O) (290) +33.2 (3 N HCl) (1)	2.02 8.80 (1)	40	3	41	42
27	Aspartic acid (α-aminosuccinic acid)	Conglutin, legumin	$HOOCCH_2CH(NH_2)COOH$	$C_4H_7NO_4$ (133.10)	270° (1)	$+5.05^{25}$ (c 2, H_2O) $+25.4^{25}$ (c 2, 5N HCl) (1)	1.88 3.65 9.60 (1)	43	3	41	41

DATA ON THE NATURALLY OCCURRING AMINO ACIDS (continued)

No.	Amino acid (synonym)	Source	Structure	Formula (mol wt)	Melting point °C[a]	$[\alpha]_D$[b]	pK_a	References: Isolation and purification	References: Chromatography	References: Chemistry	References: Spectral data
28	Ethylasparagine	*Ecballium elaterium, Bryonia dioica*	$CH_3CH_2NHCOCH_2CH(NH_2)COOH$	$C_6H_{12}N_2O_3$ (160.19)	–	–	–	45	45	45	45
29	γ-Ethylidene glutamic acid	Mimosa	CH_3CH ‖ $HOOCCCH_2CH(NH_2)COOH$	$C_7H_{11}NO_4$ (173.18)	–	$+21^{20}$ (*c* 2.8, H_2O) (47) $+38.3^{20}$ (*c* 1.4, 6 *N* HCl) (47)	–	46	46	–	46
30	*N*-Fumarylalanine	*Penicillium recticulosum*	$CH_3CHCOOH$ | $NHCOCH{=}CHCOOH$	$C_7H_9NO_5$ (187.16)	229° (48)	–	–	48	–	48	–
31	Glutamic acid (α-aminoglutaric acid)	Gluten-fibrin hydrolyzates	$HOOC(CH_2)_2CH(NH_2)COOH$	$C_5H_9NO_4$ (147.13)	249° (1)	$+12^{25}$ (*c* 2, H_2O) $+31.8^{28}$ (*c* 2, 5 *N* HCl) (1)	2.19 4.25 9.67 (1)	49	3	50	50
32	Glutamine (α-aminoglutaramic acid)	Beet juice	$H_2NCO(CH_2)_2CH(NH_2)COOH$	$C_5H_{10}N_2O_3$ (146.15)	185° (1)	$+6.3^{25}$ (*c* 2, H_2O) $+31.8^{25}$ (*c* 2, 1 *N* HCl) (1)	2.17 9.13 (1)	51	3	50	50
33	N^5-Isopropyl-glutamine	*Lunaria annua*	$HNCO(CH_2)_2CH(NH_2)COOH$ | H_3CCHCH_3	$C_8H_{16}N_2O_3$ (188.23)	–	$+7.1^{22}$ (*c* 1.7, H_2O) (53)	–	53	53	53	–
34	N^4-Methylasparagine	–	$CH_3NHCOCH_2CH(NH_2)COOH$	$C_5H_{10}N_2O_3$ (146.15)	241–244° (54)	−4.2 (*c* 5.5, H_2O) (54)	–	–	54	–	–
35	β-Methylaspartic acid	*Clostridium tetanomorphum*	$HOOCCH(CH_3)CH(NH_2)COOH$	$C_5H_9NO_4$ (147.13)	–	−10 (*c* 0.4, H_2O) +12.4 (*c* 3, 1 *N* HCl) +13.3 (*c* 3, 5 *N* HCl) (55)	3.5 9.9 (290)	55	55	55	–

No.	Amino acid (synonym)	Source	Structure	Formula (mol wt)	Melting point °C[a]	$[\alpha]_D$[b]	pK_a	References: Isolation and purification	References: Chromatography	References: Chemistry	References: Spectral data
36	γ-Methylglutamic acid	*Phyllitis scolopendrium*	$HOOCCH(CH_3)CH_2\ CH(NH_2)COOH$	$C_6H_{11}NO_4$ (141.17)	–	–	–	56	56	–	–
37	γ-Methyleneglutamic acid	*Archis hypogaea*	$HOOCC(=CH_2)CH_2CH(NH_2)COOH$	$C_6H_9NO_4$ (159.15)	196° (57)	–	–	57	57	57	–
38	γ-Methyleneglutamine	*Archis hypogaea*	$H_2NCOC(=CH_2)CH_2CH(NH_2)COOH$	$C_6H_{10}N_2O_3$ (158.17)	173–182° (57)	–	–	57	57	57	–
39	Theanine (α-amino-γ-*N*-ethylglutaramic acid)	*Xerocomus hadius*	$HOOCCH(NH_2)(CH_2)_2CONHCH_2CH_3$	$C_7H_{14}N_2O_3$ (174)	–	–	–	59	59	59	59
39a	β-*N*-Acetyl-α-, β-diaminopropionic acid (β-acetamido-L-alanine)	*Acacia armata* seeds	$CH_3CONHCH_2CH(NH_2)COOH$	$C_5H_{10}N_2O_3$ (146.17)	–	-87^{20} (*c* 8, H_2O) -35^{20} (*c* 4, 6 *N* HCl) (59a)	–	59a	59a	59a	59a
	C. L-Diamino, Monocarboxylic Acids										
40	*N*-Acetylornithine	*Asplenium* species	$CH_3CONH(CH_2)_3CH(NH_2)COOH$	$C_7H_{14}N_2O_3$ (174.11)	200° (dec) (60)	–	–	60	60	60	–
41	α-Amino-γ-*N*-acetylamino-butyric acid	Latex of *Euphorbia pulcherrima*	$CH_3CONH(CH_2)_2CH(NH_2)COOH$	$C_6H_{12}N_2O_3$ (160.18)	220–222° (dec) (61)	–	4.45 (33)	61	61	–	–
42	*N*-ε-(2-Amino-2-carboxyethyl)-lysine	Alkali-treated protein	$HOOCCH(NH_2)CH_2NH(CH_2)_4CH(NH_2)COOH$	$C_9H_{19}N_3O_4$ (233.28)	–	–	2.2 6.5 8.8 9.9 (62)	62	62	62	–
43	*N*-δ-(2-Amino-2-carboxyethyl)-ornithine	Alkali-treated wool	$HOOCCH(NH_2)CH_2NH(CH_2)_3CH(NH_2)COOH$	$C_8H_{17}N_3O_4$ (273.72)	–	–	–	63	63	63	–
44	2-Amino-3-dimethylamino-propionic acid	*Streptomyces neocaliberis*	$(CH_2)_3NCH_2CH(NH_2)COOH$	$C_5H_{12}N_2O_2$ (117.15)	–	-17.8^{25} (*c* 1, H_2O) +18.1 (*c* 1, HCl pH 3) (64)	–	64	64	64	64

DATA ON THE NATURALLY OCCURRING AMINO ACIDS (continued)

No.	Amino acid (synonym)	Source	Structure	Formula (mol wt)	Melting point °C[a]	$[\alpha]_D$[b]	pK_a	References: Isolation and purification	References: Chromatography	References: Chemistry	References: Spectral data
45	α-Amino-β-methyl-aminopropionic acid	*Cycas circinalis*	$CH_3NHCH_2CH(NH_2)COOH$	$C_4H_{11}N_2O_2$ (105.15)	165–167° (65)	–	–	65	65	65	65
45a	α-Amino-β-oxalyl-aminopropionic acid	*Crotalaria*	$HOOC–CH(NH_2)CH_2NHCOCOOH$	$C_5H_8N_2O_5$ (176.15)	–	–	–	–	65a	–	–
46	Canaline	*Canavalia ensiformis*	$H_2NO(CH_2)_2CH(NH_2)COOH$	$C_4N_{10}N_2O_3$ (134.14)	–	–	2.40 3.70 9.20 (20)	66	–	67	–
46a	Threo-α,β-diamino-butyric acid	Amphomycin hydrolyzate	$CH_3CH(NH_2)CH(NH_2)COOH$	$C_4H_{10}N_2O_2$ (118.16)	213–214° (dec) (67a)	$+27.1^{25}$ (*c* 2, 5 *N* HCl) (67a)	–	67b	67b	67a	67b
47	α,γ-Diaminobutyric acid (γ-aminobutyrine)	Glumamycin	$H_2NCH_2CH_2CH(NH_2)COOH$	$C_4H_{10}N_2O_2$ (134.14)	–	$+7.2^{25}$ (*c* 2, H_2O) $+14.6^{18}$ (*c* 3.67, H_2O)[g] (290)	1.85 8.28 10.50 (20)	68	68	68	–
48	3,5-Diaminohexanoic acid	*Clostridium stricklandii*	$CH_3CH(NH_2)CH_2CH(NH_2)CH_2COOH$	$C_6H_{14}N_2O_2$ (146.19)	204–208° (69)	–	–	69	69	69	69
48a	2,6-Diamino-7-hydroxyazelaic acid	*Bacillus brevis* (edeine A and B)	$HOOCCH_2CH(OH)CH(NH_2)(CH_2)_3CH(NH_2)COOH$	$C_9H_{18}N_2O_{15}$ (234.29)	–	–	–	69a	69a	69a	–
49	α,β-Diaminopropionic acid (β-aminoalanine)	Mimosa	$H_2NCH_2CH(NH_2)COOH$	$C_3H_8N_2O_2$ (104.11)	–	–	1.23 6.73 9.56 (20)	70	70	70	–
49a	*N*ε,*N*ε-Dimethyl-lysine	Human urine	$(H_3C)_2N(CH_2)_4CH(NH_2)COOH$	$C_8H_{18}N_2O_2$ (174.28)	214–216° (dec) (70a)	–	–	70a	70a	–	70a
49b	N^5-Iminoethyl-ornithine	*Streptomyces* broth	$HN{=}C(CH_3)NH(CH_2)_3CH(NH_2)COOH$	$C_7H_{15}N_3O_2$ (173.25)	226–229° (70b)	$+20.6^{25}$ (*c* 1, 5 *N* HCl) (70b)	1.97 8.86 11.83 (70b)	70b	70b	–	70b

DATA ON THE NATURALLY OCCURRING AMINO ACIDS (continued)

No.	Amino acid (synonym)	Source	Structure	Formula (mol wt)	Melting point °C[a]	$[\alpha]_D$[b]	pK_a	References: Isolation and purification	Chromatography	Chemistry	Spectral data
50	Lathyrus factor (β-*N*-(γ-glutamyl)-aminopropionitrile)	*Lathyrus pusillus*	$CH(CH_2)_2NHCO(CH_2)_2CH(NH_2)COOH$	$C_8H_{13}N_3O_3$ (199.22)	193–194° (72)	$+28^{18}$ (*c* 1, 6 *N* HCl) (72)	2.2 9.14	71	72	72	72
51	Lysine (α,ε-diaminocaproic acid)	Casein	$H_2NCH_2(CH_2)_3CH(NH_2)COOH$	$C_6H_{14}N_2O_2$ (146.19)	224–225° (dec) (73)	$+14.6^{20}$ (H_2O) (73)	2.16 9.18 10.79 (290)	74	3	75	–
52	β-Lysine (isolysine; β,ε-diaminocaproic acid)	Viomycin	$HOOCCH_2CH(NH_2)CH_2CH_2\ CH_2(NH_2)$	$C_6H_{14}N_2O_2$ (146.19)	240–241° (76)	–	–	76	76	76	76
53	Lysopine *N*[2]-(D-1-carboxy-ethyl)-lysine	Calf thymus histone	$H_2N(CH_2)_4CHCOOH$ | NH | $H_3CCHCOOH$	$C_9H_{18}N_2O_4$ (204.25)	157–160° (77)	+18 (*c* 1.4, H_2O) (77)	–	78	78	77	77
54	ε-*N*-Methyllysine	Calf thymus histone	$CH_3NH(CH_2)_4CH(NH_2)COOH$	$C_7H_{16}N_2O_2$ (160.23)	–	–	–	79	79	–	79
55	Ornithine (α,δ-diaminovaleric acid)	*Asplenium nidus*	$H_2N(CH_2)_3CH(NH_2)COOH$	$C_5H_{12}H_2O_2$ (132.16)	–	+12.1 (*c* 2, H_2O) +28.4 (*c* 2, 5 *N* HCl) (1)	1.71 8.69 10.76 (290)	60	60	81	81
55a	4-Oxalysine	*Streptomyces*	$H_2N{-}(CH_2)_2O(CH_2)_2CH(NH_2)COOH$	$C_5N_{12}N_2O_3$ (148.19)	–	–	–	81a	81a	–	81a
56	β-*N*-Oxalyl-α,β-diaminopropionic acid	*Lathyrus sativus*	$HOOCCONHCH_2CH(NH_2)COOH$	$C_5H_8N_2O_5$ (176.13)	206° (dec) (82)	-36.9^{27} (*c* 0.66, 4 *N* HCl) (82)	1.95 2.95 9.25 (82)	82	82	82	82
56a	β-Putreanine [*N*-(4-aminobutyl)-3-aminopropionic acid]	Bovine brain	$NH_2(CH_2)_4NH(CH_2)_2COOH$	$C_7H_{16}N_2O_2$ (160.25)	250–251° (dec) (82a)	–	3.2 9.4 11.2 (82a)	82a	82a	–	82a
	D. L-Diamino, Dicarboxylic Acids										
57	Acetylenic dicarboxylic acid diamide	*Streptomyces chibaensis*	$CCONH_2$ ||| $CCONH_2$	$C_4H_4N_2O_2$ (112.09)	216–218° (dec)	–	–	83	83	83	83

DATA ON THE NATURALLY OCCURRING AMINO ACIDS (continued)

No.	Amino acid (synonym)	Source	Structure	Formula (mol wt)	Melting point °C[a]	$[\alpha]_D$[b]	pK_a	References: Isolation and purification	References: Chromatography	References: Chemistry	References: Spectral data
58	α,ε-Diaminopimelic acid	Pine pollen	$HOOCCH(NH_2)(CH_2)_3CH(NH_2)COOH$	$C_7H_{14}N_2O_4$ (190.20)	–	$+8.1^{25}$ (c 5, H_2O) $+45^{26}$ (c 1, 1 N HCl) $+45.1^{24}$ (c 2.6, 5 N HCl) (290)	1.8 2.2 9.9 8.8 (290)	84	84	85	–
59	2,3-Diaminosuccinic acid	*Streptomyces rimosus*	$HOOCCH(NH_2)CH(NH_2)COOH$	$C_4H_8N_2O_4$ (148.10)	240–290° (dec)	–	–	86	86	86	86
	E. L-Keto, Hydroxy, and Hydroxy Substituted Amino Acids										
60	*O*-Acetylhomoserine	*Pisum*	$H_3CC(=O)O(CH_2)_3CH(NH_2)COOH$	$C_6H_{11}NO_4$ (161.17)	–	–	–	87	87	87	–
60a	Threo-α-amino-β,γ-dihydroxybutyric acid	*Streptomyces*	$HOCH_2CH(OH)CH(NH_2)COOH$	$C_4H_9NO_4$ (135.14)	210° (dec)	-13.3^{25} (c 1, H_2O) -1.1^{25} (c 1, 2.2 N HCl) (87a)	–	87a	–	–	–
61	2-Amino-4,5-dihydroxy pentanoic acid	*Lunaria annua*	$HOCH_2CH(OH)CH_2CH(NH_2)COOH$	$C_5H_{11}NO_4$ (149.15)	–	–	–	88	88	88	–
61a	2-Amino-3-formyl-3-pentenoic acid	*Bankera fuligineoalba*	$H_3CCH{=}C(CHO)CH(NH_2)COOH$	$C_6H_9NO_3$ (143.16)	–	–	–	88a	88a	88a	88a
62	α-Amino-γ-hydroxy-adipic acid	*Vibrio comma*	$HOOCCH_2CH(OH)CH_2CH(NH_2)COOH$	$C_6H_{11}NO_5$ (177.17)	–	–	–	89	–	–	–
63	2-Amino-6-hydroxy-aminohexanoic acid	*Mycobacterium phlei*	$HOCH(NH_2)(CH_2)_3CH(NH_2)COOH$	$C_6H_{14}N_2O_3$ (162.19)	–	$+6.3^{20}$ (c 5, H_2O) $+23.9^{20}$ (c 5.1, 1 N HCl)(90)	–	90	90	90	–

DATA ON THE NATURALLY OCCURRING AMINO ACIDS (continued)

No.	Amino acid (synonym)	Source	Structure	Formula (mol wt)	Melting point °C[a]	$[\alpha]_D$[b]	pK_a	References: Isolation and purification	References: Chromatography	References: Chemistry	References: Spectral data
64	α-Amino-γ-hydroxy-butyric acid	*Escherichia coli* mutants	$HOCH_2CH_2CH(NH_2)COOH$	$C_4H_9NO_3$ (119.12)	199° (91)	–	–	91	91	91	–
65	γ-Amino-β-hydroxy-butyric acid	*Escherichia coli* mutants	$CH_3CH(NH_2)CH(OH)COOH$	$C_4H_9NO_3$ (119.12)	–	–	–	–	92	–	–
66	2-Amino-6-hydroxy-4-methyl-4-hexenoic acid	*Aesculus Californica*	$(HOCH_2)(H)C{=}C(H_3C)(CH_2CH(NH_2)COOH)$	$C_7H_{13}NO_3$ (159.19)	–	-30^{20} (*c* 2.2, H_2O) $+2^{20}$ (*c* 1.1, 5 *N* HCl)(19)	–	19	19	19	19, 21
67	2-Amino-3-hydroxy-methyl-3-pentenoic acid	*Bankera fuligineoalba*	$CH_3CH{=}C(CH_2OH)CH(NH_2)COOH$	$C_6H_{11}NO_3$ (145.17)	160–161° (13)	$+182^{25}$ (*c* 0.8, H_2O) $+201^{25}$ (*c* 0.8, 0.3 *N* HCl (13)	–	13	13	13	13
68	α-Amino-γ-hydroxy-pimelic acid	*Asplenium septentrionale*	$HOOC(CH_2)_2CH(OH)CH_2CH(NH_2)COOH$	$C_7H_{13}NO_5$ (191.19)	–	–	–	96	96	–	–
69	α-Amino-δ-hydroxy-valeric acid	*Canavalia ensiformis*	$HOCH_2(CH_2)_2CH(NH_2)COOH$	$C_5H_{11}NO_3$ (133.15)	216° (dec) (97)	$+6^{25}$ (*c* 2.65, H_2O) $+2.4^{25}_{365}$ (*c* 2.65, H_2O) (97)	–	97	97	97	97
70	α-Amino-β-ketobutyric acid	Mikramycin A	$CH_3C({=}O)CH(NH_2)COOH$	$C_4H_7NO_3$ (117.11)	–	–	–	98	–	–	–
71	α-Amino-β-methyl-γ,δ-dihydroxyisocaproic acid	Phalloidin	$COOH–HCNH_2–HCCH_3–HOCCH_3–CH_2OH$	$C_7H_{15}NO_4$ (177.21)	208–210° (99)	–	–	99	99	99	99

DATA ON THE NATURALLY OCCURRING AMINO ACIDS (continued)

No.	Amino acid (synonym)	Source	Structure	Formula (mol wt)	Melting point °C[a]	$[\alpha]_D$[b]	pK_a	References: Isolation and purification	References: Chromatography	References: Chemistry	References: Spectral data
71a	2-Amino-5-methyl-6-hydroxyhex-4-enoic acid	*Blighia unijugata*	HOH_2C(H$_3$C)C=CHCH$_2$CH(NH$_2$)COOH	$C_7H_{13}NO_3$ (159.21)	–	–	–	99a	99a	–	99a
72	*O*-Butylhomoserine	Soil bacterium	$CH_3(CH_2)_3OCH_2CH_2CH(NH_2)COOH$	$C_8H_{17}NO_3$ (175.23)	267° (100)	–	–	100	100	100	100
72a	Dihydrorhizobitoxine [*O*-(2-amino-3-hydroxypropyl)-homoserine]	*Rhizobium japonicum*	$HOCH_2CH(NH_2)CH_2-O-(CH_2)_2CH(NH_2)COOH$	$C_7H_{16}N_2O_4$ (192.25)	–	–	7.2 8.6 (100b)	100a	–	–	100
73	β,γ-Dihydroxy-glutamic acid	*Rheum rhaponticum*	$HOOCCH(OH)CH(OH)CH(NH_2)COOH$	$C_5H_9NO_6$ (179.13)	–	–	–	101	101	101	–
74	β,γ-Dihydroxyiso-leucine	Thiostrepton	$CH_3CH_2C(CH_3)(OH)CH(OH)CH(NH_2)COOH$	$C_6H_{15}NO_4$ (165.20)	–	–	–	102	–	–	–
75	γ,δ-Dihydroxy-leucine	Phalloin	$HOCH_2C(CH_3)(OH)CH_2CH(NH_2)COOH$	$C_6H_{13}NO_4$ (163.18)	–	–	–	103	–	103	103
75a	δ,ε-Dihydroxy-norleucine	Bovine tendon	$HOOC-CH(NH_2)(CH_2)_2CH(OH)CH_2OH$	$C_6H_{13}NO_4$ (163.20)	–	–	–	103a	–	103a	–
76	*O*-Ethylhomo-serine	Soil bacterium	$CH_3CH_2OCH_2CH_2CH(NH_2)COOH$	$C_6H_{13}NO_3$ (147.18)	262° (100)	-14^{30} (*c* 2.5, H_2O) (100)	–	100	100	100	100
76a	β-Guanido-γ-hydroxyvaline	Viomycin	$H_2NC(=NH)NHCH_2$(HOH$_2$C)CHCH(NH$_2$)COOH	$C_6H_{14}N_4O_3$ (190.24)	182° (dec) (104a)	–	–	–	–	–	104a

DATA ON THE NATURALLY OCCURRING AMINO ACIDS (continued)

No.	Amino acid (synonym)	Source	Structure	Formula (mol wt)	Melting point °C[a]	$[\alpha]_D$[b]	pK_a	References: Isolation and purification	References: Chromatography	References: Chemistry	References: Spectral data
77	Homoserine (α-amino-γ-hydroxybutyric acid)	*Pisum sativum*	$HOCH_2CH_2CH(NH_2)COOH$	$C_4H_9NO_3$ (119.12)	–	-8.8^{25} (*c* 1–2, H_2O) $+18.3^{26}$ (*c* 2, 2 *N* HCl)(290)	2.71 9.62 (290)	105	105	–	–
78	α-Hydroxyalanine	Peptides of ergot	$CH_3C(OH)(NH_2)COOH$	$C_3H_7NO_3$ (105.10)	–	–	–	106	–	–	–
79	α-Hydroxy-γ-aminobutyric acid	*E. coli* mutants	$H_2NCH_2CH_2CH(OH)COOH$	$C_4H_9NO_3$ (119.12)	–	–	–	91	91	–	–
80	β-Hydroxy-γ-aminobutyric acid	Mammalian brain	$H_2NCH_2CH(OH)CH_2COOH$	$C_4H_9NO_3$ (119.12)	–	–	–	–	108	–	–
81	α-Hydroxy-ε-aminocaproic acid	*Neurospora crassa*	$N_2NCH_2(CH_2)_3CH(OH)COOH$	$C_6H_{13}NO_3$ (147.18)	–	–	–	109	109	–	–
82	β-Hydroxyasparagine	Human urine	$H_2NCOCH(OH)CH(NH_2)COOH$	$C_4H_8O_4N_2$ (148.12)	238–240° (dec) (110)	–	2.09 8.29 (20)	110	110	110	110
83	β-Hydroxyaspartic acid	*Azotobacter*	$HOOCCH(OH)CH(NH_2)COOH$	$C_4H_7NO_5$ (149.10)	–	+41.4 (*c* 2.42, H_2O) +53.0 (*c* 2.46, 1 *N* HCl) (290)[d]	1.91 3.51 9.11 (20)	111	111	–	–
83a	*N*-(2-Hydroxyethyl)-alanine	Rumen protozoa	$HO(CH_2)_2NHCH(CH_3)COOH$	$C_5H_{11}NO_3$ (133.17)	–	–	–	111a	–	–	111a
84	N^4-(2-Hydroxyethyl)-asparagine	*Bryonia dioica*	$HO(CH_2)_2NHCOCH_2CH(NH_2)COOH$	$C_6H_{12}N_2O_4$ (176.20)	199–200° (112)	-2.9^{20} (*c* 5, H_2O) (112)	–	112	112	112	112
85	N^5-(2-Hydroxyethyl)-glutamine	*Lunaria annua*	$HO(CH_2)_2NHCO(CH_2)_2CH(NH_2)COOH$	$C_7H_{14}N_2O_4$ (190.23)	–	$+5.8^{19}$ (*c* 1.8, H_2O) (88)	–	88	88	88	–

DATA ON THE NATURALLY OCCURRING AMINO ACIDS (continued)

No.	Amino acid (synonym)	Source	Structure	Formula (mol wt)	Melting point °C[a]	$[\alpha]_D$[b]	pK_a	References: Isolation and purification	References: Chromatography	References: Chemistry	References: Spectral data
86	β-Hydroxyglutamic acid	*Mycobacterium tuberculosis*	$HOOCCH_2CH(OH)CH(NH_2)COOH$	$C_5H_9NO_5$ (163.13)	187° (dec) (290)	$+8.6^{9}$ (H_2O) $+30.8^{20}$ (*c* 2, 20% HCl) (290)[d]	–	114	114	–	–
87	γ-Hydroxyglutamic acid	*Linaria vulgaris*	$HOOCCH(OH)CH_2CH(NH_2)COOH$	$C_5H_9NO_5$ (163.13)	–	–	–	115	115	115	–
88	γ-Hydroxyglutamine	*Phlox decussata*	$H_2NCOCH(OH)CH_2CH(NH_2)COOH$	$C_5H_{10}N_2O_4$ (162.15)	163–164° (dec) (116)	–	–	116	116	116	–
89	ε-Hydroxylamino-norleucine (α-amino-ε-hydroxy-aminohexanoic acid)	*Mycobacterium phlei*	$HONHCH_2(CH_2)_3CH(NH_2)COOH$	$C_5H_{12}N_2O_3$ (148.17)	223–225° (dec) (117)	$+6.3^{20}$ (*c* 5, H_2O) $+23.9^{18}$ (*c* 5.1, 1 *N* HCl) (117)	–	117	117	117	–
89a	4-Hydroxyiso-leucine	*Trigonella foenumgraecum*	CH_3CHOH \ $CHCH(NH_2)COOH$ / CH_3	$C_6H_{13}NO_3$ (147.20)	–	$+31^{20}$ (*c* 1, H_2O) (117a)	–	117a	117a	–	117a
90	δ-Hydroxy-γ-keto-norvaline	*Streptomyces akiyoshiensis* novo	$HOCH_2COCH_2CH(NH_2)COOH$	$C_5H_9NO_4$ (147.13)	–	-8.2^{17} (*c* 3.4, H_2O) (118)	2.0 9.1	118	118	118	118
91	δ-Hydroxyleucenine (δ-ketoleucine)	Phalloidin	$HOCH_2C(CH_3){=}CHCH(NH_2)COOH$	$C_6H_{11}NO_3$ (145.17)	–	–	–	119	119	119	119
92	β-Hydroxyleucine	Antibiotic from *Paecilomyces* strain	$(CH_3)_2CHCH(OH)CH(NH_2)COOH$	$C_6H_{13}NO_3$ (147.17)	–	–	–	16	16	16	–
93	δ-Hydroxyleucine	*Paecilomyces*	HOH_2C \ $CHCH_2CH(NH_2)COOH$ / H_3C	$C_6H_{13}NO_3$ (147.17)	–	–	–	121	121	121	121

DATA ON THE NATURALLY OCCURRING AMINO ACIDS (continued)

No.	Amino acid (synonym)	Source	Structure	Formula (mol wt)	Melting point °C[a]	$[\alpha]_D$[b]	pK_a	References: Isolation and purification	References: Chromatography	References: Chemistry	References: Spectral data
94	Threo-β-hydroxyleucine	*Deutzia gracilis*	$HOOCCH(NH_2)CH(OH)CH(CH_3)CH_3$	$C_6H_{13}NO_3$ (147.17)	–	–	–	122	122	122	122
95	α-Hydroxylysine (α,ε-diamino-α-hydroxycaproic acid)	*Salvia officinalis*	$H_2N(CH_2)_4C(OH)(NH_2)COOH$	$C_6H_{15}N_2O_3$ (162.19)	–	–	2.13 8.62 9.67 (20)	123	123	–	–
96	δ-Hydroxylysine (α,ε-diamino-δ-hydroxycaproic acid)	Fish gelatin	$H_2NCH_2CH(OH)(CH_2)_2CH(NH_2)COOH$	$C_6H_{15}N_2O_3$ (162.19)	–	–	–	124	111	–	–
96a	β-Hydroxynorvaline	*Streptomyces* species	$CH_3CH_2CH(OH)CH(NH_2)COOH$	$C_5H_{11}NO_3$ 133.17)	244° (dec) (124a)	–	–	124a	124a	–	124a
97	γ-Hydroxynorvaline	*Lathyrus odoratus*	$CH_3CH(OH)CH_2CH(NH_2)COOH$	$C_5H_{11}NO_3$ (133.15)	–	$+22^{20}$ (c 5, H_2O) +32 (c 2.5, gl acetic) (126)	–	126	126	126	–
98	γ-Hydroxyornithine	*Vicia sativa*	$H_2NCH_2CH(OH)CH_2CH(NH_2)COOH$	$C_5H_{12}N_2O_3$ (148.16)	–	–	–	127	127	–	–
99	α-Hydroxyvaline	Ergot	$(CH_3)_2CHC(OH)(NH_2)COOH$	$C_5H_{11}NO_3$ (133.15)	–	–	2.55 9.77 (20)	106	–	–	–
100	γ-Hydroxyvaline	*Kalanchoe daigremontiana*	$(H_3C)(HOH_2C)CHCH(NH_2)COOH$	$C_5H_{11}NO_3$ (133.15)	228° (dec) (129)	$+10^{20}$ (H_2O) (129)	–	129	129	–	–
100a	Hypusine	Bovine brain	$H_2N(CH_2)_2CH(OH)CH_2NH(CH_2)_4CH(NH_2)COOH$	$C_{10}H_{23}N_3O_3$ (233.27)	234–238° (dec) (129a)	–	–	129a	–	129a	129a
100b	Isoserine	*Bacillus brevis* (edeine A and B)	$H_2NCH_2CH(OH)COOH$	$C_3H_7NO_3$ (105.11)	–	–	–	129b	129b	129b	–
101	4-Ketonorleucine (2-amino-4-keto-hexanoic acid)	*Citrobacter freundii*	$HOOCCH(NH_2)CH_2COCH_2CH_3$	$C_6H_{11}NO_3$ (145.17)	142–143° (130)	–	–	131	131	130	130

DATA ON THE NATURALLY OCCURRING AMINO ACIDS (continued)

No.	Amino acid (synonym)	Source	Structure	Formula (mol wt)	Melting point °C[a]	$[\alpha]_D$[b]	pK_a	References: Isolation and purification	References: Chromatography	References: Chemistry	References: Spectral data
102	γ-Methyl-γ-hydroxy glutamic acid	*Phyllitis scolopendrium*	$HOOCC(CH_3)(OH)CH_2CH(NH_2)COOH$	$C_6H_{11}NO_5$ (157.17)	–	–	–	56	56	–	–
103	Pantonine (α-amino-β,β-dimethyl-γ-hydroxybutyric acid)	*Escherichia coli*	$HOCH_2C(CH_3)_2CH(NH_2)COOH$	$C_6H_{13}NO_3$ (147.18)	–	–	–	132	132	132	–
103a	*Threo*-β-phenylserine	*Canthium eurysides*	C_6H_5–CH(OH)–CH(NH$_2$)COOH (HO—CH—CH(NH$_2$)COOH with phenyl on CH)	$C_9H_{11}NO_3$ (181.21)	–	–	–	–	–	–	252a
103b	Pinnatanine [N^5-(2-hydroxy-methylbutadienyl)-*allo*-γ-hydroxy-glutamine]	*Staphylea pinnata*	$(HOH_2C)(H_2C{=}HC)C{=}CHNH{-}COCH(OH)CH_2CH(NH_2)COOH$	$C_{10}H_{16}N_2O_5$ (244.28)	165° (dec) (132a)	+3.2[27] (c 0.5, H_2O) (132a)	9.1 (132a)	132a	132a	–	132a
104	*O*-Propylhomoserine	Soil bacterium	$CH_3(CH_2)_2O(CH_2)_2CH(NH_2)COOH$	$C_7H_{15}NO_3$ (161.21)	265° (100)	–11[30] (c 2, H_2O) (100)	–	100	100	100	100
104a	Rhizobitoxine [2-amino-4-(2-amino-3-hydroxypropoxy)-but-3-enoic acid]	*Rhizobium japonicum*	$HOH_2CCH(NH_2)CH_2OCH{=}CHCH_2(NH_2)COOH$	$C_7H_{15}N_2O_4$ (191.24)	–	–	–	132b	–	–	132c
105	Serine (α-amino-β-hydroxypropionic acid)	Silk fibroin	$HOCH_2CH(NH_2)COOH$	$C_3H_7NO_3$ (105.09)	228° (dec) (1)	–7.5[25] (c2, H_2O) +15.1[25] (c 2, 5 *N* HCl) (290)	2.19 9.21 (290)	134	3	134	134
106	*O*-Succinylhomoserine	*Escherichia coli*	$HOOC(CH_2)_2C(=O)O(CH_2)_2CH(NH_2)COOH$	$C_8H_{13}NO_6$ (219.20)	180–181° (135)	–	4.4 9.5 (135)	135	135	135	–

DATA ON THE NATURALLY OCCURRING AMINO ACIDS (continued)

No.	Amino acid (synonym)	Source	Structure	Formula (mol wt)	Melting point °C[a]	$[\alpha]_D$[b]	pK_a	References: Isolation and purification	References: Chromatography	References: Chemistry	References: Spectral data
107	Tabtoxinine (α,ϵ-diamino-β-hydroxypimelic acid)	*Pseudomonas tabaci*	$HOOCCH(NH_2)(CH_2)_2CH(OH)CH(NH_2)COOH$	$C_7H_{14}N_2O_5$ (186.20)	–	–	–	136	137	137	–
108	Threonine (α-amino-β-hydroxybutyric acid)	Fibrin hydrolyzate	$CH_3CH(OH)CH(NH_2)COOH$	$C_4H_9NO_3$ (119.12)	253° (1)	-28^{25} (c 1–2, H_2O) -15^{25} (c 1–2, 5 N HCl) (290)	2.09 9.10 (290)	138	3	138, 139	139
			F. L-Aromatic Amino Acids								
109	α-Amino-β-phenylbutyric acid	*Streptomyces bottropensis*	–$CH(CH_3)CH(NH_2)COOH$	$C_{10}H_{13}NO_2$ (187)	176–177° (140)	–	–	140	140	140	140
109a	β-Amino-β-phenylpropionic acid	*Roccella canariensis* hydrolyzate	–$CH(NH_2)CH_2COOH$	$C_9H_{11}NO_2$ (165.21)	–	–	–	–	–	–	140a
110	3-Carboxy-4-hydroxyphenylalanine (m-carboxytyrosine)	*Reseda odorata*	COOH, HO–, –$CH_2CH(NH_2)COOH$	$C_{10}H_{11}NO_5$ (165.15)	–	-7.7^{25} (c 0.9, 1 N NaOH) -29.9^{24} (c 0.6, 0.2 M PO_4, pH 7) (297)	2 3.4 9.3 12–13 (297)	141	141	141	141
111	m-Carboxyphenylalanine	Iris bulbs	COOH, –$CH_2CH(NH_2)COOH$	$C_{10}H_{11}NO_4$ (191.26)	–	–	1.5 3.9 (142)	142	142	142	142
111a	2,3-Dihydroxy-N-benzoylserine	*Escherichia coli*	OH, OH, –$CONHCH(CH_2OH)COOH$	$C_{10}H_{11}NO_6$ (241.22)	193–194° (142a)	–	–	142a	142a	–	142a

DATA ON THE NATURALLY OCCURRING AMINO ACIDS (continued)

No.	Amino acid (synonym)	Source	Structure	Formula (mol wt)	Melting point °C[a]	$[\alpha]_D$[b]	pK_a	References: Isolation and purification	References: Chromatography	References: Chemistry	References: Spectral data
112	2,4-Dihydroxy-6-methylphenylalanine	*Agrostemma githago*	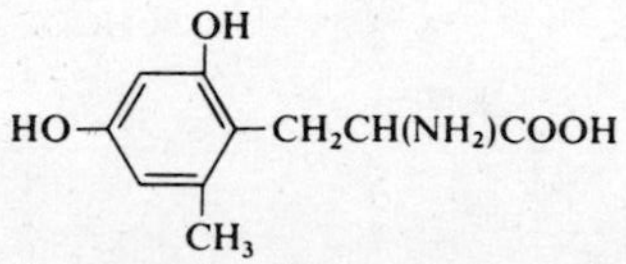 OH; HO; $CH_2CH(NH_2)COOH$; CH_3	$C_{10}H_{13}NO_4$ (211.23)	252° (144)	+19.7[18] (1 *N* HCl) (144)	–	144	144	144	144
113	3,4-Dihydroxyphenylalanine (DOPA)	*Vicia faba*	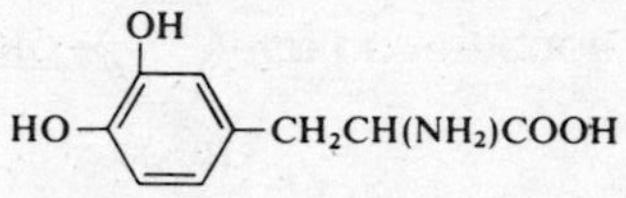 OH; HO; $CH_2CH(NH_2)COOH$	$C_9H_{11}NO_4$ (209.21)	–	–14.3 (1 *N* HCl) (146)	2.32[e] 8.68[e] 9.88[e]	145	1	146	–
114	3,5-Dihydroxyphenylglycine	*Euphorbia helioscopia*	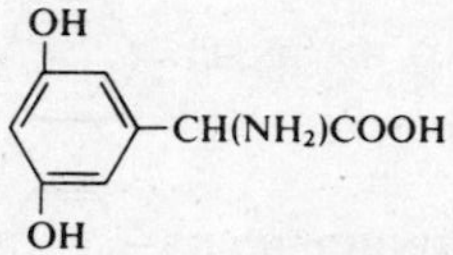 OH; $CH(NH_2)COOH$; OH	$C_8H_9NO_4$ (183.16)	230–232° (147)	–	–	147	147	147	147
114a	γ-Glutaminyl-4-hydroxybenzene	*Agarius bisporus*	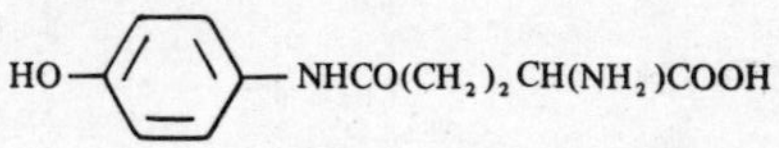 HO; $NHCO(CH_2)_2CH(NH_2)COOH$	$C_{11}H_{14}N_2O_4$ (238.27)	225–226° (147a)	+42.5[25] (*c* 0.67, 0.1 *N* NaOH) (147a)	–	147a	–	–	147a
114b	3-Hydroxymethylphenylalanine	*Caesalpinice tinctoria*	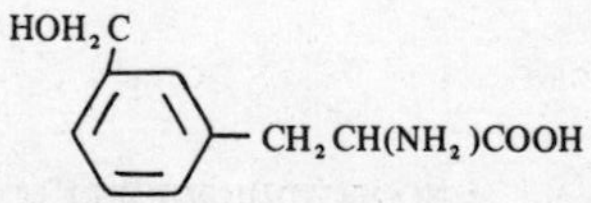 HOH_2C; $CH_2CH(NH_2)COOH$	$C_{10}H_{13}NO_3$ (195.24)	–	–	–	147b	147b	–	147b
114c	4-Hydroxy-3-hydroxymethylphenylalanine	*Caesalpinia tinctoria*	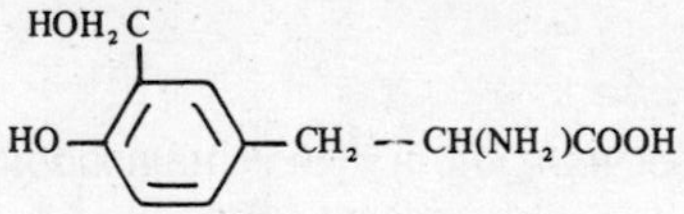 HOH_2C; HO; CH_2 – $CH(NH_2)COOH$	$C_{10}H_{13}NO_4$ (211.24)	–	–36[20] (*c* 1, H_2O) –4[20] (*c* 0.67, 1 *N* NaOH) (147b)	–	147b	147b	–	147b
115	3-Hydroxykynurenine	Human urine	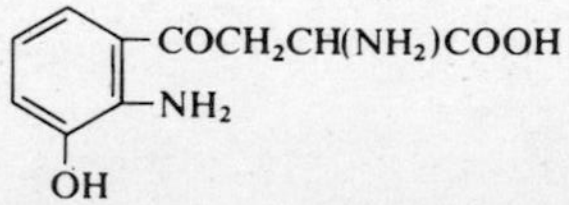 $COCH_2CH(NH_2)COOH$; NH_2; OH	$C_{10}H_{12}N_2O_4$ (224.21)	–	–	–	148	148	–	–

DATA ON THE NATURALLY OCCURRING AMINO ACIDS (continued)

No.	Amino acid (synonym)	Source	Structure	Formula (mol wt)	Melting point °C[a]	$[\alpha]_D$[b]	pK_a	References: Isolation and purification	References: Chromatography	References: Chemistry	References: Spectral data
115a	*p*-Hydroxymethylphenylalanine	*Escherichia coli*	HOH$_2$C–C$_6$H$_4$–CH$_2$CH(NH$_2$)COOH (para)	$C_{10}H_{13}NO_3$ (195.24)	231–233° (dec) (148a)	-32.5^{20} (148a)	–	–	148b	148b	148a
116	*m*-Hydroxyphenylglycine	*Euphorbia helioscopia*	OH–C$_6$H$_4$–CH(NH$_2$)COOH (meta)	$C_8H_9NO_3$ (177.16)	212–214° (147)	–	–	147	147	147	147
117	Kynurenine (β-arthraniloyl-α-aminopropionic acid)	Rabbit urine	C$_6$H$_4$(NH$_2$)–COCHCH(NH$_2$)COOH (ortho)	$C_{10}H_{12}N_2O_3$ (208.21)	191° (dec) (290)	-30.5^{25} (*c* 1, H_2O) (290)	–	148	–	148	148
118	*O*-Methyltyrosine (β-(*p*-methoxyphenyl)-alanine)	Puromycin	H$_3$CO–C$_6$H$_4$–CH$_2$CH(NH$_2$)COOH (para)	$C_{10}H_{13}NO_3$ (195.22)	191° (150)	-5.9_{546} (HCl) (150) -3.2_{546} (1 *N* NaOH) (150)	–	149	–	150	–
119	Phenylalanine	*Lupinus luteus*	C$_6$H$_5$–CH$_2$CH(NH$_2$)COOH	$C_9H_{11}NO_2$ (165.19)	284° (1)	-34.5^{25} (*c* 1–2, H_2O) (290) -4.5^{25} (*c* 1–2, 5 *N* HCl) (290)	2.16 9.18 (290)	151	3	152	152
120	Tyrosine (α-amino-β-hydroxyphenyl propionic acid)	Casein, alkaline hydrolyzate	HO–C$_6$H$_4$–CH$_2$CH(NH$_2$)COOH (para)	$C_9H_{11}NO_3$ (181.19)	344° (1)	-10^{25} (*c* 2, 5 *N* HCl)(1)	2.20 9.11 (1) 10.13 (290)	153	3	154	154
120a	β-Tyrosine	*Bacillus brevis* (edeine A and B)	HO–C$_6$H$_4$–CH(NH$_2$)CH$_2$COOH (para)	$C_9H_{11}NO_3$ (181.21)	–	–	–	129b	129b	129b	–

DATA ON THE NATURALLY OCCURRING AMINO ACIDS (continued)

No.	Amino acid (synonym)	Source	Structure	Formula (mol wt)	Melting point °C[a]	$[\alpha]_D$[b]	pK_a	References: Isolation and purification	References: Chromatography	References: Chemistry	References: Spectral data
121	*m*-Tyrosine	*Euphorbia myrsinites* L	3-hydroxyphenyl–$CH_2CH(NH_2)COOH$ (ring with OH)	$C_9H_{11}NO_3$ (181.19)	272–274° (155)	-14.5^{22} (70% EtOH) +8.9(70% EtOH, 2 *N* HCl) (155)	–	155	155	155	155
	G. L-Ureido and Guanido Amino Acids										
121a	*N*-Acetylarginine	Cattle brain	$H_2N-C(=NH)-NH(CH_2)_3CH(NHCOCH_3)COOH$	$C_7H_{16}N_4O_3$ (204.27)	270° (155a)	–	–	155a	155a	155a	155a
122	Albizziine (2-amino-3-ureido-propionic acid)	*Mimosaceae*	$HOOCCH(NH_2)CH_2NHCONH_2$	$C_4H_9NO_3$ (119.12)	–	-66^{26} (*c* 2, H_2O) (157)	–	156	157	157	157
123	Arginine (amino-δ-guanidino-valeric acid)	*Lupinus luteus*	$H_2NC(=NH)NH(CH_2)_3CH(NH_2)COOH$	$C_6H_{14}N_4O_2$	238° (1)	$+12.5^{25}$ (*c* 2, H_2O) $+27.6^{25}$ (*c* 2, 5 *N* HCl)(1)	1.82 8.99 12.48 (290)	158	3	159	159
124	Canavanine (α-amino-(*O*-guanidyl)-γ-hydroxybutyric acid)	*Canavalia ensiformis*	$H_2NC(=NH)NHOCH_2CH_2CH(NH_2)COOH$	$C_5H_{12}N_4O_3$ (176.19)	172° (160)	$+18.6^{18.5}$ (*c* 7.8, H_2O) (160)	2.50 6.60 9.25 (33)	161	160	160	162
125	Canavanosuccinic acid	*Canavalia ensiformis*	$HOOCCH_2CH(COOH)NHC(=NH)NHOCH_2CH_2CH(NH_2)COOH$	$C_9H_{16}N_4O_7$ (292.27)	–	–	–	163	163	–	–

DATA ON THE NATURALLY OCCURRING AMINO ACIDS (continued)

No.	Amino acid (synonym)	Source	Structure	Formula (mol wt)	Melting point °C[a]	$[\alpha]_D$[b]	pK_a	References: Isolation and purification	References: Chromatography	References: Chemistry	References: Spectral data
126	Citrulline	Watermelon	$H_2NCONH(CH_2)_3CH(NH_2)COOH$	$C_6H_{13}N_3O_3$	202° (164)	$+4^{25}$ (*c* 2, H_2O) $+24.2^{25}$ (*c* 2, *N* HCl) +10.8 (*c* 1.1, 0.1 *N* NaOH) (290)	2.43 9.41 (290)	164	165	164	–
127	Desaminocanavanine	Canavanine	$HN{=}CNHO(CH_2)_2CHCOOH$ (ring closed through NH: C···NH)	$C_5H_9N_3O_3$	256–257° (166)	$+26.61^{21}$ (H_2O) (166)	–	166	–	166	–
127a	N^G,N^G-Dimethyl-arginine	Bovine brain	H_3C NH $NCNH(CH_2)_3CH(NH_2)COOH$ H_3C	$C_8H_{18}N_4O_2$ (202.30)	198–201° (70a)	–	–	166a	166a	–	166a
127b	N^G,N'^G-Dimethyl-arginine	Bovine brain	NCH_3 ‖ $H_3C—NH—CNH(CH_2)_3CH(NH_2)COOH$	$C_8H_{18}N_4O_2$ (202.30)	237–239° (dec) (70a)	–	–	166a	166a	70a	166a
128	Gigartinine [α-amino-δ-(guanylureido)-valeric acid]	*Gymnogongrus flabelliformis*	NH ‖ $H_2NCNHCONH(CH_2)_3CH(NH_2)COOH$	$C_7H_{15}N_5O_3$ (217.25)	–	–	–	167	167	167	167
129	Homoarginine	*Lathyrus* species	NH ‖ $H_2NCNH(CH_2)_4CH(NH_2)COOH$	$C_7H_{16}N_4O_2$ (188.25)	–	+42.4 (*c* 0.452, 1.02 *N* HCl)[f] (168)	–	168	168	168	168
130	Homocitrulline	Human urine	$H_2NCONH(CH_2)_4CH(NH_2)COOH$	$C_7H_{14}N_3O_3$ (189.22)	–	–	–	169	169	–	–
131	γ-Hydroxyarginine	*Vicia sativa*	NH ‖ $H_2NCNHCH_2CH(OH)CH_2CH(NH_2)COOH$	$C_6H_{14}N_4O_3$ (190.20)	–	–	–	170	170	170	170

DATA ON THE NATURALLY OCCURRING AMINO ACIDS (continued)

No.	Amino acid (synonym)	Source	Structure	Formula (mol wt)	Melting point °C[a]	$[\alpha]_D$[b]	pK_a	References: Isolation and purification	References: Chromatography	References: Chemistry	References: Spectral data
131a	N^5-Hydroxyarginine	*Bacillus* species	NH ‖ $H_2NCNOH(CH_2)_3CH(NH_2)COOH$	$C_6H_{14}N_4O_3$ (190.24)	206–212° (dec) (170a)	$+21^{25}$ (*c* 1, 5 *N* HCl) (170a)	–	170a	170a	170a	170a
132	γ-Hydroxyhomo-arginine (α-amino-ε-guanidino-γ-hydroxy-hexanoic acid)	*Lathyrus tingitanus*	NH ‖ $H_2NCNH(CH_2)_2CHOHCH_2CH(NH_2)COOH$	$C_7H_{16}N_4O_3$ (204.23)	–	–	–	171	171	171	171
133	Indospicine (α-amino-ε-amidino caproic acid)	*Endigofera spicata*	$H_2NC(CH_2)_4CH(NH_2)COOH$ ‖ NH	$C_7H_{15}N_3O_2$ (173.23)	131–134° (173)	$+18^{22}$ (*c* 1.1, 5 *N* HCl)(173)	–	173	173	173	–
133a	ω-*N*-Methylarginine (guanidinomethyl-arginine)	Bovine brain	NH ‖ $H_3CNHCNH(CH_2)_3CH(NH_2)COOH$	$C_7H_{16}N_4O_2$ (188.27)	–	–	–	166a	166a	–	166a
	H. L-Amino Acids Containing Other Nitrogenous Groups										
134	Alanosine [2-amino-3-(*N*-nitrosohydroxy-amino) propionic acid]	*Streptomyces alanosinicus*	NO | $HONCH_2CH(NH_2)COOH$	$C_3H_7N_3O_4$ (149.11)	–	+8 (1 *N* HCl) −46 (0.1 *N* NaOH) (174)	(174)	174	–	174	174
135	Azaserine (*O*-diazoacetylserine)	*Streptomyces*	O ‖ $N_2CHCOCH_2CH(NH_2)COOH$	$C_5H_7N_3O_4$ (173.14)	146–162° (175)	$-0.5^{27.5}$ (*c* 8.46, H_2O) (175)	8.55 (34)	175	–	175	175
136	β-Cyanoalanine	*Vicia sativa*	$N{\equiv}CCH_2CH(NH_2)COOH$	$C_4H_6N_2O_2$ (114.11)	214.5° (176)	-2.9^{26} (*c* 1.4, 1 *N* acetic acid) (177)	1.7 7.4 (177)	176	176	177	177

DATA ON THE NATURALLY OCCURRING AMINO ACIDS (continued)

No.	Amino acid (synonym)	Source	Structure	Formula (mol wt)	Melting point °C[a]	$[\alpha]_D$[b]	pK_a	References: Isolation and purification	References: Chromatography	References: Chemistry	References: Spectral data
136a	γ-Cyano-α-amino-butyric acid	*Chromobacterium violaceum*	$N{\equiv}(CH_2)_2CH(NH_2)COOH$	$C_5H_8N_2O_2$ (128.15)	221–223° (177a)	$+32.1^{21}$ (*c* 0.38, 1 *N* HOAc) (177a)	–	177a	177a	177a	177a
137	ε-Diazo-δ-ketonorleucine	*Streptomyces*	$N_2CHCO(CH_2)_2\ CH(NH_2)COOH$	$C_6H_9N_3O_3$ (171.17)	145–155° (178)	21^{26} (*c* 5.4, H_2O) (178)	–	178	178	178	178
138	Hadacidin (*N*-formyl-*N*-hydroxyaminoacetic acid)	*Penicillium frequentans*	CH_2COOH | NOH | CHO	$C_3H_5NO_4$ (119.08)	205–210° (179)	–	–	179	–	179	179

I. L-Heterocyclic Amino Acids

No.	Amino acid (synonym)	Source	Structure	Formula (mol wt)	Melting point °C[a]	$[\alpha]_D$[b]	pK_a	References: Isolation and purification	References: Chromatography	References: Chemistry	References: Spectral data
138a	2-Alanyl-3-isoxazolin-5-one	Pea seedlings	O ‖ O N (isoxazolinone ring) | CH_2 $CHNH_2$ COOH	$C_6H_8N_2O_4$ (172.16)	203–205° (dec) (180b)	–	–	180a	180b	180b	180b
139	*Allo*hydroxyproline	*Santalum album*	OH | HC—CH_2 | | COOH H_2C C N H | H	$C_5H_9NO_3$ (131.13)	248° (180)	-57^{25} (*c* 0.65, H_2O) (180)	–	180	180	180	–
140	*Allo*kainic acid (3-carboxymethyl-4-isopropenyl proline)	*Digenea simplex*	$H_2C{=}C{-}CH{-}CHCH_2COOH$ | | | H_3C C CHCOOH N H	$C_{10}H_{15}NO_4$ (213.24)	–	$+8^{26}$ (H_2O) (184, 185)	–	184, 185	–	184, 185	184, 185

DATA ON THE NATURALLY OCCURRING AMINO ACIDS (continued)

No.	Amino acid (synonym)	Source	Structure	Formula (mol wt)	Melting point °C[a]	$[\alpha]_D$[b]	pK_a	References: Isolation and purification	References: Chromatography	References: Chemistry	References: Spectral data
140a	1-Amino-2-nitro-cyclopentane carboxylic acid	*Aspergillus wentii*	NH_2 H_2C — C — COOH H_2C $CHNO_2$ CH_2	$C_6H_{10}N_2O_4$ (174.18)	150° (dec) (185a)	–	–	185a	185a	185a	185a
141	4-Aminopipecolic acid	*Strophanthus scandens*	NH_2 C H_2C CH_2 H_2C CHCOOH N H	$C_6H_{11}N_2O_2$ (143.18)	–	–	–	186	186	–	–
141a	*cis*-3-Aminoproline	*Morchella esculenta*	NH_2 H_2C — CH — COOH H_2C C — H N H	$C_5H_{10}N_2O_2$ (130.17)	215° (dec) (186a)	$+5.8^{20}$ (c 2, H_2O) $+23.0^{20}$ (c 2, 5 N HCl) (186a)	–	186a	186a	–	186a
141b	Anticapsin	*Streptomyces griseoplanus*	O O CH_2 H_2N — CH — COOH	$C_9H_{13}NO_4$ (199.23)	240° (dec) (186b)	$+125^{25}$ (c 1, H_2O) (186b)	4.3 10.1 (186b)	186b	186b	–	186b
142	Ascorbigen	Cabbage	$CH_2OC_6H_7O_5$ N	$C_{15}H_{15}NO_6$ (305.30)	–	–	–	187	188	189	–

DATA ON THE NATURALLY OCCURRING AMINO ACIDS (continued)

No.	Amino acid (synonym)	Source	Structure	Formula (mol wt)	Melting point °C[a]	$[\alpha]_D$[b]	pK_a	References: Isolation and purification	References: Chromatography	References: Chemistry	References: Spectral data
143	Azetidine-2-carboxylic acid	*Convallaria majalis*	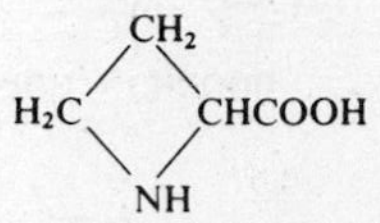 CH_2, H_2C, CHCOOH, NH	$C_4H_7NO_2$ (101.11)	–	–	–	190	190	190	190
143a	Azirinomycin (3-methyl-2-hydro-azirine carboxylic acid)	*Streptomyces aureus*	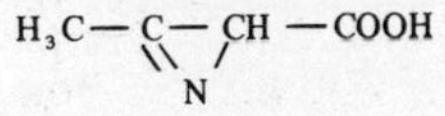 H_3C—C—CH—COOH, N	$C_4H_5NO_2$ (99.10)	–	–	–	189a	189a	–	189a
144	Baikiain 1,2,3,6-tetra-hydropyridine-α-carboxylic acid)	*Baikiaea plurijuga*	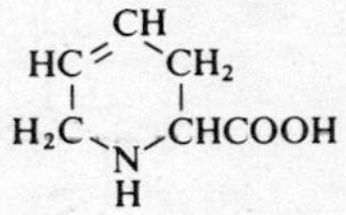 CH, HC, CH_2, H_2C, CHCOOH, N, H	$C_6H_9NO_2$ (127.15)	273–274° (183)	–	–	181	182	183	183
144a	*N*-Carbamoyl-2-(*p*-hydroxyphenyl) glycine	*Vicia faba*	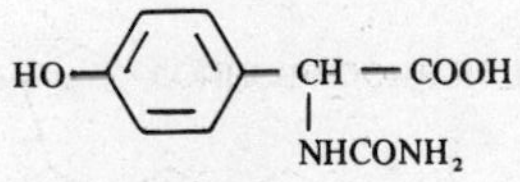 HO, CH—COOH, $NHCONH_2$	$C_9H_{10}N_2O_4$ (210.21)	194–195° (dec) (190a)	–	–	190a	–	–	190a
144b	3-Carboxy-6,7-dihydroxy-1,2,3,4-tetrahydroiso-quinoline	*Mucuna Mutisiana*	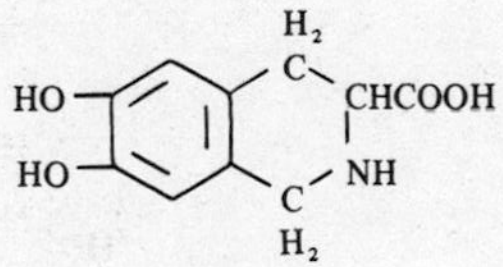 H_2, C, CHCOOH, HO, HO, NH, C, H_2	$C_{10}H_{11}NO_4$ (209.22)	286–288° (190b)	-114.9^{25} (*c* 1.65, 20% HCl) (190b)	–	190b	–	190b	190b
144c	Clavicipitic acid	*Claviceps* (ergot fungus)	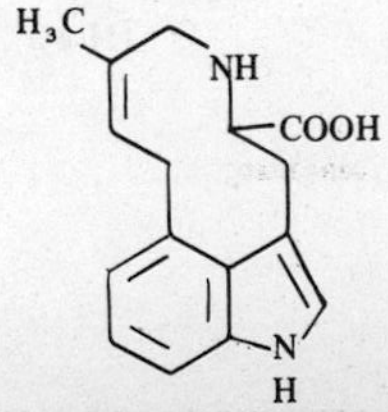 H_3C, NH, COOH, N, H	$C_{16}H_{18}N_2O_2$ (270.36)	262° (dec) (190c)	–	–	190c	–	–	190c

DATA ON THE NATURALLY OCCURRING AMINO ACIDS (continued)

No.	Amino acid (synonym)	Source	Structure	Formula (mol wt)	Melting point °C[a]	$[\alpha]_D$[b]	pK_a	References: Isolation and purification	References: Chromatography	References: Chemistry	References: Spectral data
145	Cucurbitine (3-amino-3-carboxy-pyrrolidine	*Cucurbita moschata*	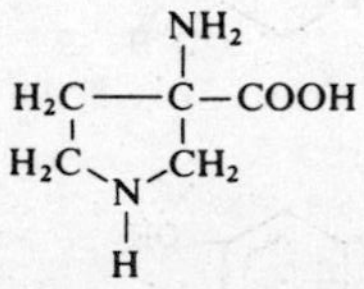 NH_2; H_2C—C—COOH; H_2C N CH_2; H	$C_5H_{10}NO_2$ (116.14)	–	-19.76^{27} (*c* 9.3, H_2O) (191)	–	191	191	191	191
145a	*N*-Dihydrojas-monoylisoleucine	*Gibberella fujikuroi*	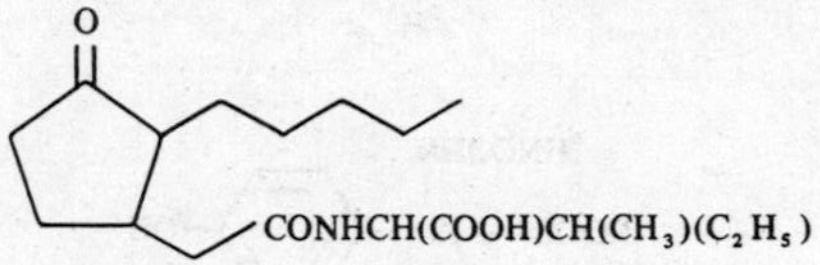 O; CONHCH(COOH)CH(CH_3)(C_2H_5)	$C_{18}H_{31}NO_4$ (325.50)	140–141° (191a)	–	–	–	191a	191a	191a
145b	2,5-Dihydro-phenylalanine (1,4-cyclohexa-diene-1-alanine)	*Streptomycete* X-13, 185	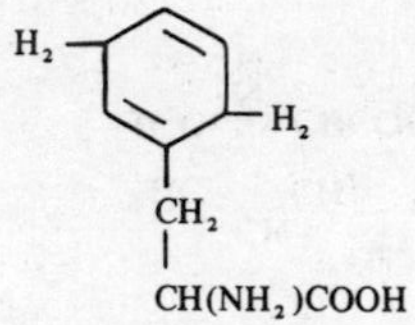 H_2; H_2; CH_2; CH(NH_2)COOH	$C_9H_{13}NO_2$ (167.23)	206–208° (191b)	-33.7^{25} (*c* 1, 5 *N* HCl) (191b)	–	191b	–	–	191b
145c	2-*N*,6-*N*-Di-(2,3-dihydroxy-benzoyl)lysine	*Azobacter vinelandii*	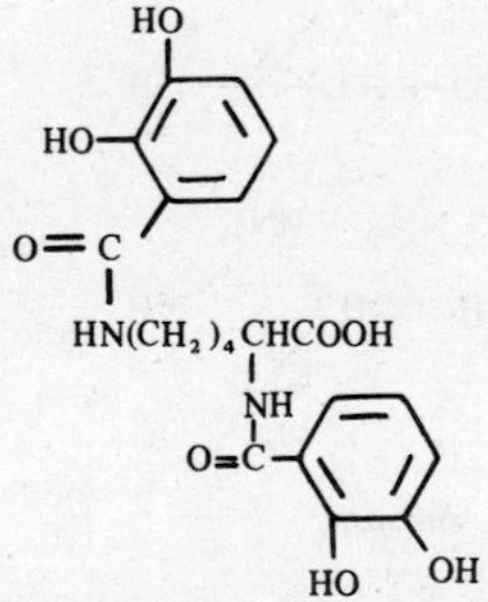 HO; HO; O=C; HN(CH_2)$_4$CHCOOH; NH; O=C; HO; OH	$C_{20}H_{22}N_2O_8$ (418.44)	–	–	4.8 9 (191c)	191c	191c	–	191c

DATA ON THE NATURALLY OCCURRING AMINO ACIDS (continued)

No.	Amino acid (synonym)	Source	Structure	Formula (mol wt)	Melting point °C[a]	$[\alpha]_D$[b]	pK_a	References: Isolation and purification	Chromatography	Chemistry	Spectral data
145d	*cis*-3,4-*trans*-3,4-Dihydroxyproline	Diatom cell walls	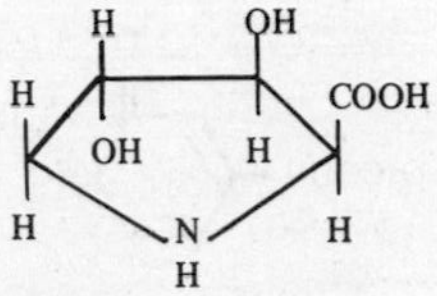 H, OH, H, COOH, OH, H, H, N, H, H	$C_5H_9NO_4$ (147.15)	262° (dec) (191d)	-61.2[20] (*c* 0.5, H_2O) (191d)	–	191d	191d	–	191e
145e	β-(2,6-Dihydroxy-pyrimidin-1-yl) alanine	Pea seedlings	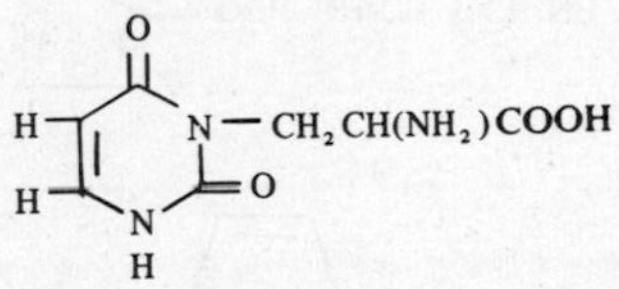 O, H, N — $CH_2CH(NH_2)COOH$, H, =O, N, H	$C_7H_9N_3O_4$ (199.19)	230° (dec) (191f)	–	–	191f	–	191f	191f
145f	4,6-Dihydroxy-quinoline-2-carboxylic acid	Tobacco leaves	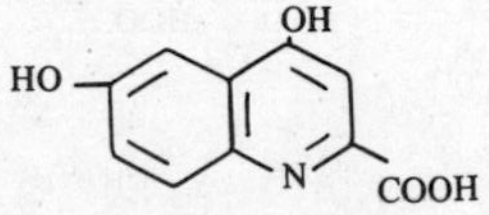 OH, HO, N, COOH	$C_{10}H_7NO_4$ (205.18)	287° (dec) (191g)	–	–	191g	191g	191g	191g
146	Dihydrozanthurenic acid (8-hydroxy-1,2,3,4-tetrahydro-4-ketoquinaldic acid)	*Lepidoptera*	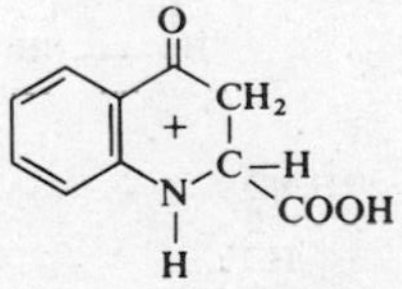 O, CH_2, +, C–H, N, COOH, H	$C_{10}H_9NO_4$ (207.19)	185–190° (192)	-45[20] (*c* 0.9, MeOH) +18[20] (*c* 0.9, MeOH-HCl) (192)	–	192	192	–	192
147	Domoic acid [2-carboxy-3-carboxymethyl-4-1-methyl-2-carboxy-1,3-hexadienyl)-pyrrolidine]	*Chondria armata*	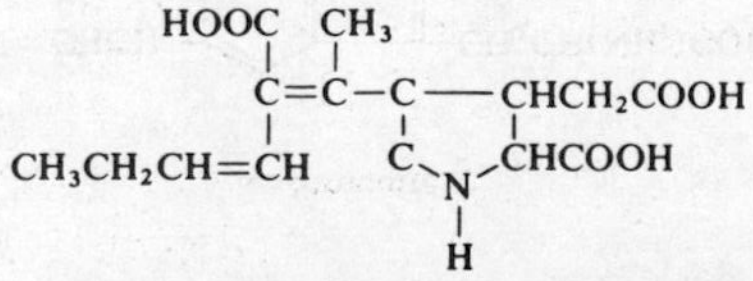 HOOC, CH_3, C=C–C — $CHCH_2COOH$, CH_3CH_2CH=CH, C, N, CHCOOH, H	$C_{15}H_{21}NO_6$ (311.35)	217° (193)	-109.6[12] (*c* 1.314, H_2O) (193)	2.20 3.72 4.93 9.82 (193)	193	193	193	193

DATA ON THE NATURALLY OCCURRING AMINO ACIDS (continued)

No.	Amino acid (synonym)	Source	Structure	Formula (mol wt)	Melting point °C[a]	$[\alpha]_D$[b]	pK_a	References: Isolation and purification	References: Chromatography	References: Chemistry	References: Spectral data
148	Echinine (2-*tert*-pentenyl-5,7-diisopentenyl-tryptophan)	*Aspergillus glaucus*	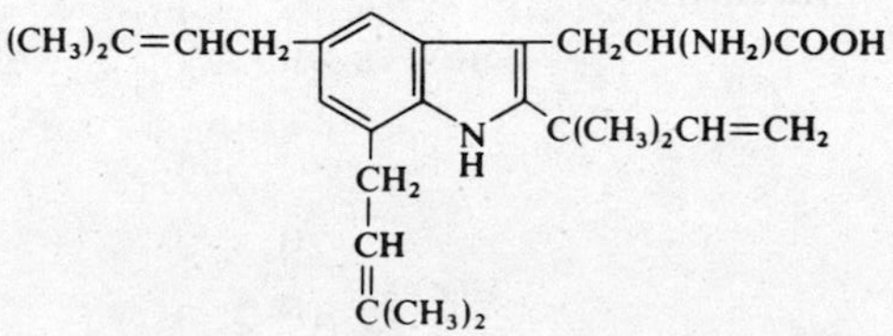 $(CH_3)_2C{=}CHCH_2$–(indole, N–H)–$CH_2CH(NH_2)COOH$; $C(CH_3)_2CH{=}CH_2$; CH_2–CH=$C(CH_3)_2$	$C_{26}H_{36}N_2O_2$ (408.50)	169–172° (194)	–	–	194	–	194	194
148a	Enduracididine [α-amino-β-(2-iminoimidazo-lidinyl)-propionic acid]	Enduracidin hydrolyzate	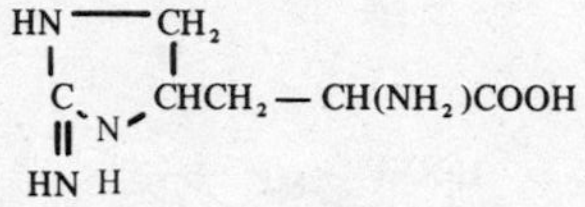 HN—CH_2; C(=NH)–N(H)–$CHCH_2$—$CH(NH_2)COOH$	$C_6H_{12}N_4O_2$ (172.22)	–	+63.3[22] (1 *M* HCl) +57.6[22] (1 *M* NaOH) (194a)	2.5 8.3 12 (94a)	–	194a	–	194a
148a′	Furanomycin [α-amino-(2,5-dihydro-5-methyl) furan-2-acetic acid]	*Streptomyces* L-803	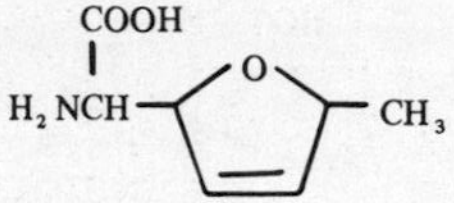 $H_2NCH(COOH)$–(2,5-dihydrofuran)–CH_3	$C_7H_{11}NO_3$ (157.19)	220–223° (dec) (194a′)	+136.1[27] (*c* 1, H_2O) (194a′)	2.4 9.1 (194a′)	194a′	194a′	194a′	194a′
148a″	Furosine [ε-*N*-(2-furoyl-methyl)-lypine]	Heated milk	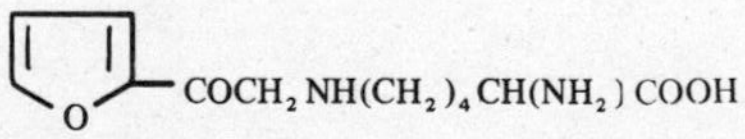 (furan)–$COCH_2NH(CH_2)_4CH(NH_2)COOH$	$C_{12}H_{18}N_2O_4$ (254.32)	–	–	–	194a″	194b	–	194a″
148b	γ-Glutaminyl-3,4-benzoquinone	*Agaricus bisporus*	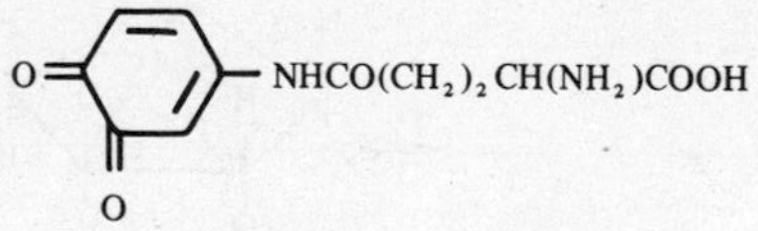 O=(benzoquinone, =O)–$NHCO(CH_2)_2CH(NH_2)COOH$	$C_{11}H_{12}N_2O_5$ (252.25)	–	–	–	194b	194b	–	194b
149	Guvacine	*Areca cathecu*	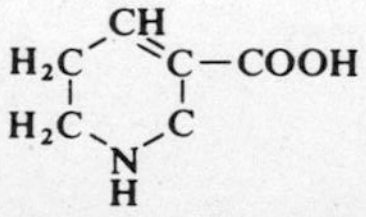 H_2C–CH=C–COOH; H_2C–N(H)–C (ring)	$C_6H_9NO_2$ (127.15)	–	–	–	195, 196	–	–	–

DATA ON THE NATURALLY OCCURRING AMINO ACIDS (continued)

No.	Amino acid (synonym)	Source	Structure	Formula (mol wt)	Melting point °C[a]	$[\alpha]_D$[b]	pK_a	References: Isolation and purification	References: Chromatography	References: Chemistry	References: Spectral data
150	Histidine	Protamine from sturgeon sperm	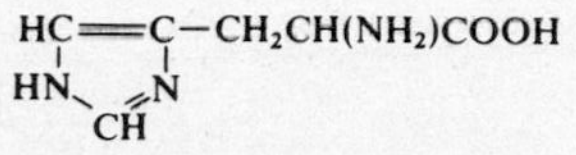 HC=C–CH₂CH(NH₂)COOH HN–CH=N (imidazole ring)	$C_6H_9N_3O_2$ (155.16)	277° (1)	-38.5^{25} (H_2O) +11.8 (5 *N* HCl)(1)	1.82 6.00 9.17 (1)	197	3	198	198
150a	β-Hydroxyhistidine	Bleomycin A_2 (antibiotic)	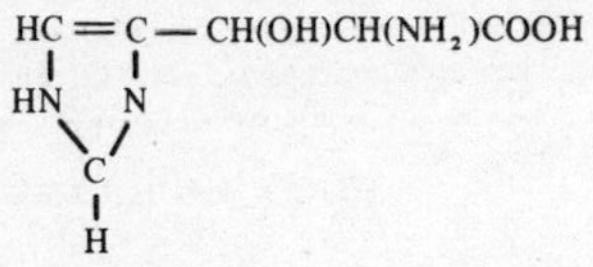 HC=C–CH(OH)CH(NH₂)COOH HN, N, C–H (imidazole ring)	$C_6H_9N_3O_3$ (171.18)	205° (dec) (198a)	$+40^{28}$ (*c* 1, H_2O) (198a)	<2.0 198a 5.5 8.8 (198a)	198a	–	198a	–
151	4-Hydroxy-4 methyl-proline	Apples	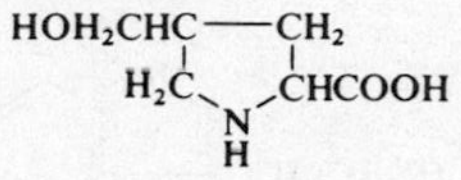 HOH₂CHC—CH₂ H₂C–N(H)–CHCOOH (ring)	$C_6H_{11}NO_2$ (145.16)	–	–	–	199	199	–	200
152	Hydroxyminaline	*Penicillium aspergillus*	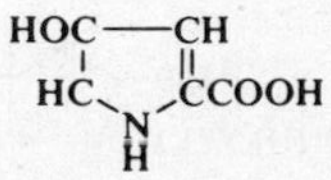 HOC—CH HC–N(H)–CCOOH (ring)	$C_5H_5NO_3$ (127.10)	–	–	–	201	–	201	201
153	4-Hydroxypipecolic acid	*Acacia pentadenia*	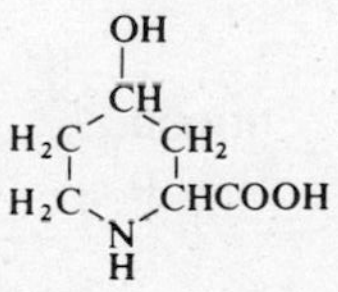 OH–CH H₂C, CH₂ H₂C–N(H)–CHCOOH (ring)	$C_6H_{11}NO_3$ (145.16)	250–270° (202)	-12.5^{21} (H_2O) $+0.34^{21}$ (1 *N* HCl) -18.5^{21} (1 *N* NaOH) (202)	–	202	202	202	203
154	5-Hydroxypipecolic acid	*Rhapis excelsa*	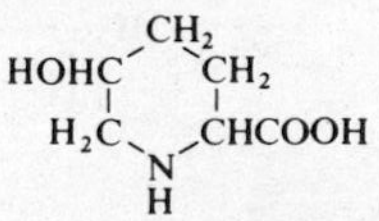 CH₂ HOHC, CH₂ H₂C–N(H)–CHCOOH (ring)	$C_6H_{11}NO_3$ (145.16)	–	–	–	204	204	–	203
154a	5-Hydroxy-piperidazine-3-carboxylic acid	Monamycin hydrolyzate	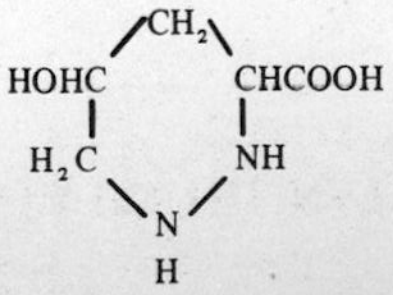 CH₂ HOHC, CHCOOH H₂C, NH N–H (ring)	$C_5H_{10}N_2O_3$ (146.17)	201–202° (DNP derv.) (204a)	$+21.4^{24}$ (*c* 0.39, H_2O) (204b)	–	204b	204b	204b	204a, 204b

DATA ON THE NATURALLY OCCURRING AMINO ACIDS (continued)

No.	Amino acid (synonym)	Source	Structure	Formula (mol wt)	Melting point °C[a]	$[\alpha]_D$[b]	pK_a	References: Isolation and purification	References: Chromatography	References: Chemistry	References: Spectral data
155	3-Hydroxyproline (3-hydroxypyrrolidine-2-carboxylic acid)	Telomycin	OH; H_2C—C—H; C, CHCOOH; N H	$C_5H_9NO_3$ (131.13)	225–235° (206)	-17.4^{20} (c 1, H_2O) $+13.3^{20}$ (c 0.5, 1 N HCl) (206a)	–	206	206	206	206a
156	4-Hydroxyproline (4-hydroxypyrrolidine-2-carboxylic acid	Gelatin hydrolyzate	HOHC—CH_2; H_2C, CHCOOH; N H	$C_5H_9NO_3$ (131.13)	273–4° (290)	-76.0^{25} (c 2, H_2O) -50.5^{25} (c 2, 5 N HCl)(290)	1.82 9.66 (290)	207	3	208	208
157	2-Hydroxytryptophan	Phalloidin	$CH_2CH(NH_2)COOH$; OH; N	$C_{11}H_{12}N_2O_3$ (220.22)	257° (210)	+40.8 (c 4.2, 1 N NaOH) (210)	–	209	–	210	–
158	5-Hydroxytryptophan	*Chromobacterium violaceum*	HO; $CH_2CH(NH_2)COOH$; N	$C_{11}H_{12}N_2O_3$ (220.22)	273° (dec) (290)	-32.5^{22} (c 1, H_2O) $+16^{22}$ (c 1, 4 N HCl) (290)	–	211	211	–	–
159	Ibotenic acid (α-amino-3-keto-4-isoxazoline-5-acetic acid)	*Amanita strobiliformis*	HC=$CCH(NH_2)COOH$; O=C, O; N H	$C_5H_6N_2O_4$ (114.10)	177–178° (212)	±0[c] (212)	5.1 8.2 (212)	212	–	212	212
160	4-Imidazoleacetic acid	*Polyporus sulfureus*	N—$CHCH_2COOH$; HC, CH; N H	$C_5H_7N_2O_2$ (125.11)	–	–	2.96 7.35 (20)	213	–	–	–
161	*N*-(Indole-3-acetyl)-aspartic acid	Magnolia	COOH; CH_2; $CH_2CONHCH$; COOH; N	$C_{14}H_{14}N_2O_5$ (290.29)	–	–	–	214	214	–	–

DATA ON THE NATURALLY OCCURRING AMINO ACIDS (continued)

No.	Amino acid (synonym)	Source	Structure	Formula (mol wt)	Melting point °C[a]	$[\alpha]_D$[b]	pK_a	References: Isolation and purification	References: Chromatography	References: Chemistry	References: Spectral data
162	Indole-3-acetyl-ε-lysine	*Pseudomonas savastanoi*	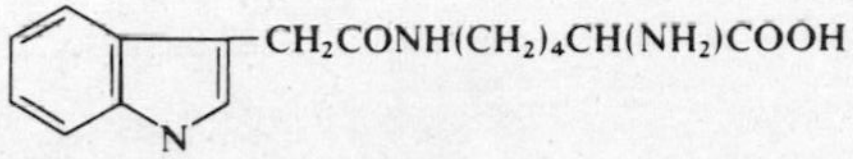 (indol-3-yl)–$CH_2CONH(CH_2)_4CH(NH_2)COOH$	$C_{16}H_{21}N_3O_3$ (303.36)	259–261° (215)	$+22^{23}$ (2 *N* HCl) (215)	–	215	215	215	215
162a	*N*-Jasmonoyliso-leucine	*Gibberella fujikuroi*	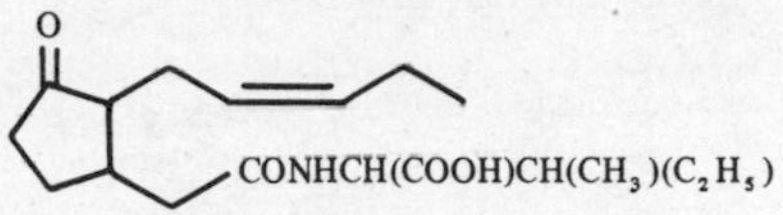 (jasmonoyl ring)–$CONHCH(COOH)CH(CH_3)(C_2H_5)$	$C_{18}H_{29}NO_4$ (323.48)	147–149° (191a)	–	–	–	191a	191a	191a
163	Kainic Acid (3-carboxymethyl-4-isopropenyl proline)	*Digenea simplex*	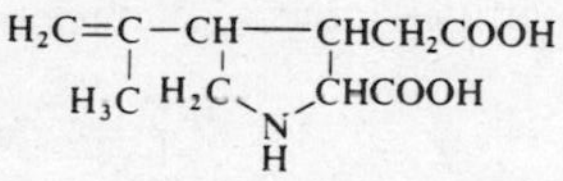 $H_2C{=}C(CH_3)$–CH–$CHCH_2COOH$; H_2C–N(H)–CHCOOH (ring)	$C_{10}H_{15}NO_4$ (213.25)	251° (216)	-15^{17} (H_2O) (185, 216)	–	216	216	–	–
164	4-Keto-5-methyl-proline	Actinomycin	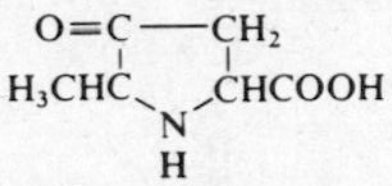 O=C—CH_2; H_3CHC–N(H)–CHCOOH (ring)	$C_6H_9NO_3$ (143.18)	215° (217)	–	–	217	217	217	217
165	4-Ketopipecolic acid	Staphylomycin	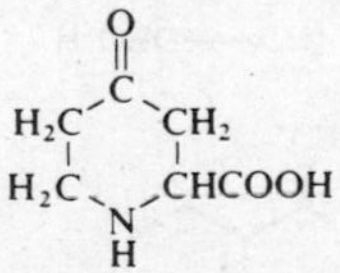 C(=O); H_2C, CH_2; H_2C–N(H)–CHCOOH (ring)	$C_6H_9NO_3$ (143.15)	–	–	–	218	–	–	–
166	4-Ketoproline	Actinomycin	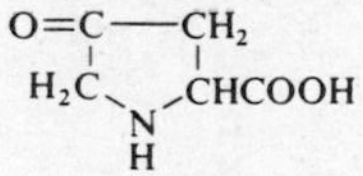 O=C—CH_2; H_2C–N(H)–CHCOOH (ring)	$C_5H_9N_4O_2$ (181.00)	–	–	–	219	–	219	–
167	Lathytine (tingitanine)	*Lathyrus tingitanus*	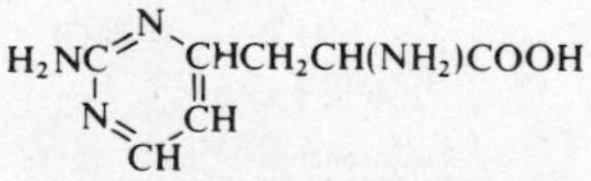 H_2NC(=N)(N=CH–CH=)–$CHCH_2CH(NH_2)COOH$ (ring)	$C_7H_{10}N_4O_2$ (182.00)	215° (220)	-55.9^{21} (H_2O) (220)	2.4 4.1 9.0 (220)	220	–	–	220
167a	exo-3,4-Methano-proline	*Aesculus parviflora*	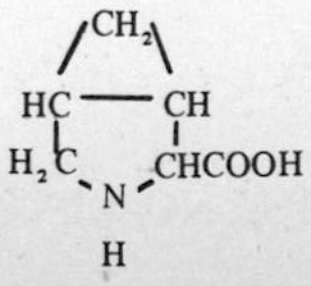 CH_2 bridge; HC—CH; H_2C–N(H)–CHCOOH (ring)	$C_6H_9NO_2$ (127.16)	–	-132^{20} (*c* 2, H_2O) -104^{20} (*c* 1, 5 *N* HCl) (23a)	–	23a	23a	–	23a

DATA ON THE NATURALLY OCCURRING AMINO ACIDS (continued)

No.	Amino acid (synonym)	Source	Structure	Formula (mol wt)	Melting point °C[a]	$[\alpha]_D$[b]	pK_a	References: Isolation and purification	References: Chromatography	References: Chemistry	References: Spectral data
168	4-Methyleneproline	*Eriobotrya japonica*	CH_2=C—CH_2 H_2C CHCOOH N H	$C_6H_9NO_2$ (127.15)	225° (221)	±0[c]	–	221	221	222	–
168a	1-Methyl-6-hydroxy-1,2,3,4-tetrahydro-isoquinoline-3-carboxylic acid	*Euphorbia myrisinites*	HO, H_2, H, COOH, NH, H, CH_3	$C_{11}H_{13}NO_3$ (207.25)	–	–	–	222a	222a	–	222a
169	4-Methylproline	Apples	H_3CHC—CH_2 H_2C CHCOOH N H	$C_6H_{11}NO_2$ (129.16)	232–234° (224)	–52 (*c* 0.3, H_2O) (224)	–	223	223	224	224
170	*N*-Methylproline (hygric acid)	Apples	H_2C—CH_2 H_2C CHCOOH N CH_3	$C_6H_{11}NO_2$	–	–	–	223	223	223	–
170a	*N*-Methylstrep-tolidine	Streptothricin	HOOCHC—CHCH(OH)—CH_2NHCH_3 N NH C NH_2	$C_7H_{14}N_4O_3$ (202.25)	–	–	–	224a	224a	–	–
171	β-Methyltryptophan	Telomycin	CH(CH_3)CH(NH_2)COOH N	$C_{12}H_{14}N_2O_2$ (218.25)	–	–	–	226	–	226	226
172	Mimosine	*Mimosa pudica*	OH C=CH O=C NCH$_2$CH(NH_2)COOH C=C H H	$C_8H_{10}N_2O_4$ (198.18)	228–229° 227	–21[22] (H_2O) +10[22] (1% HCl)(227)	–	227a	–	227	227

DATA ON THE NATURALLY OCCURRING AMINO ACIDS (continued)

No.	Amino acid (synonym)	Source	Structure	Formula (mol wt)	Melting point °C[a]	$[\alpha]_D$[b]	pK_a	References: Isolation and purification	References: Chromatography	References: Chemistry	References: Spectral data
173	Minaline (pyrrole-2-carboxylic acid)	Diastase	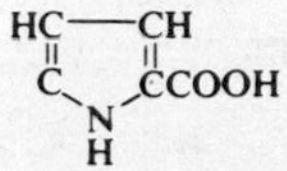 HC—CH, C, CCOOH, N, H (ring)	$C_5H_5NO_2$ (111.10)	180° (228)	–	–	228	–	228	228
174	Muscazone	*Amanita muscaria*	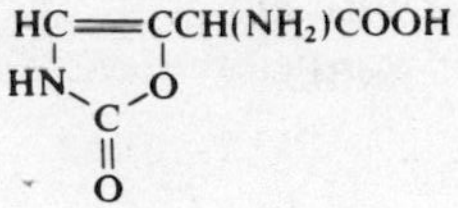 HC═CCH(NH$_2$)COOH, HN, O, C═O (ring)	$C_5H_6N_2O_5$ (174.12)	190° (229)	–	–	229	–	230	231
174a	Nicotianamine [1-(3′-(γ-amino-α-γ-dicarboxypropyl-amino)-propyl-azetidine-2-carboxylic acid]	Tobacco leaves	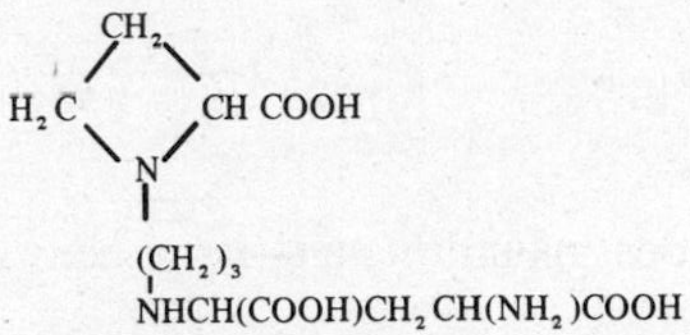 CH$_2$, H$_2$C, CH COOH, N (ring); N—(CH$_2$)$_3$NHCH(COOH)CH$_2$CH(NH$_2$)COOH	$C_{12}H_{21}N_3O_6$ (303.36)	240° (dec) (231a)	−60.5[23] (*c* 2.7, H_2O) (231a)	–	231a	–	231a	231a
175	β-3-Oxindolyl-alanine	Phalloidin	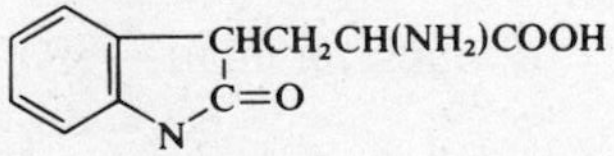 (benzene ring)—CHCH$_2$CH(NH$_2$)COOH, C═O, N	$C_{11}H_{12}N_2O_3$ (220.24)	249–253° (209)	+39.2[20] (1 *N* NaOH) (209)	–	209	–	209	–
176	Pipecolic acid	Apples	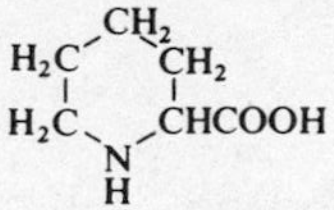 CH$_2$, H$_2$C, CH$_2$, H$_2$C, CHCOOH, N, H (ring)	$C_6H_{11}NO_2$ (129.17)	260° (dec) (290)	−25.4[18] (*c* 5, H_2O) −13.3[25] (*c*2, 5 *N* HCl)(290)	–	233	233	233	–
176a	Piperidazine-3-carboxylic acid	Monamycin hydrolyzate	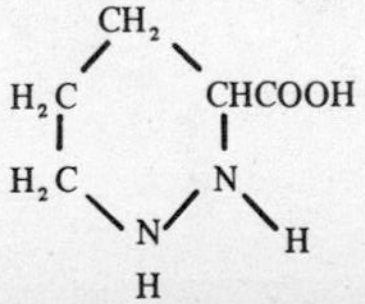 CH$_2$, H$_2$C, CHCOOH, H$_2$C, N, N, H, H (ring)	$C_5H_{10}N_2O_2$ (130.17)	153–155° (DNP derv.) (204b)	+307[25] (*c* 0.18, CH_3OH) (204b)	–	–	–	–	204a

DATA ON THE NATURALLY OCCURRING AMINO ACIDS (continued)

No.	Amino acid (synonym)	Source	Structure	Formula (mol wt)	Melting point °C[a]	$[\alpha]_D$[b]	pK_a	References: Isolation and purification	References: Chromatography	References: Chemistry	References: Spectral data
177	Proline (pyrolidine-2-carboxylic acid)	Casein hydrolyzate	H_2C—CH_2, H_2C—N(H)—CHCOOH (ring)	$C_5H_9NO_2$ (115.13)	222° (1)	-86.2^{25} (c 1–2, H_2O) −60.4 (c 1–2, 5 N HCl) (1)	1.95 10.64 (290)	234	3	235	235
178	β-Pyrazol-1-ylalanine	*Citrullus vulgaris*	HC=N—NCH₂CH(NH₂)COOH ring with C(H)=CH	$C_6H_9N_3O_2$ (155.17)	236–238° (236)	-7.3^{20} (c 3.4, H_2O) (236)	2.2 (236)	236	236	236	236
178a	Pyridosine [ε-(1,4-dihydro-6-methyl-3-hydroxy-4-oxo-1-pyridyl)-lysine]	Heated milk	O= ring (OH, CH₃) N—NH(CH₂)₄CH(NH₂)COOH	$C_{12}H_{19}N_3O_4$ (269.34)	–	–	–	236a	–	–	236a
178b	Pyrrolidone carboxylic acid (5-oxoproline)	Human plasma	H_2C—CH_2, O=C—N(H)—CHCOOH (ring)	$C_5H_7NO_3$ (129.13)	–	–	–	236b	236b	–	–
179	Roseanine	Roseothricin	N=C(H₂N)—N(H)—CH₂—CHC(OH)COOH ring with CH₂NH₂	$C_6H_{12}N_4O_3$ (188.21)	–	$+56.8^{22}$ (c 2.35, H_2O)[c] (238)	–	237	238	238	–
179a	Stendomycidine	Stendomycin (from *Streptomyces*)	CH₃N=C ring (HN, CH₂, CH₂, N-CH₃) CHCH(NH₂)COOH	$C_8H_{16}N_4O_2$ (200.28)	–	–	–	238a	238a	238a	238a, 238b

DATA ON THE NATURALLY OCCURRING AMINO ACIDS (continued)

No.	Amino acid (synonym)	Source	Structure	Formula (mol wt)	Melting point °C[a]	$[\alpha]_D$[b]	pK_a	References: Isolation and purification	References: Chromatography	References: Chemistry	References: Spectral data
180	Stizolobic acid [β-(3-carboxypyran-5-yl)alanine]	*Stizolobium hassjoo*	$HOOCCH(NH_2)CH_2C$ (ring: C=O, CCOOH, HC, CH, O)	$C_9H_9NO_6$ (227.18)	231–233° (239)	–	–	239	239	239	239
181	Stizolobinic acid [β-(6-carboxy-α-pyran-3-yl)alanine]	*Stizolobium hassjoo*	$HOOCCH(NH_2)CH_2$– (pyranone ring, O, COOH)	$C_9H_9NO_6$ (207.18)	–	–	–	239	239	240	240
181a	Streptolidine	Streptothricin	HOOCHC—CHCH(OH)CH$_2$NH$_2$ (ring: N=C(NH$_2$)–NH)	$C_6H_{12}N_4O_3$ (188.22)	215° (dec) (224a)	$+55.3^{25}$ (c 1.01, H_2O) (224a)	–	224a	224a	224a	224a
182	Tricholomic acid (α-amino-3-keto-5-isoxazolidine acetic acid)	*Tricholoma muscarium*	H_2C—CHCH(NH$_2$)COOH (ring: O=C, N–H, O)	$C_5H_8N_4O_4$ (160.13)	207° (242)	–	6.0 8.6 (242)	242	–	242	242
183	Tryptophan (α-amino-β-3-indole-propionic acid)	Casein	(indole, NH)–$CH_2CH(NH_2)COOH$	$C_{11}H_{12}N_2O_2$ (204.24)	282° (1)	-33.7^{25} (c 1–2, H_2O) $+2.8^{25}$ (c 1–2, 1 N HCl) (1)	2.43 9.44 (290)	243	3	244	244
183a	Tuberactidine (2-imino-4-hydroxyhexahydro-6-pyrimidinyl glycine)	*Streptomyces griseoverticillatus*	(ring: HO, H, N–H, =NH, NH, H, H_2) CH(NH$_2$)COOH	$C_6H_{12}N_4O_3$ (188.22)	182° (dec) (244a)	-25.8^{15} (c 0.5, H_2O) (244a)	–	244a	–	–	244a
184	Viomycidine (guanidine-1-pyrroline-5-carboxylic acid)	Viomycin	(ring: H_2C, H_2C–C, =N, CH–COOH) NHC(=NH)NH$_2$	$C_6H_{10}N_4O_2$ (170.19)	181-182° (dec) (246a)	$-151^{32.2}$ (c 1.25, H_2O) $-38^{22.2}$ (c 0.8, HCl) (246a)	1.3 5.5 12.6 (246)	76, 246a	246, 246a	246, 246a	246a

DATA ON THE NATURALLY OCCURRING AMINO ACIDS (continued)

No.	Amino acid (synonym)	Source	Structure	Formula (mol wt)	Melting point °C[a]	$[\alpha]_D$[b]	pK_a	References: Isolation and purification	References: Chromatography	References: Chemistry	References: Spectral data
185	Willardine [3-(1-uracyl)-alanine]	*Mimosa*	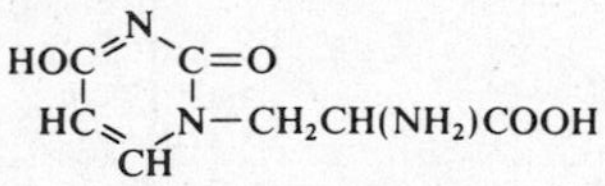 HOC=N–C=O / HC=CH–N–$CH_2CH(NH_2)COOH$	$C_7H_9N_3O_4$ (199.18)	–	-12.1^{22} (*c* 1.2, 1 *N* HCl) (248)	–	247	247	247, 248	248
	J. L-*N*-Substituted Amino Acids										
186	Abrine (*N*-methyl-tryptophan)	*Abrus precatorius*	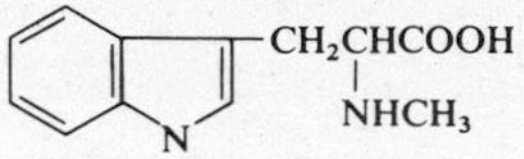 (indole, N)–$CH_2CHCOOH$ with $NHCH_3$	$C_{12}H_{14}N_2O_2$ (218.41)	297° (dec) (249)	–	–	250	–	249	–
186a	*N*-Acetylalanine	Human brain	CH_3–$CH(NHCOCH_3)COOH$	$C_5H_9NO_3$ (131.15)	–	–	–	250a	250a	250a	–
186a′	*N*-Acetylglutamic acid	Mammalian liver	HOOC–$(CH_2)_2$–$CH(NHCOCH_3)COOH$	$C_7H_{11}NO_5$ (189.19)	–	–	–	250a′	250a′	–	–
186b	*N*-Acetylaspartic acid	Extract of cat brain	COOH–CH_2–$CH(NHCOCH_3)COOH$	$C_6H_9NO_5$ (175.16)	123–125° (250b)	-11.8^{27} (2% methyl cellosolve) (250b)	–	250b	250b	–	250b
187	2,3-Dihydro-3,3-dimethyl-1 *H*-pyrrolo-[1,2-α] indole-9-alanine		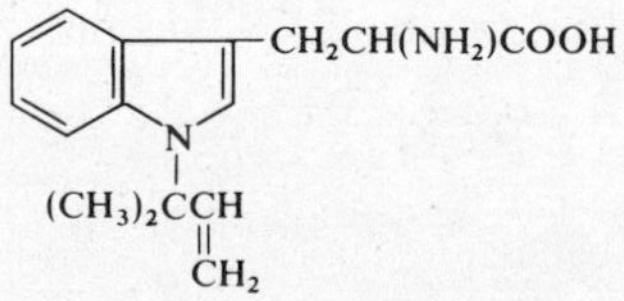 (indole)–$CH_2CH(NH_2)COOH$; N–$(CH_3)_2CCH$=CH_2	$C_{16}H_{20}N_2O_2$ (272.34)	170–175° (dec) (251)	–	–	251	251	251	251
188	β,*N*-Dimethylleucine	Etamycin	$(CH_3)_2CHCH(CH_3)CHCOOH$ with $NHCH_3$	$C_8H_{17}NO_2$ (159.23)	315–316° (dec) (252)	+33.15 (*c* 2, H_2O) $+39.2^{29}$ (*c* 2.2, 5 *N* HCl)(252)	–	252	252	252	–
188a	*N,N*-Dimethyl-phenylalanine	*Canthium eurysides*	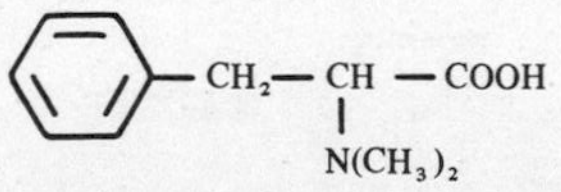 (phenyl)–CH_2–CH–COOH with $N(CH_3)_2$	$C_{11}H_{15}NO_2$ (193.27)	–	–	–	–	–	–	252a
189	γ-Formyl-*N*-methyl-norvaline	Ilamycin	CHO / $CH_3CHCH_2CHCOOH$	$C_7H_{13}NO_3$ (159.19)	–	–	1.8 10.2 (253)	253	–	253	253

DATA ON THE NATURALLY OCCURRING AMINO ACIDS (continued)

No.	Amino acid (synonym)	Source	Structure	Formula (mol wt)	Melting point °C[a]	$[\alpha]_D$[b]	pK_a	References: Isolation and purification	References: Chromatography	References: Chemistry	References: Spectral data
190	Fusarinine [δ-*N*-(*cis*-5-hydroxy-3-methylpent-2-enoyl)-δ-*N*-hydroxyornithine]	*Fusarium*	$HO(CH_3)_2C{=}C(CH_3)H{-}C(=O){-}N(OH){-}(CH_2)_3CH(NH_2)COOH$	$C_{11}H_{20}N_2O_4$ (244.30)	–	–	–	254	254	254	254
191	Homarine (*N*-methylpicolinic acid)	*Arenicola marina*	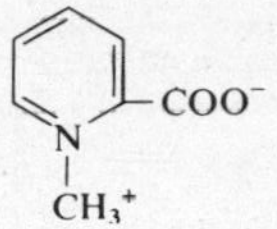 pyridinium ring with COO^- at C-2 and $N^+{-}CH_3$	$C_7H_7NO_2$ (137.15)	–	–	–	255	255	–	–
192	4-Hydroxy-*N*-methylproline	*Afrormosia elata* heartwood	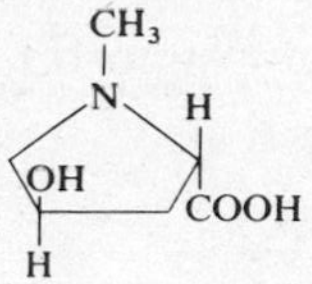 pyrrolidine ring: $N{-}CH_3$, C-2 H/COOH, C-4 OH/H	$C_6H_{11}NO_3$ (145.16)	238–240° (dec) (256)	-86.6 (*c* 1.5, H_2O) (256)	–	256	–	256	256
193	Merodesmosine	Elastin	$HOOC(H_2N)CH{-}(CH_2)_2CH{<}$ [$CHCH(CH_2)_2CH(NH_2)COOH$; $CH_2NH(CH_2)_4CH(NH_2)COOH$]	$C_{18}H_{34}N_4O_6$ (420)	–	–	–	257	257	257	257
194	*N*-Methylalanine	Dichapetalum cymosum (Gifblaar)	$CH_3CH(NHCH_3)COOH$	$C_4H_9NO_2$ (103.12)	–	–	–	258	258	258	258
195	1-Methylhistidine	Cat urine	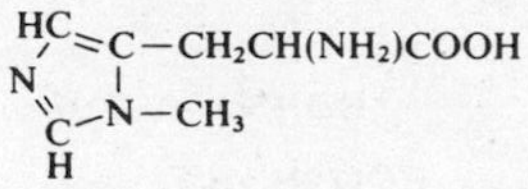 imidazole ring $HC{=}C{-}CH_2CH(NH_2)COOH$, $N{=}CH{-}N{-}CH_3$	$C_7H_{11}N_3O_2$ (169.18)	245–247° (259)	-25.8^{18} (*c* 3.9, H_2O) (290)	1.69 6.48 (imidazole) 8.85 (290)	259	259	–	–
196	3-Methylhistidine	Human urine	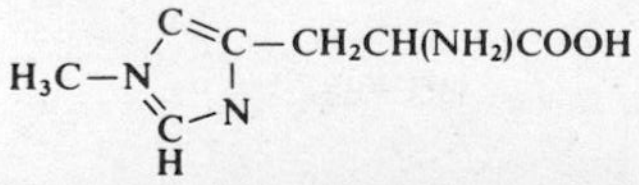 imidazole ring $H_3C{-}N{-}CH{=}C{-}CH_2CH(NH_2)COOH$, $C(H){=}N$	$C_7H_{11}N_3O_2$ (169.18)	–	-26.5^{26} (*c* 2.1, H_2O) (290) $+13.5^{27}$ (*c* 1.9, 1 *N* HCl)(260)	–	260	260	260	260

DATA ON THE NATURALLY OCCURRING AMINO ACIDS (continued)

No.	Amino acid (synonym)	Source	Structure	Formula (mol wt)	Melting point °C[a]	$[\alpha]_D$[b]	pK_a	References: Isolation and purification	References: Chromatography	References: Chemistry	References: Spectral data
197	*N*-Methylisoleucine	Enniatin A	$H_3CCH_2(H_3C)CHCH(NHCH_3)COOH$	$C_7H_{15}NO_2$ (145.21)	–	$+28.6^{22}$ (*c* 1.034, H_2O) $+44.8^{22}$ (*c* 1.162, 5 *N* HCl) (261)	–	261	261	261	–
198	*N*-Methylleucine	Enniatin A	$CH_3CH(CH_3)CH_2CH(NHCH_3)COOH$	$C_7H_{15}NO_2$ (145.21)	–	$+21.4^{15}$ (*c* 0.77, H_2O) $+31.3^{15}$ (*c* 0.86, 5 *N* HCl) (262)	–	262	262	262	–
199	*N*-Methyl-*β*-methylleucine	Etamycin	$(CH_3)_2CHCH(CH_3)CH(NHCH_3)COOH$	$C_8H_{17}NO_2$ (159.23)	–	$+26^{30}$ (*c* 1.8, H_2O) $+33.2^{30}$ (*c* 1.9, 5 *N* HCl) (252)	–	252	252	252	–
200	*N*-Methyl-*O*-methylserine	*Myobacterium butyricum*	$CH_3OCH_2CH(NHCH_3)COOH$	$C_5H_{11}NO_3$ (133.14)	203–205° (dec) (264)	–	–	264	264	264	264
201	*N*-Methylphenylglycine (α-phenylsarcosine)	Etamycin	C_6H_5–$CH(NHCH_3)COOH$	$C_9H_{12}NO_2$ (166.21)	–	$+118^{31}$ (*c* 4.8, 1 *N*, HCl) (252)	–	252	252	252	–
201a	*N*-Methylthreonine	Stendomycin hydrolyzate	$CH_3CH(OH)CH(NHCH_3)COOH$	$C_5H_{11}NO_3$ (133.17)	–	-17^{25} (*c* 2, 5 *N* HCl) (264a)	–	–	–	–	264a
202	*N*-Methyltyrosine (surinamine)	*Andira*	HO–C_6H_4–$CH(NHCH_3)COOH$	$C_{10}H_{13}NO_3$ (195.22)	280–300° (267)	−18.6 (267)	–	266	–	267	–
203	*N*-Methylvaline	Actinomycin	$(CH_3)_2CHCH(NHCH_3)COOH$	$C_6H_{13}NO_2$ (131.18)	–	+17.5 (H_2O) +30.9 (5 *N* HCl) (269)	–	268	268	269	–

DATA ON THE NATURALLY OCCURRING AMINO ACIDS (continued)

No.	Amino acid (synonym)	Source	Structure	Formula (mol wt)	Melting point °C[a]	$[\alpha]_D$[b]	pK_a	References: Isolation and purification	References: Chromatography	References: Chemistry	References: Spectral data
204	Saccharopine (N^6-(2-glutaryl)-lysine)	*Saccharomyces*	$HOOCCH(NH_2)(CH_2)_4NHCH(CH_2CH_2COOH)COOH$	$C_{11}H_{20}N_2O_6$ (276.30)	–	$+33.6^{23}$ (*c* 1, 0.5 *N* HCl) (271)	2.6, 4.1, 9.2, 10.3 (270)	270	270	271	271
205	Sarcosine (*N*-methylglycine)	*Cladonia silvatica*	CH_3NHCH_2COOH	$C_3H_7NO_2$ (89.10)	210 (dec) (290)	–	2.21, 10.20 (290)	272	272	–	–
	K. L-Sulfur and Selenium Containing-Amino Acids										
206	*N*-Acetyldjenkolic acid	*Acacia farnesiana*	$HOOCCH(NHC(=O)CH_3)CH_2SCH_2SCH_2CH(NH_2)COOH$	$C_9H_{16}N_2O_5S$ (264.32)	170° (273)	-49.0^{25} (*c* 2, 1% HCl) -60.2^{25} (*c* 1, 1 *N* HCl)(273)	–	273	273	273	273
207	Alliin	*Allium sativum*	$CH_2{=}CHCH_2S(\rightarrow O)CH_2CH(NH_2)COOH$	$C_6H_{11}NO_3S$ (177.24)	163–165° (274)	$+62.8^{21}$ (H_2O) (274)	–	274	–	356	–
208	*S*-Allylcysteine	*Allium sativum*	$CH_2{=}CHCH_2SCH_2CH(NH_2)COOH$	$C_6H_{11}NO_2S$ (161.24)	218° (275)	-8.7^{20} (275)	–	275	275	275	–
209	*S*-Allylmercapto-cysteine	*Allium sativum*	$CH_2{=}CHCH_2SSCH_2CH(NH_2)COOH$	$C_6H_{11}NO_2S_2$ (193.31)	188° (276)	$-95.3 \pm 10^{23.5}$ (*c* 0.19, 6 *N* HCl) (276)	–	276	276	276	276
210	α-Amino-β-(2-amino-2-carboxyethyl-mercaptobutyric acid)	Subtilin	$HOOCCH(NH_2)CH(CH_3)-S-CH_2CH(NH_2)COOH$	$C_7H_{14}N_2O_4S$ (222.28)	–	-34.7^{24} (*c* 5.4, 1 *N* HCl)(277)	–	277	277	277	277
211	Cystathionine *S*-(2-amino-2-carboxyethyl-homocysteine)	Human brain	$HOOCCH(NH_2)CH_2CH_2SCH_2CH_2(NH_2)COOH$	$C_7H_{14}N_2O_4S$ (222.28)	301° (279) 312° (290)	$+26.4^{25}$ (*c* 0.8, 1 *N* HCl)(279)	–	279	279	279	279
212	3-Amino-(3-carboxy-propyldimethyl-sulfonium)	Cabbage	$(CH_3)_2S^+CH_2CH_2CH(NH_2)COOH$	$C_6H_{14}NO_2S$ (164.26)	–	–	–	280	280	280	–

DATA ON THE NATURALLY OCCURRING AMINO ACIDS (continued)

No.	Amino acid (synonym)	Source	Structure	Formula (mol wt)	Melting point °C[a]	$[\alpha]_D$[b]	pK_a	References: Isolation and purification	References: Chromatography	References: Chemistry	References: Spectral data
212a	α-Amino-γ-[2-(4-carboxy)-thiazolyl butyric acid	*Xeromus subtomentosus* (mushroom)	Thiazole ring (H, S; HOOC, N) —CH_2—CH_2—$CH(NH_2)COOH$	$C_8H_{10}N_2O_4S$ (230.26)	237–238° (280a)	–	3.7 (280a)	280a	280a	280a	280a
213	Carbamyltaurine	Cat urine	$H_2NCONHCH_2CH_2SO_3H$	$C_3H_8N_2O_3S$ (152.17)	–	–	–	281	–	–	–
214	2-[*S*-(β-Carboxy-β-aminoethyl-tryptophan)]	*Amanita phalloides*	Indole ring (N, H) with —$CH_2CH(NH_2)COOH$ and —$SCH_2CH(NH_2)COOH$	$C_{14}H_{17}N_3O_4S$ (323.40)	–	–	–	–	119	–	–
215	*S*-(β-Carboxyethyl)-cysteine	*Albizia julibrissin*	$HOOC(CH_2)_2SCH_2CH(NH_2)COOH$	$C_6H_{11}NO_4S$ (193.17)	218° (282)	-9.33[20] (*c* 3, 1 *N* HCl)(283)	–	283	283	282	283
216	*N*-(1-Carboxyethyl)-taurine	Red algae	$HO_3S(CH_2)_2NHCH(CH_3)COOH$	$C_5H_{11}NO_5S$ (197.22)	258° (283)	-1.15[13] (*c* 5, 1 *N* NaOH) (284)	–	284	284	284	–
216a	*S*-(Carboxymethyl)-homocysteine	Human urine	$HOOCCH_2S(CH_2)_2CH(NH_2)COOH$	$C_6H_{11}NO_4S$ (193.24)	223–225° (dec) (284a)	–	–	284a	–	–	284a
217	*S*-(2-Carboxy-isopropyl)-cysteine	*Acacia*	CH_3CHCH_2COOH | S | $CH_2CH(NH_2)COOH$	$C_7H_{13}NO_4S$ (207.26)	202° (284)	+6.6[22] (*c* 2, H_2O) +31[25] (*c* 1.94, 1 *N* NaOH) (285)	–	285	285	285	285
218	*S*-(2-Carboxypropyl)-cysteine	Onions	CH_3 | CHCHCOOH | S | $CH_2CH(NH_2)COOH$	$C_7H_{13}NO_4S$ (207.26)	191–194° (285) (286)	-50.1[21] (H_2O) (286)	–	286, 287	286, 287	286, 287	286, 287

DATA ON THE NATURALLY OCCURRING AMINO ACIDS (continued)

No.	Amino acid (synonym)	Source	Structure	Formula (mol wt)	Melting point °C[a]	$[\alpha]_D$[b]	pK_a	References: Isolation and purification	References: Chromatography	References: Chemistry	References: Spectral data
219	Chondrine (1,4-thiazane-5-carboxylic acid 1-oxide)	*Chondria crassicaulis*	O ↑ S; H_2C–S–CH_2; HOOCHC–N(H)–CH_2 (ring)	$C_5H_9NO_3S$ (163.20)	255–257° (287)	+20.9[16] (*c* 2, H_2O) (288) +30.2[16] (*c* 2, 6 *N* HCl)(288)	–	288	288	288	–
220	Cycloallin (3-methyl-1,4-thiazane-5-carboxylic acid oxide)	Onions	O ↑ S; H_2C–S–CH_2; H_3CHC–N(H)–CHCOOH (ring)	$C_6H_{11}NO_3S$ (177.24)	–	–17.4[20] (H_2O) (289)	–	289	289	289	–
221	Cysteic acid (β-sulfo-α-aminopropionic acid)	Sheep's fleece	$HO_3SCH_2CH(NH_2)COOH$	$C_3H_7NO_5S$ (169.17)	289° (dec) (290)	+8.66 (*c* 7.4, H_2O (290)	1.3 (SO_3H) 1.9 8.7 (290)	291	292	–	–
222	Cysteine (α-amino-β-mercaptopropionic acid	Cystine	$HSCH_2CH(NH_2)COOH$	$C_3H_7NO_2S$ (121.15)	178° (1)	–16.5[25] (*c* 2, H_2O) +6.5[25] (*c* 2, 5 *N* HCl) (290)	1.92 8.35 10.46 (SH) (290)	293	3	294	294
223	Cysteine sulfinic acid (β-sulfinyl-α-aminopropionic acid)	Rat brain	$HO_2SCH_2CH(NH_2)COOH$	$C_3H_7NO_4S$ (153.17)	–	+11 (H_2O) +24 (*c* 1, 1 *N* HCl)	ca 2.1 (290)	295	295	–	–
224	Cystine [β,β′-Dithiodi-(α-aminopropionic acid)]	Urinary calculi	CH_2–S–S–CH_2; CHNH$_2$ CHNH$_2$; COOH COOH	$C_6H_{12}N_4O_4S_2$ (240.29)	261° (1)	–232[25] (*c* 1, 5 *N* HCl)(1)	<1 2.1 8.02 8.71 (290)	296	3	294	294
225	Cystine disulfoxide	Rat tissue	O O (=); CH_2–S–S–CH_2; CHNH$_2$ CHNH$_2$; COOH COOH	$C_6H_{12}N_2O_6S_2$ (272.33)	–	–	–	–	298	298	298

DATA ON THE NATURALLY OCCURRING AMINO ACIDS (continued)

No.	Amino acid (synonym)	Source	Structure	Formula (mol wt)	Melting point °C[a]	$[\alpha]_D$[b]	pK_a	References: Isolation and purification	References: Chromatography	References: Chemistry	References: Spectral data
226	*S*-(1,2-Dicarboxyethyl)-cysteine	Bovine lens	$HOOCCH_2CH(COOH)SCH_2CH(NH_2)COOH$	$C_7H_{11}NO_6S$ (237.25)	–	–	–	299	299	299	
227	Dichrostachinic acid [*S*-(β-hydroxy-β-carboxyethanesulfonyl methyl) cysteine]	*Mimosa*	$HOOCCH(OH)CH_2SO_2CH_2SCH_2CH(NH_2)COOH$	$C_7H_{13}NO_7S$ (255.27)	201° (300)	$+9.2^{24}$ (*c* 2.2, 1 *N* HCl) (300)	–	300	300	300	300
228	Dihydroallin (*S*-propylcysteine sulfoxide)	*Allium cepa*	$CH_3CH_2CH_2S(=O)CH_2CH(NH_2)COOH$	$C_6H_{13}NO_3S$ (179.26)	–	–	–	301	301	301	
229	Djenkolic acid	Djenkol beans	$CH_2(SCH_2CH(NH_2)COOH)_2$	$C_7H_{14}N_2O_4S_2$ (254.32)	300–350° (303)	$-65^{20.5}$ (*c* 1, 1 *N* HCl)(303)	–	302	–	303	–
230	Ethionine	*Escherichia coli*	$CH_3CH_2SCH_2CH_2CH(NH_2)COOH$	$C_6H_{13}NO_2S$ (163.23)	–	–	–	304	304	–	–
231	Felinine	Cat urine	$HOCH_2CH_2C(CH_3)_2SCH_2CH(NH_2)COOH$	$C_8H_{17}NO_3S$ (207.30)	177° (305)	$+23^{20}$ (*c* 2.2, H_2O) (305)	–	305	305	305	–
232	*N*-Formyl methionine	*Escherichia coli*	$CH_3SCH_2CH_2CH(NHCHO)COOH$	$C_6H_{11}NO_3$ (177.22)	–	–	–	306	306	–	–
233	Glucobrassicin [*S*-β-1-(glucopyranosyl)-3-indolylacetothiohydroximyl-*O*-sulfate]	*Brassica* species	[indol-3-yl]–$CH_2C(=NOSO_3H)SC_6H_{11}O_5$	$C_{16}H_{20}N_2O_9S_2$ (447.47)	–	-13.3^{23} (*c* 3, H_2O) (189)	–	189	189	189	189
234	Guanidotaurine	*Arenicola marina*	$H_2NC(=NH)NHCH_2CH_2SO_3H$	$C_3H_9N_3O_3S$ (167.20)	228–230° (308)	–	–	308	308	308	–

DATA ON THE NATURALLY OCCURRING AMINO ACIDS (continued)

No.	Amino acid (synonym)	Source	Structure	Formula (mol wt)	Melting point °C[a]	$[\alpha]_D$[b]	pK_a	References: Isolation and purification	References: Chromatography	References: Chemistry	References: Spectral data
235	Homocysteine	*Neurospora*	$HSCH_2CH_2CH(NH_2)COOH$	$C_4H_9NO_2S$ (135.18)	–	–	2.22 8.87 10.86 (290)	278	310	–	–
236	Homocystine	Human urine	$CH_2-S-S-CH_2$ / CH_2 CH_2 / $CHNH_2$ $CHNH_2$ / $COOH$ $COOH$	$C_8H_{16}N_2O_4S_2$ (268.36)	282–3° (dec) (290)	-16^{21} (c 0.06, H_2O) $+78^{25}$ (c 1–2, 5 N HCl) (290)	1.59 2.54 8.52 9.44 (290)	311	311	–	–
237	Homocysteinecysteine disulfide	Human urine	$CH_2-S-S-CH_2$ / CH_2 $CHNH_2$ / $CHNH_2$ $COOH$ / $COOH$	$C_7H_{14}N_2O_4S_2$ (254.35)	–	-52.2^{25} (1 N HCl) (312)	–	312	–	312	312
238	Homolanthionine	*Escherichia coli*	$HOOCCH(NH_2)(CH_2)_2S(CH_2)_2CH(NH_2)COOH$	$C_8H_{16}N_2O_4S$ (236.29)	–	$+37.3^{24}$ (c 1, 1 N HCl)(363)	–	313	313	313	–
239	Homomethionine (5-methylthionor-valine)	Cabbage	$CH_3S(CH_2)_3CH(NH_2)COOH$	$C_6H_{13}NO_2S$ (163.25)	223–225° (314)	$+21^{25.5}$ (c 0.3, 6 N HCl)(314)	–	314	314	314	314
239a	β-6-(4-Hydroxy-benzothiazolyl)-alanine	Chicken feather pigment	OH / benzothiazole ring (N, S) / H_2C / $CH(NH_2)$ / COOH	$C_{10}H_9N_2O_3S$ (237.27)	–	–	–	314a	314a	–	314a
239a′	*S*-(2-Hydroxy-2-carboxyethyl)-homocysteine	Human urine	$COOHCH(OH)CH_2S(CH_2)_2CH(NH_2)COOH$	$C_7H_{13}NO_5S$ (223.27)	–	–	–	284a	–	–	–
239a″	*S*-(2-Hydroxy-2-carboxyethylthio)-homocysteine	Human urine	$HOOCCHCH_2SS(CH_2)_2CH(NH_2)COOH$ (OH on CH)	$C_7H_{13}NO_5S_2$ (223.27)	–	–	–	377	377	377	–

DATA ON THE NATURALLY OCCURRING AMINO ACIDS (continued)

No.	Amino acid (synonym)	Source	Structure	Formula (mol wt)	Melting point °C[a]	$[\alpha]_D$[b]	pK_a	References: Isolation and purification	References: Chromatography	References: Chemistry	References: Spectral data
239b	*S*-(3-Hydroxy-3-carboxy-*n*-propyl thio)-cysteine	Human urine	HOOC—CH(OH)(CH$_2$)$_2$—SSCH$_2$CH(NH$_2$)COOH	$C_7H_{13}NO_5S_2$ (255.33)	176–177° (dec) (377)	-96.3^{26} (*c* 2.3, 1 *N* HCl) (377)	–	377	377	377	377
239c	*S*-(3-Hydroxy-3-carboxy-*n*-propyl) cysteine	Human urine	COOHCH(OH)(CH$_2$)$_2$SCH$_2$CH(NH$_2$)COOH	$C_7H_{13}NO_5S$ (223.27)	–	–	–	284a	–	–	–
239c′	*S*-(3-Hydroxy-3-carboxy-*n*-propyl-thio)-homocysteine	Human urine	HOOCCH(OH) — (CH$_2$)$_2$SS(CH$_2$)$_2$CH(NH$_2$)COOH	$C_8H_{15}NO_5S_2$ (269.36)	187–188° (dec) (377)	$+8.1^{25}$ (*c* 15.2, 1 *N* HCl) (377)	–	377	377	377	377
239d	α-Hydroxycysteine-cysteine disulfide	Human urine	CH$_2$ — S — S — CH$_2$ CHNH$_2$ COHNH$_2$ COOH COOH	$C_6H_{12}N_2O_5S_2$ (256.32)	181° (dec) (378)	-263^{23} (*c* 2.2, H_2O) (378)	–	–	378	378	–
240	Hypotaurine (2-aminoethane-sulfinic acid)	Rat brain	HO$_2$SCH$_2$CH$_2$NH$_2$	$C_2H_7NO_2S$ (109.14)	–	–	–	295	295	–	–
241	Isovalthine (isopropylcarboxy-methylcysteine)	Human urine	(CH$_3$)$_2$CHCH(COOH)SCH$_2$CH(NH$_2$)COOH	$C_8H_{15}NO_4S$ (221.28)	–	–	–	316	317	317	317
242	Lanthionine [β,β′-thiodi-(α-aminopropionic acid)]	Wool	HOOCCH(NH$_2$)CH$_2$SCH$_2$CH(NH$_2$)COOH	$C_6H_{12}N_2O_4S$ (208.23)	270–304° (318)	–	–	318	–	319	–
242a	β-Mercaptolactate-cysteine disulfide	Human urine	CH$_2$ — S — S — CH$_2$ CHOH CHNH$_2$ COOH COOH	$C_6H_{11}NO_5S_2$ (241.30)	–	–	–	–	319a	319a	–
243	*S*-Methylcysteine	*Phaseolus vulgaris*	CH$_3$SCH$_2$CH(NH$_2$)COOH	$C_4H_9NO_2S$ (135.19)	220° (320)	-26^{25} (*c* 2.5) (320)	8.75 (20)	320	320	320	320
243a	*S*-Methylcysteine sulfoxide	Cabbage	H$_3$C–S(→O)CH$_2$CH(NH$_2$)COOH	$C_4H_9NO_3S$ (151.20)	173° (320a)	–	–	–	–	–	–

DATA ON THE NATURALLY OCCURRING AMINO ACIDS (continued)

No.	Amino acid (synonym)	Source	Structure	Formula (mol wt)	Melting point °C[a]	$[\alpha]_D$[b]	pK_a	References: Isolation and purification	References: Chromatography	References: Chemistry	References: Spectral data
244	Methionine (α-amino-γ-methylthiobutyric acid)	Casein hydrolyzate	$CH_3SCH_2CH_2CH(NH_2)COOH$	$C_5H_{11}NO_2S$ (149.21)	283° (1)	-10^{25} (H_2O) $+23.2^{25}$ (5 *N* HCl) (1)	2.28 9.21 (1)	321	3	322	322
245	3,3′-(2-Methylethylene-1,2-dithio)dialanine	*Allium shoenoprasum*	$S-CH_2CH(CH_3)-S$; each S bonded to CH_2-CHNH_2-COOH	$C_9H_{17}N_2O_4S_2$ (282.40)	–	–	–	323	323	323	323
246	β-Methyllanthionine	Yeast	$HOOCCH(NH_2)CH_2SCH(CH_3)CH(NH_2)COOH$	$C_7H_{14}N_2O_4S$ (222.28)	–	–	–	277	–	277	–
247	*S*-Methylmethionine (α-aminodimethyl-γ-butyrothetin)	Asparagus	$(CH_3)_2S^+CH_2CH_2CH(NH_2)COO^-$	$C_6H_{13}NO_2S$ (164.24)	–	–	–	325	325	280	325
248	β-Methylselenoalanine (β-methylselenocysteine)	*Stanleya pinnata*	$CH_3SeCH_2CH(NH_2)COOH$	$C_4H_9NO_2Se$ (182.08)	–	–	–	327	327	–	–
249	Neoglucobrassicin	*Brassica napus*	indole (N–OCH_3)-3-$CH_2C(=NOSO_3H)SC_6H_{11}O_5$	$C_{17}H_{21}N_2O_{10}S_2$ (476.48)	175° (328)	–	–	328	328	328	328
250	*S*-(Prop-1-enyl)-cysteine	*Allium sativum*	$CH_3CH{=}CHSCH_2CH(NH_2)COOH$	$C_6H_{11}NO_2S$ (161.24)	–	–	–	329	329	–	–
251	*S*-(Prop-1-enyl)-cysteine sulfoxide	Onions	$CH_3CH{=}CHS(\rightarrow O)CH_2CH(NH_2)COOH$	$C_6H_{11}NO_3S$ (177.22)	146–148° (330)	–	–	330	330	330	330
252	*S*-*n*-Propylcysteine	*Allium sativum*	$CH_3(CH_2)_2SCH_2CH(NH_2)COOH$	$C_6H_{13}NO_2S$ (163.26)	–	–	–	331	331	–	–
253	*S*-*n*-Propylcysteine sulfoxide	Onions	$CH_3CH_2CH_2S(\rightarrow O)CH_2CH(NH_2)COOH$	$C_6H_{13}NO_3S$ (179.26)	–	–	–	301	301	–	–

DATA ON THE NATURALLY OCCURRING AMINO ACIDS (continued)

No.	Amino acid (synonym)	Source	Structure	Formula (mol wt)	Melting point °C[a]	$[\alpha]_D$[b]	pK_a	References: Isolation and purification	References: Chromatography	References: Chemistry	References: Spectral data
254	Selenocystathionine	*Stanleya pinnata*	$HOOCCH(NH_2)CH_2CH_2SeCH_2CH(NH_2)COOH$	$C_7H_{14}N_2O_4Se$ (269.07)	–	–	–	327	327	327	–
255	Selenocystine	*Astragalus pectinatus*	$CH_2-Se-Se-CH_2$ (each CH_2 bearing $CHNH_2$–$COOH$)	$C_6H_{12}N_2O_4Se_2$ (334.11)	263–265° (334)	–	–	334	–	334	–
255a	Selenomethionine	*Escherichia coli* (hydrolyzate)	$H_3CSe(CH_2)_2CH(NH_2)COOH$	$C_5H_{11}NO_2Se$ (196.13)	–	–	–	334a	334a	–	–
256	Selenomethyl-selenocysteine	*Astragalus bisulcatus*	$SeCH_2CH(NH_2)COOH$	$C_3H_6NO_2Se$ (167.03)	–	–	–	335	335	335	–
257	Taurine (2-amino-ethane-sulfonic acid)	Plant and animal tissue	$H_2NCH_2CH_2SO_3H$	$C_2H_7NO_3S$ (125.15)	320° (dec) (290)	–	-0.3 9.06 (290)	–	292	–	–
258	β-(2-Thiazole)-β-alanine	*Bottromycin*	thiazole ring (HC—N, HC–S–C) $-CH(NH_2)CH_2COOH$	$C_6H_8N_2O_2S$ (172.22)	197.5–201.5° (337)	–	–	337	337	337	337
259	Thiolhistidine	Erythrocytes and microorganisms	$HC{=}C-CH_2CH(NH_2)COOH$ (imidazole ring N=C(SH)–NH)	$C_6H_9N_2O_2S$ (173.23)	–	-10^{25} (*c* 2, 1 *N* HCl) (339)	1.84 8.47 11.4 (290)	338	–	339	–
260	Thiostreptine	Thiostrepton	$H_3CCH(OH)C(CH_3)(OH)CH(NH_2)-C$ (thiazole ring S—CH, N) $CCOOH$	$C_9H_{14}N_2O_4S$ (246.32)	–	-4^{25} (*c* 1, 1 *N* acetic acid)(102)	–	102	102	102	102
261	Tyrosine-*O*-sulfate	Human urine	$HO_3SO-C_6H_4-CH_2CH(NH_2)COOH$	$C_9H_{11}NO_6S$ (261.26)	–	–	–	341	342	–	–

DATA ON THE NATURALLY OCCURRING AMINO ACIDS (continued)

No.	Amino acid (synonym)	Source	Structure	Formula (mol wt)	Melting point °C[a]	$[\alpha]_D$[b]	pK_a	References: Isolation and purification	References: Chromatography	References: Chemistry	References: Spectral data
	L. L-Halogen-Containing Amino Acids										
262	2-Amino-4,4-dichlorobutyric acid	*Streptomyces armentosus*	$Cl_2CHCH_2CH(NH_2)COOH$	$C_4H_7Cl_2NO_2$ (172.02)	–	+6.7[25] (*c* 0.74, H_2O) +26.2[25] (*c* 0.74, 1 *N* HCl) (64)	–	64	64	64	64
262a	5-Chloropiperidazine-3-carboxylic acid	Monamycin hydrolyzate	ClHC–CH₂–CHCOOH ring: H_2C, NH, N–H (piperidazine ring)	$C_5H_9ClN_2O_2$ (164.61)	83–85° (DNP deriv) (204a)	+157[33] (*c* 0.18, $CHCl_3$) (204b)	–	204 b	204b	–	204a, 204b
263	3,5-Dibromotyrosine	*Gorgona* species	HO–(3,5-Br₂-phenyl)–$CH_2CH(NH_2)COOH$	$C_9H_9NO_3Br_2$ (339.01)	245° (344)	–	2.17 6.45 (OH) 7.60 (20)	344	–	344	–
263a	2,4-Diiodohistidine	Human urine	I–C=C–$CH_2CH(NH_2)COOH$ (imidazole ring: HN, N, C–I)	$C_6H_7I_2N_3O_2$ (406.96)	–	–	–	344a	344a	–	–
264	3,3′-Diiodothyronine	Bovine thyroid gland	HO–(3-I-phenyl)–O–(3-I-phenyl)–$CH_2CH(NH_2)COOH$	$C_{15}H_{13}I_2NO_4$ (525.11)	233–234° (dec) (345)	–	–	345	345	345	–
265	3,5-Diiodotyrosine (iodogorgoic acid)	Coral protein	HO–(3,5-I₂-phenyl)–$CH_2CH(NH_2)COOH$	$C_9H_9I_2NO_3$ (433.01)	194° (dec) (290)	+2.9 (1.1 *N* HCl)(1)	2.12 6.48 (OH) 7.82 (290)	346	347	348	–

DATA ON THE NATURALLY OCCURRING AMINO ACIDS (continued)

No.	Amino acid (synonym)*	Source	Structure	Formula (mol wt)	Melting point °C[a]	$[\alpha]_D$[b]	pK_a	References: Isolation and purification	References: Chromatography	References: Chemistry	References: Spectral data
266	3-Monobromotyrosine	Sea fans and sponges	Br; HO–, –$CH_2CH(NH_2)COOH$	$C_9H_{10}NO_3Br$ (260.10)	–	–	–	349	349	–	–
266a	3-Monobromo-5-monochlorotyrosine	*Buccinum undatum*	Br; HO–, –$CH_2CH(NH_2)COOH$; Cl	$C_9H_9ClNO_3Br$ (294.55)	–	–	–	349a	349a	–	349a
266b	5-Monochlorotyrosine	*Buccinum undatum*	HO–, –CH_2 – $CH(NH_2)COOH$; Cl	$C_9H_{10}ClNO_3$ (215.65)	–	–	–	349b	349b	–	349b
267	2-Monoiodohistidine	Rat thyroid gland	HC = C – $CH_2CH(NH_2)COOH$; HN, N, C, I	$C_6H_8IN_3O_2$ (281.02)	–	–	–	350	350	–	–
268	Monoiodotyrosine	*Nereocystis luetkeana* (an alga)	I; HO–, –$CH_2CH(NH_2)COOH$	$C_9H_{10}NO_3$ (307.11)	–	–	–	351	351	–	–
269	Thyroxine	Thyroid gland	I, I; HO–, –O–, –$CH_2CH(NH_2)COOH$; I, I	$C_{15}H_{11}I_4NO_4$ (776.88)	236° (dec) (290) 95% EtOH) (290)	+15 (*c* 5, 1 *N* HCl, (OH) 10.1 (290)	2.2 6.45	352	353	348	348
270	3,5,3′-Triiodothyronine	*Phaseolus vulgaris*	I, I; HO–, –O–, –$CH_2CH(NH_2)COOH$; I	$C_{15}H_{12}I_3NO_4$ (650.98)	233–234° (dec) (290)	+23.6[24] (*c* 5, 1 *N* HCl-EtOH) (290)	2.2 8.40 (OH) 10.1 (290)	355	347	–	–

DATA ON THE NATURALLY OCCURRING AMINO ACIDS (continued)

No.	Amino acid (synonym)	Source	Structure	Formula (mol wt)	Melting point °C[a]	$[\alpha]_D$[b]	pK_a	References: Isolation and purification	References: Chromatography	References: Chemistry	References: Spectral data
			M. L-Phosphorus-Containing Amino Acids								
271	α-Amino-β-phosphonopropionic acid	*Tetrahymena pyriformis, Zoanthus sociatus*	$H_2O_3PCH_2CH(NH_2)COOH$	$C_2H_8NO_5P$ (157.07)	228° (dec) (357)	–	2.2 4.5 8.8 11.0 (357)	358	358	357	–
272	Ciliatine (2-aminoethyl phosphonic acid)	Sea anemone	$H_2NCH_2CH_2PO_3H_2$	$C_2H_8NO_3P$ (125.07)	280–281° (dec) (359)	–	6.4 (359)	359	359	359	359
273	2-Dimethylamino-ethylphosphonic acid	Sea anemone	$(CH_3)_2NCH_2CH_2PO_3H_2$	$C_4H_{12}NO_3P$ (153.13)	249.5° (360)	–	–	360	360	360	360
273a	1-Hydroxy-2-amino ethylphosphonic acid	*Acanthamoeba castellanii* (plasma membrane)	$H_2N{-}CH_2{-}CH(OH)PO_3H_2$	$C_2H_8NO_4P$ (141.08)	–	–	–	360a	360a	–	360a
274	Lombricine (2-amino-2-carboxy-ethyl-2-guanidine-ethyl hydrogen phosphate)	Earthworm	$H_2N\overset{NH}{\overset{\Vert}{C}}NH(CH_2)_2O\overset{O}{\overset{\Vert}{\underset{OH}{\underset{\vert}{P}}}}OCH_2CH(NH_2)COOH$	$C_6H_{15}N_4O_6P$ (270.21)	223–224° (361)	$+14.5^{23.5}$ (c 0.93, H_2O) (362)	8.9 (20)	361	361	362	362
275	2-Methylamino-ethylphosphonic acid	Sea anemone	$\underset{CH_3}{\underset{\vert}{H}}NCH_2CH_2PO_3H_2$	$C_3H_{10}NO_3P$ (139.10)	291° (360)	–	–	360	360	360	360
276	*O*-Phosphohomoserine	*Lactobacillus*	$H_2O_3POCH_2CH_2CH(NH_2)COOH$ $H_2O_3POCH_2CH_2CH(NH_2)COOH$	$C_4H_{10}NO_6P$ (199.11)	178° (dec) (364)	$+6.25^{22.5}$ (c 2.4, H_2O) (364)	–	364	364	364	–
277	*O*-Phosphoserine	Casein	$H_2O_3POCH_2CH(NH_2)COOH$	$C_3H_8NO_6P$ (185.08)	–	+7.2 (H_2O) (366)	–	365	366	366	–
278	2-Trimethyl-aminoethyl-betaine phosphonic acid	Sea anemone	$(CH_3)_3{}^+NCH_2CH_2\overset{O}{\overset{\Vert}{\underset{O^-}{\underset{\vert}{P}}}}OH$	$C_5H_{14}NO_3P$ (167.16)	250–252° (360)	–	–	360	360	360	360

DATA ON THE NATURALLY OCCURRING AMINO ACIDS (continued)

No.	Amino acid (synonym)	Source	Structure	Formula (mol wt)	Melting point °C[a]	$[\alpha]_D$[b]	pK_a	References: Isolation and purification	References: Chromatography	References: Chemistry	References: Spectral data
				N. L-Betaines							
279	*N*-(3-Amino-3-carboxy-propyl)-β-carboxypyridinium betaine	Tobacco leaves	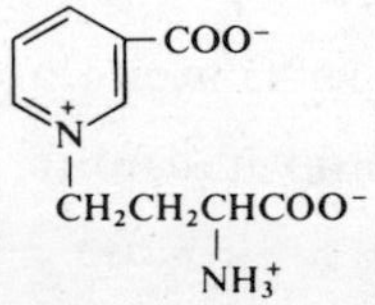 Pyridinium ring (3-COO^-), N^+–$CH_2CH_2CH(NH_3^+)COO^-$	$C_{10}H_{14}N_2O_5$ (242.24)	241–243° (dec) (368)	+24[24] (*c* 2, H_2O) (368)	–	368	368	368	368
280	Betonicine (4-hydroxyproline betaine)	*Stachys* (*Betonica*) *officinalis*	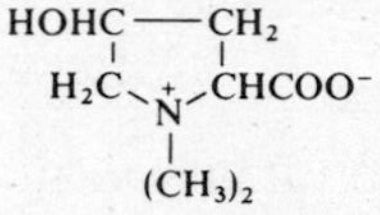 HOHC—CH_2; H_2C–$N^+(CH_3)_2$–CHCOO$^-$ (ring)	$C_7H_{13}NO_3$ (159.19)	243–244° (dec) (369)	−36.6[15] (369)	–	369	–	369	–
281	γ-Butyrobetaine (γ-aminobutyric acid betaine)	Rat brain	$(CH_3)_3\overset{+}{N}CH_2CH_2CH_2COO^-$	$C_7H_{15}NO_2$ (144.20)	180–184° (370)	–	–	370	370	–	–
282	Carnitine (γ-amino-β-hydroxybutyric acid betaine)	Vertebrate muscle	$(CH_3)_3N^+CH_2CH(OH)CH_2COO^-$	$C_7H_{15}NO_3$ (161.21)	195–197° (371)	+23.5[22] (*c* 5, H_2O) (374)	–	372	373	367, 371	–
283	Desmosine	Bovine elastin	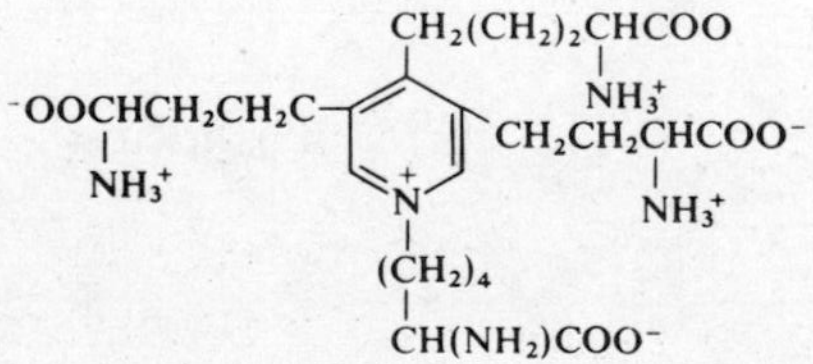 Pyridinium ring bearing $CH_2(CH_2)_2CH(NH_3^+)COO$, $^-OOCCH(NH_3^+)CH_2CH_2C$–, $CH_2CH_2CH(NH_3^+)COO^-$; N^+–$(CH_2)_4$–$CH(NH_2)COO^-$	$C_{24}H_{39}N_5O_8$ (879.43)	–	–	1.70, 2.40, 8.80, 9.85 (375)	375	–	257, 375	375
284	Ergothionine (betaine of thiol histidine)	Ergot	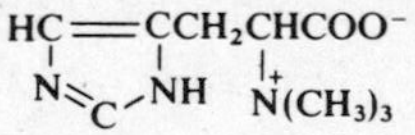 HC=C–$CH_2CH(\overset{+}{N}(CH_3)_3)COO^-$; N=C–NH (imidazole ring)	$C_9H_{15}N_3O_2$ (229.29)	290° (340)	+115[27.5] (*c* 1, H_2O) (290)	–	336	340	340	340
285	Hercynin (histidine betaine)	Mushrooms	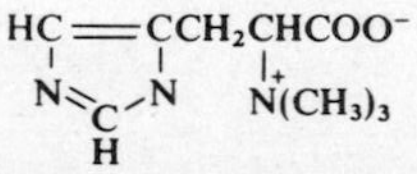 HC=C–$CH_2CH(\overset{+}{N}(CH_3)_3)COO^-$; N=C(H)–N (imidazole ring)	$C_9H_{15}N_3O_2$ (203.22)	–	–	–	332	–	333	–
286	Homobetaine (β-alanine betaine)		$(CH_3)_3\overset{+}{N}CH_2CH_2COO^-$	$C_6H_{13}NO_2$ (131.18)	–	–	–	324	–	–	–

DATA ON THE NATURALLY OCCURRING AMINO ACIDS (continued)

No.	Amino acid (synonym)	Source	Structure	Formula (mol wt)	Melting point °C[a]	$[\alpha]_D$[b]	pK_a	References: Isolation and purification	References: Chromatography	References: Chemistry	References: Spectral data
287	Homostachydrine (pipecolic acid betaine)	Alfalfa	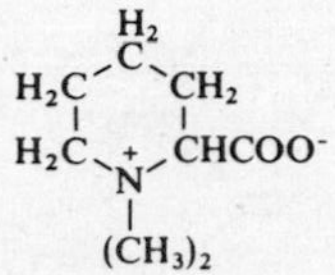 H_2C–C(H_2)–CH_2 H_2C–$\overset{+}{N}$–$CHCOO^-$ $(CH_3)_2$	$C_8H_{14}NO_2$ (156.21)	–	–	–	315	309	–	–
288	3-Hydroxystachydrine (3-hydroxyproline betaine)	*Courbonia virgata*	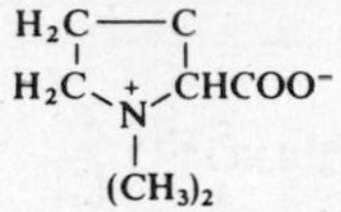 H_2C—C H_2C–$\overset{+}{N}$–$CHCOO^-$ $(CH_3)_2$	$C_7H_{13}NO_3$ (159.20)	210–212° (307)	$+10^{30}$ (c 2.9, H_2O) (307)	–	307	–	307	–
289	Hypaphorin (tryptophan betaine)	*Erythrina subumbrans*	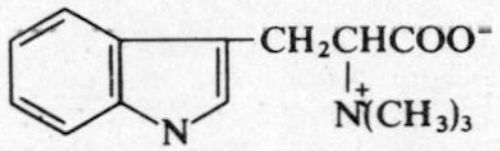 (indol-3-yl)–CH_2CHCOO^- $\overset{+}{N}(CH_3)_3$	$C_{14}H_{18}N_2O_2$ (246.29)	–	–	–	58	–	282	–
290	Isodesmosine	Bovine elastin	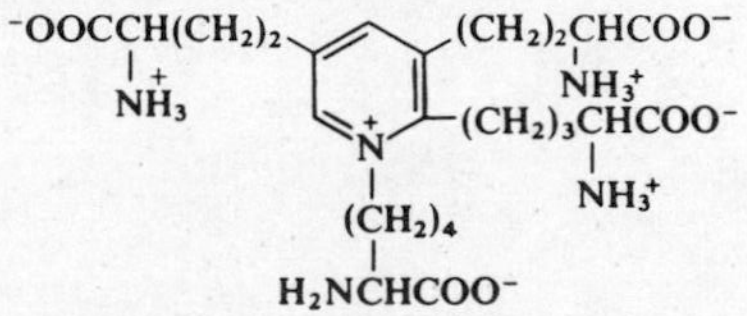 $^-OOCCH(CH_2)_2$–(pyridinium ring)–$(CH_2)_2CHCOO^-$ (NH_3^+) $\overset{+}{N}H_3$; –$(CH_2)_3CHCOO^-$ (NH_3^+) $\overset{+}{N}$–$(CH_2)_4$–$H_2NCHCOO^-$	$C_{24}H_{39}N_5O_8$ (879.43)	–	–	–	375	–	257	–
291	Laminine (α-amino-ε-trimethylamino adipic acid)	*Laminaria angustata*	$(CH_3)_3N^+(CH_2)_4CH(NH_2)COO^-$	$C_9H_{20}N_2O_4$ (220.29)	–	–	–	245	–	245	245
292	Lycin (glycine betaine)	*Lycium barbarum*	$(CH_3)_3N^+CH_2COO^-$	$C_5H_{11}NO_2$ (118.16)	–	–	–	234	–	–	–
293	Miokinine (ornithine betaine)	Human skeletal muscle	$(CH_3)_3N^+(CH_2)_3CHCOO^-$ $N^+(CH_3)_3$	$C_{11}H_{25}N_2O_2$ (217.33)	–	–	–	232	–	–	–
294	Nicotianine (N-3-amino-3-carboxypropyl-β-carboxypyridinium betaine)	Tobacco leaves	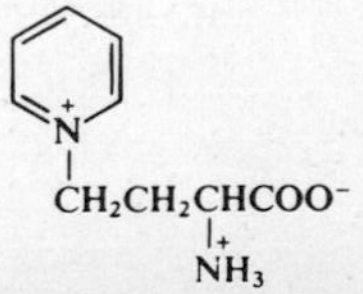 (pyridinium ring) $\overset{+}{N}$–$CH_2CH_2CHCOO^-$ $\overset{+}{N}H_3$	$C_{10}H_{14}N_2O_5$	241–243° (dec) (156)	$+28.4^{24}$ (c 5, H_2O) (156)	–	156	156	156	156

DATA ON THE NATURALLY OCCURRING AMINO ACIDS (continued)

No.	Amino acid (synonym)	Source	Structure	Formula (mol wt)	Melting point °C[a]	$[\alpha]_D$[b]	pK_a	References: Isolation and purification	References: Chromatography	References: Chemistry	References: Spectral data
295	Stachydrine (proline betaine)	*Stachys tuberifera*	H_2C—CH_2 / H_2C—$\overset{+}{N}(CH_3)_2$—$CHCOO^-$ (ring)	$C_7H_{13}NO_2$ (157.22)	–	-20.7^{25g} (H_2O) (125)	–	128	–	128	–
296	Trigonelline (coffearin)	*Foenum graceum*	pyridinium ring with COO^-, $\overset{+}{N}$—CH_3	$C_7H_7NO_2$ (137.15)	–	–	–	120	107	–	–
296a	ε-*N*-Trimethyl-δ-hydroxylysine betaine	Diatom cell walls	$(H_3C_3)N^+CH_2CH(OH)(CH_2)_2CH(NH_2)COO^-$	$C_9H_{20}N_2O_3$ (204.31)	243° (dec) (381)	$+15.0^{24}$ (*c* 0.84, H_2O) $+23.1^{24}$ (*c* 0.84, 2 *N* HCl) (381)	–	381	381	–	–
297	ε-*N*-Trimethyllysine betaine	Histone of murine ascites cells	$(CH_3)_3N^+(CH_2)_4CH(NH_2)COO^-$	$C_9H_{20}N_2O_2$ (188.28)	225–226° (dec) (70a)	$+10.8^{18}$ (*c* 5, H_2O) (379)	–	44, 379	44, 380	380	379
297a	D-Alloenduracididine [α-amino-β-(2-iminoimidazolidinyl)-propionic acid]	Enduracidin hydrolyzate	HN—CH_2 / HN=C—N(H)—$CHCH_2CH(NH_2)COOH$ (ring)	$C_6H_{12}N_4O_2$ (172.22)	–	$+8.7^{23}$ (1 *M* HCl) $+13.3^{23}$ (1 *M* NaOH) (194)	2.5 8.3 12 (194a)	–	194a	–	194a
297b	D-*Allo*isoleucine	Stendomycin hydrolyzate	CH_3CH_2 / CH_3 > $CHCH(NH_2)COOH$	$C_6H_{13}NO_2$ (131.20)	–	–	–	–	264a	–	–

O. D-Amino Acids

No.	Amino acid (synonym)	Source	Structure	Formula (mol wt)	Melting point °C[a]	$[\alpha]_D$[b]	pK_a	References: Isolation and purification	References: Chromatography	References: Chemistry	References: Spectral data
298	D-Allothreonine	*Actinomycetales* species	$CH_3CHOHCH(NH_2)COOH$	$C_4H_9NO_3$ (119.12)	–	–	2.11 9.10 (290)	52	52	–	–

DATA ON THE NATURALLY OCCURRING AMINO ACIDS (continued)

No.	Amino acid (synonym)	Source	Structure	Formula (mol wt)	Melting point °C[a]	$[\alpha]_D$[b]	pK_a	References: Isolation and purification	References: Chromatography	References: Chemistry	References: Spectral data
299	1-Amino-D-proline	Flax seed	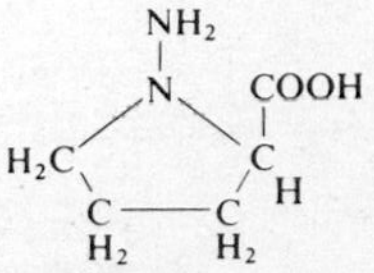 [structure: pyrrolidine ring, N bearing NH_2, C bearing COOH and H]	$C_5H_{10}N_2O_2$ (130.15)	155° (dec) (80)	$+113^{25}$ (c 2, 0.5 N HCl)(80)	–	80	80	80	80
299a	O-Carbamyl D-serine	*Streptomyces* strain	$NH_2COOCH_2CH(NH_2)COOH$	$C_4H_8N_2O_4$ (148.14)	238° (dec) (384)	−19.6 (c 2, 1 N HCl) +2 (c 2, H_2O) (384)	–	384	384	384	–
300	D-(3-Carboxy-4-hydroxyphenyl) glycine	*Reseda luteola*	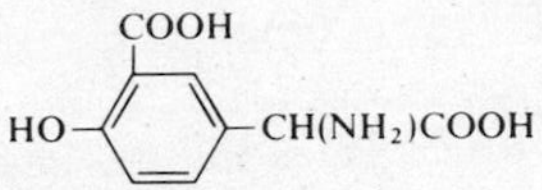 [structure: HO–benzene ring (COOH)–CH(NH₂)COOH]	$C_9H_9NO_5$ (211.18)	<250° (dec)(93)	-121^{22} (c 0.75, 1 N HCl) (93)	–	93	93	93	93
301	D-Cycloserine (oxamycin, D-4-amino-3-isoxazolinone)	*Streptomyces orchidaceus*	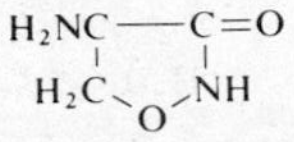 [structure: H_2NC—C=O, H_2C–O–NH ring]	$C_3H_6N_2O_2$ (101.09)	156° (dec) (94)	-89^{25} (c 0.6, H_2O)(94)	(94)	94	–	94	94
302	m-Carboxyphenyl-glycine	Iris bulbs	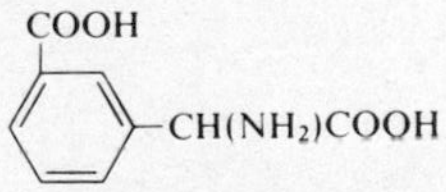 [structure: benzene ring (COOH)–CH(NH₂)COOH]	$C_9H_9NO_4$ (195.18)	215° (143)	$+112^{25}$ (c 5, 2 N HCl)(326)	4.4 7.3 (326)	143	143	143	143
302a	N-α-Malonyl-D-alanine	Pea seedlings	$H_3C-CH(HNC(=O)CH_2COOH)-COOH$	$C_6H_9NO_5$ (175.16)	138–140° (dec) (382)	$+33^{28}$ (c 0.38, H_2O) (382)	–	382	382	382	382
303	N-α-Malonyl-D-methionine	Tobacco	$CH_3SCH_2CH_2CH(HNC(=O)CH_2COOH)COOH$	$C_8H_{13}NO_5S$ (235.30)	–	–	–	95	95	95	–
304	N-α-Malonyl-D-tryptophan	Spinach	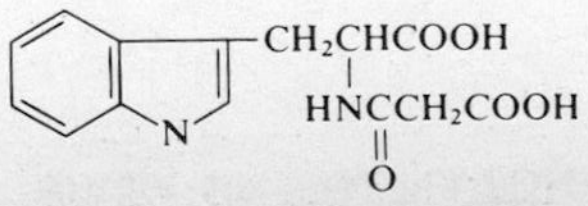 [structure: indol-3-yl–$CH_2CH(HNC(=O)CH_2COOH)COOH$]	$C_{14}H_{13}N_2O_5$ (289.27)	–	–	–	104	–	104	–

DATA ON THE NATURALLY OCCURRING AMINO ACIDS (continued)

No.	Amino acid (synonym)	Source	Structure	Formula (mol wt)	Melting point °C[a]	$[\alpha]_D$[b]	pK_a	References: Isolation and purification	References: Chromatography	References: Chemistry	References: Spectral data
304a	*N*-Methyl-D-leucine	*Griselimycin* hydrolyzate	$CH_3-CH(CH_3)-CH_2-CH(COOH)-NHCH_3$	$C_7H_{15}NO_2$ (145.23)	–	-19.6^{20} (*c* 0.9, H_2O) (376)	–	–	–	–	376
305	D-α-Methylserine	*Streptomyces*	$HOCH_2C(CH_3)(NH_2)COOH$	$C_4H_9NO_3$ (119.12)	–	–	2.3 9.4 (113)	113	–	113	113
306	D-Octopine [*N*-α-(1-Carboxyethyl-arginine]	Octopus muscle	$H_2NC(=NH)N(CH_2)_3CH(COOH)HNCHCH_3(COOH)$	$C_9H_{18}N_4O_4$ (246.28)	–	$+20.6^{24}$ (*c* 1, H_2O) 20^{25} (*c* 2, 5 *N* HCl) (290)	1.36 2.40 8.76 11.3 (290)	133	–	205	–
307	D-Penicillamine	Penicillin	$(CH_3)_2C(SH)CH(NH_2)COOH$	$C_5H_{11}NO_2S$ (149.22)	–	(225)	1.8 7.9 10.5	225	–	225	–
308	Turcine (D-allohydroxy-proline betaine)	*Stachys* (*Betonica*) *officinalis*	$HOHC-CH_2$, $H_2C-\overset{+}{N}(CH_3)_2-CHCOO^-$ (ring)	$C_7H_{13}NO_3$ (159.19)	249° (241)	$+36.26^{15}$ (241)	–	241	–	241	–

Compiled by Elizabeth Dodd Mooz.

[a]Melting point or decomposition point in degrees, C.
[b]*c*, grams/100 ml of solution at 20 to 25°C unless specified; wavelength as subscript in millimicrons; temperature as superscript. References are 1, 3, and 290 unless indicated.
[c]The isolated amino acid appears to be a racemic DL mixture.
[d]The naturally occurring isomer is the *erythro*-L-form 290.
[e]Thermodynamic values.
[f]As hydrochloride of amino acid.
[g]For cycloalliin sulfoxide.

REFERENCES

1. Meister, *Biochemistry of the Amino Acids*, 2nd ed., Academic Press, New York, 1965, 28.
2. Schutzenberger and Bourgeois, *C. R. Hebd. Seances Acad. Sci.* (Paris), 81, 1191 (1875).
3. Greenstein and Winitz, *Chemistry of the Amino Acids*, Vol. 2, John Wiley & Sons, New York, 1961, 1382.
4. Greenstein and Winitz, *Chemistry of the Amino Acids*, Vol. 2, John Wiley & Sons, New York, 1961, 1819.
5. Morris and Thompson, *Nature*, 190, 718 (1961).
6. Abderhalden and Bahn, *Hoppe Seyler's Z. Physiol. Chem.*, 245, 246 (1937).
7. Virtanen and Miettinen, *Biochim. Biophys. Acta*, 12, 181 (1953).
8. Ackerman, *Hoppe Seyler's Z. Physiol. Chem.*, 69, 273 (1910).
9. Work, *Bull. Soc. Chim. Biol.*, 31, 138 (1949).
10. Steward, *Science*, 110, 439 (1949).
11. Gabriel, *Chem. Ber.*, 23, 1767 (1890).
12. Vahatalo and Virtanen, *Acta Chem. Scand.*, 11, 741 (1957).
13. Doyle and Levenberg, *Biochemistry*, 7, 2457 (1968).
14. Steiner and Hartmann, *Biochem. Z.*, 340, 436 (1964).

14a. Chilton, Tsou, Kirk, and Benedict, *Tetrahedron Lett.*, p. 6283 (1968).

15. Takita, *J. Antibiot.* (Tokyo), 17, 264 (1964).

15a. Sung, Fowden, Millington, and Sheppard, *Phytochemistry*, 8, 1227 (1969).

15b. Millington and Sheppard, *Phytochemistry*, 7, 1027 (1968).

16. Kenner and Sheppard, *Nature*, 181, 48 (1958).
17. Asen, *J. Biol. Chem.*, 234, 343 (1959).

17a. Scannell, Pruess, Demny, Sello, Williams, and Stempel, *J. Antibiot.* (Tokyo), 25, 122 (1972).

17b. Sahm, Knobloch, and Wagner, *J. Antibiot.* (Tokyo), 26, 389 (1973).

18. Fowden and Done, *Biochem. J.*, 55, 548 (1953).
19. Fowden and Smith, *Phytochemistry*, 7, 809 (1968).

19a. Kelly, Martin, and Hanka, *Can. J. Chem.*, 47, 2504 (1969).

20. Perrin, *Dissociation Constants of Organic Bases in Aqueous Solution*, Butterworths, London, 1965.
21. Millington and Sheppard, *Phytochemistry*, 7, 1027 (1968).
22. Dardenne, Casimar, and Jadot, *Phytochemistry*, 7, 1401 (1968).

22a. Dardenne, Casimar, and Jadot, *Phytochemistry*, 7, 1401 (1968).

23. Staron, Allard, and Xuong, *C. R. Hebd. Seances Acad. Sci.* (Paris), 260, 3502 (1965).

23a. Scannell, Pruess, Demny, Weiss, Williams, and Stempel, *J. Antibiot.* (Tokyo), 24, 239 (1971).

23a′. Fowden, Smith, Millington, and Sheppard, *Phytochemistry*, 8, 437 (1969).

23b. Fowden and Smith, *Phytochemistry*, 7, 809 (1968).

24. Braconnot, *Ann. Chim. Phys.*, 13, 113 (1820).
25. Shorey, *J. Am. Chem. Soc.*, 19, 881 (1897).
26. Hassall and Keyle, *Biochem. J.*, 60, 334 (1955).
27. Carbon, *J. Am. Chem. Soc.*, 80, 1002 (1958).
28. Ehrlich, *Chem. Ber.*, 37, 1809 (1904).
29. Greenstein and Winitz, *Chemistry of the Amino Acids*, Vol. 3, John Wiley & Sons, New York, 1961, 2043.
30. Proust, *Ann. Chim. Phys.*, 10, 29 (1819).
31. Greenstein and Winitz, *Chemistry of the Amino Acids*, Vol. 3, John Wiley & Sons, New York, 1961, 2075.

31a. Walker, Bodanszky, and Perlman, *J. Antibiot.* (Tokyo), 23, 255 (1970).

32. Gray and Fowden, *Biochem. J.*, 82, 385 (1961).
33. Kortrum, Vogel, and Andrussow, *Dissociation Constants of Organic Acids in Aqueous Solutions*, Butterworths, London, 1961.
34. Yukowa, *Handbook of Organic Structural Analysis*, Benjamin, New York, 1965, 584.
35. Fischer, *Hoppe Seyler's Z. Physiol. Chem.*, 33, 151 (1901).

35a. Vervier and Casimir, *Phytochemistry*, 9, 2059 (1970).

35b. Levenberg, *J. Biol. Chem.*, 243, 6009 (1968).

36. Greenstein and Winitz, *Chemistry of the Amino Acids*, Vol. 3, John Wiley & Sons, New York, 1961, 2368.
37. Hatanaka and Virtanen, *Acta Chem. Scand.*, 16, 514 (1962).
38. Takemoto and Sai, *J. Pharm. Soc. Jap.*, 85, 33 (1965).
39. Virtanen and Berg, *Acta Chem. Scand.*, 8, 1085 (1954).
40. Vauqelin and Robiquet, *Ann. Chim.*, 57, 88 (1806).
41. Greenstein and Winitz, *Chemistry of the Amino Acids*, Vol. 3, John Wiley & Sons, New York, 1961, 1856.
42. Davies and Evans, *J. Chem. Soc.*, p. 480 (1953).
43. Ritthausen, *J. Prakt. Chem.*, 103, 233 (1868).
44. Hempel, Lange, and Berkofer, *Naturwissenschaften*, 55, 37 (1968).
45. Gray and Fowden, *Nature*, 189, 401 (1961).
46. Gmelin and Larsen, *Biochim. Biophys. Acta*, 136, 572 (1967).

46a. Nulu and Bell, *Phytochemistry*, 11, 2573 (1972).

47. Fowden, *Biochem. J.*, 98, 57 (1966).
48. Birkinshaw, Raistick, and Smith, *Biochem. J.*, 36, 829 (1942).
49. Ritthausen, *J. Prakt. Chem.*, 99, 454 (1866).
50. Greenstein and Winitz, *Chemistry of the Amino Acids,* Vol. 3, John Wiley & Sons, New York, 1961, 1929.
51. Schulze and Bosshard, *Landwirtsch. Vers. Stn.*, 29, 295 (1883).
52. Ikawa, Snell, and Lederer, *Nature,* 188, 558 (1961).
53. Larsen, *Acta Chem. Scand.*, 19, 1071 (1965).
54. Fowden and Gray, *Amino Acid Pools,* Holden, Ed., Elsevier, Amsterdam, 1962, 46.
55. Barker, Smyth, Wawszkiewicz, Lee, and Wilson, *Arch. Biochem. Biophys.*, 78, 468 (1958).
56. Virtanen and Berg, *Acta Chem. Scand.*, 9, 553 (1955).
56a. Przybylska and Strong, *Phytochemistry,* 7, 471 (1968).
57. Done and Fowden, *Biochem. J.*, 51, 451 (1952).
58. Greshoff, *Chem. Ber.*, 23, 3537 (1890).
59. Casimir, Jadot, and Renard, *Biochim. Biophys. Acta,* 39, 462 (1960).
59a. Seneviratne and Fowden, *Phytochemistry,* 7, 1030 (1968).
60. Virtanena and Linko, *Acta Chem. Scand.*, 9, 531 (1955).
61. Liss, *Phytochemistry,* 1, 87 (1962).
62. Bohak, *J. Biol. Chem.*, 239, 2878 (1964).
63. Ziegler, Melchert, and Lurken, *Nature,* 214, 404 (1967).
64. Argoudelis, Herr, Mason, Pyke, and Zieserl, *Biochemistry,* 6, 165 (1967).
65. Vega and Bell, *Phytochemistry,* 6, 759 (1967).
65a. Bell, *Nature,* 218, 197 (1968).
66. Damodaran and Narayanan, *Biochem. J.*, 34, 1449 (1940).
67. Kitagawa and Monobe, *J. Biochem.* (Tokyo), 18, 333 (1933).
67a. Bodanszky and Bodanszky, *J. Antibiot.* (Tokyo), 23, 149 (1970).
67b. Bodanszky, Chaturvedi, Scozzie, Griffith, and Bodanszky, *Antimicrob. Agents Chemother.*, p. 135 (1969).
68. Fujino, Inoue, Ueyanagi, and Miyake, *Bull. Chem. Soc. Jap.*, 34, 740 (1961).
69. Tsai and Stadtman, *Arch. Biochem. Biophys.*, 125, 210 (1968).
69a. Hettinger and Craig, *Biochemistry,* 9, 1224 (1970).
70. Gmelin, Strauss, and Hasenmaier, *Hoppe Seyler's Z. Physiol. Chem.*, 314, 28 (1959).
70a. Kakimoto and Akazawa, *J. Biol. Chem.*, 245, 5751 (1970).
70b. Scannell, Ax, Pruess, Williams, Demm, and Stempel, *J. Antibiot.* (Tokyo), 25, 179 (1972).
71. Dupuy and Lee, *J. Am. Pharm. Assoc. (Sci. Ed.)*, 43, 61 (1954).
72. Schilling and Strong, *J. Am. Chem. Soc.*, 77, 2843 (1955).
73. Vickery and Leavenworth, *J. Biol. Chem.*, 76, 437 (1928).
74. Drechsel, *Z. Prakt. Chem.*, 39, 425 (1889).
75. Greenstein and Winitz, *Chemistry of the Amino Acids,* Vol. 3, John Wiley & Sons, New York, 1961, 2097.
76. Haskell, Fusari, Frohardt, and Bartz, *J. Am. Chem. Soc.*, 74, 599 (1952).
77. Biemann, Lioret, Asselineau, Lederer, and Polonsky, *Bull. Soc. Chim. Biol.*, 42, 979; *Biochim. Biophys. Acta,* 40, 369 (1960).
78. Lioret, *C. R. Hebd Seances Acad. Sci.* (Paris), 244, 2171 (1957).
79. Murray, *Biochemistry,* 3, 10 (1964).
80. Klosterman, Lamoureux, and Parsons, *Biochemistry,* 6, 170 (1967).
81. Greenstein and Winitz, *Chemistry of the Amino Acids,* Vol. 3, John Wiley & Sons, New York, 1961, 2477.
81a. Stapley, Miller, Mata, and Hendlin, *Antimicrob. Agents Chemother.*, p. 401 (1967).
82. Rao, Adiga, and Sarma, *Biochemistry,* 3, 432 (1964).
82a. Shiba, Kubota, and Kaneto, *Tetrahedron,* 26, 4307 (1970).
83. Suzuki, Nakamura, Okuma, and Tomiyama, *J. Antibiot.* (Tokyo), 11A(81), 84 (1958).
84. Cummings and Hudgins, *Am. J. Med. Sci.*, 236, 311 (1958).
85. Sorensen and Andersen, *Hoppe Seyler's Z. Physiol. Chem.*, 56, 250 (1908).
86. Hochstein, *J. Org. Chem.*, 24, 679 (1959).
87. Grobbelaar and Steward, *Nature,* 182, 1358 (1958).
87a. Westley, Pruess, Volpe, Demny, and Stempel, *J. Antibiot.* (Tokyo), 24, 330 (1971).
88. Larsen, *Acta Chem. Scand.*, 21, 1592 (1967).
88a. Doyle and Levenberg, *Biochemistry,* 7, 2457 (1968).
89. Blass and Macheboeuf, *Helv. Chim. Acta,* 29, 1315 (1946).
90. Snow, *J. Chem. Soc.*, 2588, 4080 (1954).
91. Virtanen and Hietala, *Acta Chem. Scand.*, 9, 549 (1955).
92. Umbreit and Heneage, *J. Biol. Chem.*, 201, 15 (1953).
93. Kjaer and Larsen, *Acta Chem. Scand.*, 17, 2397 (1963).
94. Hidy, Hodge, Yound, Harned, Brewer, Phillips, Runge, Staveley, Pohland, Boaz, and Sullivan, *J. Am. Chem. Soc.*, 77, 2345 (1955).
95. Keglevic, Ladesic, and Pokorney, *Arch. Biochem. Biophys.*, 124, 443 (1968).

96. Virtanen, Uksila, and Matikkala, *Acta Chem. Scand.*, 8, 1091 (1954).
97. Thompson, Morris, and Hunt, *J. Biol. Chem.*, 239, 1122 (1964).
98. Okabe, *J. Antibiot. Ser. A*, 13, 412 (1961).
99. Wieland and Haufer, *Justus Liebigs Ann. Chem.*, 619, 35 (1968).
99a. Fowden, MacGibbon, Mellon, and Sheppard, *Phytochemistry*, 11, 1105 (1972).
99a'. Rudzats, Gellert, and Halpern, *Biochem. Biophys. Res. Commun.*, 47, 290 (1972).
100. Murooka and Harada, *Agric. Biol. Chem.*, 31, 1035 (1967).
100a. Giovanelli, Owens, and Mudd, *Biochim. Biophys. Acta*, 227, 671 (1971).
100b. Owens, Thompson, and Fennessey, *Chem. Commun.*, p. 715 (1972).
101. Virtanen and Ettala, *Acta Chem. Scand.*, 11, 182 (1957).
102. Bodanszky, Alicino, Birkhimer, and Williams, *J. Am. Chem. Soc.*, 84, 2004 (1962).
103. Wieland and Schopf, *Justus Liebigs Ann. Chem.*, 626, 174 (1959).
103a. Mechanic and Tanzer, *Biochem. Biophys. Res. Commun.*, 41, 1597 (1970).
104. Good and Andreae, *Plant Physiol.*, 32, 561 (1957).
104a. Takita and Maeda, *J. Antibiot.* (Tokyo), 22, 39 (1969).
105. Saarivirta and Virtanen, *Acta Chem. Scand.*, 19, 1008 (1965).
106. Craig, *J. Biol. Chem.*, 125, 289 (1938).
107. Joshi and Handler, *J. Biol. Chem.*, 235, 2981 (1961).
108. Setsuseo, *J. Osaka Univ.*, 7, 833 (1957).
109. Schweet, Holden, and Lowy, *J. Biol. Chem.*, 211, 517 (1954).
110. Tominaga, Hiwaki, Maekawa, and Yoshida, *J. Biochem.* (Tokyo), 53, 227 (1963).
111. Wilding and Stahlmann, *Phytochemistry*, 1, 241 (1962).
111a. Kemp and Dawson, *Biochim. Biophys. Acta*, 176, 678 (1969).
112. Fowden, *Biochem. J.*, 81, 155 (1961).
113. Flynn, Hinman, Caron, and Woolf, *J. Am. Chem. Soc.*, 75, 5867 (1953).
114. Nagao, *Bull. Fac. Fish Hokkaido Univ.*, 2, 128 (1951).
115. Hatanaka, *Acta Chem. Scand.*, 16, 513 (1962).
116. Brandner and Virtanen, *Acta Chem. Scand.*, 17, 2563 (1963).
117. Snow, *J. Chem. Soc.*, p. 2589 (1954).
117a. Fowden, Pratt, and Smith, *Phytochemistry*, 12, 1707 (1973).
118. Miyake, *Chem. Pharm. Bull.* (Tokyo), 8, 1071 (1960).
119. Wieland and Schön, *Justus Liebigs Ann. Chem.*, 593, 157 (1955).
120. Jahns, *Chem. Ber.*, 18, 2518 (1885); 20, 2840 (1887).
121. Jadot and Casimir, *Biochim. Biophys. Acta*, 48, 400 (1961).
122. Jadot, Casimir, and Alderweireldt, *Biochim. Biophys. Acta*, 78, 500 (1963).
123. Brieskorn and Glasz, *Naturwissenschaften*, 51, 216 (1964).
124. Schryver, *Proc. R. Soc.*, B98, 58 (1925).
124a. Godtfredsen, Vangedal, and Thomas, *Tetrahedron*, 26, 4931 (1970).
125. Steenbock, *J. Biol. Chem.*, 35, 1 (1918).
126. Fowden, *Nature*, 209, 807 (1966).
127. Bell and Tirimanna, *Biochem., J.*, 91, 356 (1964).
128. Planta and Schulze, *Chem. Ber.*, 26, 939 (1893).
129. Pollard, Sondheimer, and Steward, *Nature*, 182, 1356 (1958).
129a. Shiba, Mizote, Kaneko, Nakajima, and Kakimoto, *Biochim. Biophys. Acta*, 244, 523 (1971).
129b. Hettinger and Craig, *Biochemistry*, 9, 1224 (1970).
130. Barry and Roark, *J. Biol. Chem.*, 239, 1541 (1964).
131. Barry, Chen, and Roark, *J. Gen. Microbiol.*, 33, 95 (1963).
132. Ackermann and Kirby, *J. Biol. Chem.*, 175, 483 (1948).
132a. Grove, Daxenbichler, Weisleder, and van Etlen, *Tetrahedron Lett.*, 4477 (1971).
132b. Owens, Guggenheim, and Hilton, *Biochim. Biophys. Acta*, 158, 219 (1968).
132c. Owens, Thompson, Pitcher, and Williams, *Chem. Commun.*, p. 714 (1972).
133. Morizawa, *Acta Sch. Med. Univer. Kioto*, 9, 285 (1927).
134. Cramer, *J. Prakt. Chem.*, 96, 76 (1865).
135. Flavin, Delavier-Klutchko, and Slaughter, *Science*, 143, 50 (1964).
136. Wooley, *J. Biol. Chem.*, 197, 409 (1952).
137. Wooley, *J. Biol. Chem.*, 198, 807 (1953).
138. Rose, McCoy, Meyer, and Carter, *J. Biol. Chem.*, 112, 283 (1935).
139. Greenstein and Winitz, *Chemistry of the Amino Acids*, Vol. 3, John Wiley & Sons, New York, 1961, 2238.
140. Wisvisz, Van der Hoever, and Nijenhuis, *J. Am. Chem. Soc.*, 79, 4522 (1957).
140a. Bohman, *Tetrahedron Lett.*, p. 3065 (1970).
141. Olesen-Larsen, *Biochim. Biophys. Acta*, 93, 200 (1964).
142. Thompson, Morris, Asen, and Irreverre, *J. Biol. Chem.*, 236, 1183 (1961).
142a. O'Brien, Cox, and Gibson, *Biochim. Biophys. Acta*, 177, 321 (1969).

143. Morris, Thompson, Asen, and Irreverre, *J. Am. Chem. Soc.*, 81, 6069 (1959).
144. Schneider, *Biochem. Z.*, 330, 428 (1958).
145. Guggenheim, *Z. Physiol. Chem.*, 88, 276 (1913).
146. Greenstein and Winitz, *Chemistry of the Amino Acids*, Vol. 3, John Wiley & Sons, New York, 1961, 2713.
147. Muller and Schulte, *Z. Naturforsch.*, 23b, 659 (1958).
147a. Weaver, Rajagopalan, Handler, Rosenthal, and Jeffs, *J. Biol. Chem.*, 246, 2010 (1971).
147b. Watson and Fowden, *Phytochemistry*, 12, 617 (1973).
148a. Smith and Sloane, *Biochim. Biophys. Acta*, 148, 414 (1967).
148b. Sloane and Smith, *Biochim. Biophys. Acta*, 158, 394 (1968).
148. Makino, Satoh, Fujik, and Kawaguchi, *Nature*, 170, 977 (1952).
149. Waller, Fryth, Hutchings, and Williams, *J. Am. Chem. Soc.*, 75, 2025 (1953).
150. Behr and Clark, *J. Am. Chem. Soc.*, 54, 1630 (1932).
151. Schultze and Barbieri, *Chem. Ber.*, 14, 1785 (1881).
152. Greenstein and Winitz, *Chemistry of the Amino Acids*, Vol. 3, John Wiley & Sons, New York, 1961, 2156.
153. Liebig, *Justus Liebigs Ann. Chem.*, 57, 127 (1846).
154. Greenstein and Winitz, *Chemistry of the Amino Acids*, Vol. 3, John Wiley & Sons, New York, 1961, 2348.
155. Mothes, Schütte, Müller, Ardenne, and Tümmler, *Z. Naturforsch.*, 196, 1161 (1964).
155a. Ohkusu and Mori, *J. Neurochem.*, 16, 1485 (1969).
156. Noguchi, Sakuma, and Tamaki, *Phytochemistry*, 7, 1861 (1968).
157. Kjaer and Larsen, *Acta Chem. Scand.*, 13, 1565 (1959).
158. Schulze and Steiger, *Chem. Ber.*, 19, 1177 (1886); *Z. Physiol. Chem.*, 11,43 (1886).
159. Greenstein and Winitz, *Chemistry of the Amino Acids*, Vol. 3, John Wiley & Sons, New York, 1961, 1841.
160. Fearon and Bell, *Biochem., J.*, 59, 221 (1955).
161. Kitagawa and Tomiyama, *J. Biochem.* (Tokyo), 11, 265 (1929).
162. Bell, *Biochem. J.*, 75, 618 (1960).
163. Walker, *J. Biol. Chem.*, 204, 139 (1954).
164. Wada, *Biochem. Z.*, 224, 420 (1930).
165. Rogers, *Biochim. Biophys. Acta*, 29, 33 (1958).
166. Kitagawa and Tsukamoto, *J. Biochem.* (Tokyo), 26, 373 (1937).
166a. Nakajima, Matsuoka, and Kakimoto, *Biochim. Biophys. Acta*, 230, 212 (1971).
167. Ito and Hashimoto, *Nature*, 211, 417 (1966).
168. Rao, Ramachandran, and Adiga, *Biochemistry*, 2, 298 (1962).
169. Gerritsen, Vaughn, and Waisman, *Arch. Biochem. Biophys.*, 100, 298 (1963).
170. Bell and Tirimanna, *Nature*, 197, 901 (1963).
170a. Maehr, Blount, Pruess, Yarmchuk, and Kellett, *J. Antibiot.* (Tokyo), 26, 284 (1973).
171. Bell, *Biochem. J.*, 21, 358 (1964).
173. Hegarty and Pound, *Nature*, 217, 354 (1968).
174. Tamoni and Gallo, *Farmaco Ed. Sci.* 21, 269 (1966).
175. Fusari, Bartz, and Elder, *Nature*, 173, 72; *J. Am. Chem. Soc.*, 76, 2878 (1954).
176. Ressler, *J. Biol. Chem.*, 237, 733 (1962).
177. Ressler and Ratzkin, *J. Org. Chem.*, 26, 3356 (1961).
177a. Brysk and Ressler, *J. Biol. Chem.*, 245, 1156 (1970).
178. Dion, Fusari, Jakubowski, Zora, and Bartz, *J. Am. Chem. Soc.*, 78, 3075 (1956).
179. Kaczka, Gitterman, Dulaney, and Folkers, *Biochemistry*, 1, 340 (1962).
180. Radhakrishnan and Giri, *Biochem. J.*, 58, 57 (1954).
180a. Lambein and Van Parijs, *Biochem. Biophys. Res. Commun.*, 32, 474 (1968).
180b. Lambein, Schamp, Vandendriessche, and Van Parijs, *Biochem. Biophys. Res. Commun.*, 37, 375 (1969).
181. King, *J. Chem. Soc.*, p. 3590 (1950).
182. Grobbelaar, *Nature*, 175, 703 (1955).
183. Dobson and Raphael, *J. Chem. Soc.*, p. 3642 (1958).
184. Tanaka, Miyamoto, Honjo, Morimoto, Sugawa, Uchibayshi, Sanno, and Tatsuoka, *Proc. Jap. Acad.*, 33, 47, 53 (1957).
185. Tanaka, Miyamoto, Honjo, Morimoto, Sugawa, Uchibayshi, Sanno, and Tatsuoka, *Chem. Abstr.*, 51, 17881i (1957).
185a. Burrows and Turner, *J. Chem. Soc.*, p. 255 (1966).
186. Schenk and Schütte, *Naturwissenschaften*, 48, 223, (1961).
186a. Hatanaka, *Phytochemistry*, 8, 1305 (1969).
186b. Shah, Neuss, Gorman, and Boeck, *J. Antibiot.* (Tokyo), 23, 613 (1970).
187. Prochazka, *Czech. Chem. Commun.*, 22, 333, 654 (1957).
188. Pironen and Virtanen, *Acta Chem. Scand.*, 16, 1286 (1962).
189. Gmelin and Virtanen, *Ann. Acad. Sci. Fenn. (Med.)*, 107, 3 (1961).
189a. Miller, Tristram, and Wolf, *J. Antibiot.* (Tokyo), 24, 48 (1971).

190. Fowden, *Nature,* 176, 347 (1955).
190a. Eagles, Laird, Matai, Self, and Synge, *Biochem. J.,* 121, 425 (1971).
190b. Bell, Nulu, and Cone, *Phytochemistry,* 10, 2191 (1971).
190c. Robbers and Floss, *Tetrahedron Lett.,* p. 1857 (1969).
191. Fang, Li, Nin, and Tseng, *Sci. Sinica,* 10, 845 (1961).
191a. Cross and Webster, *J. Chem. Soc. (Org.),* p. 1839 (1970).
191b. Scannell, Pruess, Demny, Williams, and Stempel, *J. Antibiot.* (Tokyo), 23, 618 (1970).
191c. Corbin and Bulen, *Biochemistry,* 8, 757 (1969).
191d. Nakajima and Volcani, *Science,* 164, 1400 (1969).
191e. Karle, Daly, and Witkop, *Science,* 164, 1401 (1969).
191f. Brown and Mangat, *Biochim. Biophys. Acta,* 177, 427 (1969).
191g. Macnicol, *Biochem. J.,* 107, 473 (1968).
192. Brown, *J. Am. Chem. Soc.,* 87, 4202 (1965).
193. Daigo, *J. Pharm. Soc. Jap.,* 79, 353 (1959).
194. Casnati, Quilico, and Ricca, *Gazz. Chim. Ital.,* 93, 349 (1963).
194a. Horii and Kameda, *J. Antibiot.* (Tokyo), 21, 665 (1968).
194a′. Katagiri, Tori, Kimura, Yoshida, Nagasaki, and Minato, *J. Med. Chem.,* 10, 1149 (1967).
194a″. Finot, Bricout, Viani and Mauron, *Experimentia,* 24, 1097 (1968).
194b. Weaver, Rajagopalan, and Handler, *J. Biol. Chem.,* 246, 2015 (1971).
195. Jahns, *Chem. Ber.,* 24, 2615 (1891).
196. Freidenberg, *Chem. Ber.,* 51, 976 (1918).
197. Kossel, *Hoppe Seyler's Z. Physiol. Chem.,* 22, 176 (1896).
198. Greenstein and Winitz, *Chemistry of the Amino Acids,* Vol. 3, John Wiley & Sons, New York, 1961, 1971.
198a. Takita, Yoshioka, Muraoka, Maeda, and Umezawa, *J. Antibiot.* (Tokyo), 24, 795 (1971).
199. Hulme, *Nature,* 174, 1055 (1954).
200. Biemann and Deffner, *Nature,* 191, 380 (1961).
201. Minagawa, *Proc. Imp. Acad.* (Tokyo), 21, 33, 37 (1945).
202. Virtanen and Gmelin, *Acta Chem. Scand.,* 13, 1244 (1959).
203. Schoolery and Virtanen, *Acta Chem. Scand.,* 16, 2457 (1962).
203. Virtanen and Kari, *Acta Chem. Scand.,* 8, 1290 (1954).
204. Virtanen and Kari, *Actu. Chem. Scand.,* 8, 1290 (1954).
204a. Hassall, Morton, Ogihara, and Thomas, *Chem. Commun.,* p. 1079 (1969).
204b. Bevan, Davies, Hassall, Morton, and Phillips, *J. Chem. Soc. (Org.),* p. 514 (1971).
205. Irvine and Wilson, *J. Biol. Chem.,* 127, 555 (1939).
206. Sheehan and Whitney, *J. Am. Chem. Soc.,* 84, 3980 (1962).
206a. Sung and Fowden, *Phytochemistry,* 7, 2061 (1968).
207. Fischer, *Chem. Ber.,* 35, 2660 (1902).
208. Greenstein and Winitz, *Chemistry of the Amino Acids,* Vol. 3, John Wiley & Sons, New York, 1961, 2018.
209. Wieland and Witkop, *Justus Liebigs Ann. Chem.,* 543, 171 (1940).
210. Kotake, Sakan, and Miwa, *Chem. Ber.,* 85, 690 (1952).
211. Mitoma, Weissbach, and Udenfriend, *Nature,* 175, 994 (1955).
212. Takemoto, Nakajima, and Yokobe, *J. Pharm. Soc. Jap.,* 84, 1232 (1964).
213. List, *Planta Med.,* 6, 424 (1958).
214. von Klämbt, *Naturwissenschaften,* 47, 398 (1960).
215. Hutzinger and Kosuge, *Biochemistry,* 7, 601 (1968).
216. Murakami, Takemoto, and Shimzu, *J. Pharm. Soc. Jap.,* 73, 1026 (1953).
217. Beockman and Staehler, *Naturwissenschaften,* 52, 391 (1965).
218. Vanderhaeghe and Parmetier, *17th Congr. Pure Appl. Chem.,* Butterworths, London, 1959, 56.
219. Brockmann, *Ann. N.Y. Acad. Sci.,* 89, 323 (1960).
220. Bell, *Biochim. Biophys. Acta,* 47, 602 (1961).
221. Gray and Fowden, *Nature,* 193, 1285 (1962).
222. Bethell, Kenner, and Shepperd, *Nature,* 194, 864 (1962).
222a. Müller and Schütte, *Z. Naturforsch.,* 236, 491 (1968).
223. Hulme and Arthington, *Nature,* 173, 588 (1954).
224. Burroughs, Dalby, Kenner, and Sheppard, *Nature,* 189, 394 (1961).
224a. Borders, Sax, Lancaster, Hausmann, Mitscher, Wetzel and Patterson, *Tetrahedron,* 26, 3123 (1970).
225. Chain, *Ann. Rev. Biochem.,* 17, 657 (1948).
226. Sheehan, Drummond, Gardner, Maeda, Mania, Nakamura, Sen, and Stock, *J. Am. Chem. Soc.,* 85, 2867 (1963).
227. Adams and Johnson, *J. Am. Chem. Soc.,* 71, 705 (1949).
227a. Murakoshi, Kuramoto, Ohmiya and Haginiwa, *Chem. Pharm. Bull.* (Japan), 20, 855 (1972).
228. Minagawa, *Proc. Jap. Acad.,* 22, 130 (1946).
229. Eugster, Muller, and Good, *Tetrahedron Lett.,* 23, 1813 (1965).
230. Reiner and Eugster, *Helv. Chim. Acta,* 50, 128 (1967).

231. Fritz, Gagnent, Zbinden, Geigy, and Eugster, *Tetrahedron Lett.*, 25, 2075 (1965).
231a. Noma, Noguchi, and Tamaki, *Tetrahedron Lett.*, p. 2017 (1971).
232. Engeland and Biehler, *Hoppe Seyler's Z. Physiol. Chem.*, 123, 290 (1922).
233. Hulme and Arthington, *Nature*, 170, 659 (1952).
234. Husemann and Marme, *Justus Liebigs Ann. Chem.*, Suppl. 2, 382 (1863).
235. Greenstein and Winitz, *Chemistry of the Amino Acids*, Vol. 3, John Wiley & Sons, New York, 1961, 2178.
236. Noe and Fowden, *Biochem. J.*, 77, 543 (1960).
236a. Finot, Viani, Bricout, and Mauron, *Experimentia*, 25, 134 (1969).
236b. Wolfersberger and Tabachnik, *Experimenta*, 29, 346 (1973).
237. Nakanishi, Ito, and Hirata, *J. Am. Chem. Soc.*, 76, 2845 (1954).
238. Carter, Sweeley, Daniels, McNary, Schaffner, West, Tamelen, Dyer, and Whaley, *J. Am. Chem. Soc.*, 83, 4296 (1961).
238a. Bodanszky, Marconi, and Bodanszky, *J. Antiobiot.*, (Tokyo), 22, 40 (1969).
238b. Marconi and Bodanszky, *J. Antibiot.* (Tokyo), 23, 120 (1970).
239. Hattori and Komamine, *Nature*, 183, 1116 (1959).
240. Senoh, Imamato, Maeno, Tokyama, Sakan, Komamine, and Hattori, *Tetrahedron Lett.*, 46, 3431 (1964).
241. Schulze and Trier, *Hoppe Seyler's Z. Physiol. Chem.*, 76, 258; 79, 235 (1912).
242. Takemoto and Nakajima, *J. Pharm. Soc. Jap.*, 84, 1230 (1964).
243. Hopkins and Cole, *J. Physiol.* (Lond.), 27, 418 (1901).
244. Greenstein and Winitz, *Chemistry of the Amino Acids*, Vol. 3, John Wiley & Sons, New York, 1961, 2316.
224a. Nakamiya, Shiba, Kaneko, Sakakibara, Take, and Abe, *Tetrahedron Lett.*, p. 3497 (1970).
245. Takemoto, Diago, and Takagi, *J. Pharm. Soc. Jap.*, 84, 1176 (1964).
246. Dyer, Hayes, Miller, and Nassar, *J. Am. Chem. Soc.*, 86, 5363 (1964).
246a. Büchi and Raleigh, *J. Org. Chem.*, 36, 873 (1971).
247. Gmelin, *Hoppe Seyler's Z. Physiol. Chem.*, 316, 164 (1959).
248. Kjaer, Knudsen, and Larsen, *Acta Chem. Scand.*, 15, 1193 (1961).
249. Gordon and Jackson, *J. Biol. Chem.*, 110, 151 (1935).
250. Ghatak and Kaul, *Chem. Zentralbl.*, p. 3730 (1932).
250a. Auditore and Wade, *J. Neurochem.*, 18, 2389 (1971).
250a′. Hall, Metzenberg, and Cohen, *J. Biol. Chem.*, 230, 1013 (1958).
250b. Tallan, Moore, and Stein, *J. Biol. Chem.*, 219, 257 (1956).
251. Takita, Naganawa, Maeda, and Umezawa, *J. Antibiot.* (Tokyo), 17, 90 (9164).
252. Sheehan, Zachau, and Lawson, *J. Am. Chem. Soc.*, 80, 3349 (1958).
252a. Bouloin, Ottinger, Pais, and Chiurdoglu, *Bull. Soc. Chim. Belges*, 78, 583 (1969).
253. Takita, *J. Antibiot.* (Tokyo), 16, 175 (1963).
254. Emery, *Biochemistry*, 4, 1410 (1965).
255. Ackerman, *Hoppe Seyler's Z. Physiol. Chem.*, 302, 80 (1955).
256. Morgan, *Chem. Ind.* (Lond.), p.542 (1964).
257. Starcher, Partridge, and Elsden, *Biochemistry*, 6, 2425 (1967).
258. Eloff and Grobbelaar, *J. S. Afr. Chem. Inst.*, 20, 190 1967).
259. Searle and Westall, *Biochem. J.*, 48, 1 (P), (1951).
260. Tallan, Stein, and Moore, *J. Biol. Chem.*, 206, 825 (1954).
261. Plattner and Nager, *Helv. Chim. Acta*, 31, 665 (1948).
262. Plattner and Nager, *Helv. Chim. Acta*, 31, 2192 (1948).
263. Corrigan, *Science*, 164, 142 (1969).
264. Vilkas, Rojas, and Lederer, *C.R. Hebd. Seances Acad. Sci.* (Paris), 261, 4258 (1965).
264a. Bodanszky, Muramatsu, Bodanszky, Lukin, and Doubler, *J. Antibiot.* (Tokyo), 21, 77 (1968).
265. Thompson, Morris, and Smith, *Ann. Rev. Biochem.*, 38, 137 (1969).
266. Hiller-Bombien, *Arch. Pharm.*, 230, 513 (1892).
267. Winterstein, *Hoppe Seyler's Z. Physiol. Chem.*, 105, 20 (1919).
268. Brockmann and Grobhofer, *Naturwissenschaften*, 36, 376 (1949).
269. Plattner and Nager, *Helv. Chim. Acta*, 31, 2203 (1948).
270. Darling and Larsen, *Acta Chem. Scand.*, 15, 743 (1961).
271. Kjaer and Larsen, *Acta Chem. Scand.*, 15, 750 (1961).
272. Linko, Alfthan, Miettinen, and Virtanen, *Acta Chem. Scand.*, 7, 1310 (1953).
273. Gmelin, Kjaer, and Larsen, *Phytochemistry*, 1, 233 (1962).
274. Stoll and Seebeck, *Helv. Chem. Acta*, 31, 189 (1948).
275. Suzuki, *Chem. Pharm. Bull.* (Tokyo), 9, 251 (1961).
276. Sugii, *Chem. Pharm. Bull.* (Tokyo), 12, 1114 (1964).
277. Alderton, *J. Am. Chem. Soc.*, 75, 2391 (1953).
278. Horowitz, *J. Biol. Chem.*, 171, 255 (1947).
279. Tallan, Moore and Stein, *J. Biol. Chem.*, 230, 707 (1958).
280. McRorie, Sutherland, Lewis, Barton, Glazener, and Shive, *J. Am. Chem. Soc.*, 76, 115 (1954).

280a. Jadot, Casimir, and Warin, *Bull. Soc. Chim. Belges,* 78, 299 (1969).
281. Salkowski, *Chem. Ber.,* 6, 744 (1873).
282. Romburgh and Barger, *J. Chem. Soc.* (Lond.), 99, 2068 (1911).
283. Gmelin, Strauss, and Hasenmaler, *Z. Naturforsch.,* 13b, 252 (1958).
284. Kuriyama, *Nature,* 192, 969 (1961).
284a. Kodama, Yao, Kobayashi, Hirayama, Fujii, Mizuhara, Haraguchi, and Hirosawa, *Physiol. Chem. Phys.,* 1, 72 (1969).
285. Gmelin and Hietala, *Hoppe Seyler's Z. Physiol. Chem.,* 322, 278 (1960).
286. Virtanen and Matikkala, *Hoppe Seyler's Z. Physiol. Chem.,* 322, 8 (1960).
287. Mizuhara and Ohmori, *Arch. Biochem. Biophys.,* 92, 53 (1961).
288. Kuriyama, Takagi, and Murata, *Bull. Fac. Fish Hokkaido Univ.,* 11, 58 (1960).
289. Virtanen and Matikkala, *Acta Chem. Scand.,* 13, 623 (1959).
290. Dawson, Elliott, Elliott, and Jones, *Data for Biochemical Research.* Oxford University Press, Oxford, 1969, 2.
291. Martin and Synge, *Adv. Protein Chem.,* 2, 7 (1945).
292. Dent, *Biochem. J.,* 43, 169 (1948).
293. Baumann, *Hoppe Seyler's Z. Physiol. Chem.,* 8, 299 (1884).
294. Greenstein and Winitz, *Chemistry of the Amino Acids,* Vol. 3, John Wiley & Sons, New York, 1961, 1879.
295. Bergeret and Chatagner, *Biochim. Biophys. Acta,* 14, 297 (1954).
296. Wollaston, *Ann. Chim. Phys.,* 76, 21 (1810).
297. Larsen and Kjaer, *Acta Chem. Scand.,* 16, 142 (1962).
298. Sweetman, *Nature,* 183, 744 (1959).
299. Calam and Waley, *Biochem. J.,* 86, 226 (1963).
300. Gmelin, *Hoppe Seyler's Z. Physiol. Chem.,* 327, 186 (1962).
301. Virtanen and Matikkala, *Acta Chem. Scand.,* 13, 1898 (1959).
302. van Veen and Hijman, *Rec. Trav. Chem. Pays-Bas,* 54, 493 (1935).
303. Armstrong and du Vigneaud, *J. Biol. Chem.,* 168, 373 (1947).
304. Fisher and Mallette, *J. Gen. Physiol.,* 45, 1 (1961).
305. Westall, *Biochem. J.,* 55, 244 (1953).
306. Adams and Capecchi, *Proc. Natl. Acad. Sci. USA,* 55, 147 (1966).
307. Cornforth and Henry, *J. Chem. Soc.,* p. 597 (1952).
308. Thoai and Robin, *Biochim. Biophys. Acta,* 13, 533 (1954).
309. Robertson and Marion, *Can. J. Chem.,* 37, 1043 (1959).
310. Strack, Friedel, and Hambsch, *Hoppe Seyler's Z. Physiol. Chem.,* 305, 166 (1956).
311. Gerritson, Vaughn, and Waisman, *Biochem. Biophys. Res. Commun.,* 9, 493 (1962).
312. Frimpter, *J. Biol. Chem.,* 236, PC51 (1961).
313. Huang, *Biochemistry,* 2, 296 (1963).
314. Sugii, Suketa, and Suzuchi, *Chem. Pharm. Bull.* (Tokyo), 12, 1115 (1964).
314a. Minale, Fattorusso, Cimino, DeStefano, and Nicolaus, *Gazzetta,* 97, 1636 (1967).
315. Wiehler and Marion, *Can. J. Chem.,* 36, 339 (1958).
316. Ohmori and Mizuhara, *Arch. Biochem. Biophys.,* 96, 179 (1962).
317. Ohmori, *Arch. Biochem. Biophys.,* 104, 509 (1964).
318. Horn, Jones, and Ringel, *J. Biol. Chem.,* 138, 141 (1941).
319. du Vigneaud and Brown, *J. Biol. Chem.,* 138, 151 (1941).
319a. Ampola, Bixby, Crawhall, Efron, Parker, Sneddon, and Young, *Biochem. J.,* 107, 16P (1968).
320. Thompson, *Nature,* 178, 593 (1956).
320a. Fujiwara, Itokawa, Uchino, and Inoue, *Experimenta,* 28, 254 (1972).
321. Mueller, *Proc. Soc. Exp. Biol. Med.,* 19, 161 (1922).
322. Greenstein and Winitz, *Chemistry of the Amino Acids,* Vol. 3, John Wiley & Sons, New York, 1961, 2125.
323. Matikkala and Virtanen, *Acta Chem. Scand.,* 17, 1799 (1963).
324. Guggenheim, *Die Biogenen. Amine,* S. Karger, Basel, (1951).
325. Challenger and Hayward, *Biochem. J.,* 58, 10 (1954).
326. Friis and Kjaer, *Acta Chem. Scand.,* 17, 2391 (1963).
327. Shrift and Virupaksha, *Biochim. Biophys. Acta,* 100, 65 (1965).
328. Gmelin and Virtanen, *Suomen Kemistilehti,* B35, 34 (1962); *Acta Chem. Scand.,* 16, 1378 (1962).
329. Matikkala and Virtanen, *Acta Chem. Scand.,* 16, 2461 (1962).
330. Spare and Virtanen, *Acta Chem. Scand.,* 17, 641 (1963).
331. Virtanen, Hatanaka, and Berlin, *Suomen Kemistilehti,* B35, 52 (1962).
332. Kutscher, *Zentralbl. Physiol.,* 24, 775 (1910).
333. Barger and Ewins, *Biochem.,* 7, 204, (1913).
334. Horn and Jones, *J. Biol. Chem.,* 139, 649 (1941).
334a. Tuve and Williams, *J. Biol. Chem.,* 236, 597 (1961).
335. Trelease, DiSomma, and Jacobs, *Science,* 132, 618 (1960).
336. Tanret, *C.R. Hebd. Seances Acad. Sci.* (Paris), 149, 222 (1909).
337. Waisvisz, van der Hoeven, and Rijenhuis, *J. Am. Chem. Soc.,* 79, 4524 (1957).

338. **Behre and Benedict,** *J. Biol. Chem.,* 82, 11 (1929).
339. **Greenstein and Winitz,** *Chemistry of the Amino Acids,* Vol. 3, John Wiley & Sons, New York, 1961, 2671.
340. **Heath,** *Nature,* 166, 106 (1950).
341. **Tallan, Bella, Stein, and Moore,** *J. Biol. Chem.,* 217, 703 (1955).
342. **Bettelheim,** *J. Am. Chem. Soc.,* 76, 2838 (1954).
343. **Partridge, Elsden, and Thomas,** *Nature,* 197, 1297 (1963).
344. **Morner,** *Hoppe Seyler's Z. Physiol. Chem.,* 88, 138 (1913).
344a. **Savoie, Massin, and Savoie,** *J. Clin. Invest.,* 52, 116 (1973).
345. **Gross and Pitt-Rivers,** *Biochem. J.,* 53, 645 (1950).
346. **Drechsel,** *Z. Biol.,* 33, 96 (1896).
347. **Greenstein and Winitz,** *Chemistry of the Amino Acids,* Vol. 3, John Wiley & Sons, New York, 1961, 1426.
348. **Greenstein and Winitz,** *Chemistry of the Amino Acids,* Vol. 3, John Wiley & Sons, New York, 1961, 2259.
349. **Low,** *J. Mar. Res.,* 10, 239 (1951).
349a. **Hunt and Breuer,** *Biochim. Biophys. Acta,* 252, 401 (1971).
349b. **Hunt,** *FEBS Lett.,* 24, 109 (1972).
350. **Roche, Lissitzky, and Michel,** *C.R. Hebd. Seances Acad. Sci.* (Paris), 232, 2047 (1951).
351. **Roche and Yagi,** *C. R. Soc. Biol.,* 146, 642 (1952).
352. **Kendall,** *J. Biol. Chem.,* 39, 125 (1919).
353. **Coulson,** *J. Sci. Food Agric. Abstr.,* 6, 674 (1955).
354. **Butenandt, Weidel, Weicher, and Von Derjugen,** *Hoppe Seyler's Z. Physiol. Chem.,* 279, 27 (1937).
355. **Fowden,** *Physiol. Plant,* 12, 657 (1959).
356. **Stoll and Seebeck,** *Helv. Chim. Acta,* 34, 481 (1951).
357. **Chambers and Isbell,** *J. Org. Chem.,* 29, 832 (1964).
358. **Kittredge and Hughes,** *Biochemistry,* 3, 991 (1964).
359. **Kittredge, Roberts, and Simonsen,** *Biochemistry,* 1, 624 (1962).
360. **Kittredge, Isbell, and Hughes,** *Biochemistry,* 6, 289 (1967).
360a. **Korn, Dearborn, Fales, and Sokoloski,** *J. Biol. Chem.,* 248, 2257 (1973).
361. **Thoai and Robin,** *Biochim. Biophys. Acta,* 14, 76 (1954).
362. **Beatty, Magrath, and Ennor,** *J. Am. Chem. Soc.,* 82, 4983 (1960); *J. Biol. Chem.,* 236, 1028 (1961).
363. **Weiss and Stekol,** *J. Am. Chem. Soc.,* 73, 2497 (1951).
364. **Agren,** *Acta Chem. Scand.,* 16, 1607 (1962).
365. **Lipmann,** *Biochem. Z.,* 262, 3 (1933).
366. **Greenstein and Winitz,** *Chemistry of the Amino Acids,* Vol. 3, John Wiley & Sons, New York, 1961, 2208.
367. **Tomita and Sendju,** *Hoppe Seyler's Z. Physiol. Chem.,* 169, 263 (1927).
368. **Noguchi, Sakuma, and Tamaki,** *Arch. Biochem. Biophys.,* 125, 1017 (1968).
369. **Küng and Trier,** *Hoppe Seyler's Z. Physiol. Chem.,* 85, 209 (1913).
370. **Hosein and Proulx,** *Nature,* 187, 321 (1960).
371. **Carter and Bhattacharyya,** *J. Am. Chem. Soc.,* 75, 2503 (1953).
372. **Gulewitsch and Krimberg,** *Hoppe Seyler's Z. Physiol. Chem.,* 45, 326 (1905).
373. **Friedman, McFarlane, Bhattacharyya, and Fraenkel,** *Arch. Biochem. Biophys.,* 59, 484 (1955).
374. **Carter, Bhattacharyya, Weidman, and Fraenkel,** *Arch. Biochem. Biophys.,* 38, 405 (1952).
375. **Thomas, Elsden, and Partridge,** *Nature,* 200, 651 (1963).
376. **Terlain and Thomas,** *Bull. Soc. Chim. Fr.,* p. 2349 (1971).
377. **Kodama, Ohmori, Suzuki, Mizuhara, Oura, Isshiki, and Uemura,** *Physiol. Chem. Phys.,* 3, 81 (1971).
378. **Wälti and Hope,** *J. Chem. Soc. (Org.),* p. 2326 (1971).
379. **Larsen,** *Acta Chem. Scand.,* 22, 1369 (1968).
380. **Delange, Glazer, and Smith,** *J. Biol. Chem.* 244, 1385 (1969).
381. **Nakajima and Volcani,** *Biochem. Biophys. Res. Commun.,* 39, 28 (1970).
382. **Ogawa, Fukuda, and Sasaoka,** *Biochim. Biophys. Acta,* 297, 60 (1973).
383. **Zygmunt and Martin,** *J. Med. Chem.,* 12, 953 (1969).
384. **Hagemann, Pénasse, and Teillon,** *Biochim. Biophys. Acta,* 17, 240 (1955).

α,β-UNSATURATED AMINO ACIDS

α,β-Unsaturated amino acids with free α-amino groups are not stable; *N*-acylated α,β-unsaturated amino acids are stable compounds. The α,β-unsaturated amino acids listed in this table are present in natural products in which they are stabilized by peptide bond formation. The addition of mercaptans to α,β-unsaturated amino acids and the reversible conversion to keto acids and amides are of biological significance.[1,2,11]

α,β-UNSATURATED AMINO ACIDS

No.	Amino acid/ synonym	Source	Structure	Formula (mol wt)	References: Chemistry	References: Spectral data
1	**Dehydroalanine**	Nisin Subtilin	$H_2C{=}C(NH_2){-}COOH$	$C_3H_5NO_2$ 87.08	1, 2 2	1, 2 2
2	**β-Methyldehydro-alanine**	Nisin, Subtilin Stendomycin	$H_3C{-}H_2C{=}C(NH_2){-}COOH$	$C_4H_7NO_2$ 101.10	1, 2 3	1, 2 3
3	**Dehydroserine**	Viomycin Capreomycin	$HOHC{=}C(NH_2){-}COOH$	$C_3H_5NO_3$ 103.08	4 4	4 4
4	**Dehydroleucine**	Albonoursin	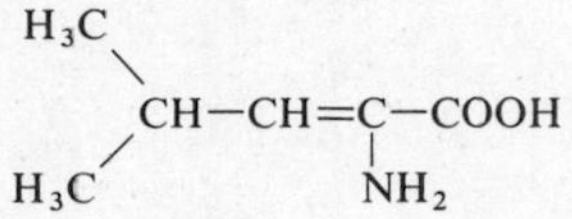 $(H_3C)_2CH{-}CH{=}C(NH_2){-}COOH$	$C_6H_{11}NO_2$ 129.16	5	
5	**Dehydrophenyl-alanine**	Albonoursin	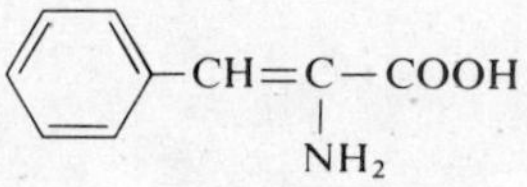 $C_6H_5{-}CH{=}C(NH_2){-}COOH$	$C_9H_9NO_2$ 163.17	5	
6	**Dehydrotryptophan**	Telomycin	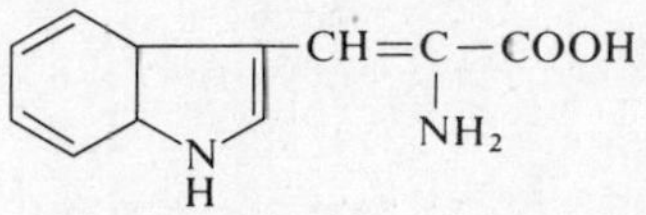 (indol-3-yl)$-CH{=}C(NH_2){-}COOH$	$C_{11}H_{10}N_2O_2$ 202.21	6	6
7	**Dehydroarginine**	Viomycin Capreomycin	$H_2N{-}C({=}NH_2){-}NH{-}CH_2{-}CH_2{-}CH{=}C(NH_2){-}COOH$	$C_6H_{12}N_4O_2$ 172.19	4 4	4 4
8	**Dehydroproline**	Ostreogrycin A	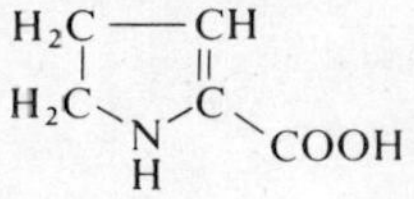 $H_2C{-}CH{=}C(COOH){-}NH{-}CH_2$ (ring)	$C_5H_7NO_2$ 113.11	7	7
9	**Dehydrovaline**	Penicillin	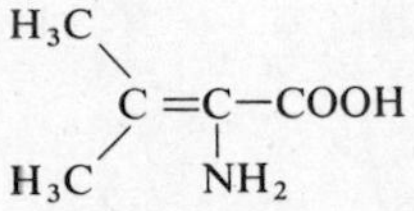 $(H_3C)_2C{=}C(NH_2){-}COOH$	$C_5H_9NO_2$ 115.13	8	
10	**Dehydrocysteine**	Micrococcin Thiostrepton	$HSHC{=}C(NH_2){-}COOH$	$C_3H_5NO_2S$	9, 10	10

Compiled by Erhard Gross.

α,β-UNSATURATED AMINO ACIDS (continued)

REFERENCES

1. **Gross and Morell,** *J. Am. Chem. Soc.,* 89, 2791 (1967).
2. **Gross and Morell,** *Fed. Eur. Biochem. Soc. Lett.,* 2, 61 (1968).
3. **Bodanszky, Izdebski, and Muramatsu,** *J. Am. Chem. Soc.,* 91, 2351 (1969).
4. **Bycroft, Cameron, Croft, Hassanali-Walji, Johnson, and Webb,** *Tetrahedron Lett.,* p. 5901 (1968).
5. **Khokhlov and Lokshin,** *Tetrahedron Lett.,* p. 1881 (1963).
6. **Sheehan, Mani, Nakamura, Stock, and Maeda,** *J. Am. Chem. Soc.,* 90, 462 (1968).
7. **Delpierre, Eastwood, Gream, Kingston, Sarin, Todd, and Williams,** *J. Chem. Soc.,* p. 1653 (1966).
8. **Abraham and Newton,** in *Antibiotics,* Gottlieb and Shaw, Eds., Springer-Verlag, New York, 1967.
9. **Brookes, Clarke, Majhofer, Mijovic, and Walker,** *J. Chem. Soc.,* p. 925 (1960).
10. **Bodanszky, Sheehan, Fried, Williams, and Birkhimer,** *J. Am. Chem. Soc.,* 82, 4747 (1960).
11. **Gross, Morell, and Craig,** *Proc. Natl. Acad. Sci. U.S.A.,* 62, 953 (1969).

This table originally appeared in Sober, Ed., *Handbook of Biochemistry and selected data for Molecular Biology,* 2nd ed., Chemical Rubber Co., Cleveland, 1970.

AMINO ACID ANTAGONISTS

Amino acid	Analog	System	Reference
α-Alanine	α-Aminoethanesulfonic acid	Bacteria	1
		Mouse tumor	2
	Glycine	Bacteria	3
	α-Aminoisobutyric acid	Bacteria	4
	Serine	Bacteria	3
D-Alanine	D-Cycloserine	Bacterial cell wall	5-9, 10
	O-Carbamyl-D-serine	Bacterial cell wall	11
	D-α-Aminobutyric acid	Bacterial cell wall	12
β-Alanine	β-Aminobutyric acid	Yeast	13
	Propionic acid	Bacteria	14
	Asparagine	Yeast	15
	D-Serine	Bacteria	16
Arginine	Canavanine	Yeast, *Neurospora*, Bacteria	17, 18-22
		Carcinosarcoma	23
		Animals	24, 25
		Plants	26, 27
		Tissue Culture	28, 29
	Lysine	Arginase	30
	Ornithine	Arginase	31
	Homoarginine	Bacteria	22, 32
Aspartic acid	Cysteic acid	Bacteria	1, 33, 34
		Bacteria	35
	β-Hydroxyaspartic acid	Bacteria	36, 34,37
	Diaminosuccinic acid	Bacteria	36
	Aspartophenone	Bacteria, yeast	4
	α-Aminolevulinic acid	Bacteria, yeast	4
	α-Methylaspartic acid	Bacteria	38
	β-Aspartic acid hydrazide	Bacteria	38
	S-Methylcysteine sulfoxide	Bacteria	39
	β-Methylaspartic acid	Bacteria	40
	Hadacidin	Purine biosynthesis	41
Asparagine	2-Amino-2-carboxyethane-sulfonamide	*Neurospora*	42
Cysteine	Allylglycine	Bacteria, yeast	43
α, ε-Diaminopimelic acid	α,α-Diaminosuberic acid	Bacteria	44
	α,α-Diaminosebacic acid	Bacteria	44
	β-Hydroxy-α,ε-diaminopimelic acid	*Escherichia coli*	45
	γ-Methyl-α,ε-diaminopimelic acid	*E. coli*	46
	Cystine	*E. coli*	47
Glutamic acid	Methionine sulfoxide	Bacteria	48, 49
		Glutamine synthesis	50
	γ-Glutamylethylamide	Bacteria	51
	β-Hydroxyglutamic acid	Bacteria	52, 53
	Methionine sulfoximine	Bacteria	54, 55
	α-Methylglutamic acid	Enzymes	56, 57
	γ-Phosphonoglutamic acid	Glutamine synthesis	58
	P-Ethyl-γ-phosphonoglutamic acid	Glutamine synthesis	58
	γ-Fluoroglutamic acid[a]	Glutamine synthesis	59
Glutamine	*S*-Carbamylcysteine	Bacteria	60
		Ascites cells	61
	O-Carbamylserine	Bacteria	62
	O-Carbazylserine	Bacteria	63
	3-Amino-3-carboxypropane-sulfonamide	*E. coli*	64
	N-Benzylglutamine	*Streptococcus lactis*	65
	Azaserine	Enzymes	66
	6-Diazo-5-oxonorleucine	Enzymes	66
	γ-Glutamylhydrazide	*S. faecalis*	67
Glycine	α-Aminomethanesulfonic acid	Bacteriophage	68
		Vaccinia virus	69
		Bacteria	1
		E. coli	4

[a]Some of the observed inhibition may be due to fluoride ion present in the amino acid preparation or formed during incubation.[208]

AMINO ACID ANTAGONISTS (continued)

Amino acid	Analog	System	Reference
Histidine	D-Histidine	Histidase	70
	Imidazole		70
	2-Thiazolealanine	*E. coli*	71
	1,2,4-Triazolealanine	*E. coli*	71, 72
		Salmonella	73
Isoleucine	Leucine	Bacteria	74
		Rats	75
	Methallylglycine	Bacteria, yeast	4, 43
	ω-Dehydroisoleucine	Bacteria	76
	3-Cyclopentene-1-glycine	Bacteria	77
	Cyclopentene glycine	Bacteria	78
	2-Cyclopentene-1-glycine	Bacteria	79
	O-Methylthreonine	Tumor cells	80
	β-Hydroxyleucine	Bacteria	37
Leucine	D-Leucine	Bacteria	81
	α-Aminoisoamylsulfonic acid	Bacteria	1, 33
		Mouse tumor	2
	Norvaline	Bacteria	4
	Norleucine	Bacteria	1, 4, 82
	Methallylglycine	Yeast, bacteria	4
	α-Amino-β-chlorobutyric acid	Yeast, bacteria	4
	Valine	Bacteria	83
	δ-Chloroleucine	*Neurospora*	84
	Isoleucine	Bacteria	85
	β-Hydroxynorleucine	Bacteria	86
	β-Hydroxyleucine	Bacteria	86
	Cyclopentene alanine	Bacteria	87
	3-Cyclopentene-1-alanine	Bacteria	88
	2-Amino-4-methylhexenoic acid	Bacteria	89
	5′,5′,5′-Trifluoroleucine	*E. coli*	90
	4-Azaleucine	*E. coli*	91
Lysine	α-Amino-ε-hydroxycaproic acid	Rat	92
	Arginine	*Neurospora*	93
	2,6-Diaminoheptanoic acid	Bacteria	94
	Oxalysine	Bacteria	95
	3-Aminomethylcyclohexane glycine	Bacteria	96
	3-Aminocyclohexane alanine	Bacteria	97
	trans-4-Dehydrolysine	Bacteria	98
	S-(β-Aminoethyl)-cysteine	Bacteria	99
	4-Azalysine	Bacteria	
Methionine	2-Amino-5-heptenoic acid (crotylalanine)	*E. coli*	100
	2-Amino-4-hexenoic acid (crotylglycine)	*E. coli*	101
	Methoxinine	Bacteria	102
		Vaccinia virus	69
		Rats	103
	Norleucine	Bacteria	82, 104, 105
		Animal tissues	106
		Casein	107
	Ethionine	Bacteria, animals	28, 102, 105, 108-120
		Amylase	121, 122
		Yeast	123
		Tumors	124
		Pancreatic proteins	125
	Methionine sulfoximine	Bacteria	126
	Threonine	*Neurospora*	127
	Selenomethionine	*Chlorella*	128
		E. coli, yeast	129-132
Ornithine	α-Amino-δ-hydroxyvaleric acid	Bacteria	21
	Canaline	Bacteria	21

AMINO ACID ANTAGONISTS (continued)

Amino acid	Analog	System	Reference
Phenylalanine	α-Amino-β-phenylethanesulfonic acid	Mouse tumor	2
	Tyrosine	Bacteria	133
	β-Phenylserine	Bacteria	134, 135, 136
	Cyclohexylalanine	Rats	137
	o-Aminophenylalanine	*E. coli*	138
	p-Aminophenylalanine	Bacteria	139, 140
	Fluorophenylalanines	Fungi, bacteria	141-146, 140, 147, 148
		Lysozyme, albumin	149
		Muscle enzymes	150
		Amylase	151
		Hemoglobin	152
		Rats	153
Phenylalanine— (Continued)	Chlorophenylalanines	Fungi	147
	Bromophenylalanines	Fungi	147
	β-2-Thienylalanine	Rat, bacteria, yeast	136, 154-164
		β-Galactosidase	165
	β-3-Thienylalanine	Bacteria, yeast	166
	β-2-Furylalanine	Bacteria, yeast	4
	β-3-Furylalanine	Bacteria, yeast	4
	β-2-Pyrrolealanine	Bacteria, yeast	167
	1-Cyclopentene-1-alanine	Bacteria	87
	1-Cyclohexene-1-alanine	Bacteria	87
	2-Amino-4-methyl-4-hexenoic acid	Bacteria	89
	S-(1,2-Dichlorovinyl)-cysteine	*E. coli*	168
	β-4-Pyridylalanine	Bacteria	169
	Tryptophan	Bacteria	170
	β-2-Pyridylalanine	Bacteria	171
	β-4-Pyrazolealanine	Bacteria	171
	β-4-Thiazolealanine	Bacteria	171
	p-Nitrophenylalanine	Bacteria	140
Proline	Hydroxyproline	Fungi	172
	3,4-Dehydroproline	Bacteria, beans	173, 174
	Azetidine-2-carboxylic acid	Bacteria, beans	175, 176
		Actinomycin	177, 178
Serine	α-Methylserine	Bacteria	4
	Homoserine	Bacteria	4
	Threonine	Bacteria	179, 180
	Isoserine	Enzymes	181
Threonine	Serine	Bacteria	179, 180, 182
	β-Hydroxynorvaline	Bacteria	86, 183
	β-Hydroxynorleucine	Bacteria	86, 183
Thyroxine	Ethers of 3,5-diiodotyrosine	Tadpoles	184
Tryptophan	Methyltryptophans	Bacteria	141, 185, 186-188
		Bacteriophage	189
	Naphthylalanines	Bacteria	190, 191
		Rat	4
	Indoleacrylic acid	Bacteria	192
	Naphthylacrylic acid	Bacteria	193
	β-(2-Benzothienyl)alanine	Bacteria	194
	Styrylacetic acid	Bacteria	193
	Indole	Bacteriophage	195
	α-Amino-β-3(indazole)-propionic acid (Tryptazan)	Yeast	196
		Enzyme	197
		E. coli	198
	5-Fluorotryptophan	Enzyme	199, 200
	6-Fluorotryptophan	Enzyme	201
	7-Azatryptophan	*E. coli*	202-204, 205

AMINO ACID ANTAGONISTS (continued)

Amino acid	Analog	System	Reference
Tyrosine	Fluorotyrosines	Fungi	147, 206
		Rat	206
	p-Aminophenylalanine	Fungi	139
	m-Nitrotyrosine	Bacteria	140
	β-(5-Hydroxy-2-pyridyl)-alanine	Bacteria	207
Valine	α-Aminoisobutanesulfonic acid	Bacteria	1, 2, 33
		Vaccinia	69
	α-Aminobutyric acid	Bacteria	182, 4
	Norvaline	Bacteria	4
	Leucine, isoleucine	Bacteria	182, 83
	Methallylglycine	Bacteria, yeast	4
	β-Hydroxyvaline	Bacteria	86, 183
	ω-Dehydroalloisoleucine	Bacteria	76

From Meister, *Biochemistry of the Amino Acids,* 2nd ed., I, Academic Press, New York, 1965, 233. With permission of copyright owners.

PROPERTIES OF THE α-KETO ACID ANALOGS OF AMINO ACIDS

α-Keto acid	α-Amino acid analog	2,4-Dinitrophenylhydrazone: Crystallization M.p. (°C)	Crystallization Solvent[a]	Amino acids after hydrogenation (14)	Reduction by lactic dehydrogenase[b,g]	Decarboxylation by yeast decarboxylase[c]
Pyruvic	Alanine	216	h	Alanine	26,800	1,200
α-Ketoadipamic	α-Aminoadipamic acid (homoglutamine)					
α-Ketoadipic	α-Aminoadipic acid	208	h	α-Aminoadipic acid		
α-Ketobutyric	α-Aminobutyric acid	198	h	α-Aminobutyric acid	21,000	
α-Ketoheptylic	α-Aminoheptylic acid	130	e, l	α-Aminoheptylic acid	483	
α-Keto-ε-hydroxycaproic	α-Amino-ε-hydroxy-caproic acid	183	h	α-Amino-ε-hydroxy-caproic acid	181	25
Mesoxalic	α-Aminomalonic acid	205	hc	α-Aminomalonic acid, glycine		
α-Ketophenylacetic	α-Aminophenyl-acetic acid	193	h	α-Aminophenyl-acetic acid, cyclohexyl-glycine	2.6	0
DL-Oxalosuccinic	α-Aminotri-carballylic acid					
α-Keto-δ-guanidinovaleric	Arginine	216, 267 (1)		Arginine	9.0	0
α-Ketosuccinamic	Asparagine	183		Asparagine, aspartic acid	8,930	
Oxalacetic	Aspartic acid	218	h	Aspartic acid, alanine, β-alanine	12.8	
α-Keto-δ-carbamidovaleric	Citrulline	190	h	Citrulline	4.3	0
β-Cyclohexylpyruvic	β-Cyclohexyl-alanine	189	h	β-Cyclohexyl-alanine	14.6	0
β-Sulfopyruvic	Cysteic acid	210	a	Cysteic acid, alanine	89.6	
β-Mercaptopyruvic	Cysteine	195–200 (2) 161–162 (3)		Alanine	27,000	750
α-Keto-γ-ethiolbutyric	Ethionine	131	h	Ethionine	1,650	121
α-Ketoglutaric	Glutamic acid	220	h	Glutamic acid	9.2	0
α-Ketoglutaric γ-ethyl ester	Glutamic acid γ-ethyl ester				49.0	721
α-Ketoglutaramic	Glutamine			Glutamine, glutamic acid		
Glyoxylic	Glycine	203	h	Glycine	21,100	0
β-Imidazolylpyruvic	Histidine	190–192, 240 (1)	hc, e, l	Histidine		
α-Keto-γ-hydroxybutyric	Homoserine					
DL-α-Keto-β-methylvaleric	DL-Isoleucine (or DL-alloisoleucine)	169	h	Isoleucine		
L-α-Keto-β-methylvaleric	L-Isoleucine (or D-alloisoleucine)	176	h	Isoleucine	5.0	1,000
D-α-Keto-β-methylvaleric	D-Isoleucine (or L-alloisoleucine)	176	h	Isoleucine	1.9	280

[a]h = water; e = ethyl acetate; l = ligroin; hc = hydrochloric acid; ac = glacial acetic acid; a = ethanol.

[b]Mole × 10^{-8} of DPNH oxidized per mg of enzyme per minute at 26° (9, 10).

[c]μl. CO_2 per hour (10).

[e]Originally designated *d* (11). Originally designated *l* (11).

[g]Additional data have been published on the reduction of α-keto acids by lactic dehydrogenase (12).

acids by lactic dehydrogenase (12).

PROPERTIES OF THE α-KETO ACID ANALOGS OF AMINO ACIDS (continued)

α-Keto acid	α-Amino acid analog	2,4-Dinitrophenylhydrazone: Crystallization: M.p. (°C)	Solvent[a]	Amino acids after hydrogenation (14)	Reduction by lactic dehydrogenase[b,g]	Decarboxylation by yeast decarboxylase[c]
α-Ketoisocaproic	Leucine	162	h	Leucine	3.2	306
Trimethylpyruvic	*tert*-Leucine	180	h	*tert*-Leucine		
α-Keto-ε-aminocaproic	Lysine	212	h	Lysine, pipecolic acid		
α-Keto-γ-methiolbutyric	Methionine	150	h	Methionine	1,550	125
α-Keto-γ-methylsulfonylbutyric	Methionine sulfone	175	h	Methione sulfone		
α-Keto-δ-nitroguanidinovaleric	Nitroarginine	225	ac	Nitroarginine, arginine	42.6	0
α-Ketocaproic	Norleucine	153	h	Norleucine	560	
α-Ketovaleric	Norvaline	167	h	Norvaline	1,470	
α-Keto-δ-aminovaleric	Ornithine, proline	232–242 (4) 219 (5) 211–212 (6)		Ornithine, proline, pentahomoserine[d]		
Phenylpyruvic	Phenylalanine	162–164, 192–194 (7)		Phenylalanine	755	0
***S*-Benzyl-β-mercaptopyruvic**	*S*-Benzylcysteine	150	a			
β-Hydroxypyruvic	Serine	162	e	Serine, alanine	26,000	0[h]
***N*-Succinyl-α-amino-ε-ketopimelic** (15)	*N*-Succinyl-α,ε-diaminopimelic acid	137–143	h	*N*-Succinyl-α,ε-diaminopimelic acid		
DL-α-Keto-β-hydroxybutyric	DL-Threonine (or DL-allothreonine)	157–158 (8)		Threonine, α-aminobutyric acid	20,000	
β-[3,5-Diiodo-4-(3′,5′-diiodo 4′-hydroxyphenoxy)phenyl]	Thyroxine					
β-Indolylpyruvic	Tryptophan	169 (1)		Tryptophan	670	0
***p*-Hydroxyphenylpyruvic**	Tyrosine	178	h	Tyrosine	345	0
α-Ketoisovaleric	Valine	196	h	Valine	103	922

[d]α-Amino-δ-hydroxy-*n*-valeric acid.
[h]This keto acid has been reported to be decarboxylated by yeast preparations; the reaction is much more rapid at pH 6.5 than at 5 (13).

From Meister (1965) *Biochemistry of the Amino Acids,* 2nd ed. Academic, New York, pp 162-4. Permission of copyright owners.

REFERENCES

1. **Stumpf and Green,** *J. Biol. Chem.,* 153, 387 (1944).
2. **Schneider and Reinefeld,** *Biochem. Z.,* 318, 507 (1948).
3. **Meister, Fraser, and Tice,** *J. Biol. Chem.,* 206, 561 (1954).
4. **Krebs,** *Enzymologia,* 7, 53 (1939).
5. **Blanchard, Green, Nocito, and Ratner,** *J. Biol. Chem.,* 155, 421 (1944).
6. **Meister,** *J. Biol. Chem.,* 206, 579 (1954).
7. **Fones,** *J. Org. Chem.,* 17, 1534 (1952).
8. **Sprinson and Chargaff,** *J. Biol. Chem.,* 164, 417 (1947).
9. **Meister,** *J. Biol. Chem.,* 197, 309 (1953).
10. **Meister,** *J. Biol. Chem.,* 184, 117 (1950).
11. **Meister,** *J. Biol. Chem.,* 190, 269 (1951).
12. **Czok and Büchler,** *Advanc. Protein. Chem.,* 15, 315 (1960).
13. **Dickens and Williamson,** *Nature,* 178, 1349 (1956).
14. **Meister and Abendschein,** *Anal. Chem.,* 28, 171 (1956).
15. **Gilvarg,** *J. Biol. Chem.,* 236, 1429 (1961).

FAR ULTRAVIOLET ABSORPTION SPECTRA OF AMINO ACIDS

The absorption of a chromophoric amino acid in this spectral region is due to the combined absorptions of the side chain chromophore and of the carboxylate group. Because the carboxylate group is consumed in polymerizing amino acids to polypeptides, an amino acid residue absorbs less intensely than a free amino acid. The magnitude of this difference can be estimated from the spectra of the nonchromophoric amino acids – leucine, proline, alanine, serine and threonine – whose total absorption is due only to the carboxylate group. The variations between the absorptions of these amino acids reflect the variability of carboxylate absorption in slightly different environments.

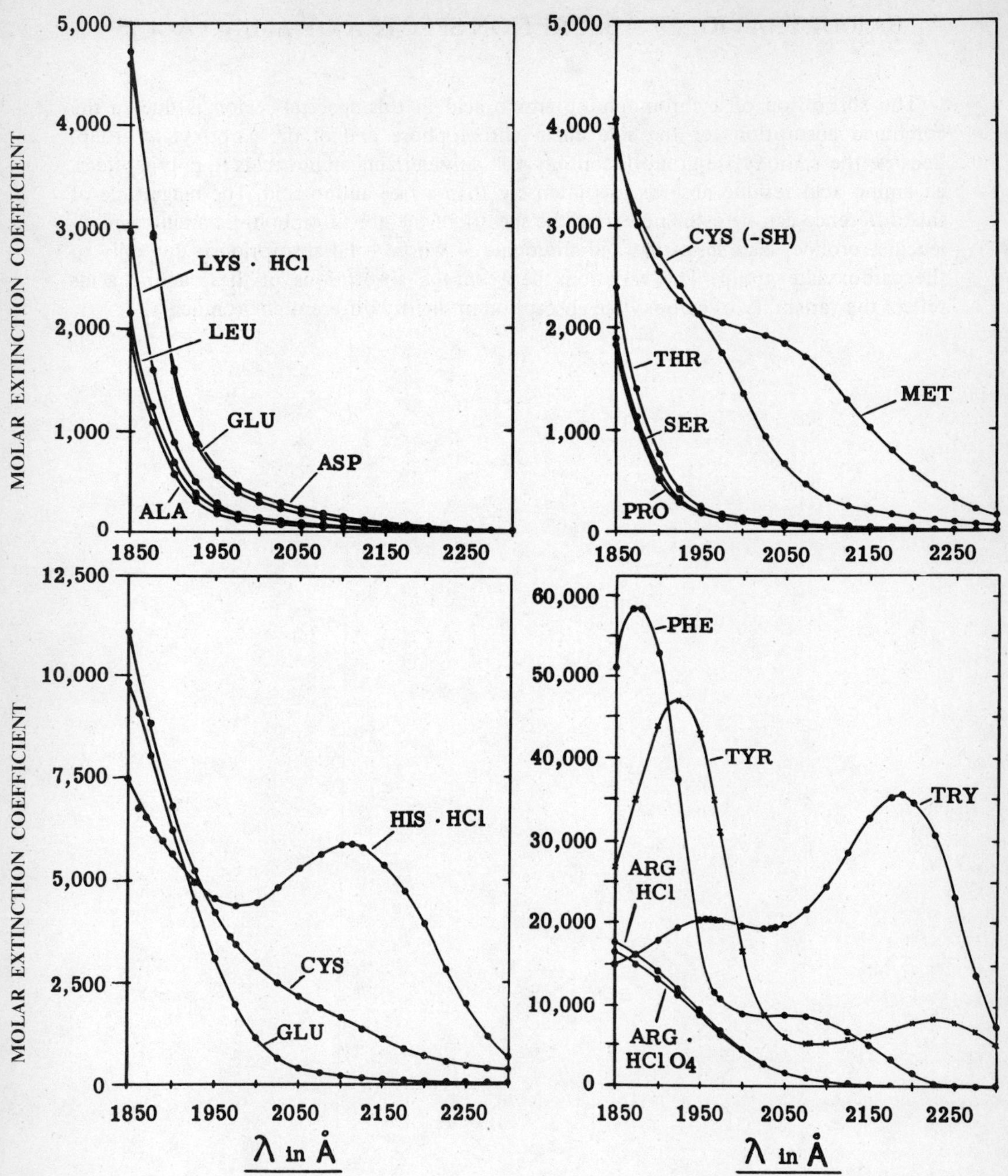

Far ultraviolet spectra of amino acids. All amino acids were in aqueous solution at pH 5, except cystine (pH 3). The dibasic amino acids were measured as hydrochlorides and the absorbance corrected by subtracting the absorbance contribution of chloride ion. Taken from Wetlaufer, *Advanc. Protein Chem.*, 17, 320 (1962). With the permission of the copyright owners. Academic Press, New York.

FAR ULTRAVIOLET ABSORPTION SPECTRA OF AMINO ACIDS

Amino acid	Feature[a]								
	$\lambda_{190.0}$	$\lambda_{197.0}$	$\lambda_{205.0}$	Maxima		Minima		Shoulder	
				λ	ε	λ	ε	λ	ε
In Neutral Water									
Tryptophan	17.60	20.50	19.60	196.7	20.60	203.3	19.40	—	—
	—	—	—	218.6	46.70	—	—	—	—
Tyrosine	42.80	35.50	5.60	192.5	47.50	208.0	4.88	—	—
	—	—	—	223.2	8.26	—	—	—	—
Phenylalanine	54.50	12.30	9.36	187.7	59.60	202.5	8.96	—	—
	—	—	—	206.0	9.34	—	—	—	—
Histidine[b]	5.57	4.35	5.17	211.3	5.86	198.4	4.22	—	—
Cysteine[c]	2.82	1.94	0.730	—	—	—	—	195.2	2.18
1/2 Cystine	3.25	1.76	1.05	—	—	—	—	207.0	0.96
Methionine	2.69	2.11	1.86	—	—	—	—	204.7	1.89
Arginine[c]	13.1	6.61	1.36	—	—	—	—	—	—
Acids[d]	1.61	0.460	0.230	—	—	—	—	—	—
Amides[d]	6.38	2.06	0.400	—	—	—	—	—	—
Lysine[b]	0.890	0.200	0.110	—	—	—	—	—	—
Leucine[d]	0.670	0.190	0.100	—	—	—	—	—	—
Alanine[d]	0.570	0.150	0.070	—	—	—	—	—	—
Proline[d]	0.540	0.150	0.070	—	—	—	—	—	—
Serine[d]	0.610	0.160	0.080	—	—	—	—	—	—
Threonine[d]	0.750	0.180	0.100	—	—	—	—	—	—
In 0.1 *M* Sodium Dodecyl Sulfate[d]									
Tryptophan	16.70	19.70	19.00	197.3	19.80	204.1	18.90	—	—
	—	—	—	220.0	46.60	—	—	—	—
Tyrosine	39.10	36.60	5.87	193.2	45.10	208.5	4.88	—	—
	—	—	—	223.7	7.86	—	—	—	—
Phenylalanine	54.10	11.30	8.45	188.3	57.0	201.5	7.79	—	—
	—	—	—	207.2	8.66	—	—	—	—
Histidine[b]	5.88	4.48	5.03	212.0	5.88	199.0	4.25	—	—
Cysteine[c]	2.66	1.79	0.650	—	—	—	—	194.5	2.15
1/2 Cystine	2.62	1.61	0.92	—	—	—	—	—	—
Methionine	2.67	2.10	1.84	—	—	—	—	204.1	1.89
Arginine[c]	12.50	5.70	0.94	—	—	—	—	—	—

Compiled by Ruth McDiarmid.

[a]Molar extinctions, $\epsilon \times 10^{-3}$. Wavelength, λ, in millimicrons.
[b]The absorptions of lysine and histidine were determined for the hydrochlorides and corrected for the absorption of the chloride ion. $\epsilon_{Cl^-} = 0.740$ ($\lambda = 190.0$), 0.050 ($\lambda = 197.0$) and 0(λ = 205.0).
[c]The absorptions of cysteine and arginine were determined for the $HClO_4$ salt.
[d]The spectra of the carboxylic acids, the amides, the aliphatic and the hydroxy amino acids and lysine are unchanged from those obtained in neutral water.

This figure and table originally appeared in Sober, Ed., *Handbook of Biochemistry and selected data for Molecular Biology*, 2nd ed., Chemical Rubber Co., Cleveland, 1970.

UV ABSORPTION CHARACTERISTICS OF *N*-ACETYL METHYL ESTERS OF THE AROMATIC AMINO ACIDS, CYSTINE AND OF *N*-ACETYLCYSTEINE

	Water[a]		Ethanol[a]	
	λ	ε	λ	ε
Phenylalanine				
Inflection	(208)	10.20	(208)	10.40
Inflection	(217)	5.00	(217)	5.30
Minimum	240	0.080	244	0.088
Maximum	241.2	0.086	242.0	0.093
Maximum	246.5	0.115	247.3	0.114
Maximum	251.5	0.157	252.3	0.158
Maximum	**257.4**	**0.197**	**258.3**	**0.195**
Inflection	(260.7)	—	(261.2)	—
Maximum	263.4	0.151	264.2	0.155
Maximum	267.1	0.091	267.8	0.096
Tyrosine				
Maximum	193	51.70	—	—
Minimum	212	7.00	212	6.20
Maximum	224	8.80	227	10.20
Minimum	247	0.176	246	0.174
Maximum	**274.6**	**1.420**	**278.4**	**1.790**
Inflection	281.9	—	285.7	—

	Water[a]		Ethanol[a]	
	λ	ε	λ	ε
Tryptophan				
Minimum	205	21.40	206	21.30
Maximum	219	35.00	221	37.20
Minimum	245	1.900	245	1.560
Maximum	**279.8**	**5.600**	**282.0**	**6.170**
Maximum	288.5	4.750	290.6	5.330
Cystine				
Inflection	(250)	0.360	(253)	0.372
Inflection	260	0.280	260	0.320
Inflection	280	0.110	280	0.135
Inflection	300	0.025	300	0.035
Inflection	320	0.006	320	0.007
***N*-Acetylcysteine**				
	250	0.015	250	0.020
	280	0.005	280	0.005
	320	(nil)	320	(nil)

[a] λ, wavelength in millimicrons; $\epsilon \times 10^{-3}$, molar extinctions. Inflection denotes unresolved inflection.

Compiled by W. B. Gratzer.

From J. E. Bailey, Ph.D. Thesis, London University (1966).

NUMERICAL VALUES OF THE ABSORBANCES OF THE AROMATIC AMINO ACIDS IN ACID, NEUTRAL AND ALKALINE SOLUTIONS

Table 1
MOLECULAR ABSORBANCES OF TYROSINE

nm[c]	Neutral[a]	Alkaline[b]	nm[c]	Neutral[a]	Alkaline[b]
230	4980	7752 ± 108	276	1367 ± 0	1206 ± 4
232	3449	8667 ± 38	278	1260 ± 2	1344 ± 5
234	1833 ± 14	9634 ± 19	280	1197 ± 0	1507 ± 5
236	1014 ± 43	10440 ± 20	282	1112 ± 2	1675 ± 5
238	571 ± 36	11000 ± 10	284	845 ± 8	1850 ± 6
240	349 ± 34	11300 ± 20	286	506 ± 7	2024 ± 4
240.5 ↑	—	11340 ± 30	288	248 ± 8	2179 ± 5
242	252 ± 20	11230 ± 40	290	113 ± 0	2300 ± 7
244	209 ± 18	10760 ± 50	292	50 ± 1	2367 ± 5
245.3 ↓	202 ± 20	—	293.2 ↑	—	2381 ± 6
246	205 ± 17	9918 ± 78	294	23 ± 1	2377 ± 8
248	218 ± 15	8734 ± 72	296	13 ± 0	2317 ± 10
250	246 ± 14	7382 ± 56	298	8 ± 1	2195 ± 16
252	287 ± 13	5844 ± 77	300	6 ± 0	2006 ± 23
254	341 ± 14	4471 ± 55	302	5 ± 1	1747 ± 29
256	401 ± 12	3360 ± 46	304	3 ± 0	1445 ± 27
258	485 ± 10	2476 ± 20	306	2 ± 1	1107 ± 35
260	582 ± 9	1883 ± 17	308	1 ± 0	800 ± 27
262	693 ± 13	1467 ± 7	310	1 ± 0	547 ± 21
264	821 ± 13	1204 ± 14	312	—	346 ± 15
266	960 ± 14	1054 ± 16	314	—	206 ± 12
268	1083 ± 13	985 ± 13	316	—	118 ± 9
269.3 ↓	—	974 ± 8	318	—	67 ± 5
270	1197 ± 9	979 ± 9	320	—	32 ± 3
272	1310 ± 9	1019 ± 8	322	—	15 ± 3
274	1394 ± 6	1094 ± 8	324	—	6 ± 2
274.8 ↑	1405 ± 7	—	326	—	1 ± 1

Compiled by Elmer Mihalyi.

[a]0.1 *M* phosphate buffer, pH 7.1.
[b]0.1 *N* KOH.
[c]Maxima, minima, and inflection points are indicated by ↑, ↓, and ~.

From *J. Chem. Eng. Data,* 13, 179 (1969). With permission of the author and copyright owner.

Table 2
MOLECULAR ABSORBANCES OF TRYPTOPHAN

nm[c]	Neutral[a]	Alkaline[b]	nm[c]	Neutral[a]	Alkaline[b]
230	6818	13200	279.0 ↑	5579 ± 14	—
232	4037 ± 60	7470	280	5559 ± 12	5377 ± 43
234	2772 ± 71	4354 ± 81	280.4 ↑	—	5385 ± 34
236	2184 ± 64	2951 ± 50	282	5323 ± 10	5302 ± 34
238	1904 ± 55	2282 ± 29	284	4762 ± 11	4962 ± 42
240	1764 ± 52	1959 ± 30	285.8 ↓	4471 ± 6	—
242.0 ↓	1737 ± 49	1813 ± 25	286	4482 ± 11	4596 ± 22
244	1772 ± 48	1773 ± 29	286.8 ↓	—	4565 ± 27
244.4 ↓	—	1763 ± 29	287.8 ↑	4650 ± 12	—
246	1869 ± 40	1792 ± 27	288	4646 ± 16	4634 ± 19
248	2018 ± 35	1877 ± 23	288.3 ↑	—	4639 ± 28
250	2217 ± 32	2013 ± 25	290	3935 ± 5	4393 ± 32
252	2462 ± 19	2187 ± 37	292	2732 ± 5	3551 ± 46
254	2760 ± 27	2410 ± 38	294	1824 ± 5	2666 ± 27
256	3087 ± 20	2664 ± 25	296	1211 ± 10	1990 ± 24
258	3422 ± 18	2953 ± 39	298	797 ± 4	1472 ± 19
260	3787 ± 17	3261 ± 34	300	510 ± 1	1064 ± 19
262	4142 ± 14	3586 ± 46	302	314 ± 3	755 ± 16
264	4472 ± 10	3895 ± 32	304	184 ± 2	517 ± 10
266	4777 ± 14	4212 ± 48	306	112 ± 4	333 ± 6
268	5020 ± 15	4481 ± 46	308	55 ± 9	217 ± 4
270	5220 ± 8	4742 ± 37	310	27 ± 11	129 ± 5
272	5331 ± 5	4933 ± 45	312	11 ± 8	84 ± 8
272.1 ↑	5344 ± 5	—	314	3 ± 2	53 ± 7
273.6 ↓	5329 ± 10	—	316	—	31 ± 7
274	5341 ± 8	5025 ± 34	318	—	17 ± 4
274.5 ~	—	5062 ± 38	320	—	8 ± 2
276	5431 ± 8	5108 ± 39	322	—	3 ± 4
278	5554 ± 12	5275 ± 46			

Compiled by Elmer Mihalyi.

[a]0.1 *M* phosphate buffer, pH 7.1.
[b]0.1 *N* KOH.
[c]Maxima, minima, and inflection points are indicated by ↑, ↓, and ~.

From *J. Chem. Eng. Data,* 13, 179 (1969). With permission of the author and copyright owner.

Table 3
MOLECULAR ABSORBANCES OF PHENYLALANINE

nm[c]	Neutral[a]	nm[c]	Alkaline[b]
230	32.8 ± 1.5	230	161.9 ± 1.9
232	32.1 ± 1.6	232	99.2 ± 1.9
234	35.6 ± 2.1	234	70.7 ± 2.4
236	42.8 ± 2.1	236	63.3 ± 2.7
238	48.5 ± 2.3	238	62.3 ± 2.6
240	59.4 ± 2.0	240	68.9 ± 3.2
242 ~	72.2 ± 2.3	242	83.0 ± 2.8
		243 ~	85.4 ± 2.9
244	80.1 ± 2.1	244	89.0 ± 3.0
246	102.0 ± 0.6	246	108.9 ± 2.8
247.4 ↑	110.7 ± 2.2	247	120.9 ± 1.5
248	109.8 ± 1.9	248.0 ↑	126.1 ± 1.4
248.3 ↓	109.5 ± 2.0	248.7 ↓	125.1 ± 1.7
250	123.5 ± 2.6	250	132.7 ± 1.8
251	143.0 ± 2.8	251	149.3 ± 1.9
252	153.9 ± 1.0	252	167.0 ± 1.1
252.2 ↑	154.1 ± 1.0	252.9 ↑	171.5 ± 1.3
254	139.6 ± 1.0	254	166.3 ± 0.8
254.5 ↓	138.5 ± 1.4	254.9 ↓	162.8 ± 1.7
256	156.5 ± 2.2	256	168.4 ± 1.9
257.6 ↑	195.1 ± 1.5	257	188.4 ± 2.8
258	193.4 ± 1.3	258	209.1 ± 0.3
259	171.9 ± 1.0	258.2 ↑	209.6 ± 0.2
260	147.0 ± 0.6	260	184.2 ± 1.0
261.9 ↓	127.7 ± 1.5	260.7 ~	178.6 ± 0.3
262	128.1 ± 1.4	262	157.8 ± 0.9
263.7 ↑	151.5 ± 0.6	262.7 ↓	105.5 ± 1.3
264	148.7 ± 0.4	263.9 ↑	161.2 ± 1.0
265	119.8 ± 1.3	264	160.0 ± 2.1
266	91.8 ± 1.4	266	114.3 ± 1.6
266.8 ~	85.6 ± 1.5	266.5 ↓	109.7 ± 1.8
		267.7 ↑	117.7 ± 1.8
268	74.7 ± 1.0	268	115.0 ± 1.0
270	30.0 ± 1.8	270	50.2 ± 2.0
272	14.3 ± 1.0	272	18.7 ± 1.1
274	5.4 ± 0.3	274	7.4 ± 0.3
276	2.2 ± 0.4	276	2.6 ± 0.4
278	1.1 ± 0.5	278	0.7 ± 0.3
280	0.7 ± 0.3	280	0.4 ± 0.2

Compiled by Elmer Mihalyi.

[a] 0.1 *M* phosphate buffer, pH 7.1.
[b] 0.1 *N* KOH.
[c] Maxima, minima, and inflection points are indicated by ↑, ↓, and ~.

From *J. Chem. Eng. Data,* 13, 179 (1969). With permission of the author and copyright owner.

Table 4
ALKALINE[b] VS. NEUTRAL[a] DIFFERENCE SPECTRA OF TYROSINE, TRYPTOPHAN, AND PHENYLALANINE

nm	Tyrosine	Tryptophan	Phenylalanine	nm	Tyrosine	Tryptophan	Phenylalanine
230	3041	4135	123.9	280	315 ± 1	−191 ± 18	—
232	5440	3213	66.0	282	558 ± 1	− 24 ± 4	—
234	7608	1621	35.4	284	994 ± 10	194 ± 3	—
236	9415	732 ± 35	20.7	286	1513 ± 11	110 ± 10	—
238	10490	345 ± 23	13.9	288	1936 ± 1	11 ± 3	—
240	11060	149 ± 21	9.9	290	2196 ± 14	467 ± 8	—
242	11090	45 ± 15	11.1	292	2331 ± 15	802 ± 3	—
244	10660	− 40 ± 15	8.8	294	2357 ± 7	830 ± 8	—
246	9844	−104 ± 11	9.0	296	2307 ± 6	755 ± 10	—
248	8567	−172 ± 10	16.4	298	2194 ± 7	652 ± 13	—
250	7205	−233 ± 9	8.3	300	2002 ± 3	527 ± 5	—
252	5671	−298 ± 16	15.3	302	1754 ± 2	413 ± 11	—
254	4344	−371 ± 10	25.8	304	1437 ± 2	300 ± 14	—
256	3127	−435 ± 14	11.3	306	1097 ± 9	205 ± 8	—
258	2142	−490 ± 19	19.1	308	792 ± 14	137 ± 5	—
260	1368 ± 37	−535 ± 13	37.8	310	526 ± 13	88 ± 3	—
262	820 ± 36	−564 ± 8	26.9	312	334 ± 9	55 ± 8	—
264	420 ± 25	−580 ± 10	16.1	314	221 ± 19	22 ± 14	—
266	125 ± 20	−573 ± 12	22.4	316	101 ± 7	16 ± 10	—
268	− 78 ± 18	−539 ± 14	42.8	318	62 ± 2	5 ± 2	—
270	−225	−486 ± 12	17.4	320	28 ± 4	0	—
272	−296	−394 ± 12	4.0	322	12 ± 5	—	—
274	−299	−308 ± 7	1.7	324	3 ± 4	—	—
276	−158	−312 ± 16	0.3	326	1 ± 1	—	—
278	89 ± 5	−278 ± 13	0	328	0	—	—

Compiled by Elmer Mihalyi.

[a]0.1 *M* phosphate buffer, pH 7.1.
[b]0.1 *N* KOH.

From *J. Chem. Eng. Data,* 13, 179 (1969). With permission of the author and copyright owner.

Table 5
ACID[b] VS. NEUTRAL[a] DIFFERENCE SPECTRA OF TYROSINE, TRYPTOPHAN, AND PHENYLALANINE

nm	Tyrosine	Tryptophan	Phenylalanine	nm	Tyrosine	Tryptophan	Phenylalanine
230	—	—	46.7	276	−40	128 ± 16	0
232	576	421 ± 49	34.1	278	−45	110 ± 10	—
234	441	610 ± 37	23.3	280	−34	71 ± 11	—
236	346	590 ± 31	16.0	282	−46	− 23 ± 7	—
238	218	512 ± 29	10.8	284	−73	− 92 ± 9	—
240	108	432 ± 23	6.9	286	−71	− 3 ± 9	—
242	40	358 ± 19	3.9	288	−49	− 26 ± 4	—
244	4	305 ± 20	3.3	290	−31	−250 ± 5	—
246	−13	263 ± 21	1.2	292	−20	−317 ± 9	—
248	−18	240 ± 16	−0.3	294	−16	−276 ± 9	—
250	−20	223 ± 14	2.5	296	−14	−227 ± 5	—
252	−16	216 ± 17	−1.8	298	−12	−177 ± 7	—
254	−12	219 ± 15	−1.9	300	−10	−131 ± 9	—
256	− 7	222 ± 11	4.0	302	− 9	− 88 ± 6	—
258	− 5	223 ± 14	−3.3	304	− 7	− 59 ± 8	—
260	− 3	225 ± 14	−4.3	306	− 6	− 40 ± 10	—
262	0	232 ± 13	2.5	308	− 4	− 24 ± 10	—
264	0	232 ± 18	−3.1	310	− 3	− 13 ± 10	—
266	− 4	224 ± 15	−3.4	312	− 2	− 7 ± 7	—
268	− 8	214 ± 7	−4.4	314	− 1	− 4 ± 4	—
270	−11	190 ± 9	−2.0	316	− 1	0	—
272	−13	159 ± 12	−0.7	318	0	—	—
274	−20	127 ± 15	−0.3				

Compiled by Elmer Mihalyi.

[a]0.1 *M* phosphate buffer, pH 7.1.
[b]0.1 *N* HCl.

From *J. Chem. Eng. Data,* 13, 179 (1969). With permission of the author and copyright owner.

ULTRAVIOLET SPECTRA OF DERIVATIVES OF CYSTEINE, CYSTINE, HISTIDINE, PHENYLALANINE, TYROSINE, AND TRYPTOPHAN

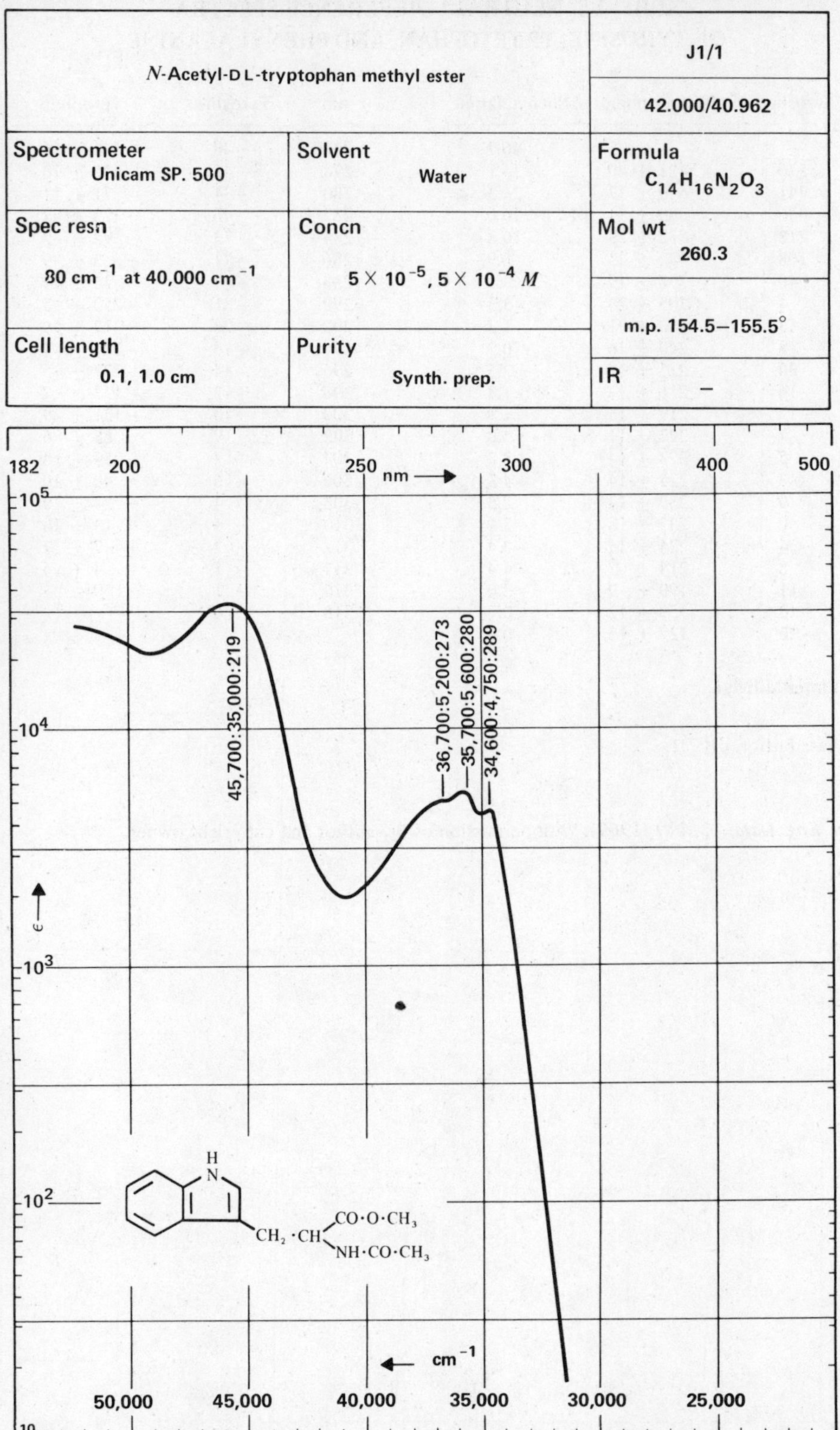

FIGURE 1. *N*-Acetyl-DL-tryptophan methyl ester. (From Bailey, in *DMS UV Atlas of Organic Compounds,* Vol. 2, Verlag Chemie, Weinheim and Butterworths, London, 1966. With permission.)

N-Acetyl-DL-β-phenylalanine methyl ester		J1/2 30.000/00.262
Spectrometer Unicam SP. 500	Solvent Water	Formula $C_{12}H_{15}NO_3$
Spec resn 80 cm^{-1} at 40,000 cm^{-1}	Concn $4 \times 10^{-4}, 4 \times 10^{-3}$ *M*	Mol wt 221.3 m.p. 60.5–61.5°
Cell length 0.1, 1.0 cm	Purity Synth. prep.	IR —

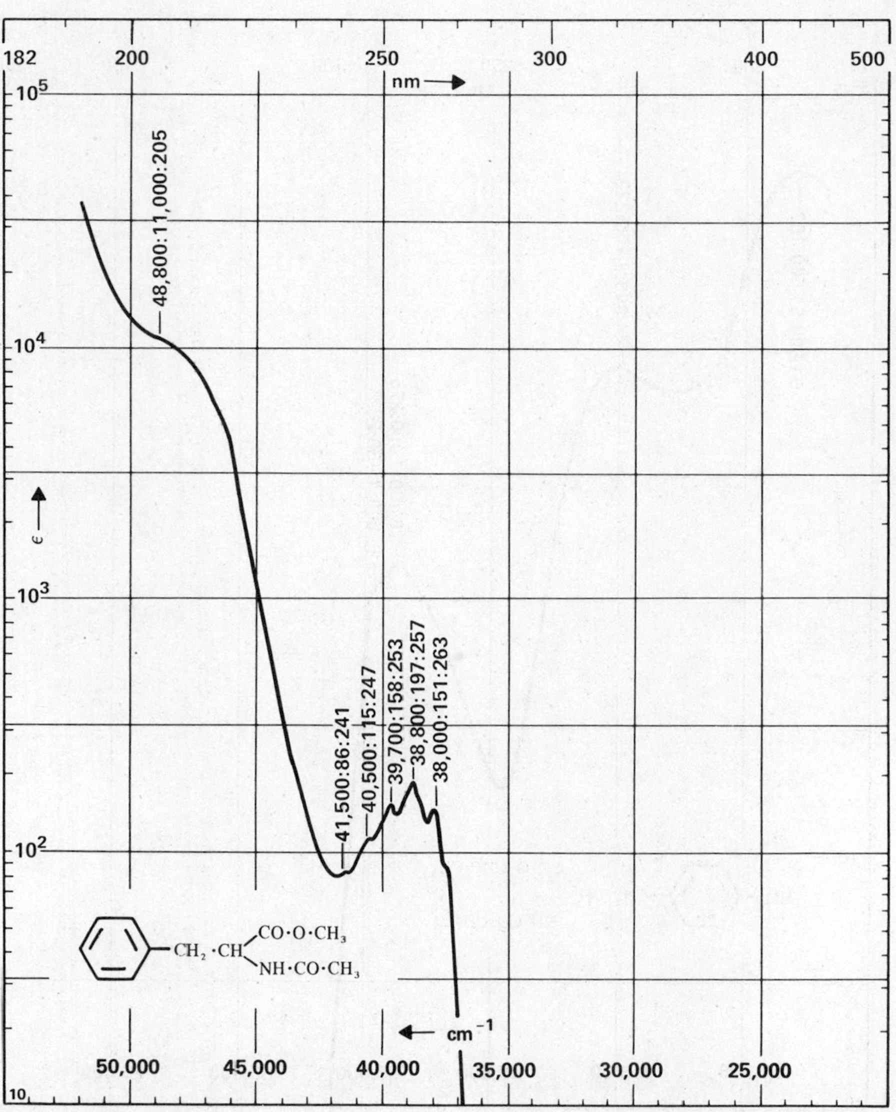

FIGURE 2. *N*-Acetyl-DL-β-phenylalanine methyl ester. (From Bailey, in *DMS UV Atlas of Organic Compounds,* Vol. 2, Verlag Chemie, Weinheim and Butterworths, London, 1966. With permission.)

N-Acetyl-L-tyrosine ethyl ester		J1/3
		30.002/00.262
Spectrometer Unicam SP. 500	Solvent Water	Formula $C_{13}H_{17}NO_4$
Spec resn $80\ cm^{-1}$ at $40{,}000\ cm^{-1}$	Concn $5 \times 10^{-5}, 5 \times 10^{-4}\ M$	Mol wt 251.3
		m.p. 97–97.5°
Cell length 0.1, 1.0 cm	Purity Synth. prep.	IR —

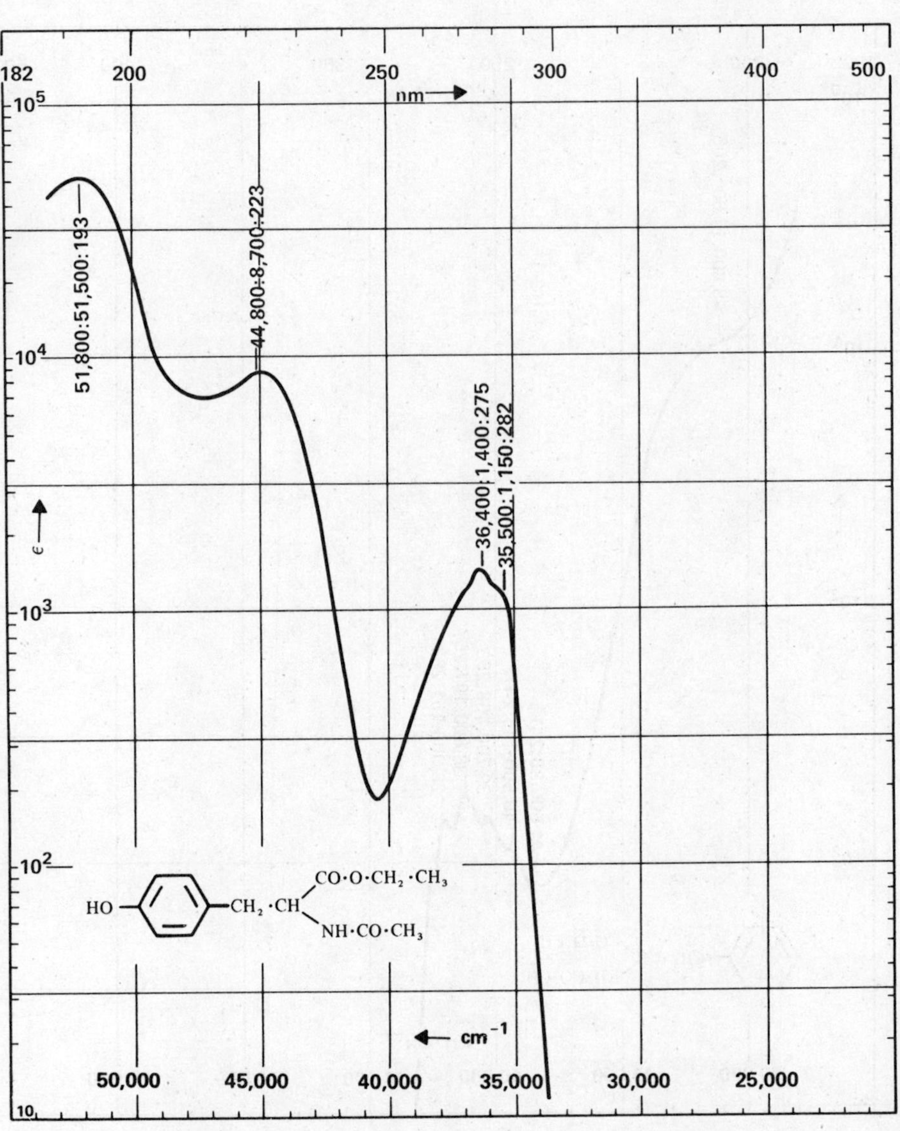

FIGURE 3. *N*-Acetyl-L-tyrosine ethyl ester. (From Bailey, in *DMS UV Atlas of Organic Compounds,* Vol. 2, Verlag Chemie, Weinheim and Butterworths, London, 1966. With permission.)

N-Acetyl-L-tyrosine ethyl ester anion		J1/4 30.002/00.262
Spectrometer Unicam SP.500	Solvent Water + 10^{-1} *M* NaOH or 10^{-1} *M* glycine buffer pH 12.4	Formula $C_{13}H_{16}NO_4$
Spec resn 80 cm^{-1} at 40,000 cm^{-1}	Concn 4×10^{-5}, 4×10^{-4} *M*	Mol wt 250.3 m.p. 97–97.5°
Cell length 0.1, 0.2, 1.0 cm	Purity Synth. prep.	IR –

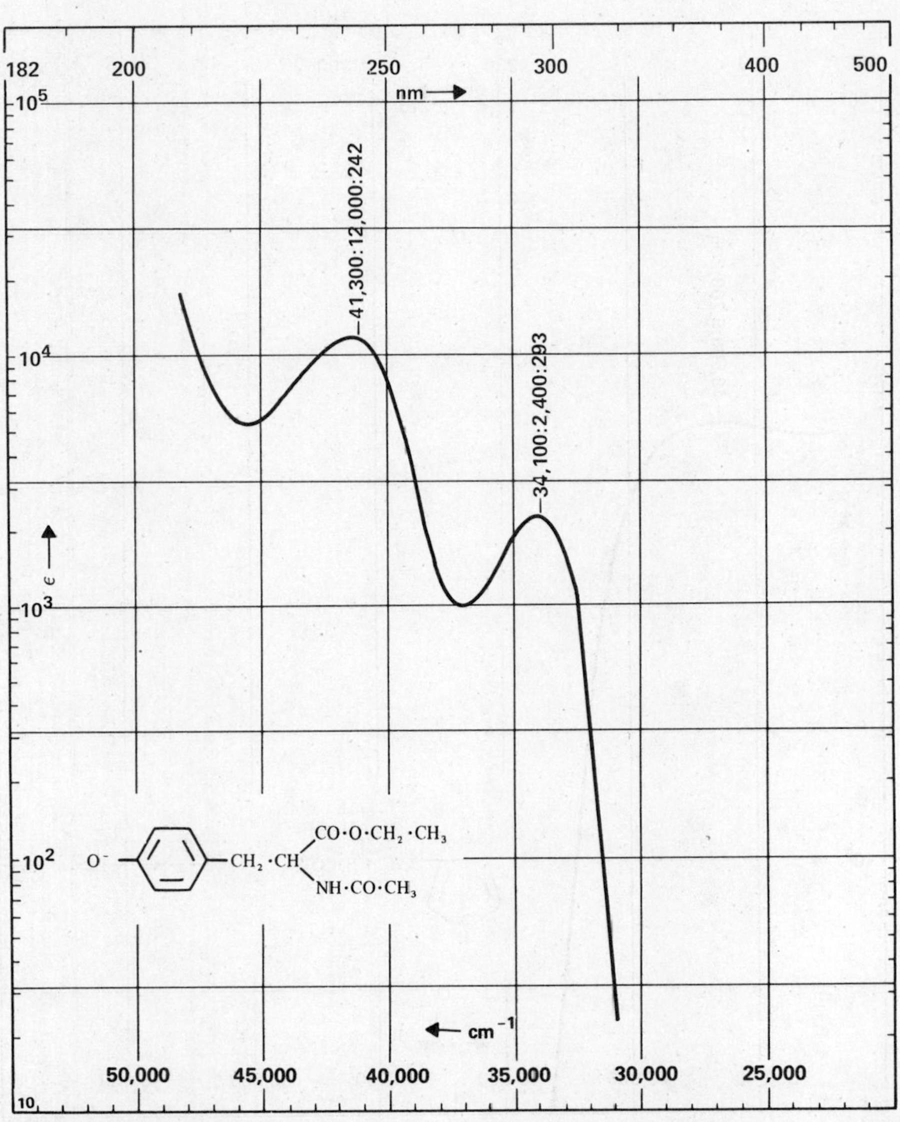

FIGURE 4. *N*-Acetyl-L-tyrosine ethyl ester anion. (From Bailey, in *DMS UV Atlas of Organic Compounds,* Vol. 2, Verlag Chemie, Weinheim and Butterworths, London, 1966. With permission.)

Histidine		J1/5 40.000/43.360
Spectrometer Unicam SP. 500 (N_2 flushed), Hilger Ultrascan	**Solvent** Aqueous phosphate buffer pH 7.4	**Formula** $C_6H_9N_3O_2$
Spec resn 100 cm^{-1}	**Concn** 10^{-3}, 10^{-2} *M*	**Mol wt** 155.2
		—
Cell length 0.1, 1.0 cm	**Purity** Recryst.	**IR** 4673

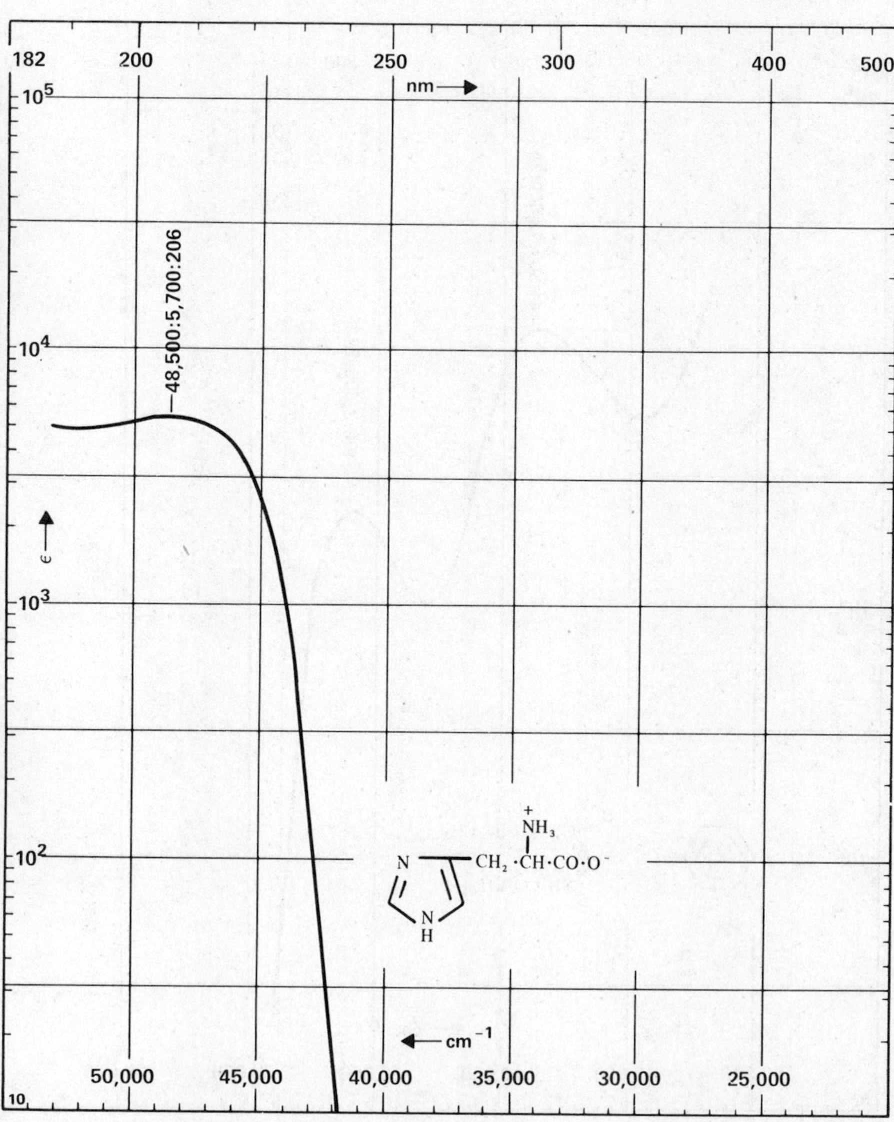

FIGURE 5. Histidine. (From Johnson, in *DMS UV Atlas of Organic Compounds*, Vol. 5, Verlag Chemie, Weinheim and Butterworths, London, 1971. With permission.)

Cysteine		J2/3
		00.000/00.060
Spectrometer Unicam SP. 500 (N_2 flushed), Hilger Ultrascan	**Solvent** Water	**Formula** $C_3H_7NO_2S$
Spec resn 100 cm^{-1} at 50,000 cm^{-1}	**Concn** 10^{-3}, 10^{-2}, 10^{-1} *M*	**Mol wt** 121.2
		–
Cell length 0.1, 1.0 cm	**Purity** Recryst.	**IR** 1497

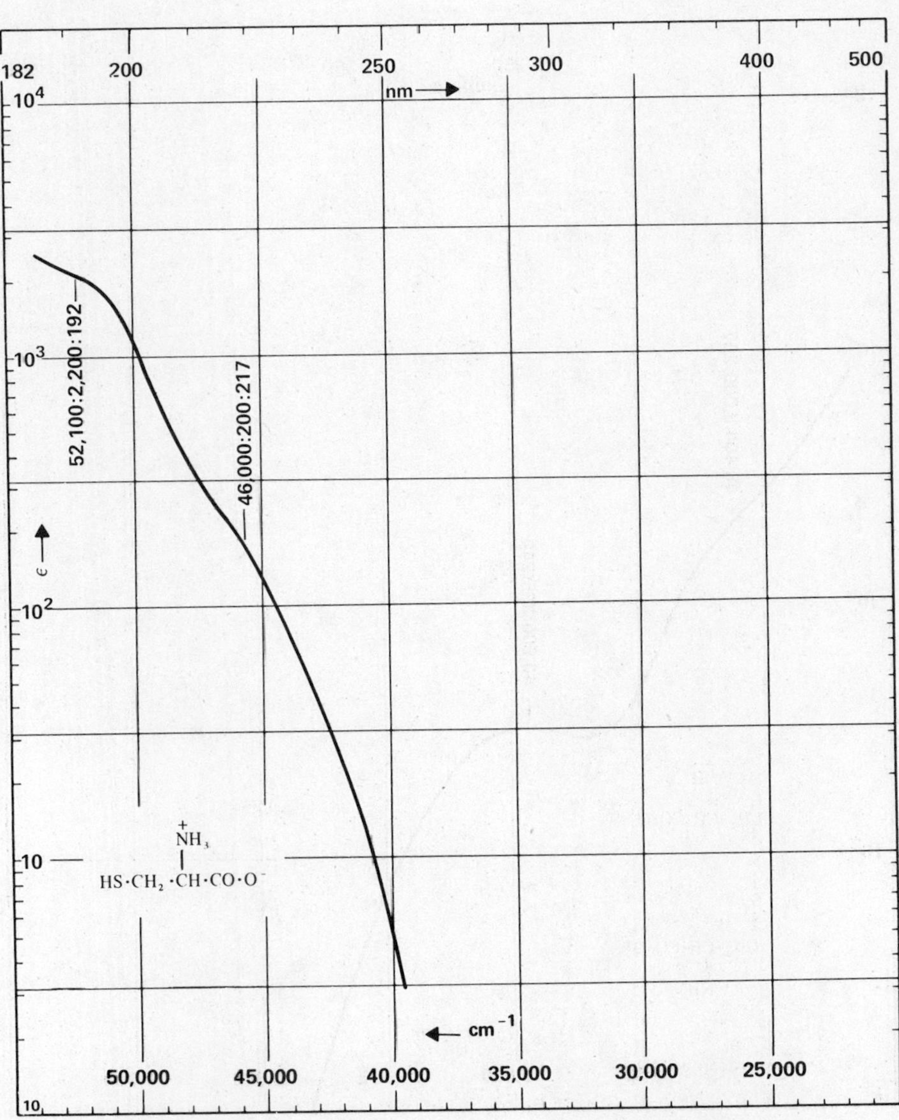

FIGURE 6. Cysteine. (From Johnson, in *DMS UV Atlas of Organic Compounds,* Vol. 5, Verlag Chemie, Weinheim and Butterworths, London, 1971. With permission.)

Cystine		J2/2
		00.000/00.062
Spectrometer Unicam SP, 500 (N_2 flushed), Hilger Ultrascan	**Solvent** Water	**Formula** $C_6H_{12}N_2O_4S_2$
Spec resn 120 cm^{-1} at 52,000 cm^{-1} 50 cm^{-1} at 40,000 cm^{-1}	**Concn** 4×10^{-4} *M*	**Mol wt** 240.3
		—
Cell length 0.2, 1.0, 5.0 cm	**Purity** Recryst.	**IR** 4675, 11038

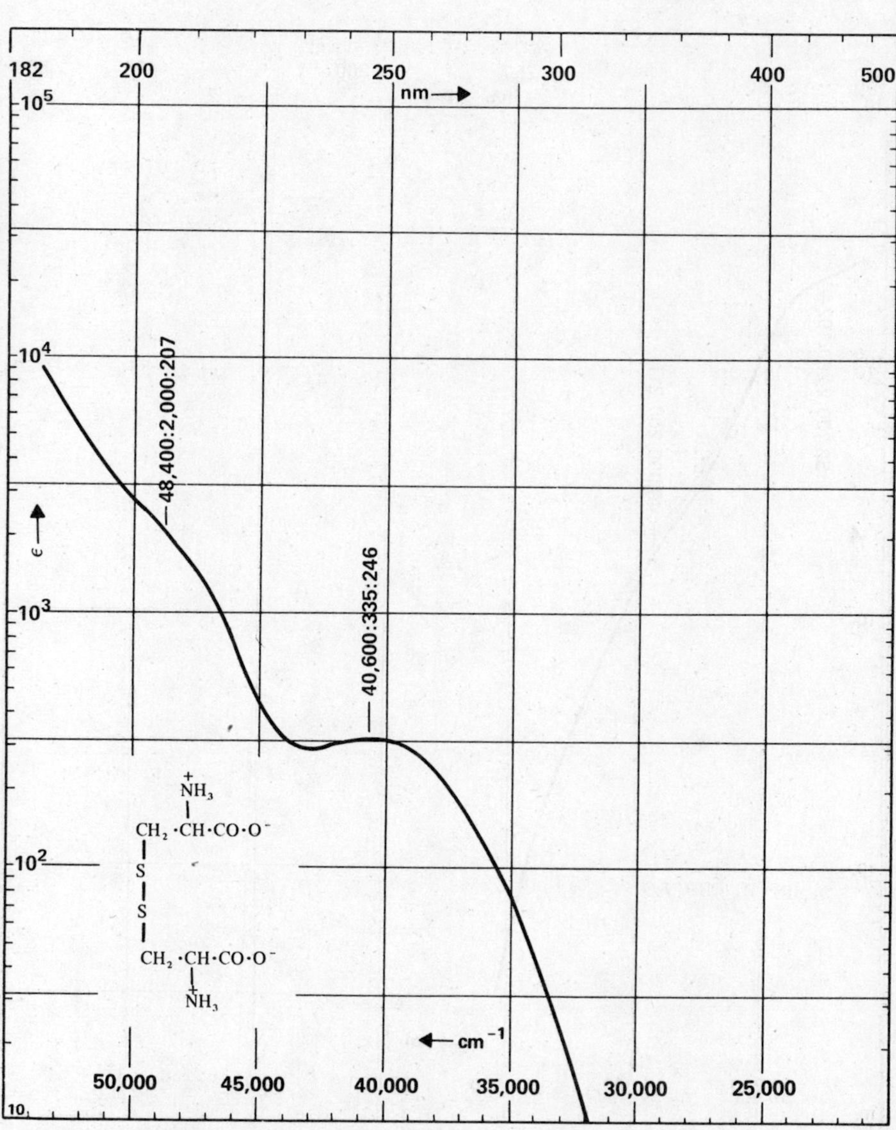

FIGURE 7. Cystine. (From Johnson, in *DMS UV Atlas of Organic Compounds*, Vol. 5, Verlag Chemie, Weinheim and Butterworths, London, 1971. With permission.)

N,N'-Diacetyl-L-cystine dimethyl ester		J2/1 00.005/00.064
Spectrometer Unicam SP. 500	**Solvent** Water + 10% v/v ethanol	**Formula** $C_{12}H_{20}N_2O_6S_2$
Spec resn 175 cm^{-1} at 50,000 cm^{-1} 90 cm^{-1} at 35,700 cm^{-1}	**Concn** 2×10^{-4}, 2×10^{-3} *M*	**Mol wt** 352.4 m.p. 128–129°
Cell length 0.1–1.0 cm	**Purity** Paper electrophoretic anal.	**IR** –

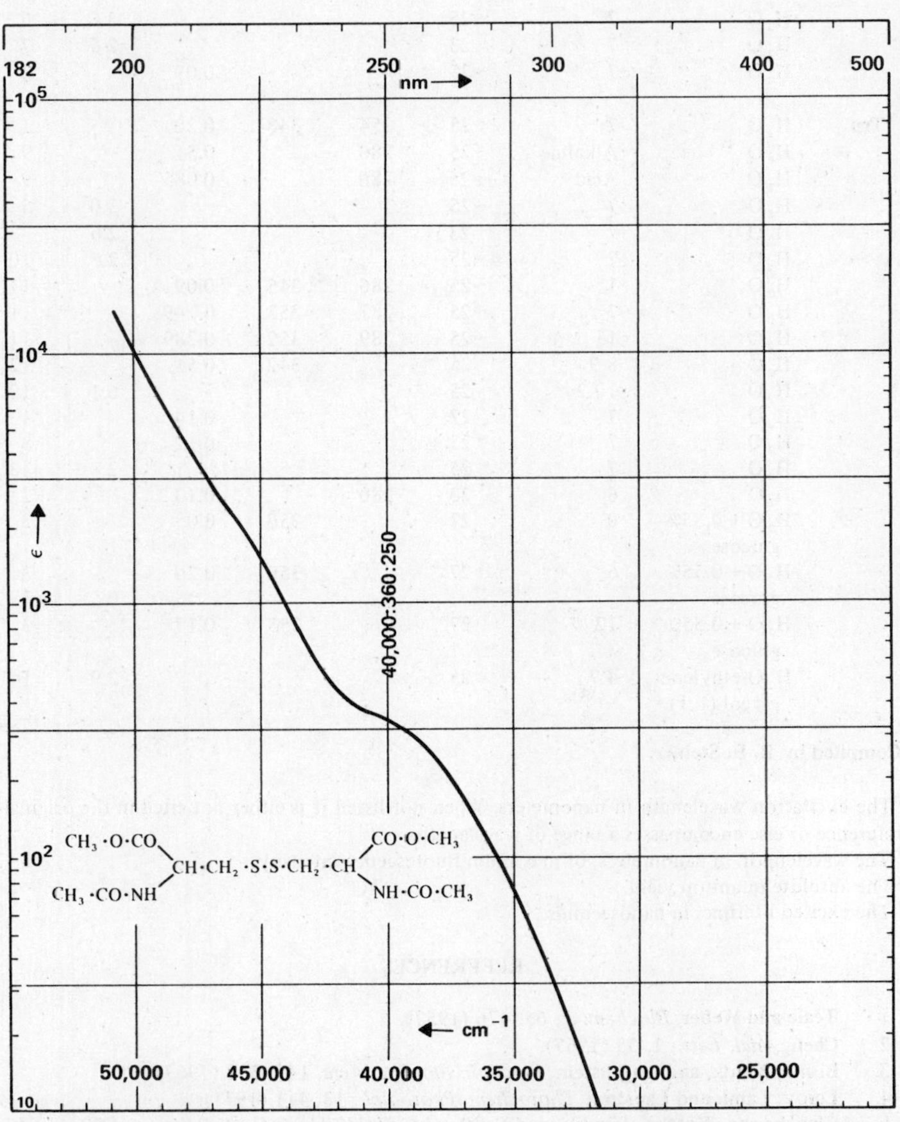

FIGURE 8. *N,N'*-Diacetyl-L-cystine dimethyl ester. (From Bailey, in *DMS UV Atlas of Organic Compounds*, Vol. 4, Verlag Chemie, Weinheim and Butterworths, London, 1968. With permission.)

Table 1
LUMINESCENCE OF THE AROMATIC AMINO ACIDS

Amino acid	Solvent	pH	Temp.	λ_{ex}[a]	$\lambda_{f,m}$[b]	Q[c]	τ[d] (ns)	Ref.
Phe	H_2O	7	~25	254	282	0.04		1
	H_2O	6	23	260		0.024		2
	H_2O + 0.55% glucose	0	27		285	0.02		3
	H_2O + 0.55% glucose	6	27		285	0.03		3
	H_2O	7	20	248	282	0.025	6.8	4
	H_2O	7	25				6.4	5
Tyr	H_2O	7	~25	254	303	0.21		1
	H_2O	7	~25				3.2	5
	H_2O	6	23	275		0.14		2
	H_2O	7	~25				3.6	6
	H_2O	7	23				2.6	7
	H_2O	7	~25			0.09		8
Trp	H_2O	7	~25	254	348	0.20		1
	H_2O	Alkaline	~25	280		0.51		9
	H_2O	Acid	~25	280		0.085		9
	H_2O	7	~25				3.0	5
	H_2O	7	23				2.6	7
	H_2O	7	~25				2.5	10
	H_2O	1	~25	286	345	0.091		11
	H_2O	7	~25	287	352	0.149		11
	H_2O	11	~25	289	359	0.289		11
	H_2O	8.9	25		342	0.51		12
	H_2O	10.2	25				6.1	10
	H_2O	7	27			0.14		13
	H_2O	7	25			0.12		8
	H_2O	7	25				2.8	13
	H_2O	6	23	280		0.13		2
	H_2O + 0.55% glucose	0	27		350	0.05		3
	H_2O + 0.55% glucose	6	27		350	0.20		3
	H_2O + 0.55% glucose	12	27		355	0.17		3
	H_2O-ethylene glycol (1:1)	4.7	25				2.9	14

Compiled by R. F. Steiner.

[a]The excitation wavelength in nanometers. When not listed it is either not cited in the original reference or else encompasses a range of wavelengths.
[b]The wavelength, in nanometers, of maximum fluorescence intensity.
[c]The absolute quantum yield.
[d]The excited lifetime, in nanoseconds.

REFERENCES

1. **Teale and Weber,** *Biochem. J.,* 65, 476 (1957).
2. **Chen,** *Anal. Lett.,* 1, 35 (1967).
3. **Bishai, Kuntz, and Augenstein,** *Biochim. Biophys. Acta,* 140, 381 (1967).
4. **Leroy, Lami, and Laustriat,** *Photochem. Photobiol.,* 13, 411 (1971).
5. **Gladchenko, Kostko, Pikulik, and Sevchenko,** *Dokl. Akad. Nauk Belorrus. SSR,* 9, 647 (1965).

Table 1 (continued)
LUMINESCENCE OF THE AROMATIC AMINO ACIDS

6. Blumberg, Eisinger, and Navon, *Biophys. J.*, 8, A-106 (1968).
7. Chen, Vurek, and Alexander, *Science*, 156, 949 (1967).
8. Borresen, *Acta Chem. Scand.*, 21, 920 (1967).
9. Cowgill, *Arch. Biochem. Biophys.*, 100, 36 (1963).
10. Badley and Teale, *J. Mol. Biol.*, 44, 71 (1969).
11. Bridges and Williams, *Biochem. J.*, 107, 225 (1968).
12. Longworth, *Biopolymers*, 4, 1131 (1966).
13. Eisinger and Navon, *J. Chem. Phys.*, 50, 2069 (1969).
14. Weinryb and Steiner, *Biochemistry*, 7, 2488 (1968).

Table 2
LUMINESCENCE OF DERIVATIVES OF THE AROMATIC AMINO ACIDS

Compound	Solvent	pH	Temp.	λ_{ex}[a]	$\lambda_{f,m}$[b]	Q_{rel}	τ[c] (ns)	Ref.
				Tryptophan[d]				
L-Trp	H_2O	7	23	290		1.00	3.0	10
L-Trp	H_2O-Ethylene glycol (1:1)	4.7	25	290	357	1.00	2.9	1
Acetyl-DL-Trp	H_2O-Ethylene glycol (1:1)	7.5	25	290	361	1.76	5.2	1
Acetyl-Trp	H_2O	5	25	290		1.6	4.8	4
Acetyl-L-Trp	H_2O	7	~25	280	355	1.40		2
Acetyl-L-Trp-amide	H_2O-Ethylene glycol (1:1)	7.5	25	290	356	0.87	3.8	1
Acetyl-Trp-amide	H_2O	7	23	290			3.0	10
Acetyl-Trp-amide	H_2O	5	25	290		1.10	2.6	4
L-Trp-amide	H_2O-Ethylene glycol (1:1)	4.7	25	290	351	0.59	1.4	1
DL-Trp-amide	H_2O	7	25	280	355	1.00		2
Trp-amide	H_2O	7	25	280		0.70		5
Acetyl-Trp-methyl ester	H_2O	7	25	280	350		0.55	5
Trp-methyl ester	H_2O	7	25	280		0.25		5
L-Trp-ethyl ester	H_2O	7	25	280	355	0.16		2
L-Trp-ethyl-ester	H_2O-Ethylene glycol (1:1)	4.7	25	290	349	0.17	0.50	1
L-Trp-Gly	H_2O	7	25	280		0.70		2
L-Trp-Gly	H_2O-Ethylene glycol (1:1)	4.7	25	290	350	0.65	1.60	1
L-Trp-Gly	H_2O	7	23	290			2.2	10
L-Trp-Gly-Gly	H_2O-Ethylene glycol (1:1)	4.7	25	290	348	0.48	1.8	1
Gly-L-Trp	H_2O-Ethylene glycol (1:1)	4.7	25	290	356	0.49	1.4	1
Gly-L-Trp	H_2O	7	25	280	355	0.29		2
Gly-Trp	H_2O	7	25	280		0.25		5

[a]The excitation wavelength in nanometers. When not listed it is either not cited in the original reference, or else encompasses a range of wavelengths.
[b]The wavelength, in nanometers, of maximum fluorescence intensity.
[c]The excited lifetime, in nanoseconds.
[d]Q_{rel} = quantum yield, relative to that of tryptophan.

Table 2 (continued)
LUMINESCENCE OF DERIVATIVES OF THE AROMATIC AMINO ACIDS

Compound	Solvent	pH	Temp.	λ_{ex}[a]	$\lambda_{f,m}$[b]	Q_{rel}[c]	τ[c] (ns)	Ref.
			Tryptophan[d] (continued)					
Gly-L-Trp	H_2O	7	23	290			1.5	10
Gly-Trp	H_2O	9.8	25	280		0.70		5
Gly-Gly-L-Trp	H_2O-Ethylene glycol (1:1)	4.7	25	290	362	0.65	2.0	1
Gly-Gly-Trp	H_2O	7	25	280		0.40		5
Gly-Gly-Trp	H_2O	9.8	25	280		0.50		5
Gly-Gly-Gly-L-Trp	H_2O-Ethylene glycol (1:1)	4.7	25	290	360	0.76	2.4	1
Trp-Gly-Trp-Gly	H_2O-Ethylene glycol (1:1)	4.7	25	290	352	0.69	2.0	1
L-Trp-L-Phe	H_2O-Ethylene glycol (1:1)	4.7	25	290	350	0.63	2.0	1
L-Trp-L-Trp	H_2O	7	~25	280		0.45		2
L-Trp-L-Trp	H_2O	7	23	290			1.6	10
L-Trp-L-Trp	H_2O-Ethylene glycol (1:1)	4.7	25	290	361	0.50	1.5	1
L-Trp-L-Tyr	H_2O-Ethylene glycol (1:1)	4.7	25	290	351	0.75	1.9	1
L-Trp-L-Tyr	H_2O	7	~25	280		0.60		2
Pro-Trp	H_2O	7	25	280		0.25		5
Pro-Trp	H_2O	10.2	25	280		0.95		5
Tryptamine	H_2O	5	25	290		2.10	6.0	4
			Tyrosine[e]					
DL-Tyr	H_2O	7	25	280	320	1.00		2
Tyramine	H_2O	7	25	280	320	0.88		2
L-Tyr-ethyl ester	H_2O	5.5	25	275		0.12		7
Tyr-amide	H_2O	5.5	25	275		0.25		7
N-Methyl-Tyr	H_2O	5.5	25	275		0.38		7
Acetyl-L-Tyr	H_2O	5.5	25	275		0.88		7
Acetyl-Tyr-amide	H_2O	5.5	25	275		0.45		7
L-Tyr-Gly	H_2O	7	25	280	320	0.35		2
L-Tyr-Gly	H_2O	5.5	25	275		0.33		7
L-Tyr-Gly-Gly	H_2O	5.5	25	275		0.22		7
L-Cystinyl-bis-L-Tyr	H_2O	8.5	25	270		0.08		11
Tyr-Cys-S-S-Cys	H_2O	8.5	25	270		0.08		11
Tyr-Ala	H_2O	6	25	270		0.43		3
Tyr-Phe	H_2O	6	25	270		0.38		3
Tyr-Tyr	H_2O	6	25	270		0.38		3
Gly-L-Tyr	H_2O	7	25	280	320	0.33		2
Gly-Tyr	H_2O	5.5	25	280		0.38		8
Gly-Gly-L-Tyr	H_2O	5.5	25	275		0.54		7
Gly-Gly-Gly-L-Tyr	H_2O	5.5	25	275		0.58		7
Gly-L-Tyr-Gly	H_2O	5.5	25	275		0.22		7
Gly-L-Tyr-Gly-amide	H_2O	7	25	280	320	0.17		2

[e] Q_{rel} = quantum yield, relative to that of tyrosine.

Table 2 (continued)
LUMINESCENCE OF DERIVATIVES OF THE AROMATIC AMINO ACIDS

Compound	Solvent	pH	Temp.	λ_{ex}[a]	$\lambda_{f,m}$[b]	Q_{rel}	τ[c] (ns)	Ref.
				Tyrosine[e] (continued)				
Gly-Tyr-Gly-amide	H_2O	6	25	270		0.18		3
Leu-L-Tyr	H_2O	7	25	280	320	0.50		2
Leu-Tyr	H_2O	5.5	25	280		0.48		12
Glu-Tyr	H_2O	5.5	25	280		0.45		8
Met-Tyr	H_2O	5.5	25	280		0.45		8
Arg-Tyr	H_2O	5.5	25	280		0.50		8
His-Tyr	H_2O	5.5	25	270		0.50		8
HS-Cys-Tyr	H_2O	8.5	25	270		0.28		11
				Phenylalanine[f]				
L-Phe	H_2O-Ethylene glycol (1:1)	4.7	25	250	287	1.00		1
Gly-DL-Phe	H_2O-Ethylene glycol (1:1)	4.7	25	250	287	1.15		1
DL-Ala-DL-Phe	H_2O-Ethylene glycol (1:1)	4.7	25	250	285	1.10		1
L-Lys-L-Phe	H_2O-Ethylene glycol (1:1)	4.7	25	250	287	0.98		1
L-His-L-Phe	H_2O-Ethylene glycol (1:1)	4.7	25	250	288	0.82		1
L-Arg-L-Phe	H_2O-Ethylene glycol (1:1)	4.7	25	250	286	0.77		1
L-Met-L-Phe	H_2O-Ethylene glycol (1:1)	4.7	25	250	287	0.44		1

[f] Q_{rel} = quantum yield, relative to that of phenylalanine.

Compiled by R. F. Steiner.

REFERENCES

1. **Weinryb and Steiner,** *Biochemistry,* 7, 2488 (1968).
2. **Cowgill,** *Arch. Biochem. Biophys.,* 100, 36 (1963).
3. **Cowgill,** *Biochim. Biophys. Acta,* 75, 272 (1963).
4. **Kirby and Steiner,** *J. Phys. Chem.,* 74, 4480 (1970).
5. **Cowgill,** *Biochim. Biophys. Acta,* 133, 6 (1967).
6. **Edelhoch, Brand, and Wilchek,** *Biochemistry,* 6, 547 (1967).
7. **Edelhoch, Perlman, and Wilchek,** *Biochemistry,* 7, 3893 (1968).
8. **Russell and Cowgill,** *Biochim. Biophys. Acta,* 154, 231 (1968).
9. **Cowgill,** *Biochim. Biophys. Acta,* 100, 37 (1967).
10. **Chen,** *Arch. Biochem. Biophys.,* 158, 605 (1973).
11. **Cowgill,** *Biochim. Biophys. Acta,* 140, 37 (1967).
12. **Cowgill,** *Arch. Biochem. Biophys.,* 104, 84 (1964).

Table 3
LUMINESCENCE OF PROTEINS LACKING TRYPTOPHAN

Protein	Solvent	pH	Temp.	λ_{ex}[a]	$\lambda_{f,m}$[b]	Q_{rel}[c]	τ[d] (ns)	Ref.
Angiotensin amide	H_2O	7	25			0.36 (0.075)		2
Insulin (bovine)	H_2O	7	25	277	304	0.19	1.4	1
Insulin B chain	H_2O	7	25			0.26 (0.055)		2
Oxytocin	H_2O	7.8	25			0.26		2, 3
Ribonuclease (bovine)	H_2O	7	25	277	304	0.10 (0.02)	1.9	1

Compiled by R. F. Steiner.

[a]The excitation wavelength in nanometers.
[b]The wavelength of maximum fluorescence.
[c]The quantum yield relative to that of tyrosine (absolute yield in parentheses).
[d]The excited lifetime, in nanoseconds.

REFERENCES

1. **Longworth,** in *Excited States of Proteins and Nucleic Acids,* Steiner and Weinryb, Eds., Plenum, New York, 1971.
2. **Cowgill,** *Biochim. Biophys. Acta,* 133, 6 (1967).
3. **Cowgill,** *Arch Biochem. Biophys.,* 104, 84 (1964).

Table 4
LUMINESCENCE OF PROTEINS CONTAINING TRYPTOPHAN

Protein	Solvent	pH	Temp.	λ_{ex}[a]	$\lambda_{f,m}$[b]	Q[c]	τ[d] (ns)	Ref.
F-Actin	H_2O	7	~25	297	332	0.24		1
Actomyosin	H_2O	7	~25	297	337	0.20		1
Albumin (bovine	H_2O	5.5–7	~25	297	343	0.39		1
serum)	H_2O	7	23		270		4.6	2
	H_2O		~25	280	342	0.15		3
	H_2O	7	25	295	343	0.26		4
	H_2O	7	25		270	0.51		5
	H_2O	7	25	280	342	0.21	4.6	5
Albumin (human	H_2O	5.5–7	~25	297	342	0.31		1
serum)	H_2O		~25	280	339	0.07		3
	H_2O	7	25	295	341	0.155		4
	H_2O					0.24	4.3	5
	H_2O					0.22	4.1	6
	H_2O	5.5	20	295	343	0.21	4.8	7
	H_2O	7	23				4.5	2
	H_2O	7	25	280	339	0.11	3.3	5
Aldolase	H_2O	7	25	280		0.10		4
	H_2O	7	25	280	328	0.10		5
	H_2O	7	25	295	328	0.12		4
β-Amylase	H_2O	7	~25	297	335	0.32		1
Arginase (bovine liver)	H_2O	7	~25	297	337	0.18		1
Avidin	H_2O	7	25	280	338			5
Azurin	H_2O	7	25	280	308	0.10		5
Carbonic anhydrase	H_2O	7	25	280	336	0.17	2.6	5
Carboxypeptidase A	H_2O	7	25	280	327	0.12		4
α-Chymotrypsin	H_2O	7	~25	297	336	0.14		1
	H_2O	7		280	334	0.095		3
	H_2O	7	25	295	332	0.144		4
	H_2O					0.13	3.0	5
	H_2O						3.0	6
	H_2O	7	23	270			3.4	2
	H_2O	7	25	280		0.13		4
Chymotryspino-	H_2O	7	~25	297	333	0.127		1
gen A	H_2O	7	~25	280	331	0.07		3
	H_2O	7	25	295	331	0.124		4
	H_2O					0.10	1.6	5
	H_2O						1.6	6
	H_2O						1.9	8
	H_2O	7	23	290			2.9	2
	H_2O	7	25	280		0.11		4
Corticotropin	H_2O	7	25	280	350	0.08		5
Elastase	H_2O	7	25	280	335			5
Endonuclease	H_2O	7	25	280	334			5
γI-Globulin	H_2O	6	25	297	335	0.08		1
(human)	H_2O	7	23				3.2	2
Glutamate dehydrogenase	H_2O	7	25	280	332	0.300	4.6	5
Glucagon	H_2O	7	25	280	345	0.14		5

[a]The excitation wavelength in nanometers.
[b]The wavelength of maximum fluorescence.
[c]The absolute quantum yield.
[d]The excited lifetime, in nanoseconds.

Table 4 (continued)
LUMINESCENCE OF PROTEINS CONTAINING TRYPTOPHAN

Protein	Solvent	pH	Temp.	λ_{ex}[a]	$\lambda_{f,m}$[b]	Q[c]	τ[d] (ns)	Ref.
Glyceraldehyde-3-phosphate dehydrogenase	H_2O	7	25	297	335.5	0.135		1
Growth hormone	H_2O	7	25	280	325	0.15		5
Hemoglobin	H_2O	7	25	280	335	0.001		5
Hyaluronidase (bovine testicle)	H_2O	7	~25	297	336	0.14		1
α-Lactalbumin	H_2O	7	25	295	328	0.06		4
	H_2O	7	25	280		0.05		4
Lactate dehydrogenase	H_2O	7	25	280	345	0.38		5
β-Lactoglobulin A	H_2O	7	25	280	330	0.08		5
β-Lactoglobulin AB (bovine)	H_2O	6–8	~25	297	332	0.12		1
	H_2O					0.15		5
	H_2O	7	25	295	333	0.082		4
	H_2O	7	25	280		0.08		4
Lysozyme (egg white)	H_2O	7.5	25	295	337			9
	H_2O	7	25	295	338	0.079		4
	H_2O		~25	280	341	0.06		3
	H_2O						2.6	5
	H_2O	7	23	270			2.0	2
	H_2O	7	25	280		0.07		4
Apomyoglobin (sperm whale)	H_2O	8.5	~25	297	335.5	0.16		1
	H_2O					0.15	2.9	10
	H_2O	7	15	288	328	0.12	2.8	11
	H_2O					0.16		5
	H_2O	7	23	290			3.0	2
Myosin (rabbit)	H_2O	7	~25	297	338	0.22		1
Ovalbumin	H_2O	7	~25	280	332	0.19	4.5	5
	H_2O	7	25	295	334	0.25		4
Papain	H_2O	9.5–10	~25	297	350	0.16		1
	H_2O	7.5–7.8	~25	297	347	0.16		1
	H_2O	7	25	295	342	0.16		4
	H_2O			280		0.15		12
	H_2O			280		0.13	4.6	13
	H_2O	7	25	280		0.14		4
	H_2O	4.5–5.3	~25	297	340	0.082		1
	H_2O					0.11		5
	H_2O					0.079		4
	H_2O	4.5	23	288		0.08		14
	H_2O			280		0.10	3.0	12
	H_2O			280		0.08	3.4	13
Pepsin	H_2O	5.2–5.5	~25	297	343	0.26		1
	H_2O		~25	280	342	0.13		3
	H_2O					0.22	4.6	15
	H_2O	7	25	295	339	0.185		4
	H_2O					0.31		5
	H_2O	4.5	23	288		0.22		14
	H_2O						4.5	8
Acid phosphatase (wheat bran)	H_2O	7	~25	297	337	0.16		1
Phosphocreatine kinase	H_2O	7	~25	297	333.5	0.11		1

Table 4 (continued)
LUMINESCENCE OF PROTEINS CONTAINING TRYPTOPHAN

Protein	Solvent	pH	Temp.	λ_{ex}[a]	$\lambda_{f,m}$[b]	Q[c]	τ[d] (ns)	Ref.
Pseudoacetyl cholinesterase (human)	H_2O	7	~25	297	334			1
Pyruvate kinase (rabbit)	H_2O	7	~25	297	339	0.44		1
	H_2O					0.20		5
Staphylococcus aureus nuclease	H_2O	5	~25	270		0.46		18
Subtilisin carlsberg	H_2O	7	~25	280	305			5
Tobacco mosaic virus protein (depolymerized)	H_2O	7.4	~25	297	332	0.40		1
Tobacco mosaic virus protein (polymerized)	H_2O	6.4	~25	297	331.5	0.37		1
Trypsin (bovine)	H_2O	7	~25	297	334.5	0.11		1
	H_2O		~25	280	332	0.08		3
	H_2O	7	25	295	335	0.13		4
	H_2O					0.08		5
	H_2O					0.15		16
	H_2O						2.0	6
	H_2O						2.4	8
	H_2O	7	25	280		0.13		4
Trypsinogen	H_2O	7	25	295	332	0.14		4
	H_2O	7	25	280		0.12		4
Tryptophanyl tRNA synthetase	H_2O	5	~25	270		0.46		18

Compiled by R. F. Steiner.

REFERENCES

1. **Burstein, Vedenkina, and Ivkova,** *Photochem. Photobiol.,* 18, 263 (1973).
2. **Chen, Vurek, and Alexander,** *Science,* 156, 949 (1967).
3. **Teale,** *Biochem. J.,* 76, 381 (1960).
4. **Kronman and Holmes,** *Photochem. Photobiol.,* 14, 113 (1971).
5. **Longworth,** in *Excited States of Proteins and Nucleic Acids,* Steiner and Weinryb, Eds., Plenum, New York (1971).
6. **Konev, Kostko, Pikulik, and Chernitski,** *Biofizika,* 11, 965 (1966).
7. **DeLauder and Wahl,** *Biochem. Biophys. Res. Commun.,* 42, 398 (1971).
8. **Konev, Kostvo, Pikulik, and Volotovski,** *Dokl. Akad. Nauk Beloruss. SSR,* 10, 500 (1966).
9. **Lehrer,** *Biochemistry,* 10, 3254 (1971).
10. **Kirby and Steiner,** *J. Biol. Chem.,* 245, 6300 (1970).
11. **Anderson, Brunori, and Weber,** *Biochemistry,* 9, 4723 (1970).
12. **Steiner,** *Biochemistry,* 10, 771 (1971).
13. **Weinryb and Steiner,** *Biochemistry,* 9, 135 (1970).
14. **Shinitzki and Goldman,** *Eur. J. Biochem.,* 3, 139 (1967).
15. **Badley and Teale,** *J. Mol. Biol.,* 44, 71 (1969).
16. **Barenboim, Sokolenko, and Turoverov,** *Cytologiya,* 10, 636 (1968).
17. **Edelhoch, Condliffe, Lippoldt, and Burger,** *J. Biol. Chem.,* 241, 5205 (1966).
18. **Edelhoch, Perlman, and Wilchek,** *Ann. N. Y. Acad. Sci.,* 158, 391 (1969).

Table 5
COVALENT PROTEIN CONJUGATES

Label	Protein	Groups/ molecule	Solvent	pH	Temp.	λ_{ex}[a]	$\lambda_{f,m}$[b]	Q[c]	τ[d] (ns)	Ref.
Fluorescamine	Bovine plasma albumin	11.2	H_2O	7.4	25	390	490	0.088	10.3	1
		9.3						0.144	10.2	
		3.6						0.103	9.7	
	Ovalbumin	2.2	H_2O	7.4	25	390	490	0.125	9.0	1
		3.9						0.123	9.0	
	Lysozyme	1.3	H_2O	7.4	25	390	490	0.099	8.0	1
		2.2						0.094	7.5	
Fluorescein isothiocyanate	γ-Globulin	0.2	H_2O	9.0	25	490	550	0.5	4.2	2
Anthracene 2-isocyanate	γ-Globulin	1	H_2O	8.0	25	362	460	0.6	29	2
Rhodamine B isothiocyanate	γ-Globulin	1	H_2O	8.0	25	550	585	0.7	3	2
Dansyl	γ-Globulin	2.5	H_2O	8.0	25	345	540–545	0.2	11	2
Dansyl	Bovine plasma albumin	1.2	H_2O	7.4	25	320–360	500	0.6	22	3
		2.5	H_2O	7.4	25	320–360	540	0.6	22	3
		16	H_2O	7.4	25	320–360	540	0.6	18	3
Dansyl	Bovine thyroglobulin	10	H_2O	7.0	25	345			8.1	4
Pyrenebutyrate	Human thyroglobulin	0.84	H_2O	7.0	25	346–347			125	4

Compiled by R. F. Steiner.

[a]The excitation wavelength in nanometers.
[b]The wavelength of maximum fluorescence.
[c]The absolute quantum yield.
[d]The excited lifetime in nanoseconds.

REFERENCES

1. **Chen,** *Anal. Lett.*, 7, 65 (1974).
2. **Chen,** *Arch. Biochem. Biophys.*, 133, 263 (1969).
3. **Chen,** *Arch. Biochem. Biophys.*, 128, 163 (1968).
4. **Rawitch, Hudson, and Weber,** *J. Biol. Chem.*, 244, 6543 (1969).

HYDROPHOBICITIES OF AMINO ACIDS AND PROTEINS

C. C. Bigelow and M. Channon

Table 1
HYDROPHOBICITIES OF AMINO ACIDS

The hydrophobicity of an amino acid side chain is derived from the measurement of the solubility of the amino acid in water and an organic solvent.[1,2] The data are converted into the molar free energy of transfer for the amino acid from an aqueous solution to a solution in the organic solvent at the same mole fraction at the limit of infinite dilution:

$$\Delta F_t = RT\ln \frac{N_w \gamma_w}{N_{org} \gamma_{org}}$$

The hydrophobicity of the side chain, HΦ, is then determined by subtracting ΔF_t for glycine. The thermodynamic interpretation of the hydrophobicity has been discussed by Nozaki and Tanford.[2]

Amino acid	HΦ (cal/mol)
Trp	3,400
Ile	2,950
Pro	2,600
Phe	2,500
Tyr	2,300
Leu	1,800
Val	1,500
Lys	1,500
Met	1,300
Cys/2	1,000
Arg	750
His	500
Ala	500
Thr	400
Gly	(0)
Asp	0[a]
Glu	0[a]
Ser	−300

Compiled by C. C. Bigelow and M. Channon.

[a] Assumed values.

REFERENCES

1. **Tanford,** *J. Am. Chem. Soc.,* 184, 4240 (1962).
2. **Nozaki and Tanford,** *J. Biol. Chem.,* 246, 2211 (1971).

Table 2
AVERAGE HYDROPHOBICITIES OF SELECTED PROTEINS

The values in the table have been calculated for proteins from amino acid compositions in the literature. Proteins have been included only if the content of all the amino acids is known. The selection has been guided by a desire to include proteins of different properties (globular and fibrous, aggregating and monomeric) as well as members of some families of related proteins.

$$H\Phi_{ave} = \frac{\sum_i n_i H\Phi_i}{\sum_i n_i}$$

where

n_i = the number of residues of the ith amino acid;
$H\Phi_i$ = its hydrophobicity from Table 1.[475,476]

	Protein	Average hydrophobicity	Reference
1	Acetoacetic acid decarboxylase (*Clostridium acetobutylicum*)	1,140	1
2	Acetylcholine receptor (*Electrophorus electricus*)	1,060	2
3	Acetylcholinesterase (*Electrophorus electricus*)	990	3
4	*N*-Acetyl-β-D-glucosaminidase (*Aspergillus oryzae*)	1,050	4
5	*N*-Acetyl-β-D-glucosaminidase-A (beef spleen)	930	5
6	*N*-Acetyl-β-D-glucosaminidase-B (beef spleen)	930	5
7	*N*-Acetyl-β-D-glucosaminidase-A (porcine kidney)	1,040	6
8	*N*-Acetyl-β-D-glucosaminidase-B (porcine kidney)	1,030	6
9	*N*-Acetyl-β-D-glucosaminidase (human plasma)	960	7
10	Acetylornithine γ-transaminase (*Escherichia coli*)	930	8
11	O-Acetylserine sulfhydrylase A (*Salmonella typhimurium*)	1,020	9
12	α_1-Acid glycoprotein (human plasma)	1,040	10
13	Actin (*Acanthamoeba castellani*)	1,030	11
14	Actin (beef carotid)	1,000	12
15	Actin (beef heart muscle)	980	13

Table 2 (continued)
AVERAGE HYDROPHOBICITIES OF SELECTED PROTEINS

	Protein	Average hydrophobicity	Reference
16	Actin (beef skeletal muscle)	980	13
17	Actin (brown trout)	1,000	14
18	Actin (chicken)	1,010	13
19	Actin (fish)	980	13
20	Actin (frog)	980	13
21	Actin (human platelet)	950	15
22	Actin (human uterus)	980	16
23	Actin (lamb)	980	13
24	Actin (mollusc)	910	13
25	Actin (pig)	990	13
26	Actin (rabbit muscle)	1,010	17
27	Actin (rabbit muscle)	1,000	18
28	Actin (sheep uterus)	990	16
29	Acylcarrier protein (*Escherichia coli*)	830	19
30	Acylphosphatase (bovine brain)	980	20
31	Acylphosphatase (horse muscle)	900	21
32	Adenosine and adenosine monophosphate deaminase (*Aspergillus oryzae*)	890	22
33	Adenosine deaminase (calf intestinal muscle)	1,010	23
34	Adenosine deaminase (calf spleen)	990	24
35	Adenosine monophosphate deaminase (rabbit muscle)	1,060	22
36	Adenosine monophosphate nucleosidase (*Azotobacter vinelandii*)	770	25
37	Adenosine triphosphatase, ($Na^+ + K^+$), large chain (canine renal medulla)	1,040	26
38	Adenosine triphosphatase, ($Na^+ + K^+$), small chain (canine renal medulla)	1,130	26
39	Adenosine triphosphate-creatine transphosphorylase (rabbit)	980	27
40	Adenylate cyclase (*Brevibacterium liquefaciens*)	880	28

Table 2 (continued)
AVERAGE HYDROPHOBICITIES OF SELECTED PROTEINS

	Protein	Average hydrophobicity	Reference
41	Adenylosuccinate – AMP-lyase (*Neurospora* sp.)	940	29
42	Adrenodoxin (bovine)	850	30
43	Adrenodoxin reductase (bovine adrenal gland)	940	31
44	Aequorin (*Aequorea*)	990	32
45	Agglutinin (wheat germ)	680	33
46	Alanine amino transferase (rat liver)	990	34
47	Albumin (bovine plasma)	1,000	35
48	Albumin (dog plasma)	990	36
49	Albumin (human serum)	960	37
50	Albumin (rat serum)	980	38
51	Alcohol dehydrogenase (horse liver)	1,030	39
52	Aldehyde dehydrogenase, protein A (Baker's yeast)	1,020	40
53	Aldolase B (rabbit liver)	940	41
54	Aldolase C (rabbit brain)	980	41
55	Aldolase, fructose-1,6-diphosphate (*Boa constrictor constrictors*)	970	42
56	Aldolase, fructose-1,6-diphosphate (*Discostichus mawsonii*)	980	43
57	Aldolase, fructose-1,6-diphosphate (*Gallus domesticus*, brain)	920	44
58	Aldolase, fructose-1,6-diphosphate (*Gallus domesticus*, breast muscle)	960	45
59	Aldolase, fructose-1,6-diphosphate (*Gallus domesticus*, liver)	970	46
60	Aldolase, fructose diphosphate (*Micrococcus aerogenes*)	940	47
61	Aldolase, fructose-1,6-diphosphate (ox)	970	48
62	Aldolase, fructose-1,6-diphosphate (porcine)	960	48
63	Aldolase, fructose-1,6-diphosphate (rabbit liver)	960	49
64	Aldolase, fructose-1,6-diphosphate (rabbit muscle)	960	48
65	Aldolase, fructose-1,6-diphosphate (sturgeon)	990	48

Table 2 (continued)
AVERAGE HYDROPHOBICITIES OF SELECTED PROTEINS

	Protein	Average hydrophobicity	Reference
66	Aldolase, fructose-1,6-diphosphate (*Trematomus borchgrevinki*)	980	43
67	Aldolase, fructose-1,6-diphosphate (yeast)	990	50
68	Aldolase, 2-keto-3-deoxy-6-phosphogluconate (*Pseudomonas putida*)	1,090	51
69	Alkaline phosphatase (*Bacillus licheniformis* MC 14)	940	52
70	Alkaline phosphatase (*Escherichia coli*)	810	53
71	Alkaline proteinase-B (*Streptomyces rectus proteolyticus*)	820	54
72	ω-Amidase (rat liver)	980	55
73	Amino acid decarboxylase, aromatic (porcine kidney)	1,050	56
74	D-Amino acid oxidase (porcine kidney)	1,050	57
75	Aminopeptidase (*Aeromonas proteolytica*)	860	58
76	Aminotripeptidase, TP-2 (porcine kidney)	1,090	59
77	Amylase (*Bacillus macerans*)	910	60
78	Amylase (rat pancreas)	970	61
79	Amylase PI (rabbit pancreas)	940	62
80	Amylase PII (rabbit pancreas)	950	62
81	Amylase PIII (rabbit pancreas)	930	62
82	Amylase (rabbit parotid)	940	62
83	α-Amylase (*Aspergillus oryzae*)	990	63
84	α-Amylase (*Bacillus stearothermophilus*)	970	64
85	α-Amylase (*Bacillus subtilis*)	970	65
86	α-Amylase (porcine pancreas)	1,040	66
87	α-Amylase (human saliva)	940	67
88	α-Amylase-I (porcine pancreas)	960	68
89	α-Amylase-II (porcine pancreas)	970	68
90	Anthranylate synthetase, component I (*Pseudomonas putida*)	1,020	69

Table 2 (continued)
AVERAGE HYDROPHOBICITIES OF SELECTED PROTEINS

	Protein	Average hydrophobicity	Reference
91	Anthranylate synthetase, component II (*Pseudomonas putida*)	950	69
92	Apo-high density lipoprotein (bovine plasma)	960	70
93	Apolipoprotein – alanine-I (human plasma)	830	71
94	Apolipoprotein – alanine-II (human plasma)	830	71
95	Apolipoprotein – glutamine-I (human plasma)	870	72
96	Apolipoprotein – glutamine-I (porcine plasma)	890	72
97	Apolipoprotein – glutamine-II (human plasma)	970	73
98	Apolipoprotein-valine (human plasma)	920	74
99	Apovitellinin-I (*Dromaeus novaehollandiae*)	1,210	75
100	α-L-Arabinofuranosidase (*Aspergillus niger* K1)	900	76
101	L-Arabinose-binding protein (*Escherichia coli* B/r)	1,020	77
102	L-Arabinose isomerase (*Escherichia coli*)	990	78
103	Arginine decarboxylase (*Escherichia coli*)	980	79
104	Arginine kinase (*Callinectus sapidus*)	940	80
105	Arginine kinase (*Homarus americanus*)	980	81
106	Arginine kinase, negative (*Limulus polyphemus*)	950	80
107	Arginine kinase, neutral (*Limulus polyphemus*)	970	80
108	Arginine kinase (*Pagurus bernhardus*)	970	80
109	Asparaginase (*Escherichia coli* B)	870	82
110	Asparaginase (guinea pig)	970	83
111	L-Asparaginase (*Proteus vulgaris*)	950	84
112	Aspartate aminotransferase (ox heart)	980	85
113	Aspartate aminotransferase (rat brain, cytoplasm)	1,050	86
114	Aspartate aminotransferase (rat brain, mitochondria, fraction II)	990	86
115	Aspartate aminotransferase (rat brain, mitochondria, fraction III)	990	86

Table 2 (continued)
AVERAGE HYDROPHOBICITIES OF SELECTED PROTEINS

	Protein	Average hydrophobicity	Reference
116	Aspartate-β-decarboxylase (*Alcaligenes faecalis*)	1,010	87
117	Aspartate-β-decarboxylase (*Pseudomonas dacunhae*)	1,020	88
118	Aspartate transcarbamylase, catalytic chain (*Escherichia coli*)	990	89
119	Aspartate transcarbamylase, regulatory chain (*Escherichia coli*)	980	89
120	Aspartokinase, lysine sensitive (*Escherichia coli* K-12, HfrH)	940	90
121	Aspergillopeptidase B (*Aspergillus oryzae*)	810	91
122	Avidin (chicken egg)	930	92
123	Azurin (*Pseudomonas fluorescens*)	880	93
124	Azurin (*Pseudomonas polymyxa*)	970	94
125	Biotin carboxyl carrier protein (*Escherichia coli*)	1,300	95
126	Blastokinin (rabbit)	1,030	96
127	Bradykininogen (bovine)	980	97
128	α-Bungarotoxin (*Bungarus multicintus*)	1,040	98
129	C_1-inactivator (human plasma)	1,020	99
130	Calcitonin (bovine)	1,080	100
131	Calcitonin (human)	1,100	101
132	Calcitonin (porcine)	1,100	102
133	Calcitonin (salmon)	860	103
134	Calcium-binding protein (bovine adrenal medulla)	780	104
135	Calcium-binding protein (chick duodenal mucosa)	930	105
136	Calcium-binding protein-B (parvalbumin) (*Cyprinus carpio* muscle)	890	106
137	Calcium-binding protein (porcine brain)	740	107
138	Carbamyl phosphate synthetase (*Escherichia coli* B)	960	108
139	Carbonic anhydrase (bovine)	1,000	109
140	Carbonic anhydrase (*Gallus domesticus*)	1,000	110

Table 2 (continued)
AVERAGE HYDROPHOBICITIES OF SELECTED PROTEINS

	Protein	Average hydrophobicity	Reference
141	Carbonic anhydrase (parsley)	1.100	111
142	Carbonic anhydrase (sheep)	990	112
143	Carbonic anhydrase B (bovine)	950	113
144	Carbonic anhydrase B (equine)	940	114
145	Carbonic anhydrase B (*Macaca mulata*)	960	115
146	Carbonic anhydrase B (porcine)	1,000	116
147	Carbonic anhydrase C (equine)	960	114
148	Carbonic anhydrase C (porcine)	1,010	117
149	β-Carboxy-cis,cis-muconate lactonizing enzyme (*Pseudomonas putida*)	940	118
150	γ-Carboxymuconolactone decarboxylase (*Pseudomonas putida*)	1,170	119
151	Carboxypeptidase A (bovine pancreas)	1,010	120
152	Carboxypeptidase A (*Penaeus setiferus*)	910	121
153	Carboxypeptidase B (*Penaeus setiferus*)	940	121
154	Carboxypeptidase B (*Protopterus aethiopicus*)	1,010	122
155	Procarboxypeptidase B (*Protopterus aethiopicus*)	990	122
156	Carboxypeptidase B (dogfish pancreas)	1,070	123
157	Carboxypeptidase Y (yeast)	1,050	124
158	Carboxytransphosphorylase (propionic acid bacteria)	950	125
159	Catalase (bovine liver)	1,110	126
160	Colchicine-binding protein (sea urchin)	930	127
161	Colchicine-binding protein (porcine brain)	930	128
162	Chorionic gonadotropin-α (human)	960	129
163	Chorionic gonadotropin-β (human)	1,120	130
164	Chorismate mutase-prephenate dehydratase (*Escherichia coli* K-12)	990	131
165	Chymoelastase, lysine free (*Streptomyces griseus* K-1)	720	132

Table 2 (continued)
AVERAGE HYDROPHOBICITIES OF SELECTED PROTEINS

	Protein	Average hydrophobicity	Reference
166	Chymoelastase, guanidine stable (*Streptomyces griseus* K-1)	760	132
167	Chymotrypsin, anionic (fin whale)	1,060	133
168	Chymotrypsin II (human)	940	134
169	Chymotrypsinogen A (dogfish)	1,010	135
170	Chymotrypsinogen B (bovine pancreas)	950	136
171	Chymotrypsinogen B (porcine pancreas)	1,010	137
172	Chymotrypsinogen C (porcine pancreas)	970	137
173	Cocoonase (*Antheraea pernyi*)	840	138
174	Cocoonase (*Antheraea polyphemus*)	860	138
175	Prococoonase (*Antheraea polyphemus*)	840	138
176	Colicin I_a-CA53 (*Escherichia coli*)	880	139
177	Colicin I_b-P9 (*Escherichia coli*)	940	139
178	Collagen (chicken tendon)	880	140
179	Collagen, α-fraction (calf skin)	880	141
180	Collagen, spongin B (sponge)	760	142
181	Collagen (sturgeon swim bladder)	770	143
182	Collagenase (*Uca pugilator*)	920	144
183	Conalbumin (chicken)	980	145
184	Corrinoid protein (*Clostridium thermoaceticum*)	890	146
185	Creatine kinase (*Cyprinus carpio* L.)	960	147
186	Creatine kinase (human)	970	148
187	Crotonase (*Clostridium acetobutylicum*)	970	149
188	Crotoxin (*Crotalus terrificus terrificus*)	970	150
189	Cystathionase (rat liver)	980	151
190	Cytochrome b_5 (bovine liver)	1,000	152

Table 2 (continued)
AVERAGE HYDROPHOBICITIES OF SELECTED PROTEINS

	Protein	Average hydrophobicity	Reference
191	Cytochrome b_5 (equine liver)	970	152
192	Cytochrome b_5 (porcine liver)	900	152
193	Cytochrome c (horse heart)	1,050	153
194	Cytochrome c (chicken)	1,050	154
195	Cytochrome c (cow, pig, sheep)	1,020	154
196	Cytochrome c (dog)	1,070	154
197	Cytochrome c (kangaroo)	1,130	154
198	Cytochrome c (king penguin)	1,060	154
199	Cytochrome c (*Macaca mulata*)	1,060	154
200	Cytochrome c (moth)	1,060	154
201	Cytochrome c (pigeon)	1,070	154
202	Cytochrome c (rabbit)	1,070	154
203	Cytochrome c (rattlesnake)	1,050	154
204	Cytochrome c (*Saccharomyces*)	1,040	154
205	Cytochrome c (tuna)	1,050	154
206	Cytochrome c (turkey)	1,050	154
207	Cytochrome C_A (*Humicola lanuginosa*)	910	155
208	Cytochrome C_3 (*Desulfovibrio desulfuricans*)	920	156
209	Cytochrome C_3 (*Desulfovibrio gigas*)	870	157
210	Cytochrome C_3 (*Desulfovibrio salexigens*)	840	157
211	Cytochrome C_3 (*Desulfovibrio vulgaris*)	840	158
212	Cytochrome 553 (*Monochysis lutheri*)	730	159
213	Daunorubicin reductase (rat liver)	1,100	160
214	3-Deoxy-D-arabinoheptulosonate 7-phosphate synthetase-chorismate mutase (*Bacillus subtilis*)	1,000	161
215	Deoxycytidylate deaminase (T_2 r^+ bacteriophage)	970	162

Table 2 (continued)
AVERAGE HYDROPHOBICITIES OF SELECTED PROTEINS

	Protein	Average hydrophobicity	Reference
216	Deoxyribonuclease (bovine pancreas)	930	163
217	Deoxyribonuclease A (bovine pancreas)	970	164
218	Deoxyribonuclease B (bovine pancreas)	980	164
219	Deoxyribonuclease C (bovine pancreas)	980	164
220	Deoxyribonuclease II (porcine spleen)	1,050	165
221	Deoxyribonucleic acid ligase (*Escherichia coli*)	970	166
222	Deoxyribonucleotidase inhibitor II (calf spleen)	1,010	167
223	α-Dialkylamino transamidase (*Pseudomonas cepacia*)	930	168
224	Dihydrofolate reductase (T_4 bacteriophage)	1,030	169
225	Dihydrofolate reductase (*Escherichia coli* MB 1428)	1,030	170
226	Dihydrofolate reductase (*Lactobacillus casei*)	1,010	171
227	Dihydrofolate reductase (*Streptococcus faecium*)	1,060	172
228	3,4-Dihydroxyphenylacetate-2,3-oxygenase (*Pseudomonas ovalis*)	980	173
229	Dopamine-β-hydroxylase (bovine adrenal)	950	174
230	Elastin (bovine aorta)	940	175
231	Elastin (bovine ear cartilage)	990	175
232	Elastin (bovine ligamentum nuchae)	980	175
233	Encephalitogenic A-1 protein (bovine brain)	850	176
234	Encephalitogenic A-1 protein (human brain)	820	176
235	Endopolygalacturonase (*Verticillium albo-atrum*)	850	177
236	Enolase (monkey muscle)	970	178
237	Enolase (*Onkorhynchus keta*)	980	179
238	Enolase (*Onkorhynchus kisutch*)	930	179
239	Enolase (rabbit muscle)	990	180
240	Enolase (*Thermus* X-1)	900	181

Table 2 (continued)
AVERAGE HYDROPHOBICITIES OF SELECTED PROTEINS

	Protein	Average hydrophobicity	Reference
241	Enolase (*Thermus aquaticus* YT-1)	770	182
242	Enolase (yeast)	890	183
243	Enterotoxin C (*Staphylococcus* 137)	990	184
244	Enterotoxin E (*Staphylococcus aureus* FRI-326)	940	185
245	Enterotoxin A (*Staphylococcus aureus* 13N-2909)	980	186
246	Erythrocuprein (human)	740	187
247	Esterase (pig liver)	1,100	188
248	Estradiol dehydrogenase (human placenta)	940	189
249	Factor III lac (*Staphylococcus aureus*)	850	190
250	Factor VIII (bovine plasma)	1,030	191
251	Factor VIII (human plasma)	990	192
252	Ferredoxin (*Chlorobium thiosulfatophilum*)	910	193
253	Ferredoxin (*Clostridium pasteurianium*)	920	194
254	Ferredoxin (*Chromatium*)	870	195
255	Ferredoxin (*Cyperus rotundus*)	900	196
256	Ferredoxin (*Desulfovibrio gigas*)	930	196
257	Ferredoxin (*Spinacea*)	850	197
258	Ferredoxin (*Scenedesmus*)	800	198
259	Ferredoxin I (*Azotobacter vinelandii*)	1,050	199
260	Ferredoxin I (*Bacillus polymxa*)	920	200
261	Ferredoxin II (*Bacillus polymxa*)	920	200
262	Ferritin (horse spleen)	900	201
263	Ferritin (tadpole red blood cell)	850	202
264	Fibrinogen B (lobster)	990	203
265	Fibroin (*Bombyx mori*)	330	204

Table 2 (continued)
AVERAGE HYDROPHOBICITIES OF SELECTED PROTEINS

	Protein	Average hydrophobicity	Reference
266	Fibroin (*Tussah*)	430	204
267	Ficin II (*Ficus glabrata*)	930	205
268	Ficin III (*Ficus glabrata*)	960	205
269	Flagellin (*Proteus vulgaris*)	730	206
270	Flagellin (*Salmonella* SJ 25)	680	207
271	Flagellin (*Salmonella typhimurium*)	700	208
272	Flavodoxin (*Clostridium* MP)	1,010	209
273	Flavodoxin (*Clostridium pasteurianium*)	820	210
274	Flavodoxin (*Desulfovibrio gigas*)	890	211
275	Flavodoxin (*Desulfovibrio vulgaris*)	870	212
276	Flavodoxin (*Peptostreptococcus elsdenii*)	870	213
277	Follicle stimulating hormone (ovine)	900	214
278	*N*-formimino-L-glutamate iminohydrolase (*Pseudomonas* ATCC 11299b)	930	215
279	Fructose-1,6-diphosphatase (porcine kidney)	1,010	216
280	β-Galactosidase (*Escherichia coli*)	970	217
281	Gastricin (human)	920	218
282	Gastricin (porcine)	1,100	219
283	Gliadin SP 2-2 (Cappelle 1966 flour)	1,300	220
284	Gliadin SP 2-1 (wheat, Wichita 1963)	1,250	220
285	Gliadin SP 2-2 (wheat, Wichita 1963)	1,180	220
286	Gliadin SP 2-3 (wheat, Wichita 1963)	1,180	220
287	Globulin, 0.6 S γ_2 (human plasma)	1,090	221
288	Glucagon (bovine)	810	222
289	Glutamate decarboxylase (*Escherichia coli*)	1,080	223
290	Glutamate dehydrogenase (bovine liver)	1,000	224

Table 2 (continued)
AVERAGE HYDROPHOBICITIES OF SELECTED PROTEINS

	Protein	Average hydrophobicity	Reference
291	Glutamate dehydrogenase (*Neurospora crassa*)	880	225
292	Glutaminase-asparaginase (*Achromobacteraceae*)	960	226
293	Glutamine synthetase (rat liver)	1,010	227
294	γ-Glutamyl cyclotransferase (porcine liver)	960	228
295	Glyceraldehyde-3-phosphate dehydrogenase (*Bacillus stearothermophilus*)	940	229
296	Glyceraldehyde-3-phosphate dehydrogenase (bovine liver)	980	230
297	Glycerol-3-phosphate dehydrogenase (rabbit muscle)	1,040	231
298	Glycocyamine kinase (*Nepthys coeca*)	990	232
299	Glycogen phosphorylase (Baker's yeast)	1,050	233
300	Glycogen phosphorylase (*Carcharhinus falciformis*)	1,080	234
301	Glycogen phosphorylase (human)	1,060	235
302	Glycogen phosphorylase (rabbit liver)	1,060	236
303	Glycogen phosphorylase (rat)	1,070	237
304	Glycogen synthetase (porcine kidney)	870	238
305	γ-Glycoprotein (human plasma)	1,070	239
306	Growth hormone (human)	950	240
307	Haptoglobin, α_2-chain (human plasma)	960	241
308	Haptoglobin, α_{1S}-chain (human plasma)	990	241
309	Haptoglobin, β-chain (human plasma)	1,020	242
310	Hemagglutinin (*Lens esculenta Muench*)	1,000	243
311	Hemagglutinin I (*Pisum sativum L. var. pyram*)	990	244
312	Hemagglutinin II (*Pisum sativum L. var. pyram*)	930	244
313	Hemagglutinin (*Robina pseudoaccacia*)	990	245
314	Hemagglutinin II (snail)	1,070	246
315	Hemagglutinin (soy)	990	247

Table 2 (continued)
AVERAGE HYDROPHOBICITIES OF SELECTED PROTEINS

	Protein	Average hydrophobicity	Reference
316	Hemagglutinin (*Ulex europeus*)	950	248
317	Hemerythrin (*Golfingia gouldii*)	1,170	249
318	Hemocyanin (*Cancer magister*)	1,000	250
319	Hemocyanin (*Crustaceae C. sapidus*)	980	251
320	Hemocyanin (*Crustaceae E. spinifrons*)	980	251
321	Hemocyanin (*Crustaceae H. vulgaris*)	980	251
322	Hemocyanin (*Crustaceae P. vulgaris*)	980	251
323	Hemocyanin (*Mollusca E. moschata*)	1,050	251
324	Hemocyanin (*Mollusca M. brandaris*)	1,000	251
325	Hemocyanin (*Mollusca M. tranculus*)	1,010	251
326	Hemocyanin (*Mollusca O. macropus*)	1,050	251
327	Hemocyanin (*Mollusca O. vulgaris*)	1,080	251
328	Hemocyanin (*Xiphosura L. polyphemus*)	980	251
329	Hemoglobin (*Ascaris* body wall)	980	252
330	Hemoglobin (*Entosphenus japonicus*)	1,010	253
331	Hemoglobin (*Glycera dibranchiata*)	870	254
332	Hemoglobin, α-chain (*Catostomus clarkii*)	1,100	255
333	Hemoglobin, α-chain (human)	960	256
334	Hemoglobin, γ-chain (bovine)	960	256
335	Hemoglobin, γ-chain (human)	960	256
336	Hemoglobin, F-1 (*Eptatretus burgeri*)	1,170	258
337	Hemoglobin, F-2 (*Eptatretus burgeri*)	1,160	258
338	Hemoglobin, F-3 (*Eptatretus burgeri*)	1,140	258
339	Hemoglobin, F-4 (*Eptatretus burgeri*)	1,040	258
340	Hemopexin (human)	1,020	259

Table 2 (continued)
AVERAGE HYDROPHOBICITIES OF SELECTED PROTEINS

	Protein	Average hydrophobicity	Reference
341	Hemopexin (rabbit)	1,010	259
342	High potential iron sulfur protein (*Chromatium vinosum* D)	880	260
343	High potential iron sulfur protein (*Thiocapsa pfennigii*)	930	261
344	Histidine: ammonia lyase (*Pseudomonas* ATCC 11299b)	890	262
345	Histidine-binding protein (*Salmonella typhimurium*)	970	263
346	Histidine decarboxylase (*Lactobacillus* 30a)	980	264
347	L-Histidinol phosphate aminotransferase (*Salmonella typhimurium*)	970	265
348	Histidyl-tRNA synthetase (*Salmonella typhimurium*)	940	266
349	Histone III (calf thymus)	990	267
350	Histone III (*Letiobus bubalus*)	980	268
351	Histone III (pea)	980	269
352	Histone III (rainbow trout)	950	270
353	Hyaluronidase (bovine testicular)	1,030	271
354	Hydrogenase (*Clostridium pasteurianium* W5)	1,160	272
355	3-Hydroxyacyl coenzyme A dehydrogenase (porcine heart muscle)	960	273
356	β-Hydroxybutyrate dehydrogenase (bovine heart mitochondria)	930	274
357	α^1-Hydroxysteroid dehydrogenase (*Pseudomonas testosteroni*)	900	275
358	17-β-Hydroxysteroid dehydrogenase (human placenta)	910	276
359	Immunoglobulin, Eu heavy chain (human plasma)	980	277
360	Immunoglobulin, Eu light chain (human plasma)	860	277
361	Immunoglobulin, New λ-chain (human plasma)	870	278
362	Inorganic pyrophosphatase (yeast)	1,130	279
363	Insulin (bovine)	1,020	280
364	Insulin (cod)	1,110	281
365	Invertase (*Saccharomyces* FH4C)	990	282

Table 2 (continued)
AVERAGE HYDROPHOBICITIES OF SELECTED PROTEINS

	Protein	Average hydrophobicity	Reference
366	Isoamylase I (porcine pancreas)	960	283
367	Isoamylase II (porcine pancreas)	970	283
368	Isocitrate dehydrogenase (*Azotobacter vinelandii* ATCC 9104)	980	284
369	Isocitrate dehydrogenase (porcine liver)	990	285
370	Isomerase I (rabbit)	1,080	286
371	Isomerase II (rabbit)	1,100	286
372	Keratinase (*Trichophyton mentagrophytes*)	880	287
373	Δ^5-3-Ketosteroid isomerase (*Pseudomonas testosteronii*)	910	288
374	α-Lactalbumin (bovine milk)	1,050	289
375	Lactoferrin (bovine)	920	290
376	β-Lactoglobulin (caprid)	1,040	291
377	β-Lactoglobulin A (bovine)	1,070	291
378	β-Lactoglobulin A (ovine)	1,050	291
379	β-Lactoglobulin B (bovine)	1,060	291
380	β-Lactoglobulin B (ovine)	1,050	291
381	β-Lactoglobulin C (bovine)	1,070	292
382	Lactoperoxidase (bovine milk)	1,080	292
383	Lectin, α-D-galactosyl-binding (*Bandeiraea simplicifola*)	980	293
384	Leghemoglobin I (soy bean root nodule)	1,010	294
385	Leghemoglobin II (soy bean root nodule)	1,010	294
386	Leucine aminopeptidase (bovine lens)	1,010	295
387	Lipase (rat pancreas)	980	61
388	β-Lipolytic hormone (porcine pituitary)	950	296
389	γ-Lipolytic hormone (sheep)	780	297
390	α-Lipovitellin (chicken egg yolk)	980	298

Table 2 (continued)
AVERAGE HYDROPHOBICITIES OF SELECTED PROTEINS

	Protein	Average hydrophobicity	Reference
391	Luciferase (*Renilla reniformis*)	940	299
392	Luciferase-α (MAV)	880	300
393	Luciferase-β (MAV)	960	300
394	Luciferase-α (*Photobacterium fischeri*)	960	300
395	Luciferase-β (*Photobacterium fischeri*)	990	300
396	Luteinizing hormone-α chain (equine)	1,070	301
397	Luteinizing hormone-β chain (equine)	1,120	301
398	Lysine decarboxylase (*Escherichia coli*)	1,080	302
399	L-Lysine monooxygenase (*Pseudomonas fluorescens*)	1,040	303
400	Lysostaphin (*Staphylococcus aureus*)	950	304
401	Lysozyme (*Chalaropsis*)	960	305
402	Lysozyme, chick type (black swan)	970	306
403	Lysozyme, goose type (black swan)	930	306
404	Lysozyme (chicken)	890	307
405	Lysozyme (papaya)	1,080	308
406	Lysozyme (turkey)	910	309
407	Lysyl: tRNA ligase (*Escherichia coli*)	990	310
408	Lysyl: tRNA ligase (yeast)	1,040	310
409	α-Lytic protease (*Sorangium* sp.)	780	311
410	β-Lytic protease (*Sorangium* sp.)	820	311
411	Malate dehydrogenase (*Bacillus subtilis*)	970	312
412	Malate-lactate transhydrogenase (*Micrococcus lactilyticus*)	960	313
413	Malate-vitamin k reductase (*Mycobacterium phleii*)	890	314
414	L-Malic enzyme (*Escherichia coli*)	1,020	315
415	β-Melanocyte stimulating hormone (bovine)	1,180	316

Table 2 (continued)
AVERAGE HYDROPHOBICITIES OF SELECTED PROTEINS

	Protein	Average hydrophobicity	Reference
416	Melilotate hydroxylase (*Pseudomonas* sp.)	970	317
417	Methionyl-tRNA synthetase (*Escherichia coli*)	1,040	318
418	β-Methyl aspartase (*Clostridium tetanomorphum*)	980	319
419	β-Methylcrotonylcoenzyme A carboxylase (*Achromobacter*)	890	320
420	Methylmalonatesemialdehydedehydrogenase (*Pseudomonas aeruginosa*)	1,030	321
421	Molydbenum-iron protein (soy bean root nodule bacteroid)	1,010	322
422	Monellin (*Dioscoreophyllum cumminsii*)	1,140	323
423	Motilin (porcine intestinal mucosa)	1,020	324
424	cis,cis-Muconate lactonizing enzyme (*Pseudomonas putida*)	990	325
425	Muconolactone isomerase (*Pseudomonas putida*)	1,070	325
426	Mutarotase (bovine kidney cortex)	850	326
427	Myoglobin (human)	1,000	327
428	Myoglobin (whale)	1,040	328
429	Myoglobin (*Zalophus californianus*)	1,030	329
430	Myohemerythrin (*Dendrostamum pyroides*)	1,090	330
431	Myosin (bovine heart)	880	331
432	Myosin (cod)	860	332
433	Myosin (rabbit)	890	333
434	Myosin (tuna)	880	333
435	Neocarzinostatin (*Streptomyces carzinostaticus* F-41)	750	334
436	Neurophysin I (bovine pituitary)	850	335
437	Neurophysin II (bovine pituitary)	830	335
438	Neurophysin III (bovine pituitary)	820	336
439	Neurotoxin I (*Androctonus australis*)	1,070	337
440	Neurotoxin II (*Androctonus australis*)	990	337

Table 2 (continued)
AVERAGE HYDROPHOBICITIES OF SELECTED PROTEINS

	Protein	Average hydrophobicity	Reference
441	Neurotoxin III (*Androctonus australis*)	1,120	337
442	Neurotoxin I (*Leiurus quinquestriatus quinquestriatus*)	1,040	337
443	Neurotoxin II (*Leiurus quinquestriatus quinquestriatus*)	970	337
444	Neurotoxin III (*Leiurus quinquestriatus quinquestriatus*)	1,030	337
445	Neurotoxin IV (*Leiurus quinquestriatus quinquestriatus*)	1,050	337
446	Neurotoxin V (*Leiurus quinquestriatus quinquestriatus*)	1,020	337
447	Nuclease (*Micrococcus sodonencis* ATCC 11880)	880	338
448	Nuclease (*Staphylococcus*)	980	339
449	Nucleoside diphosphokinase (Brewer's yeast)	1,030	340
450	Nucleoside phosphotransferase-A-chain (carrot)	920	341
451	Nucleoside phosphotransferase-B-chain (carrot)	860	341
452	5′-Nucleotidase (*Escherichia coli*)	930	342
453	Ornithine transcarbamylase (bovine liver)	1,040	343
454	Ornithine transcarbamylase (*Streptococcus faecalis*)	990	343
455	Ovalbumin (chicken egg)	980	145
456	Ovomacroglobulin (chicken egg)	1,040	344
457	Ovomucin (chicken egg)	990	345
458	Ovomucoid (chicken egg)	830	145
459	Ovotransferrin (chicken egg)	960	346
460	2-Oxoglutarate dehydrogenase (porcine heart)	1,010	347
461	Oxytocin (mammalian)	1,290	348
462	Papain (papaya)	1,030	349
463	Parathyroid hormone (bovine)	900	350
464	Parvalbumin (rabbit skeletal muscle)	920	351
465	Penicillocarboxypeptidase-S (*Penicillium janthinellum*)	980	352

Table 2 (continued)
AVERAGE HYDROPHOBICITIES OF SELECTED PROTEINS

	Protein	Average hydrophobicity	Reference
466	Pepsin (bovine)	940	353
467	Pepsin (human)	930	218
468	Pepsin I (*Rhizopus chinensis*)	910	354
469	Pepsin II (*Rhizopus chinensis*)	910	354
470	Pepsinogen A (bovine)	960	355
471	Pepsinogen A (*Mustelus canis*)	920	356
472	Pepsinogen A (porcine)	970	357
473	Pepsinogen C (porcine)	1,020	358
474	Phenylalanine ammonia lyase (*Solanum tuberosum*)	1,030	359
475	Phenylalanine ammonia lyase (*Zea mays* L.)	940	359
476	Phenylalanine hydroxylase (rat liver)	930	360
477	Phosphatidyl serine decarboxylase (*Escherichia coli*)	1,050	361
478	Phosphoenol pyruvate carboxykinase (yeast)	1,030	362
479	Phosphofructokinase (chicken liver)	940	363
480	Phosphoglucomutase (rabbit muscle)	1,020	364
481	Phosphoglucose isomerase (human)	1,020	365
482	Phosphoglucose isomerase (rabbit muscle)	1,010	366
483	Phosphoglycerate kinase (human erythrocyte)	980	367
484	Phospholipase A (*Bitis gabonica* venom)	910	368
485	Phospholipase A_1 (*Crotalus adamanteus*)	1,030	369
486	Phospholipase A_2 (*Crotalus atrox*)	920	370
487	Phospholipase A (*Laticuda semifasciata*)	950	371
488	Phosphorylase (frog skeletal muscle)	1,070	372
489	Phycocyanin (*P. calothricoides*)	910	93
490	Phycocyanin (*S. lividicus*)	1,040	93

Table 2 (continued)
AVERAGE HYDROPHOBICITIES OF SELECTED PROTEINS

	Protein	Average hydrophobicity	Reference
491	Phycoerythrin (*P. tenera*)	830	373
492	Phytochrome (oat)	1,010	374
493	Plastocyanin (parsley)	880	375
494	Prealbumin (human serum)	990	376
495	Prealbumin, Pt 1-1 (monkey)	990	377
496	Prealbumin, Pt 2-2 (monkey)	1,010	377
497	Procarboxypeptidase A (bovine)	1,010	378
498	Procarboxypeptidase A (*Squalus acanthias*)	1,010	379
499	Progesterone-binding protein (guinea pig plasma)	1,040	380
500	Prohistidine decarboxylase (*Lactobacillus* 30a)	990	381
501	Prolactin (ovine pituitary)	950	382
502	Protease (*Bacillus thermoprotolyticus*)	940	383
503	Protease (*Mucor miehei* CBS 370.65)	940	384
504	Protease (*Staphylococcus aureus* V8)	900	385
505	Protease – A_1 (*Streptomyces griseus*)	700	386
506	Protease-acid (*Rhizopus chinensis*)	950	387
507	Protein kinase modulator (lobster tail muscle)	720	388
508	Prothrombin (bovine)	960	389
509	Protyrosinase (*Rana pipiens pipiens*)	1,100	390
510	Putidaredoxin (*Pseudomonas putida*)	910	391
511	Putrescine oxidase (*Micrococcus rubens*)	860	392
512	Pyruvate kinase (bovine skeletal muscle)	1,010	393
513	Quinonoid dihydropterin reductase (sheep liver)	890	394
514	Retinol-binding protein (human)	960	395
515	Retinol-binding protein (monkey)	960	396

Table 2 (continued)
AVERAGE HYDROPHOBICITIES OF SELECTED PROTEINS

	Protein	Average hydrophobicity	Reference
516	Rhodopsin (bovine)	1,120	397
517	Riboflavin-binding protein (chicken)	910	398
518	Ribonuclease (*Aspergillus fumigatus*)	830	399
519	Ribonuclease (*Bacillus subtilis*)	1,000	400
520	Ribonuclease (bovine pancreas)	780	401
521	Rubonuclease (ovine pancreas)	750	402
522	Ribonuclease Ch. (*Chalaropsis sp.*)	940	403
523	Ribonuclease N (*Neurospora crassa*)	890	404
524	Ribonuclease R_1 (*Rhizopus oligosporus*)	970	405
525	Ribonuclease R_2 (*Rhizopus oligosporus*)	960	405
526	Ribonuclease T_1 (*Aspergillus oryzae*)	740	406
527	Ribonuclease U_1 (*Ustilago sphaerogena*)	760	407
528	Ribonucleotide reductase (*Lactobacillus leichmanii*)	960	408
529	Ribulose diphosphate carboxylase (*Hydrogenomonas eutropha*)	980	409
530	Ribulose diphosphate carboxylase (*Hydrogenomonas facilis*)	980	409
531	Ribulose diphosphate carboxylase (*Rhodospirillum rubrum*)	940	410
532	Rubredoxin (*Clostridium pasteurianum*)	990	411
533	Rubredoxin (*Micrococcus aerogenes*)	1,120	412
534	Rubredoxin (*Pseudomonas oleovorans*)	960	413
535	D-Serine dehydratase (*Escherichia coli*)	920	414
536	Streptokinase (human plasma)	980	415
537	Subtilisin BPN (*Bacillus subtilis*)	810	416
538	Succinyl coenzyme A synthetase (*Escherichia coli*)	970	417
539	Sucrose synthetase (*Phaseolus aureus*)	1,040	418
540	Superoxide dismutase (*Neurospora crassa*)	760	419

Table 2 (continued)
AVERAGE HYDROPHOBICITIES OF SELECTED PROTEINS

	Protein	Average hydrophobicity	Reference
541	Thermolysin (*Bacillus thermoproteolyticus*)	890	420
542	Thioredoxin (T-4 bacteriophage)	1,150	421
543	Thioredoxin II (yeast)	960	422
544	Thrombin (bovine)	1,040	423
545	Thymidylate synthetase (T-2 bacteriophage)	1,140	424
546	Thyroglobulin (calf)	950	425
547	Thyroglobulin (human)	950	425
548	Thyroglobulin (porcine)	950	425
549	Thyroglobulin (rabbit)	920	425
550	Thyroglobulin (sheep)	950	425
551	Thyroid stimulating hormone – α-chain (human)	1,010	426
552	Thyroid stimulating hormone – β-chain (human)	1,170	426
553	Toxin FVII (*Dendroaspis angusticeps*)	890	427
554	Toxin α (*Dendroaspis polylepsis*)	940	428
555	Toxin γ (*Dendroaspis polylepsis*)	1,000	428
556	Toxin 4 (*Enhydrina schistosa*)	760	429
557	Toxin 5 (*Enhydrina schistosa*)	710	429
558	Toxin (*Laticauda colubrina*)	790	430
559	Toxin (*Laticauda laticauda*)	790	430
560	Toxin a (*Laticauda semifasciata*)	880	431
561	Toxin b (*Laticauda semifasciata*)	930	431
562	Toxin (*Naja haje haje*)	830	432
563	Toxin b (*Naja melanoleuca*)	1,090	433
564	Toxin d (*Naja melanoleuca*)	830	433
565	Toxin (*Naja naja atra*)	680	434

Table 2 (continued)
AVERAGE HYDROPHOBICITIES OF SELECTED PROTEINS

	Protein	Average hydrophobicity	Reference
566	Toxin (*Naja nigricollis*)	930	435
567	α-Toxin A (*Staphylococcus aureus* Woods 46)	890	436
568	α-Toxin B (*Staphylococcus aureus* Woods 46)	890	436
569	Transaminase B (*Salmonella typhimurium*)	980	437
570	Transcortin (guinea pig)	960	438
571	Transcortin (human)	970	438
572	Transcortin (rabbit)	960	440
573	Transcortin (rat)	1,030	440
574	Transferrin (bovine)	930	441
575	Transferrin (equine)	970	441
576	Transferrin (human)	930	442
577	Transferrin (porcine serum)	970	441
578	Transferrin (rabbit serum)	970	441
579	Triose phosphate dehydrogenase (*Bombus nevadensis*)	1,040	443
580	Triose phosphate dehydrogenase (honey bee)	1,020	443
581	Triose phosphate dehydrogenase (lobster)	980	443
582	Triose phosphate dehydrogenase (porcine)	990	443
583	Triose phosphate isomerase–I (human erythrocyte)	920	444
584	Triose phosphate isomerase-III (human erythrocyte)	940	444
585	Triose phosphate isomerase (rabbit muscle)	930	445
586	Triose phosphate isomerase (yeast)	970	446
587	Troponin-C (rabbit muscle)	780	447
588	Troponin-T (rabbit muscle)	1,040	447
589	Trypsin (*Evasterias trochelli*)	1,000	448
590	Trypsin (lungfish)	920	449

Table 2 (continued)
AVERAGE HYDROPHOBICITIES OF SELECTED PROTEINS

	Protein	Average hydrophobicity	Reference
591	Trypsin (human)	870	450
592	Trypsin, anionic (human)	930	451
593	Trypsin (porcine)	970	452
594	Trypsin (sheep)	850	453
595	Trypsin (shrimp)	840	454
596	Trypsin (*Streptomyces griseus*)	820	455
597	Trypsin inhibitor (*Ascaris lumbricoides suis*)	970	456
598	Trypsin inhibitor (bovine pancreas)	1,150	457
599	Trypsin inhibitor (porcine pancreas)	900	458
600	Trypsin inhibitor (bovine pancreas)	1,070	459
601	Trypsin inhibitor I (*Phaseolus vulgaris*)	800	460
602	Trypsin inhibitor II (*Phaseolus vulgaris*)	780	460
603	Trypsinogen (bovine)	910	461
604	Trypsinogen (dogfish)	950	462
605	Trypsinogen (lungfish)	930	449
606	Trypsinogen, anionic (porcine)	980	463
607	Trypsinogen (sheep)	900	464
608	Tryptophan synthetase A (*Escherichia coli*)	1,060	465
609	Tryptophan synthetase–α-chain (*Aerobacter aerogenes*)	1,060	466
610	Tryptophan synthetase – α-chain (*Bacillus subtilis*)	1,020	467
611	Tryptophan synthetase – α-chain (*Salmonella typhimurium*)	980	466
612	Tryptophan synthetase – β-chain (*Bacillus subtilis*)	940	468
613	Tryptophanase (*Bacillus alvei*)	980	469
614	Tryptophanase (*Escherichia coli* B/1t7-A)	1,030	470
615	Tyrosyl-tRNA synthetase (*Bacillus stearothermophilus*)	1,060	471

Table 2 (continued)
AVERAGE HYDROPHOBICITIES OF SELECTED PROTEINS

	Protein	Average hydrophobicity	Reference
616	Urease (jackbean meal)	990	472
617	Urokinase S-1 (human urine)	1,030	473
618	Urokinase S-2 (human urine)	990	473
619	Valine:tRNA ligase (yeast)	1,080	310
620	Vasopressin, lysine (mammalian)	1,210	474

Compiled by C. C. Bigelow and M. Channon.

REFERENCES

1. **Lederer, Coutts, Laursen, and Westheimer,** *Biochemistry,* 5, 823 (1966).
2. **Klett, Fulpius, Cooper, Smith, Reich, and Possani,** *J. Biol. Chem.,* 248, 6841 (1973).
3. **Rosenberry, Chang, and Chen,** *J. Biol. Chem.,* 247, 1555 (1972).
4. **Mega, Ikenaka, and Matsushima,** *J. Biochem.* (Tokyo), 68, 109 (1970).
5. **Verpoorte,** *J. Biol. Chem.,* 247, 4787 (1972).
6. **Wetmore and Verpoorte,** *Can. J. Biochem.,* 50, 563 (1972).
7. **Verpoorte,** *Biochemistry,* 13, 793 (1974).
8. **Forsyth, Theil, and Jones,** *J. Biol. Chem.,* 245, 5354 (1970).
9. **Becker, Kredich, and Tomkins,** *J. Biol. Chem.,* 244, 2418 (1969).
10. **Ikenaka, Ishiguro, Emura, Kaufman, Isemura, Bauer, and Schmid,** *Biochemistry,* 11, 3817 (1972).
11. **Weihing and Korn,** *Biochemistry,* 10, 590 (1971).
12. **Gosselin-Rey, Gerady, Gaspar-Godfroid, and Carsten,** *Biochim. Biophys. Acta,* 175, 165 (1969).
13. **Carsten and Katz,** *Biochim. Biophys. Acta,* 90, 534 (1964).
14. **Bridgen,** *Biochem. J.,* 123, 591 (1971).
15. **Booyse, Hoveke, and Rafelson, Jr.,** *J. Biol. Chem.,* 248, 4083 (1973).
16. **Carsten,** *Biochemistry,* 4, 1049 (1965).
17. **Adelstein and Kuehl,** *Biochemistry,* 9, 1355 (1970).
18. **Elzinga,** *Biochemistry,* 9, 1365 (1970).
19. **Vanaman, Wakil, and Hill,** *J. Biol. Chem.,* 243, 6420 (1968).
20. **Diederich and Grisolia,** *J. Biol. Chem.,* 244, 2412 (1969).
21. **Ramponi, Guerritore, Treves, Nassi, and Baccari,** *Arch. Biochem. Biophys.,* 130, 362 (1969).
22. **Wolfenden, Tomozawa, and Bamman,** *Biochemistry,* 7, 3965 (1968).
23. **Phelan, McEvoy, Rooney, and Brady,** *Biochim. Biophys. Acta,* 200, 370 (1970).
24. **Pfrogner,** *Arch. Biochem. Biophys.,* 119, 147 (1967).
25. **Schramm and Hochstein,** *Biochemistry,* 11, 2777 (1972).
26. **Kyte,** *J. Biol. Chem.,* 247, 7642 (1972).
27. **Noltmann, Mahowald, and Kuby,** *J. Biol. Chem.,* 237, 1146 (1962).
28. **Takai, Kurashina, Suzuki-Hori, Okamoto, and Hayashi,** *J. Biol. Chem.,* 249, 1965 (1974).
29. **Woodward and Braymer,** *J. Biol. Chem.,* 241, 580 (1966).
30. **Tanaka, Haniu, and Yasunobu,** *J. Biol. Chem.,* 248, 1141 (1973).
31. **Chu and Kimura,** *J. Biol. Chem.,* 248, 2089 (1973).
32. **Shimomura and Johnson,** *Biochemistry,* 8, 3991 (1969).
33. **Nagata and Burger,** *J. Biol. Chem.,* 249, 3116 (1974).
34. **Matsuzawa and Segal,** *J. Biol. Chem.,* 243, 5929 (1968).
35. **Pederson and Foster,** *Biochemistry,* 8, 2357 (1969).
36. **Allerton, Elwyn, Edsall, and Spahr,** *J. Biol. Chem.,* 237, 85 (1962).
37. **McMenamy, Dintzis, and Watson,** *J. Biol. Chem.,* 246, 4744 (1971).
38. **Peters, Jr.,** *J. Biol. Chem.,* 237, 2182 (1962).
39. **Jornvall,** *Eur. J. Biochem.,* 16, 25 (1970).
40. **Steinman and Jakoby,** *J. Biol. Chem.,* 243, 730 (1968).

Table 2 (continued)
AVERAGE HYDROPHOBICITIES OF SELECTED PROTEINS

41. Penhoet, Kochman, and Rutter, *Biochemistry,* 8, 4396 (1969).
42. Schwartz and Horecker, *Arch. Biochem. Biophys.*, 115, 407 (1966).
43. Komatsu and Feeney, *Biochim. Biophys. Acta,* 206, 305 (1970).
44. Marquardt, *Can. J. Biochem.*, 48, 322 (1970).
45. Marquardt, *Can. J. Biochem.*, 47, 527 (1969).
46. Marquardt, *Can. J. Biochem.*, 49, 658 (1971).
47. Lebherg, Bradshaw, and Rutter, *J. Biol. Chem.*, 248, 1660 (1973).
48. Anderson, Gibbons, and Perham, *Eur. J. Biochem.*, 11, 503 (1969).
49. Rutter, Woodfin, and Blostein, *Acta Chem. Scand.*, 17, S226 (1963).
50. Harris, Kobes, Teller, and Rutter, *Biochemistry,* 8, 2442 (1969).
51. Robertson, Hammerstedt, and Wood, *J. Biol. Chem.*, 246, 2075 (1971).
52. Hulett-Cowling and Campbell, *Biochemistry,* 10, 1364 (1971).
53. Christen, Vallee, and Simpson, *Biochemistry,* 10, 1377 (1971).
54. Mizusawa and Yoshida, *J. Biol. Chem.*, 247, 6978 (1972).
55. Hersh, *Biochemistry,* 10, 2884 (1971).
56. Christenson, Dairman, and Udenfriend, *Arch. Biochem. Biophys.*, 141, 356 (1970).
57. Tu and McCormick, *J. Biol. Chem.*, 248, 6339 (1973).
58. Prescott, Wilkes, Wagner, and Wilson, *J. Biol. Chem.*, 246, 1756 (1971).
59. Chenoweth, Brown, Valenzuela, and Smith, *J. Biol. Chem.*, 248, 1684 (1972).
60. DePinto and Campbell, *Biochemistry,* 7, 114 (1968).
61. Vandermeers and Christophe, *Biochim. Biophys. Acta,* 154, 110 (1968).
62. Malacinski and Rutter, *Biochemistry,* 8, 4382 (1969).
63. Stein, Junge, and Fischer, *J. Biol. Chem.*, 235, 371 (1960).
64. Pfueller and Elliott, *J. Biol. Chem.*, 244, 48 (1967).
65. Junge, Stein, Neurath, and Fischer, *J. Biol. Chem.*, 234, 556 (1959).
66. Caldwell, Dickey, Hanrahan, Kung, Kung, and Misko, *J. Am. Chem. Soc.*, 76, 143 (1954).
67. Muus, *J. Am. Chem. Soc.*, 76, 5163 (1954).
68. Cozzone, Paséro, Beaupoil, and Marchis-Mouren, *Biochim. Biophys. Acta,* 207, 490 (1970).
69. Queener, Queener, Meeks, and Gunsalus, *J. Biol. Chem.*, 248, 151 (1973).
70. Jonas, *J. Biol. Chem.*, 247, 7767 (1972).
71. Morrisett, David, Pownall, and Gotto, Jr., *Biochemistry,* 12, 1290 (1973).
72. Jackson, Baker, Taunton, Smith, Garner, and Gotto, Jr., *J. Biol. Chem.*, 248, 2639 (1973).
73. Lux, John, and Brewer, Jr., *J. Biol. Chem.*, 247, 7510 (1972).
74. Brown, Levy, and Fredrickson, *J. Biol. Chem.*, 245, 6588 (1970).
75. Burley, *Biochemistry,* 12, 1464 (1973).
76. Kaji and Tagawa, *Biochim. Biophys. Acta,* 207, 456 (1970).
77. Parsons and Hogg, *J. Biol. Chem.*, 249, 3602 (1974).
78. Patrick and Lee, *J. Biol. Chem.*, 244, 4277 (1969).
79. Boeker, Fischer, and Snell, *J. Biol. Chem.*, 244, 5239 (1969).
80. Blethen and Kaplan, *Biochemistry,* 7, 2123 (1968).
81. Blethen and Kaplan, *Biochemistry,* 6, 1413 (1967).
82. Ho, Milikin, Bobbitt, Grinnon, Burch, Frank, Boeck, and Squires, *J. Biol. Chem.*, 245, 3708 (1970).
83. Yellin and Wriston, Jr., *Biochemistry,* 5, 1605 (1966).
84. Tosa, Sano, Yamamoto, Nakamura, and Chibata, *Biochemistry,* 12, 1075 (1973).
85. Marino, Scardi, and Zito, *Biochem. J.*, 99, 595 (1966).
86. Magee and Phillips, *Biochemistry,* 10, 3397 (1971).
87. Tate and Meister, *Biochemistry,* 7, 3240 (1968).
88. Tate and Meister, *Biochemistry,* 9, 2626 (1970).
89. Weber, *J. Biol. Chem.*, 243, 543 (1968).
90. Lafuma, Gros, and Patte, *Eur. J. Biochem.*, 15, 111 (1970).
91. Subramanian and Kalnitsky, *Biochemistry,* 3, 1868 (1964).
92. Green and Toms, *Biochem. J.*, 118, 67 (1970).
93. Berns, Scott, and O'Reilly, *Science,* 145, 1054 (1964).
94. Ambler and Brown, *J. Mol. Biol.*, 9, 825 (1964).
95. Fall and Vagelos, *J. Biol. Chem.*, 247, 8005 (1972).
96. Krishnan and Daniel, Jr., *Biochim. Biophys. Acta,* 168, 579 (1968).
97. Nagawa, Mizushima, Sato, Iwanaga, and Suziki, *J. Biochem.* (Tokyo), 60, 643 (1966).
98. Clark, Macmurchie, Elliott, Wolcott, Landel, and Raftery, *Biochemistry,* 11, 1663 (1972).

Table 2 (continued)
AVERAGE HYDROPHOBICITIES OF SELECTED PROTEINS

99. Haupt, Heimburger, Kranz, and Schwick, *Eur. J. Biochem.*, 17, 254 (1970).
100. Brewer, Jr., Schlueter, and Aldred, *J. Biol. Chem.*, 245, 4232 (1970).
101. Neher, Riniker, Maier, Byfield, Gudmundsson, and MacIntyre, *Nature* (Lond.), 220, 984 (1968).
102. Brewer, Jr., Keutman, Potts, Jr., Reisfeld, Schlueter, and Munson, *J. Biol. Chem.*, 243, 5739 (1968).
103. O'Dor, Parkes, and Copp, *Can. J. Biochem.*, 47, 823 (1969).
104. Brooks and Siegel, *J. Biol. Chem.*, 248, 4189 (1973).
105. Bredderman and Wasserman, *Biochemistry*, 13, 1687 (1974).
106. Coffee and Bradshaw, *J. Biol. Chem.*, 248, 3305 (1973).
107. Wolff and Siegel, *J. Biol. Chem.*, 247, 4180 (1972).
108. Foley, Poon, and Anderson, *Biochemistry*, 10, 4562 (1971).
109. Nyman and Lindskog, *Biochim. Biophys. Acta*, 85, 141 (1964).
110. Bernstein and Schraer, *J. Biol. Chem.*, 247, 1306 (1972).
111. Tobin, *J. Biol. Chem.*, 245, 2656 (1970).
112. Tanis and Tashian, *Biochemistry*, 10, 4852 (1971).
113. Wong and Tanford, *J. Biol. Chem.*, 248, 8518 (1973).
114. Furth, *J. Biol. Chem.*, 243, 4832 (1968).
115. Duff and Coleman, *Biochemistry*, 5, 2009 (1966).
116. Ashworth, Spencer, and Brewer, *Arch. Biochem. Biophys.*, 142, 122 (1971).
117. Tanis, Tashian, and Yu, *J. Biol. Chem.*, 245, 6003 (1970).
118. Patel, Meagher, and Ornston, *Biochemistry*, 12, 3531 (1973).
119. Parke, Meagher, and Ornston, *Biochemistry*, 12, 3537 (1973).
120. Smith and Stockell, *J. Biol. Chem.*, 207, 501 (1954).
121. Gates and Travis, *Biochemistry*, 12, 1867 (1973).
122. Reeck and Neurath, *Biochemistry*, 11, 3947 (1972).
123. Prahl and Neurath, *Biochemistry*, 5, 4137 (1966).
124. Haberland, Willard, and Wood, *Biochemistry*, 11, 712 (1972).
125. Hayashi, Moore, and Stein, *J. Biol. Chem.*, 248, 2296 (1973).
126. Schnuchel, *Z. Physiol. Chem.*, 303, 91 (1956).
127. Shelanski and Taylor, *J. Cell Biol.*, 38, 304 (1968).
128. Weisenbug, Borisy, and Taylor, *Biochemistry*, 7, 4466 (1968).
129. Bellisario, Carlsen, and Bahl, *J. Biol. Chem.*, 248, 6796 (1973).
130. Carlsen, Bahl, and Swaminathan, *J. Biol. Chem.*, 248, 6810 (1973).
131. Davidson, Blackburn, and Dopheide, *J. Biol. Chem.*, 247, 4441 (1972).
132. Siegel and Awad, Jr., *J. Biol. Chem.*, 248, 3233 (1973).
133. Koide and Matsuoka, *J. Biochem.* (Tokyo), 68, 1 (1970).
134. Coan, Roberts, and Travis, *Biochemistry*, 10, 2711 (1971).
135. Prahl and Neurath, *Biochemistry*, 5, 2131 (1966).
136. Smillie, Enenkel, and Kay, *J. Biol. Chem.*, 241, 2097 (1966).
137. Gratecos, Guy, Rovery, and Desnuelle, *Biochim. Biophys. Acta*, 175, 82 (1969).
138. Kramer, Felsted, and Law, *J. Biol. Chem.*, 248, 3021 (1973).
139. Konisky, *J. Biol. Chem.*, 247, 3750 (1972).
140. Leach, *Biochem. J.*, 67, 83 (1957).
141. Piez, Weiss, and Lewis, *J. Biol. Chem.*, 235, 1987 (1960).
142. Piez and Gross, *Biochim. Biophys. Acta*, 34, 24 (1959).
143. Eastoe, *Biochem. J.*, 65, 363 (1957).
144. Eisen, Henderson, Jeffrey, and Bradshaw, *Biochemistry*, 12, 1814 (1973).
145. Lewis, Snell, Hirschman, and Fraenkel-Conrat, *J. Biol. Chem.*, 186, 23 (1950).
146. Ljungdahl, LeGall, and Lee, *Biochemistry*, 12, 1802 (1973).
147. Gosselin-Rey and Gerday, *Biochim. Biophys. Acta*, 221, 241 (1970).
148. Kumdavalli, Moreland, and Watts, *Biochem. J.*, 117, 513 (1970).
149. Waterson, Castellino, Hass, and Hill, *J. Biol. Chem.*, 247, 5266 (1972).
150. Fischer and Dorfel, *Z. Physiol. Chem.*, 297, 278 (1954).
151. Loiselet and Chatagner, *Biochim. Biophys. Acta*, 130, 180 (1966).
152. Ozols, *Biochemistry*, 13, 426 (1974).
153. Stellwagen and Rysary, *J. Biol. Chem.*, 247, 8074 (1972).
154. Margoliash and Schejter, *Adv. Protein Chem.*, 21, 113 (1966).
155. Morgan, Hensley, Jr., and Riehm, *J. Biol. Chem.*, 247, 6555 (1972).
156. Drucker, Trousil, Campbell, Barlow, and Margoliash, *Biochemistry*, 9, 1515 (1970).

Table 2 (continued)
AVERAGE HYDROPHOBICITIES OF SELECTED PROTEINS

157. Drucker, Trousil, and Campbell, *Biochemistry,* 9, 3395 (1970).
158. Trousil, and Campbell, *J. Biol. Chem.,* 249, 386 (1974).
159. Laycock and Cragie, *Can. J. Biochem.,* 49, 641 (1971).
160. Felsted, Gee, and Bachur, *J. Biol. Chem.,* 249, 3672 (1974).
161. Huang, Nakatsukasa, and Nester, *J. Biol. Chem.,* 249, 4467 (1974).
162. Maley, Guarino, and Maley, *J. Biol. Chem.,* 247, 931 (1972).
163. Gehrmann, and Okada, *Biochim. Biophys. Acta,* 23, 621 (1957).
164. Salnikow, Moore, and Stein, *J. Biol. Chem.,* 245, 5685 (1970).
165. Oshima and Price, *J. Biol. Chem.,* 248, 7522 (1973).
166. Modrich, Anraku, and Lehaman, *J. Biol. Chem.,* 248, 7495 (1973).
167. Lindberg, *Biochemistry,* 6, 323 (1967).
168. Lamartiniere, Itoh, and Dempsey, *Biochemistry,* 10, 4783 (1971).
169. Erickson and Mathews, *Biochemistry,* 12, 372 (1973).
170. Poe, Greenfield, Hirshfield, Williams, and Hoogsteen, *Biochemistry,* 11, 1023 (1972).
171. Gundersen, Dunlap, Harding, Freisheim, Otting, and Huennekens, *Biochemistry,* 11, 1018 (1972).
172. D'Souza, Warwick, and Freisheim, *Biochemistry,* 11, 1528 (1972).
173. Senoh, Kita, and Kamimoto, in *Biological and Chemical Aspects of Oxygenases,* Bloch and Hayaishi, Eds. Maruzen, Tokyo, 1966, 378.
174. Craine, Daniels, and Kaufman, *J. Biol. Chem.,* 248, 7838 (1973).
175. Gotte, Stern, Elsden, and Partridge, *Biochem. J.,* 87, 344 (1963).
176. Oshiro and Eylar, *Arch. Biochem. Biophys.,* 138, 606 (1970).
177. Wang and Keen, *Arch. Biochem. Biophys.,* 141, 749 (1970).
178. Winstead, *Biochemistry,* 11, 1046 (1972).
179. Ruth, Soja, and Wold, *Arch. Biochem. Biophys.,* 140, 1 (1970).
180. Holt and Wold, *J. Biol. Chem.,* 236, 3227 (1961).
181. Barnes and Stellwagen, *Biochemistry,* 12, 1559 (1973).
182. Stellwagen, Cronlund, and Barnes, *Biochemistry,* 12, 1552 (1973).
183. Malmström, Kimmel, and Smith, *J. Biol. Chem.,* 234, 1108 (1959).
184. Huang, Shih, Borja, Avena, and Bergdoll, *Biochemistry,* 6, 1480 (1967).
185. Borja, Fanning, Huang, and Bergdoll, *J. Biol. Chem.,* 247, 2456 (1972).
186. Schantz, Roessler, Woodburn, Lynch, Jacoby, Silverman, Gorman, and Spero, *Biochemistry,* 11, 360 (1972).
187. Kimmel, Markowitz, and Brown, *J. Biol. Chem.,* 234, 46 (1959).
188. Barker and Jencks, *Biochemistry,* 8, 3879 (1969).
189. Burns, Engle, and Bethune, *Biochemistry,* 11, 2699 (1972).
190. Hays, Simoni, and Roseman, *J. Biol. Chem.,* 248, 941 (1973).
191. Schmer, Kirby, Teller, and Davie, *J. Biol. Chem.,* 247, 2512 (1972).
192. Legaz, Schmer, Counts, and Davie, *J. Biol. Chem.,* 248, 3946 (1973).
193. Buchanan, Matsubara, and Evans, *Biochim. Biophys. Acta,* 189, 46 (1969).
194. Tanaka, Nakashima, Benson, Mower, and Yasunobu, *Biochemistry,* 5, 1666 (1966).
195. Sasaki and Matsubara, *Biochem. Biophys. Res. Commun.,* 28, 467 (1967).
196. Lee, Travis, and Black, *Arch. Biochem. Biophys.,* 141, 676 (1970).
197. Matsubara, Sasaki, and Chain, *Proc. Natl. Acad. Sci. USA,* 57, 439 (1967).
198. Matsubara, *J. Biol. Chem.,* 243, 370 (1968).
199. Yoch and Arnon, *J. Biol. Chem.,* 247, 4514 (1972).
200. Stombaugh, Burris, and Orme-Johnson, *J. Biol. Chem.,* 248, 7951 (1973).
201. Harrison, Hofmann, and Mainwaring, *J. Mol. Biol.,* 4, 251 (1962).
202. Theil, *J. Biol. Chem.,* 248, 622 (1973).
203. Fuller and Doolittle, *Biochemistry,* 10, 1305 (1971).
204. Lucas, Shaw, and Smith, *J. Mol. Biol.,* 2, 339 (1960).
205. Kortt, Hamilton, Webb, and Zener, *Biochemistry,* 13, 2023 (1974).
206. Kobayashi, Rinker, and Koffler, *Arch, Biochem. Biophys.,* 84, 342 (1959).
207. Hotani, Ooi, Kagawa, Asakura, and Yamaguchi, *Biochim. Biophys. Acta,* 214, 206 (1970).
208. Joys and Rankis, *J. Biol. Chem.,* 247, 5180 (1972).
209. Tanaka, Haniu, and Yasunobu, *J. Biol. Chem.,* 249, 4393 (1974).
210. Knight, Jr. and Hardy, *J. Biol. Chem.,* 242, 1370 (1967).
211. Dubourdieu and LeGall, *Biochem. Biophys. Res. Commun.,* 38, 965 (1970).
212. Tanaka, Haniu, Matsueda, Yasunobu, Mayhew, and Massey, *Biochemistry,* 10, 3041 (1971).
213. Mayhew and Massey, *J. Biol. Chem.,* 244, 794 (1969).

Table 2 (continued)
AVERAGE HYDROPHOBICITIES OF SELECTED PROTEINS

214. Cahill, Shetlar, Payne, Endecott, and Li, *Biochim. Biophys. Acta,* 154, 40 (1968).
215. Wickner and Tabor, *J. Biol. Chem.,* 247, 1605 (1972).
216. Mendicino, Kratowich, and Oliver, *J. Biol. Chem.,* 247, 6643 (1972).
217. Fowler and Zabin, *J. Biol. Chem.,* 245, 5032 (1970).
218. Mills and Tang, *J. Biol. Chem.,* 242, 3093 (1967).
219. Tauber and Madison, *J. Biol. Chem.,* 240, 645 (1965).
220. Booth and Ewart, *Biochim. Biophys. Acta,* 181, 226 (1969).
221. Nimberg and Schmid, *J. Biol. Chem.,* 247, 5056 (1972).
222. Behrens and Bromer, *Vitam. Horm.* (New York), 16, 263 (1958).
223. Strausbauch and Fischer, *Biochemistry,* 9, 226 (1970).
224. Moon and Smith, *J. Biol. Chem.,* 248, 3082 (1973).
225. Jacobson, Strickland, and Barratt, *Biochim. Biophys. Acta,* 188, 283 (1969).
226. Roberts, Holcenberg, and Dolowy, *J. Biol. Chem.,* 247, 84 (1972).
227. Tate, Leu, and Meister, *J. Biol. Chem.,* 247, 5312 (1972).
228. Adamson, Sewczuk, and Connell, *Can. J. Biochem.,* 49, 218 (1971).
229. Singleton, Kimmel, and Amelunxen, *J. Biol. Chem.,* 244, 1623 (1969).
230. Heinz and Kulbe, *Z. Physiol. Chem.,* 351, 249 (1970).
231. Fondy, Ross, and Sollohub, *J. Biol. Chem.,* 244, 1631 (1969).
232. Pradel, Kassab, Conlay, and Thoai, *Biochim. Biophys. Acta,* 154, 305 (1968).
233. Fosset, Muir, Nielsen, and Fischer, *Biochemistry,* 10, 4105 (1971).
234. Assaf and Yunis, *Biochemistry,* 12, 1423 (1973).
235. Appelman, Yunis, Krebs, and Fischer, *J. Biol. Chem.,* 238, 1358 (1963).
236. Wolf, Fischer, and Krebs, *Biochemistry,* 9, 1923 (1970).
237. Sevilla and Fischer, *Biochemistry,* 8, 2161 (1969).
238. Issa and Mendicino, *J. Biol. Chem.,* 248, 685 (1973).
239. Boenisch and Alper, *Biochim. Biophys. Acta,* 214, 135 (1970).
240. Dixon and Li, *J. Gen. Physiol.,* Suppl. 45, 176 (1962).
241. Black and Dixon, *Nature* (Lond.), 218, 736 (1968).
242. Barnett, Lee, and Bowman, *Biochemistry,* 11, 1189 (1972).
243. Ticha, Entlicher, Kostir, and Kocourek, *Biochim. Biophys. Acta,* 221, 282 (1970).
244. Entlicher, Kostir, and Kocourek, *Biochim. Biophys. Acta,* 221, 272 (1970).
245. Bourillon and Fort, *Biochim. Biophys. Acta,* 154, 28 (1968).
246. Hammarström and Kabat, *Biochemistry,* 8, 2696 (1969).
247. Wada, Pallansch, and Liener, *J. Biol. Chem.,* 233, 395 (1958).
248. Matsumoto and Osawa, *Biochim. Biophys. Acta,* 194, 180 (1969).
249. Groskopf, Holleman, Margoliash, and Klotz, *Biochemistry,* 5, 3779 (1966).
250. Carpenter and van Holde, *Biochemistry,* 12, 2231 (1973).
251. Ghiretti-Magaldi, Nuzzolo, and Ghiretti, *Biochemistry,* 5, 1943 (1966).
252. Okazaki, Wittenberg, Briehl, and Wittenberg, *Biochim. Biophys. Acta,* 140, 258 (1967).
253. Dohi, Sugita, and Yoneyama, *J. Biol. Chem.,* 248, 2354 (1973).
254. Imamura, Baldwin, and Riggs, *J. Biol. Chem.,* 247, 2785 (1972).
255. Powers and Edmundson, *J. Biol. Chem.,* 247, 6694 (1972).
256. Schroeder, Shelton, Shelton, Cormick, and Jones, *Biochemistry,* 2, 992 (1963).
257. Babin, Schroeder, Shelton, Shelton, and Robberson, *Biochemistry,* 5, 1297 (1966).
258. Bannai, Sugita, and Yoneyama, *J. Biol. Chem.,* 247, 505 (1972).
259. Hrkal and Muller-Eberhard, *Biochemistry,* 10, 1746 (1971).
260. Dus, Tedro, and Bartsch, *J. Biol. Chem.,* 248, 7318 (1973).
261. Tedro, Meyer, and Kamen, *J. Biol. Chem.,* 249, 1182 (1974).
262. Klee and Gladner, *J. Biol. Chem.,* 247, 8051 (1972).
263. Lever, *J. Biol. Chem.,* 247, 4317 (1972).
264. Chang and Snell, *Biochemistry,* 7, 2012 (1968).
265. Henderson and Snell, *J. Biol. Chem.,* 248, 1906 (1973).
266. DeLorenzo, Di Natale, and Schechter, *J. Biol. Chem.,* 249, 908 (1974).
267. Delange, Hooper, and Smith, *J. Biol. Chem.,* 248, 3261 (1973).
268. Hooper, Smith, Sommer, and Chalkley, *J. Biol. Chem.,* 248, 3275 (1973).
269. Patthy, Smith, and Johnson, *J. Biol. Chem.,* 248, 6834 (1973).
270. Bailey and Dixon, *J. Biol. Chem.,* 248, 5463 (1973).
271. Brunish and Hogberg, *C.R. Trav. Lab. Carlsberg,* 32, 35 (1960).

Table 2 (continued)
AVERAGE HYDROPHOBICITIES OF SELECTED PROTEINS

272. Nakos and Mortenson, *Biochemistry,* 10, 2442 (1971).
273. Noyes and Bradshaw, *J. Biol. Chem.,* 248, 3052 (1973).
274. Menzel and Hammes, *J. Biol. Chem.,* 248, 4885 (1973).
275. Squire, Delin, and Porath, *Biochim. Biophys. Acta,* 89, 409 (1964).
276. Jarabak and Street, *Biochemistry,* 10, 3831 (1971).
277. Edelman, *Biochemistry,* 9, 3197 (1970).
278. Chen and Poljak, *Biochemistry,* 13, 1295 (1974).
279. Heinrikson, Sterner, Noyes, Cooperman, and Bruchman, *J. Biol. Chem.,* 248, 4235 (1973).
280. Corfield and Robson, *Biochem. J.,* 84, 146 (1962).
281. Grant and Reid, *Biochem. J.,* 106, 531 (1968).
282. Neumann and Lampen, *Biochemistry,* 6, 468 (1967).
283. Cozzone, Paséro, and Marchis-Mouren, *Biochim. Biophys. Acta,* 200, 590 (1970).
284. Chung and Franzen, *Biochemistry,* 8, 3175 (1969).
285. Illingworth and Tipton, *Biochem. J.,* 118, 253 (1970).
286. Yoshida and Carter, *Biochim. Biophys. Acta,* 194, 151 (1969).
287. Yu, Harmon, Wachter, and Blank, *Arch. Biochem. Biophys.,* 135, 363 (1969).
288. Kawahara, Wang, and Talalay, *J. Biol. Chem.,* 237, 1500 (1962).
289. Gordon and Ziegler, *Arch. Biochem. Biophys.,* 57, 80 (1955).
290. Castellino, Fish, and Mann, *J. Biol. Chem.,* 245, 4269 (1970).
291. Bell, McKenzie, and Shaw, *Biochim. Biophys. Acta,* 154, 284 (1968).
292. Rombauts, Schroeder, and Morrison, *Biochemistry,* 6, 2965 (1967).
293. Hayes and Goldstein, *J. Biol. Chem.,* 249, 1094 (1974).
294. Elfolk, *Acta Chem. Scand.,* 15, 545 (1961).
295. Carpenter and Vahl, *J. Biol. Chem.,* 248, 294 (1972).
296. Gilardeau and Chretien, *Can. J. Biochem.,* 48, 1017 (1970).
297. Gráf, Cseh, and Medzihradszky-Schweigher, *Biochim. Biophys. Acta,* 175, 444 (1969).
298. Bernardi and Cook, *Biochim. Biophys. Acta,* 44, 96 (1960).
299. Karkhanis and Cormier, *Biochemistry,* 10, 317 (1971).
300. Hastings, Weber, Friedland, Eberbard, Mitchell, and Gunsalus, *Biochemistry,* 8, 4681 (1969).
301. Landefeld and McShan, *Biochemistry,* 13, 1389 (1974).
302. Sabo and Fischer, *Biochemistry,* 13, 670 (1974).
303. Flashner and Massey, *J. Biol. Chem.,* 248, 2579 (1974).
304. Trayer and Buckley, *J. Biol. Chem.,* 245, 4842 (1970).
305. Shih and Hash, *J. Biol. Chem.,* 246, 996 (1971).
306. Arnheim, Hindenburg, Begg, and Morgan, *J. Biol. Chem.,* 248, 8036 (1973).
307. Canfield and Anfinsen, *J. Biol. Chem.,* 238, 2684 (1963).
308. Smith, Kimmel, Brown, and Thompson, *J. Biol. Chem.,* 215, 67 (1955).
309. Larue and Speck, Jr., *J. Biol. Chem.,* 245, 1985 (1970).
310. Rymo, Lundvik, and Lagerkvist, *J. Biol. Chem.,* 247, 3888 (1972).
311. Jurášek and Whitaker, *Can. J. Biochem.,* 45, 917 (1967).
312. Yoshida, *J. Biol. Chem.,* 240, 1113 (1965).
313. Allen and Patil, *J. Biol. Chem.,* 247, 909 (1972).
314. Imai and Brodie, *J. Biol. Chem.,* 248, 7487 (1973).
315. Spina, Jr., Bright, and Rosenbloom, *Biochemistry,* 9, 3794 (1970).
316. Li, *Adv. Protein Chem.,* 12, 270 (1957).
317. Strickland and Massey, *J. Biol. Chem.,* 248, 2944 (1973).
318. Lawrence, *Eur. J. Biochem.,* 15, 436 (1970).
319. Hsiang, Myrtle, and Bright, *J. Biol. Chem.,* 242, 3079 (1967).
320. Apitz-Castro, Rehn, and Lynen, *Eur. J. Biochem.,* 16, 71 (1970).
321. Bannerjee, Sanders, and Sokatch, *J. Biol. Chem.,* 245, 1828 (1970).
322. Israel, Howard, Evans, and Russel, *J. Biol. Chem.,* 249, 500 (1974).
323. Morris, Mortenson, Deibler, and Cagan, *J. Biol. Chem.,* 248, 534 (1973).
324. Brown, Cook, and Dryburgh, *Can. J. Biochem.,* 51, 533 (1973).
325. Meagher and Ornston, *Biochemistry,* 12, 3523 (1973).
326. Fishman, Pentchev, and Bailey, *Biochemistry,* 12, 2490 (1973).
327. Perkoff, Hill, Brown, and Tyler, *J. Biol. Chem.,* 237, 2820 (1962).
328. Edmundson, *Nature,* (Lond.), 205, 883 (1965).
329. Vigna, Gurd, and Gurd, *J. Biol. Chem.,* 249, 4144 (1974).
330. Klippenstein, Riper, and Oosterom, *J. Biol. Chem.,* 247, 5956 (1972).

Table 2 (continued)
AVERAGE HYDROPHOBICITIES OF SELECTED PROTEINS

331. Tada, Bailin, Bárány, and Bárány, *Biochemistry,* 8, 4842 (1969).
332. Connell and Howgate, *Biochem. J.,* 71, 83 (1959).
333. Chung, Richards, and Olcott, *Biochemistry,* 6, 3154 (1967).
334. Samy, Atreyi, Maeda, and Meienhofer, *Biochemistry,* 13, 1007 (1974).
335. Hollenberg and Hope, *Biochem. J.,* 106, 557 (1968).
336. Furth and Hope, *Biochem. J.,* 116, 545 (1970).
337. Miranda, Kupeyan, Rochat, Rochat, and Lissitzky, *Eur. J. Biochem.,* 16, 514 (1970).
338. Berry, Johnson, and Campbell, *Biochim. Biophys. Acta,* 220, 269 (1970).
339. Omenn, Ontjes, and Anfinsen, *Biochemistry,* 9, 304 (1970).
340. Palmieri, Yue, Jacobs, Maland, Yu, and Kuby, *J. Biol. Chem.,* 248, 4486 (1973).
341. Rodgers and Chargaff, *J. Biol. Chem.,* 247, 5448 (1972).
342. Neu, *J. Biol. Chem.,* 242, 3896 (1967).
343. Marshall and Cohen, *J. Biol. Chem.,* 247, 1641 (1972).
344. Donovan, Mapes, Davis, and Hamburg, *Biochemistry,* 8, 4190 (1969).
345. Donovan, Davis, and White, *Biochim. Biophys. Acta,* 207, 190 (1970).
346. Phillips and Azari, *Biochemistry,* 10, 1160 (1971).
347. Koike, Hamada, Tanaka, Otsuka, Ogasahara, and Koike, *J. Biol. Chem.,* 249, 3836 (1974).
348. DuVigneaud, Ressler, Swan, Roberts, and Katsoyannis, *J. Am. Chem. Soc.,* 76, 3115 (1954).
349. Smith, Stockell, and Kimmel, *J. Biol. Chem.,* 207, 551 (1954).
350. Rasmussen and Craig, *J. Biol. Chem.,* 236, 759 (1961).
351. Lehky, Blum, Stein, and Fischer, *J. Biol. Chem.,* 249, 4332 (1974).
352. Jones and Hofmann, *Can. J. Biochem.,* 50, 1297 (1972).
353. Lang and Kassell, *Biochemistry,* 10, 2296 (1971).
354. Graham, Šodek, and Hofmann, *Can. J. Biochem.,* 51, 789 (1973).
355. Chow and Kassell, *J. Biol. Chem.,* 243, 1718 (1968).
356. Merrett, Bar-Eli, and Van Vunakis, *Biochemistry,* 8, 3696 (1969).
357. Rajagopalan, Moore, and Stein, *J. Biol. Chem.,* 241, 4940 (1966).
358. Ryle and Hamilton, *Biochem. J.,* 101, 176 (1966).
359. Havir and Hanson, *Biochemistry,* 12, 1583 (1973).
360. Fisher, Kirkwood, and Kaufman, *J. Biol. Chem.,* 247, 5161 (1972).
361. Dowhan, Wickner, and Kennedy, *J. Biol. Chem.,* 249, 3079 (1974).
362. Cannata, *J. Biol. Chem.,* 245, 792 (1970).
363. Kono, Uyeda, and Oliver, *J. Biol. Chem.,* 248, 8592 (1973).
364. Harshman and Six, *Biochemistry,* 8, 3423 (1969).
365. Tilley and Gracy, *J. Biol. Chem.,* 249, 4571 (1974).
366. Pon, Schnackerz, Blackburn, Chatterjee, and Noltmann, *Biochemistry,* 9, 1506 (1970).
367. Yoshida and Watanabe, *J. Biol. Chem.,* 247, 440 (1972).
368. Botes and Viljoen, *J. Biol. Chem.,* 249, 3827 (1974).
369. Wells and Hanahan, *Biochemistry,* 8, 414 (1969).
370. Hachimori, Wells, and Hanahan, *Biochemistry,* 10, 4084 (1971).
371. Tu, Passey, and Toom, *Arch. Biochem. Biophys.,* 140, 96 (1970).
372. Metzger, Glaser, and Helmreich, *Biochemistry,* 7, 2021 (1968).
373. Kimmel and Smith, *Bull. Soc. Chem. Biol.,* 40, 2049 (1958).
374. Mumford and Jenner, *Biochemistry,* 5, 3657 (1966).
375. Graziani, Agrò, Rotilio, Barra, and Mondovi, *Biochemistry,* 13, 804 (1974).
376. Peterson, *J. Biol. Chem.,* 246, 34 (1971).
377. van Jaarsveld, Branch, Robbins, Morgan, Kanda, and Canfield, *J. Biol. Chem.,* 248, 7898 (1973).
378. Freisheim, Walsh, and Neurath, *Biochemistry,* 6, 3010 (1967).
379. Lacko and Neurath, *Biochemistry,* 9, 4680 (1970).
380. Milgrom, Allouch, Atger, and Baulieu, *J. Biol. Chem.,* 248, 1106 (1973).
381. Recsei and Snell, *Biochemistry,* 12, 365 (1973).
382. Ma, Brovetto-Cruz, and Li, *Biochemistry,* 9, 2302 (1970).
383. Ohta, *J. Biol. Chem.,* 242, 509 (1967).
384. Rickert and Elliot, *Can. J. Biochem.,* 51, 1638 (1973).
385. Drapeau, Boily, and Houmard, *J. Biol. Chem.,* 247, 6720 (1972).
386. Johnson and Smillie, *Can. J. Biochem.,* 50, 589 (1972).
387. Tsuru, Hattori, Tsuji, and Fukumoto, *J. Biochem.* (Tokyo), 67, 415 (1970).
388. Donnelly, Jr., Kuo, Reyes, Liu, and Greengard, *J. Biol. Chem.,* 248, 190 (1973).

Table 2 (continued)
AVERAGE HYDROPHOBICITIES OF SELECTED PROTEINS

389. Heldebrant, Butkowski, Bajaj, and Mann, *J. Biol. Chem.*, 248, 7149 (1973).
390. Barisas and McGuire, *J. Biol. Chem.*, 249, 3151 (1974).
391. Tanaka, Haniu, Yasunobu, Dus, and Gunsalus, *J. Biol. Chem.*, 249, 3689 (1974).
392. DeSa, *J. Biol. Chem.*, 247, 5527 (1972).
393. Cardenas, Dyson, and Strandholm, *J. Biol. Chem.*, 248, 6931 (1973).
394. Cheema, Soldin, Knapp, Hofmann, and Scrimgeour, *Can. J. Biochem.*, 51, 1229 (1973).
395. Rask, Vahlquist, and Peterson, *J. Biol. Chem.*, 246, 6638 (1971).
396. Vahlquist and Peterson, *Biochemistry*, 11, 4526 (1972).
397. Shields, Dinovo, Henriksen, Kimbel, Jr., and Millar, *Biochim. Biophys. Acta*, 147, 238 (1967).
398. Farrell, Jr., Mallette, Buss, and Clagett, *Biochim. Biophys. Acta*, 194, 433 (1969).
399. Glitz, Angel, and Eichler, *Biochemistry*, 11, 1746 (1972).
400. Lees and Hartley, Jr., *Biochemistry*, 5, 3951 (1966).
401. Yankeelov, Jr., *Biochemistry*, 9, 2433 (1970).
402. Becker, Halbrook, and Hirs, *J. Biol. Chem.*, 248, 7826 (1973).
403. Fletcher, Jr. and Hash, *Biochemistry*, 11, 4281 (1972).
404. Uchida and Egami, in *The Enzymes*, 3rd ed., Boyer, Ed., Academic Press, New York, 1971, 205.
405. Woodroof and Glitz, *Biochemistry*, 10, 1532 (1971).
406. Takahashi, *J. Biol. Chem.*, 240, PC 4117 (1965).
407. Kenney and Dekker, *Biochemistry*, 10, 4962 (1971).
408. Panagou, Orr, Dunstone, and Blakley, *Biochemistry*, 11, 2378 (1972).
409. Kuehn and McFadden, *Biochemistry*, 8, 2403 (1969).
410. Tabita and McFadden, *J. Biol. Chem.*, 249, 3459 (1974).
411. Lovenburg and Williams, *Biochemistry*, 8, 141 (1969).
412. Bachmayer, Benson, Yasunobu, Garrard, and Whiteley, *Biochemistry*, 7, 986 (1968).
413. Lode and Coon, *J. Biol. Chem.*, 246, 791 (1971).
414. Dowhan, Jr. and Snell, *J. Biol. Chem.*, 245, 4618 (1970).
415. Brockway and Castellino, *Biochemistry*, 13, 2063 (1974).
416. Matsubara, Kaspar, Brown, and Smith, *J. Biol. Chem.*, 240, 1125 (1965).
417. Leitzmann, Wu, and Boyer, *Biochemistry*, 8, 2338 (1970).
418. Delmer, *J. Biol. Chem.*, 247, 3822 (1972).
419. Misra and Fridovich, *J. Biol. Chem.*, 247, 3410 (1972).
420. Titani, Hermodson, Ericsson, Walsh, and Neurath, *Biochemistry*, 11, 2427 (1972).
421. Sjöberg, *J. Biol. Chem.*, 247, 8058 (1972).
422. Gonzalez, Baldesten, and Reichard, *J. Biol. Chem.*, 245, 2363 (1970).
423. Batt, Mihula, Mann, Guarracino, Altiere, Graham, Quigley, Wolf, and Zafonte, *J. Biol. Chem.*, 245, 4857 (1970).
424. Galivan, Maley, and Maley, *Biochemistry*, 13, 2282 (1974).
425. Spiro, *J. Biol. Chem.*, 245, 5820 (1970).
426. Cornell and Pierce, *J. Biol. Chem.*, 248, 4327 (1973).
427. Viljoen and Botes, *J. Biol. Chem.*, 248, 4915 (1973).
428. Strydom, *J. Biol. Chem.*, 247, 4029 (1972).
429. Karlsson, Eaker, Fryklund, and Kadin, *Biochemistry*, 11, 4628 (1972).
430. Sato, Abe, and Tamiya, *Biochem. J.*, 115, 85 (1969).
431. Tu, Hong, and Solie, *Biochemistry*, 10, 1295 (1971).
432. Botes and Strydom, *J. Biol. Chem.*, 244, 4147 (1969).
433. Botes, *J. Biol. Chem.*, 247, 2866 (1972).
434. Chang and Hayashi, *Biochem. Biophys. Res. Commun.*, 37, 841 (1969).
435. Karlsson, Eaker, and Porath, *Biochim. Biophys. Acta*, 127, 505 (1966).
436. Six and Harshman, *Biochemistry*, 12, 2672 (1973).
437. Lipscomb, Horton, and Armstrong, *Biochemistry*, 13, 2070 (1974).
438. Schneider and Slaunwhite, Jr., *Biochemistry*, 10, 2086 (1971).
439. Chader and Westphal, *J. Biol. Chem.*, 243, 928 (1968).
440. Chader and Westphal, *Biochemistry*, 7, 4272 (1968).
441. Hudson, Ohno, Brockway, and Castellino, *Biochemistry*, 12, 1047 (1973).
442. Mann, Fish, Cox, and Tanford, *Biochemistry*, 9, 1348 (1970).
443. Carlson and Brosemer, *Biochemistry*, 10, 2113 (1971).
444. Sawyer, Tilley, and Gracy, *J. Biol. Chem.*, 247, 6499 (1972).
445. Norton, Pfuderer, Stringer, and Hartman, *Biochemistry*, 9, 4952 (1970).
446. Krietsch, Pentchev, Klingenburg, Hofstatter, and Bucher, *Eur. J. Biochem.*, 14, 289 (1970).

Table 2 (continued)
AVERAGE HYDROPHOBICITIES OF SELECTED PROTEINS

447. Greaser and Gergely, *J. Biol. Chem.*, 248, 2125 (1973).
448. Winter and Neurath, *Biochemistry*, 9, 4673 (1970).
449. Reeck and Neurath, *Biochemistry*, 11, 503 (1972).
450. Travis and Roberts, *Biochemistry*, 8, 2884 (1969).
451. Mallery and Travis, *Biochemistry*, 12, 2847 (1973).
452. Travis and Liener, *J. Biol. Chem.*, 240, 1967 (1965).
453. Travis, *Biochem. Biophys Res. Commun.*, 30, 730 (1968).
454. Gates and Travis, *Biochemistry*, 8, 4483 (1969).
455. Jurášek and Smillie, *Can. J. Biochem.*, 51, 1077 (1973).
456. Kucich and Peanasky, *Biochim. Biophys. Acta*, 200, 47 (1970).
457. Sherman and Kassell, *Biochemistry*, 7, 3634 (1968).
458. Tschesche and Wachter, *Eur. J. Biochem.*, 16, 187 (1970).
459. Kassell, Radicevic, Berlow, Peanasky, and Laskowski, Sr., *J. Biol. Chem.*, 238, 3274 (1966).
460. Wilson and Laskowski, Sr., *J. Biol. Chem.*, 248, 756 (1973).
461. Walsh and Neurath, *Proc. Natl. Acad. Sci. USA*, 52, 884 (1964).
462. Bradshaw, Neurath, Tye, Walsh, and Winter, *Nature* (Lond.), 226, 237 (1970).
463. Voytek and Gjessing, *J. Biol. Chem.*, 246, 508 (1971).
464. Schyns, Bricteaux-Grégoire, and Florkin, *Biochim. Biophys. Acta*, 175, 97 (1969).
465. Henning, Helinski, Chao, and Yanofsky, *J. Biol. Chem.*, 237, 1523 (1962).
466. Li and Yanofsky, *J. Biol. Chem.*, 248, 1830 (1973).
467. Hoch, *J. Biol. Chem.*, 248, 2999 (1973).
468. Hoch, *J. Biol. Chem.*, 248 2992 (1973).
469. Hoch and DeMoss, *J. Biol. Chem.*, 247, 1750 (1972).
470. Kamamiyama, Wada, Matsubara, and Snell, *J. Biol. Chem.*, 247, 1571 (1972).
471. Koch, *Biochemistry*, 13, 2307 (1974).
472. Milton and Taylor, *Biochem. J.*, 113, 678 (1969).
473. White, Barlow, and Mozen, *Biochemistry*, 5, 2160 (1966).
474. DuVigneaud, Bartlett, and Johl, *J. Am. Chem. Soc.*, 79, 5572 (1957).
475. Tanford, *J. Am. Chem. Soc.*, 84, 4240 (1962).
476. Bigelow, *J. Theor. Biol.*, 16, 187 (1967).

SPECIFIC ROTATORY DISPERSION CONSTANTS FOR 0.1 *M* AMINO ACID SOLUTIONS

Amino acid	Solution acidity	Drude parameters[a] A	λ_a^2	B	λ_b^2	RMS[e] deviation (deg)
Alanine	2 *M* HCl	298.5195	0.04220239	294.7037	0.04170058	±0.17
	pH 6.5	360.1463	0.04069301	360.1262	0.04047647	±0.24
	0.25 *M* NaOH	344.4703	0.04989568	343.1793	0.04989547	±0.13
Valine	2 *M* HCl	438.4119	0.03926320	431.6605	0.03866173	±0.46
	pH 6.06	460.4233	0.03465918	459.4160	0.03421852	±0.30
	0.25 *M* NaOH	319.3442	0.03118233	315.3691	0.03050604	±0.39
Leucine	2 *M* HCl	377.4991	0.04044465	374.1618	0.03980577	±0.17
	pH 6.18	274.5246	0.03802343	278.2870	0.03743202	±0.16
	0.25 *M* NaOH	314.6331	0.03777259	312.5013	0.03727019	±0.17
Serine	2 *M* HCl	338.1789	0.04230850	334.9272	0.04186100	±0.24
	pH 5.73	318.8132	0.03560563	321.5245	0.03501944	±0.26
	0.25 *M* NaOH	382.4514	0.03479284	383.6311	0.03466110	±0.14
Aspartic Acid	2 *M* HCl	360.9105	0.04260890	353.2918	0.04253996	±0.22
	pH 2.93[b]	446.1446	0.03954388	444.9421	0.03935718	±0.29
	pH 7.8	374.4645	0.03857365	380.6139	0.03817383	±0.19
	pH 12.1	347.8941	0.03823479	349.8450	0.03755913	±0.12
Glutamic Acid	2 *M* HCl	361.7220	0.03488210	352.7541	0.03430429	±0.57
	pH 3.25[c]	414.0331	0.03695381	410.9175	0.03669701	±0.29
	pH 7.94	393.2113	0.03567558	395.3116	0.03533174	±0.13
	pH 11.6	336.5308	0.03729399	333.6660	0.03702243	±0.24
Asparagine	pH 1.68	398.0964	0.04233022	391.5547	0.04225800	±0.39
	pH 5.2	253.8327	0.03896011	256.2606	0.03865567	±0.12
	pH 10.42	417.5699	0.03106792	421.1466	0.03100037	±0.22
Lysine	2 *M* HCl	1179.298	0.04094295	1172.252	0.04083325	±0.24
	pH 4.83	441.1781	0.03954097	437.5107	0.03937528	±0.16
	pH 9.93	431.4914	0.03917215	427.8984	0.03901931	±0.14
	0.25 *M* NaOH	183.0545	0.03726084	179.6172	0.03679289	±0.14
Ornithine	2 *M* HCl[d]	1350.506	0.04207689	1342.576	0.04200871	±0.21
	pH 5.62	432.2491	0.03778435	428.066	0.03761066	±0.09
	pH 9.80	525.6925	0.04052814	522.2509	0.04041411	±0.24
	0.25 *M* NaOH[d]	486.3102	0.03752583	482.6644	0.03735010	±0.33
Proline	2 *M* HCl	445.3052	0.04125130	461.8143	0.04066942	±0.34
	pH 6.08	403.8729	0.04138662	430.2741	0.04088043	±0.51
	0.25 *M* NaOH	1719.7714	0.01997902	1749.4527	0.02039450	±0.56

Compiled by L. I. Katzin and E. Gulyas.

From Katzin and Gulyas, *J. Amer. Chem. Soc.*, 86, 1655 (1964). With permission of copyright owners.

[a] $[\alpha]_\lambda = \frac{A}{\lambda^2 - \lambda_a^2} - \frac{B}{\lambda^2 - \lambda_b^2}$ with λ in microns. Parameters valid for 650–270 mμ.

[b] Amino acid 0.0375 *M*.

[c] Amino acid 0.05 *M*.

[d] Amino acid 0.08 *M*.

[e] Root-mean-square deviation between measured and calculated $[\alpha]_\lambda$, for 42 wavelengths between 650 and 270 mμ.

This table originally appeared in Sober, Ed., *Handbook of Biochemistry and selected data for Molecular Biology*, 2nd ed., Chemical Rubber Co., Cleveland, 1970.

CIRCULAR DICHROISM (CD) SPECTRA OF METAL COMPLEXES OF AMINO ACIDS AND PEPTIDES[a]

L. I. Katzin

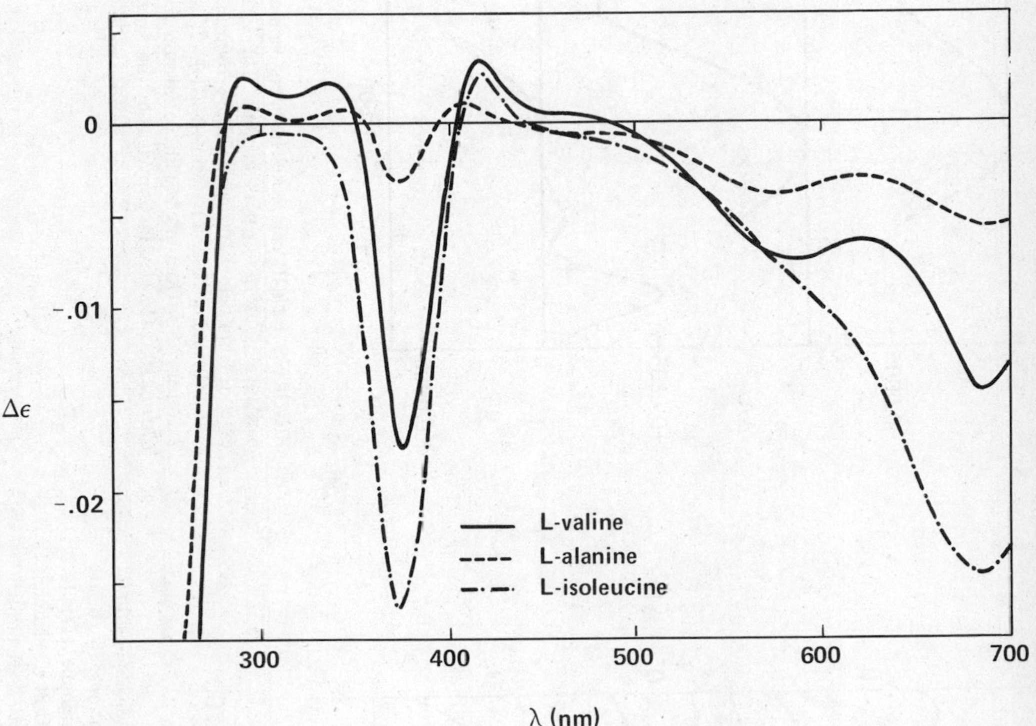

FIGURE 1. Nickel(II) complexes, NiA_2, of simple amino acids. (Reprinted with permission from Haines and Reimer, *Inorg. Chem.*, 12, 1483 (1973). Copyright by the American Chemical Society.)

[a] $\Delta\epsilon = \epsilon_1 - \epsilon_r$.

FIGURE 2. Ni(II)-amino acid systems. Vertical bars indicate positions of maxima in the absorption spectra of the given solutions: (b′, b, a) valine-Ni(II), 1:1, at pH 3, 5, and 6, respectively; (c) arginine, 1:1, pH 6; (d) glutamic acid, 1:1, at pH 6; (e) alanine, 2:1, pH 8; (f) alanine, 3:1, pH 8; (g) arginine, 3:1, pH 7; (h) ornithine, 3:1, pH 7. (Reprinted with permission from Katzin and Gulyas, *J. Am. Chem. Soc.*, 91, 6941 (1969). Copyright by the American Chemical Society.)

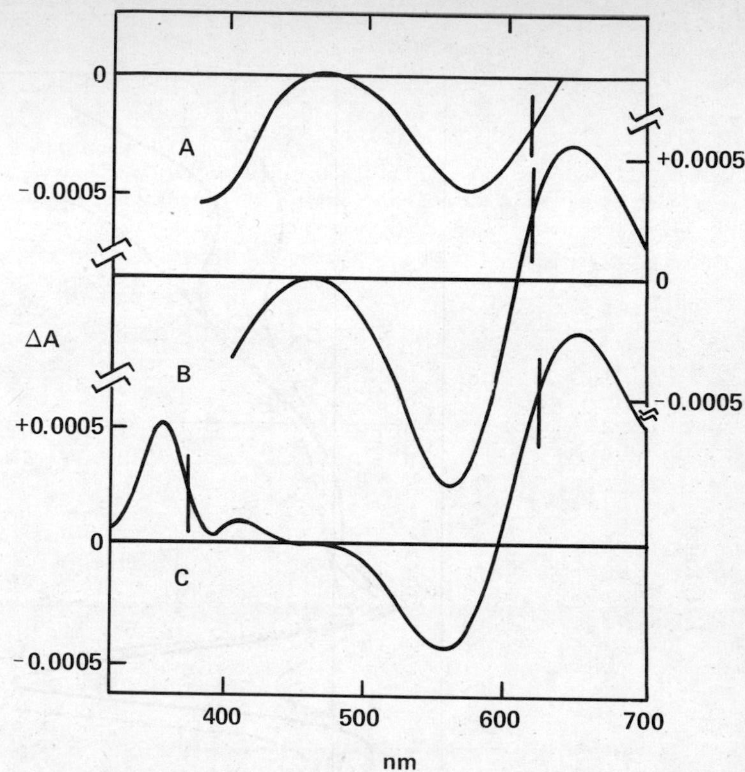

FIGURE 3. Ni(II) complexes involving valine (50-nm path length). Vertical bars locate absorption peak maxima. The absorption of the peak in the red extends from below 500 nm to above 700 nm. (A) Valine-Ni(acac)$_2$, 2:1, pH 8.70; (B) valine-Ni(acac)$_2$-(valine-Ni(II)), 2:1:1, pH 8.01; (C) valine-NiCl$_2$, 2:1, pH 7.50. (Reprinted with permission from Katzin and Gulyas, *Inorg. Chem.*, 10, 2412 (1971). Copyright by the American Chemical Society.)

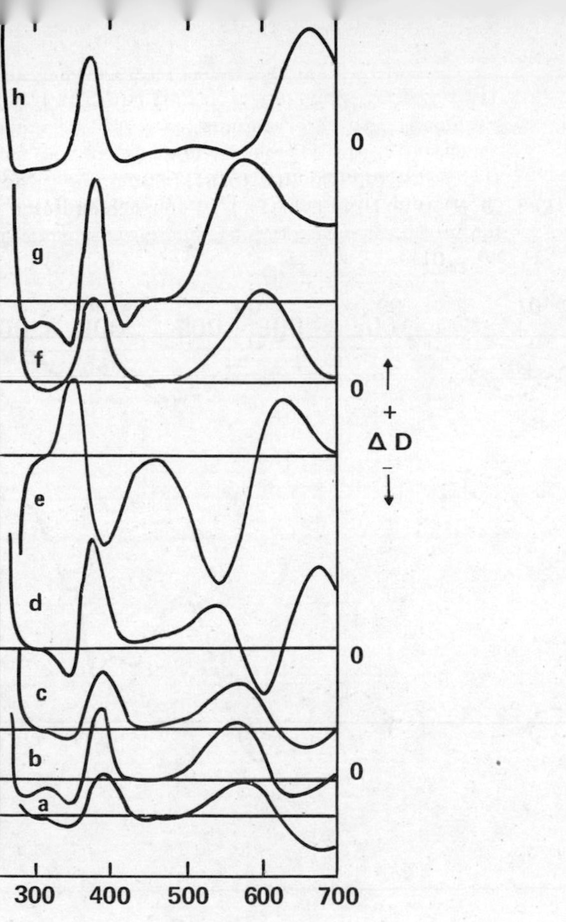

FIGURE 4. Ni(II) with asparagine and aspartic acid. Vertical bars indicate positions of maxima in the absorption spectra of the given solutions: (a) asparagine-Ni(II), 1:1, pH 4; (b) aspartic acid, 1:1, pH 6; (c) asparagine, 1:1, pH 8; (d) aspartic acid, 1:1, pH 9; (e) asparagine, 3:1, pH 8; (f) asparagine, 2:1, pH 7; (g) aspartic acid, 2:1, pH 6; (h) aspartic acid, pH 10. (Reprinted with permission from Katzin and Gulyas, *J. Am. Chem. Soc.*, 91, 6941 (1969). Copyright by the American Chemical Society.)

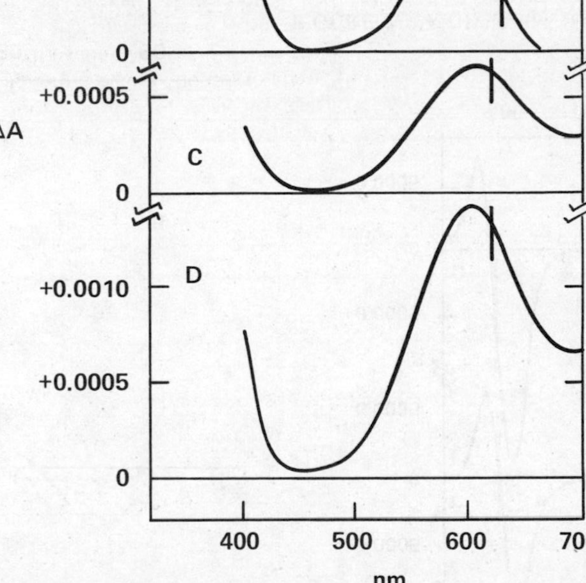

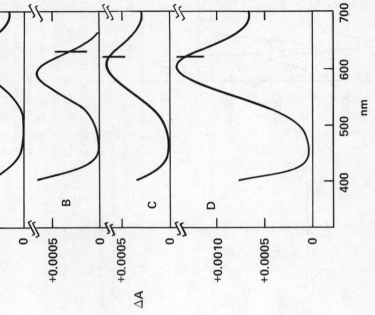

FIGURE 5. Ni(II) complexes involving asparagine and aspartic acid (50-nm path length): (A) asparagine-Ni(acac)$_2$, 1:1, pH 8.36; (B) asparagine- Ni(acac)$_2$-(asparagine-Ni(II)), 1:1:1, pH 6.29; (C) aspartic acid-Ni(acac)$_2$, 1:1, pH 8.97; (D) aspartic acid-Ni(acac)$_2$-(aspartate-Ni(II)), 1:1:1, pH 8.66. (Reprinted with permission from Katzin and Gulyas, *Inorg. Chem.*, 10, 2413 (1971). Copyright by the American Chemical Society.)

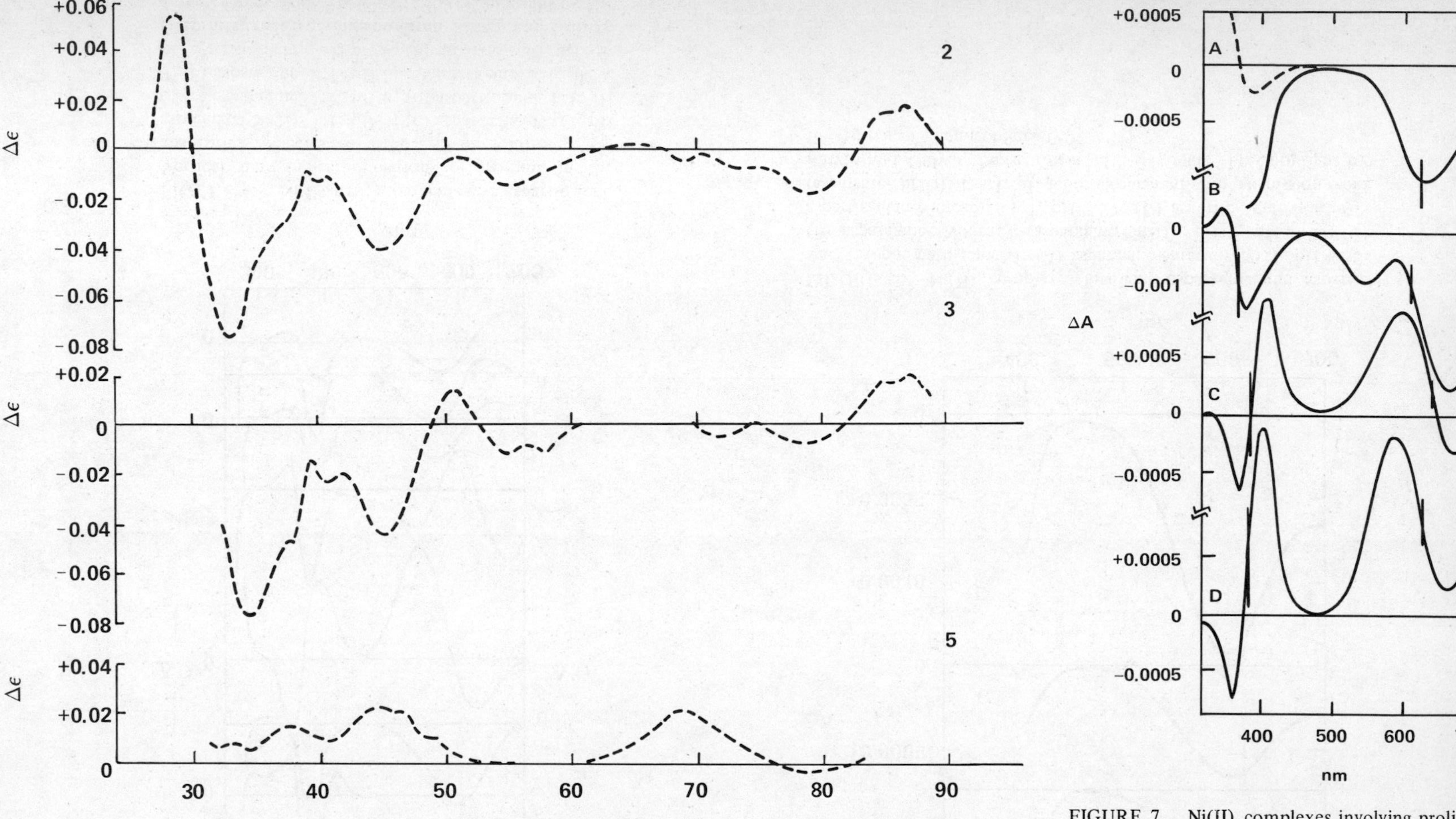

FIGURE 6. K[Ni(L-pro)$_3$] (2), K[Ni(L-hypro)$_3$] (3), and [Ni(L-pro)$_2$ (H$_2$O)$_2$] (5) in aqueous solutions. (From Hidaka and Shimura, *Bull. Chem. Soc. Jap.*, 43, 3000 (1970). With permission.)

FIGURE 7. Ni(II) complexes involving proline: (A) proline-Ni(acac)$_2$, 1:1, pH 8.33 (solid curve, 50-mm path, dotted, 10 mm); (B) proline-NiCl$_2$, 5:1, pH 9 (5-mm path); (C) proline-NiCl$_2$, 1:1, pH 6 (10-mm path); (D) proline-NiCl$_2$, 1:1, pH 7.5 (10-mm path). (Reprinted with permission from Katzin and Gulyas, *Inorg. Chem.*, 10, 2413 (1971). Copyright by the American Chemical Society.)

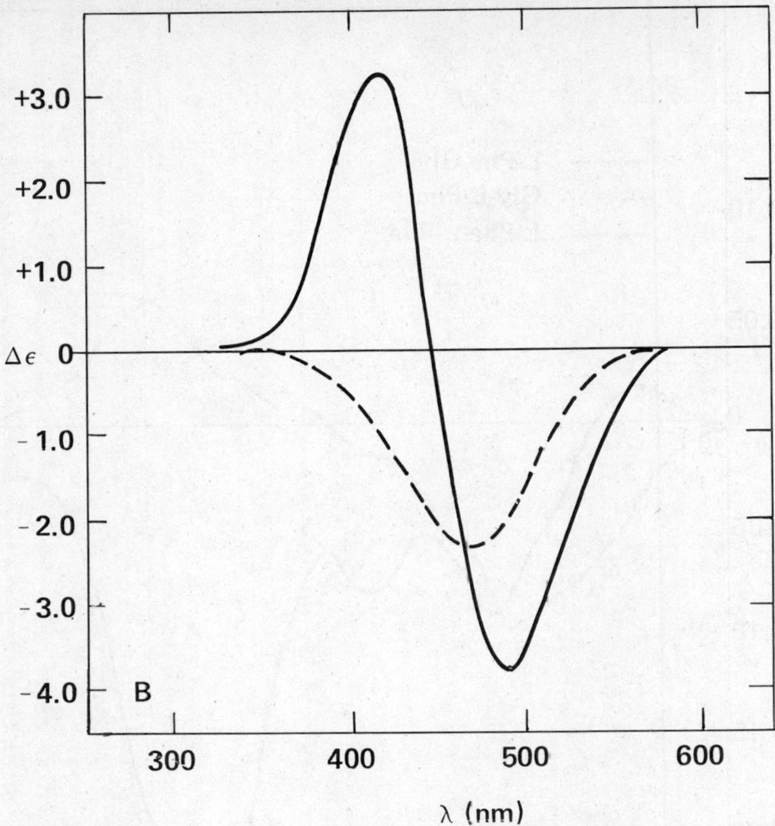

FIGURE 8. NiL^{2-} (–) species of nickel(II) and acetylglycylglycyl-L-histidylglycine; NiL^{-}(- - -) species of nickel(II) and di-L-alanyl-L-alanine. (From Bryce, Roeske, and Gurd, *J. Biol. Chem.*, 241, 1078 (1966). With permission.)

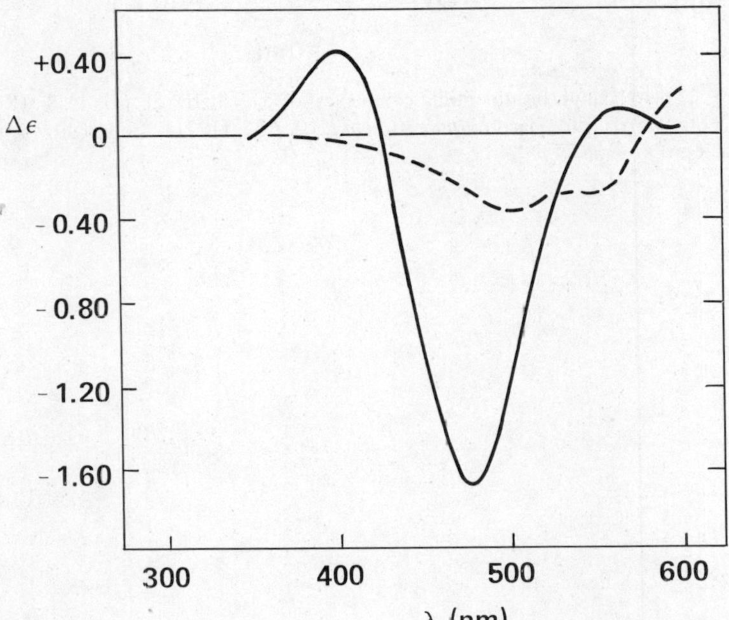

FIGURE 9. Apomyoglobin complexes; (–) 4:1 nickel(II) complex, pH 11.0; (- - -) 1:1 copper(II) complex, pH 11.0. (From Bryce, Roeske, and Gurd, *J. Biol. Chem.*, 241, 1078 (1966). With permission.)

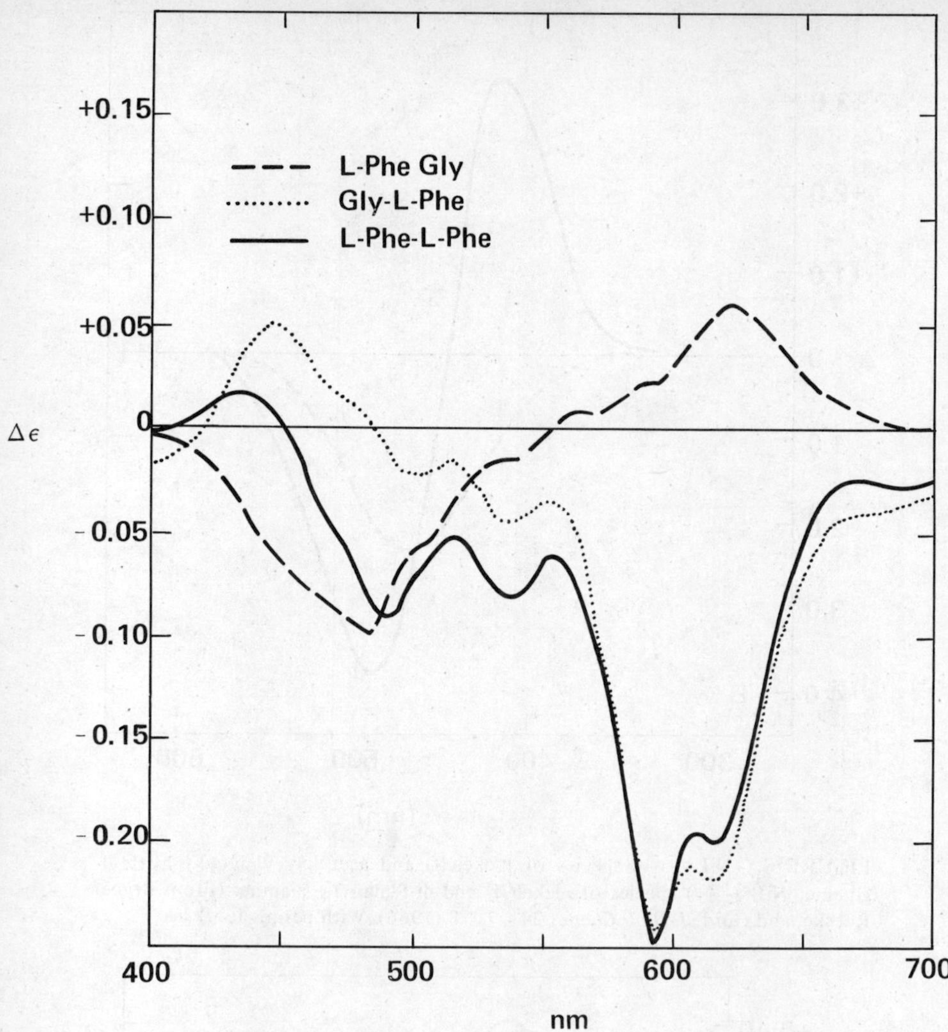

FIGURE 10. Phenylalanyl bis-dipeptide complexes of cobalt(II) at pH 11.8. (Reprinted with permission from Morris and Martin, *Inorg. Chem.*, 10, 965 (1971). Copyright by the American Chemical Society.)

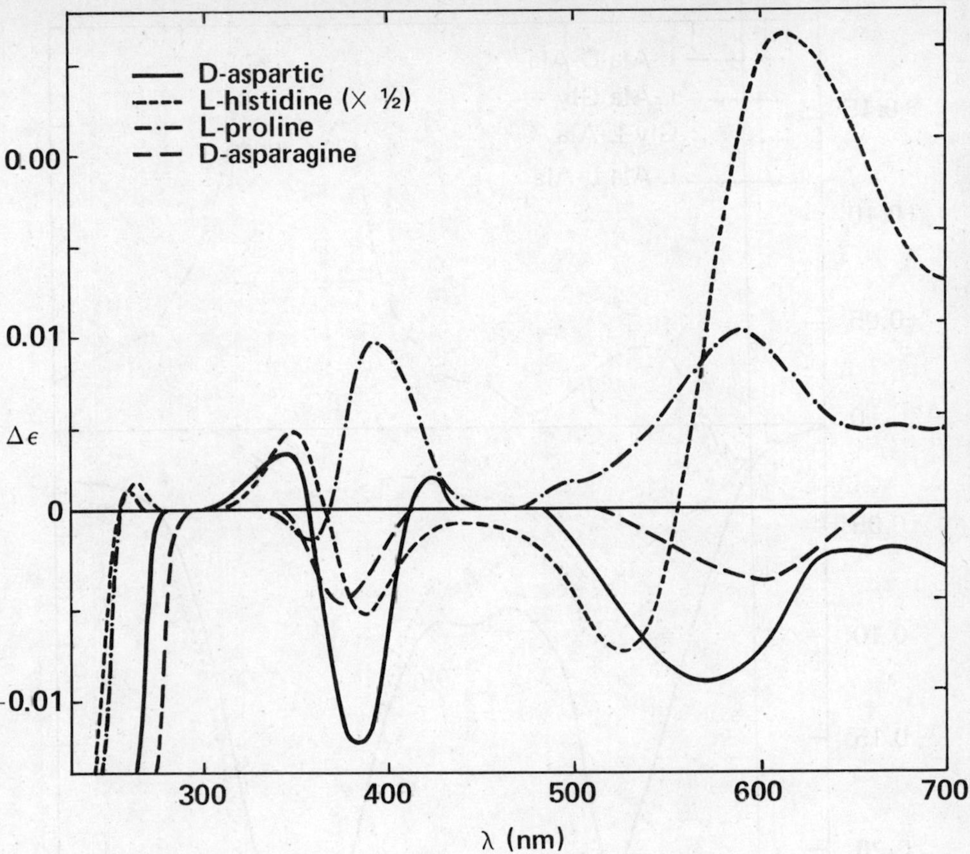

FIGURE 11. Nickel(II) complexes of proline and some polyfunctional amino acids. (Reprinted with permission from Haines and Reimer, *Inorg. Chem.*, 12, 1485 (1973). Copyright by the American Chemical Society.)

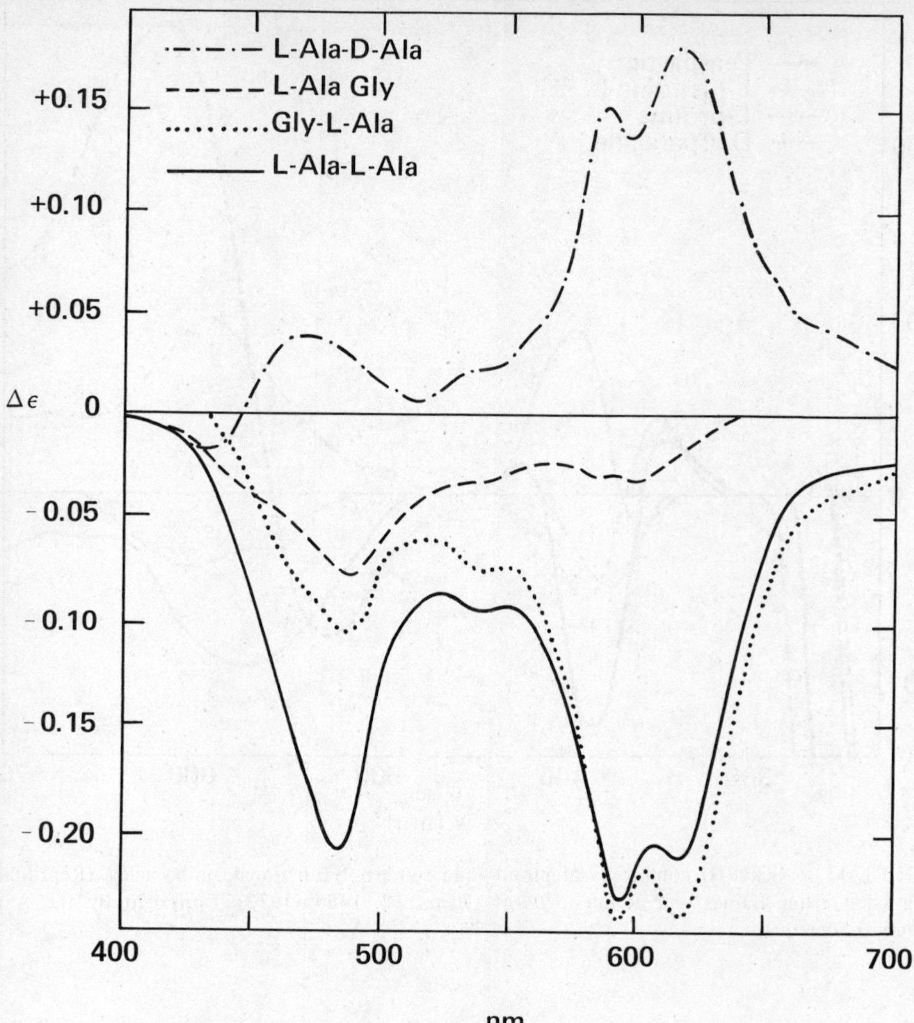

FIGURE 12. Alanyl bis-dipeptide complexes of cobalt(II) at pH 11.8 (Reprinted with permission from Morris and Martin, *Inorg. Chem.*, 10, 965 (1971). Copyright by the American Chemical Society.

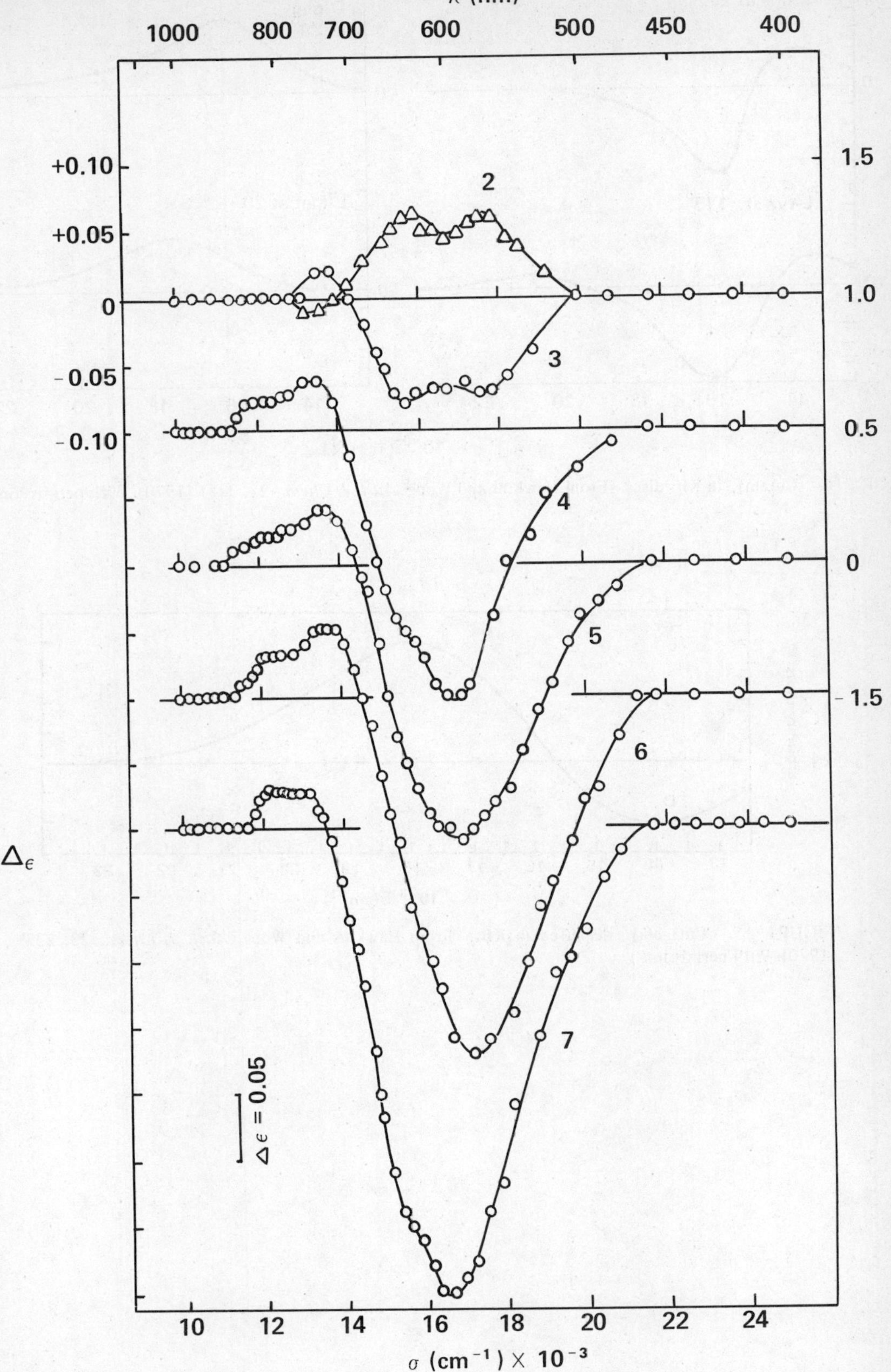

FIGURE 13. $[Cu(D\text{-}ala)_2]$ (2), $[Cu(L\text{-}ala)_2]$ (3), $[Cu(L\text{-}ser)_2]$ (4), $[Cu(L\text{-}thr)_2]\cdot H_2O$ (5), $[Cu(L\text{-}val)_2]\cdot H_2O$ (6), and $[Cu(L\text{-}allothr)_2]\cdot H_2O$ (7) in aqueous solutions. (Reprinted with permission from Yasui, Hidaka, and Shimura, *J. Am. Chem. Soc.*, 87, 2763 (1965). Copyright by the American Chemical Society.)

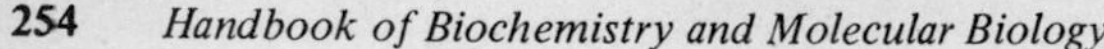

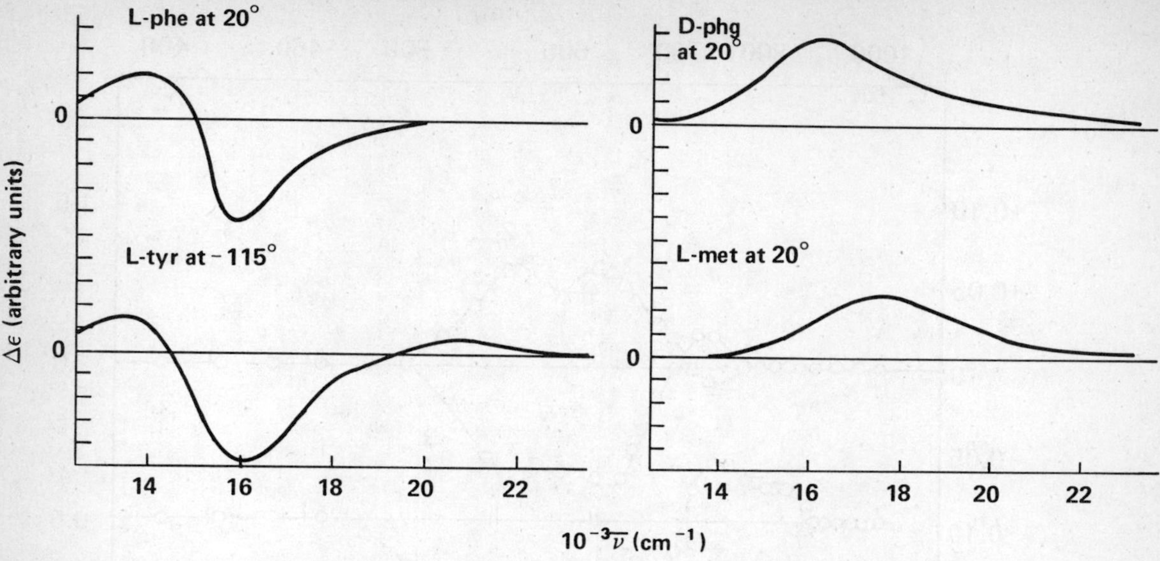

FIGURE 14. Cu(am)$_2$ in KBr discs. (From Hawkins and Wong, *Aust. J. Chem.*, 23, 2241 (1970). With permission.)

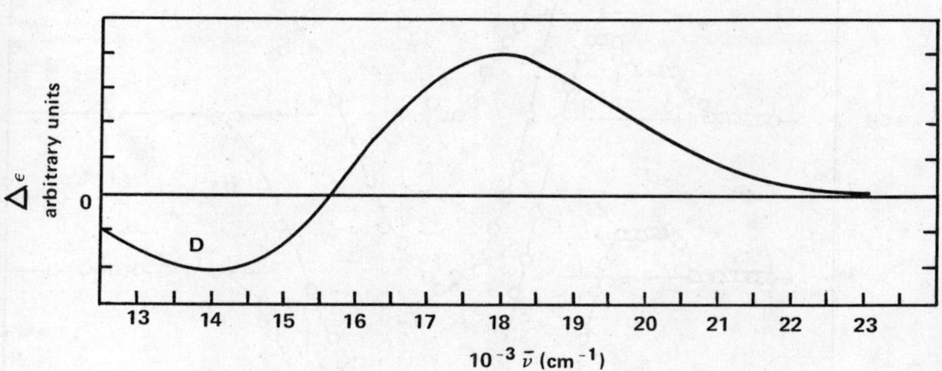

FIGURE 15. Cu(L-ala)$_2$ dispersed in KBr. (From Hawkins and Wong, *Aust. J. Chem.*, 23, 2239 (1970). With permission.)

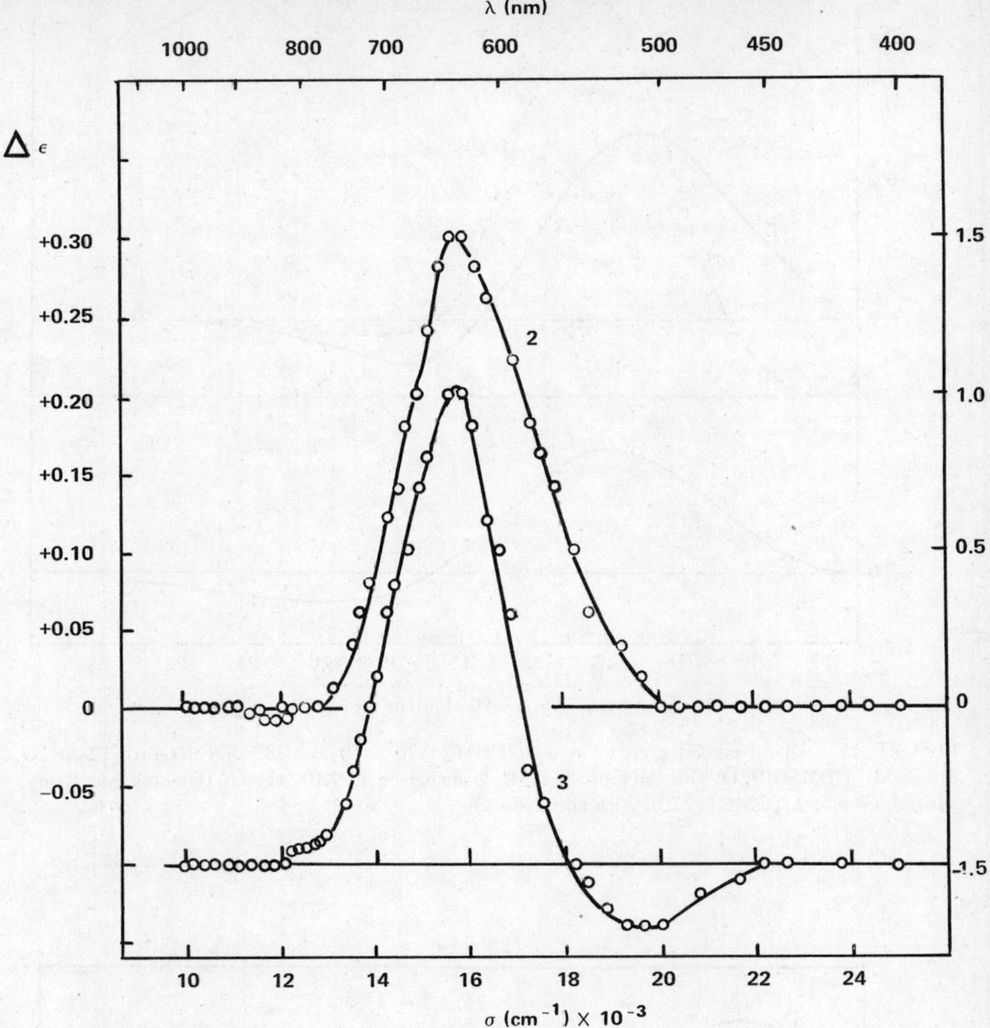

FIGURE 16. $[Cu(L\text{-hydprol})_2]\cdot 3H_2O$ (2) and $[Cu(L\text{-prol})_2]\cdot 2H_2O$ (3) in aqueous solutions. (Reprinted with permission from Yasui, Hidaka, and Shimura, *J. Am. Chem. Soc.*, 87, 2763 (1965). Copyright by the American Chemical Society.)

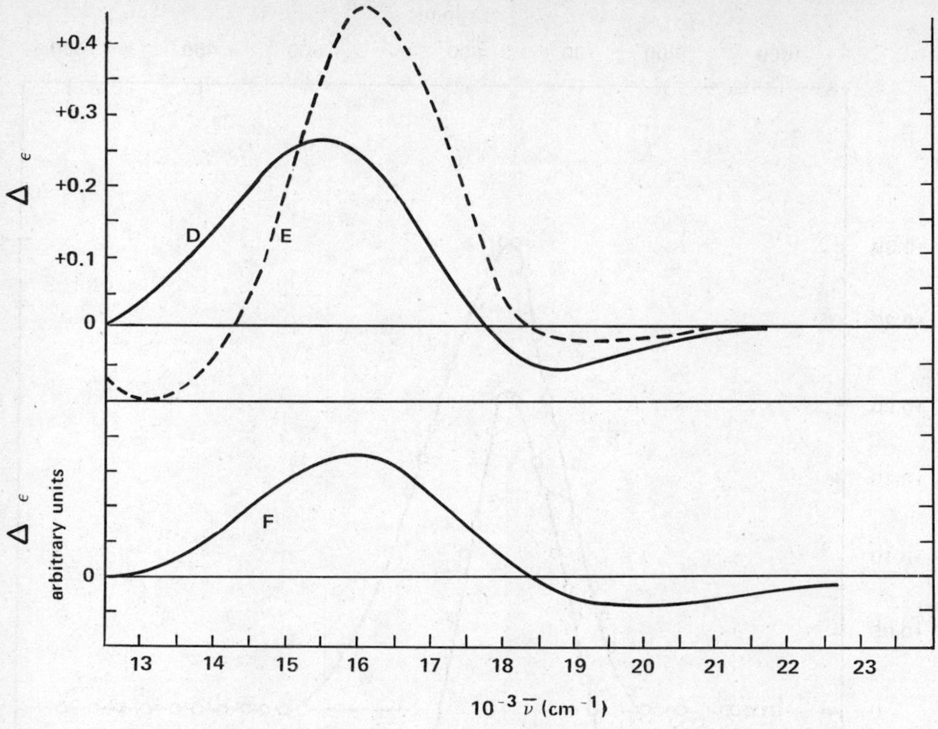

FIGURE 17. Dissolved Cu(L-pro)$_2$ in water: (D) at 20°; (E) at −90° of Cu(L-pro)$_2$,2·5H_2O dissolved in DMSO-H_2O; (F) Cu(L-pro)$_2$,2·5H_2O dispersed in KBr. (From Hawkins and Wong, *Aust. J. Chem.*, 23, 2240 (1970). With permission.)

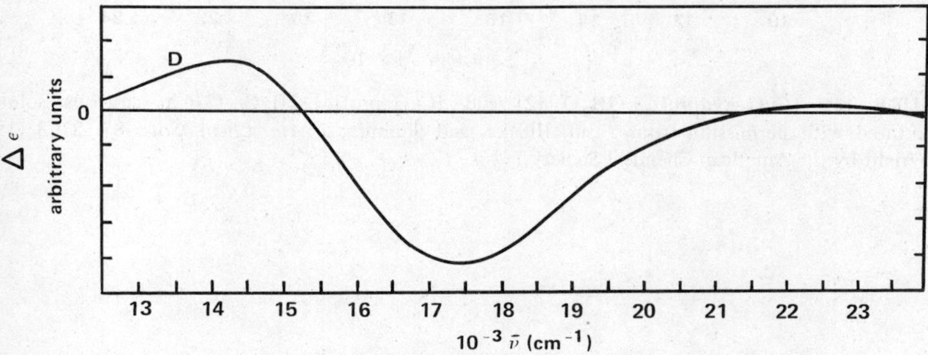

FIGURE 18. Cu(L-val)$_2$ dispersed in KBr. (From Hawkins and Wong, *Aust. J. Chem.*, 23, 2240 (1970). With permission.)

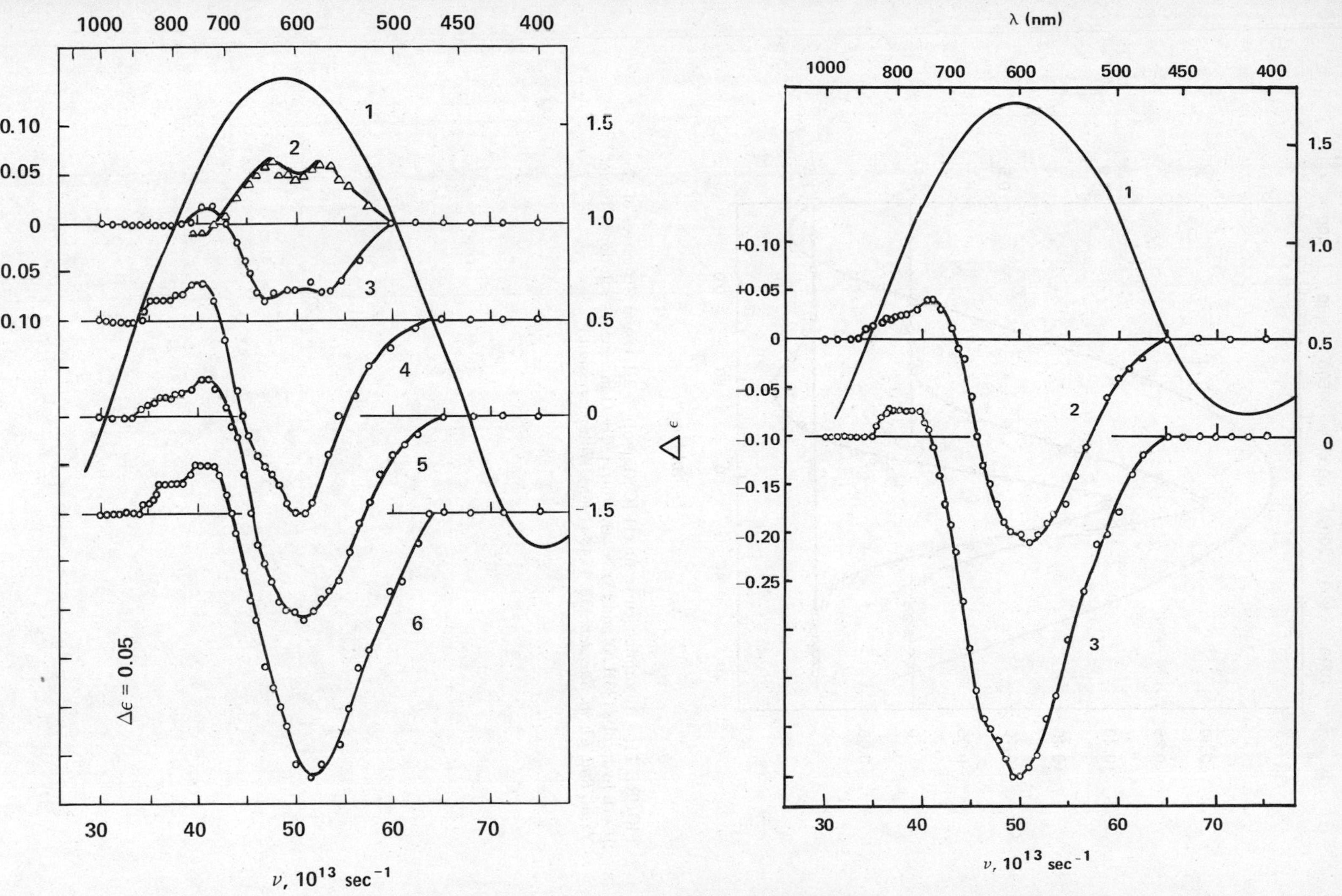

FIGURE 19. Absorption curve of (1) [Cu(L-ala)$_2$] and CD curves of (2) [Cu(D-ala)$_2$], (3) [Cu(L-ala)$_2$], (4) [Cu(L-ser)$_2$], (5) [Cu(L-thr)$_2$]·H_2O and (6) [Cu(L-val)$_2$]·H_2O in aqueous solutions. (From Yasui, *Bull. Chem. Soc. Jap.*, 38, 1747 (1965). With permission.)

FIGURE 20. Absorption curve of (1) [Cu(L-thr)$_2$]·H_2O and CD curves of (2) [Cu(L-thr)$_2$]·H_2O and (3) [Cu(L-allothr)$_2$]·H_2O in aqueous solutions. (From Yasui, *Bull. Chem. Soc. Jap.*, 38, 1748 (1965). With permission.)

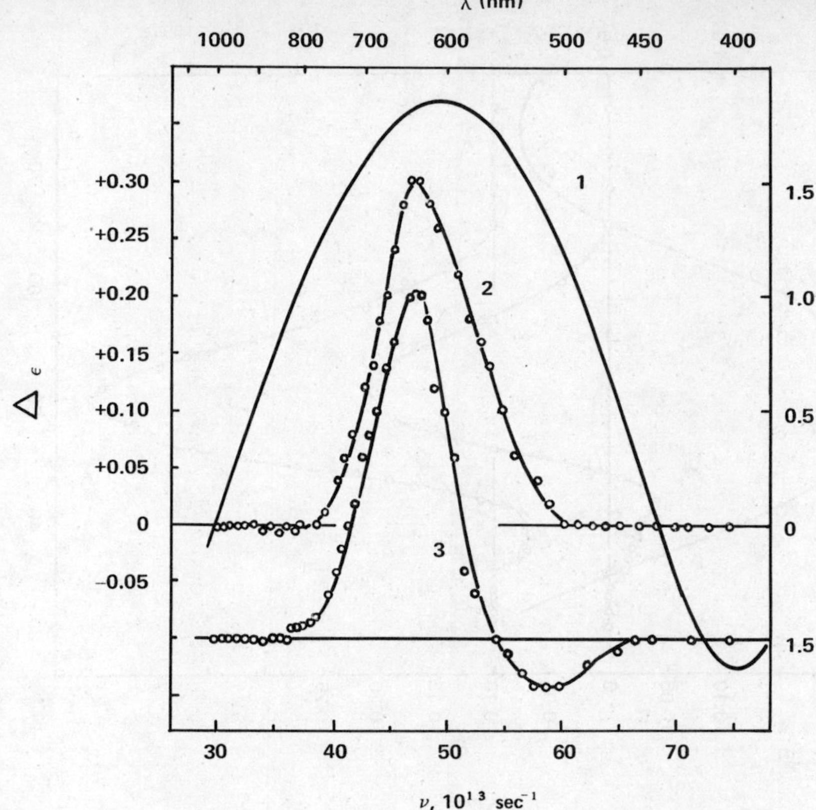

FIGURE 21. Absorption curve of (1) $[Cu(L\text{-}prol)_2]\cdot 2H_2O$ and CD curves of (2) $[Cu(L\text{-}hydprol)_2]\cdot 3H_2O$ and (3) $[Cu(L\text{-}prol)_2]\cdot 2H_2O$ in aqueous solutions. (From Yasui, *Bull. Chem. Soc. Jap.*, 38, 1748 (1965). With permission.)

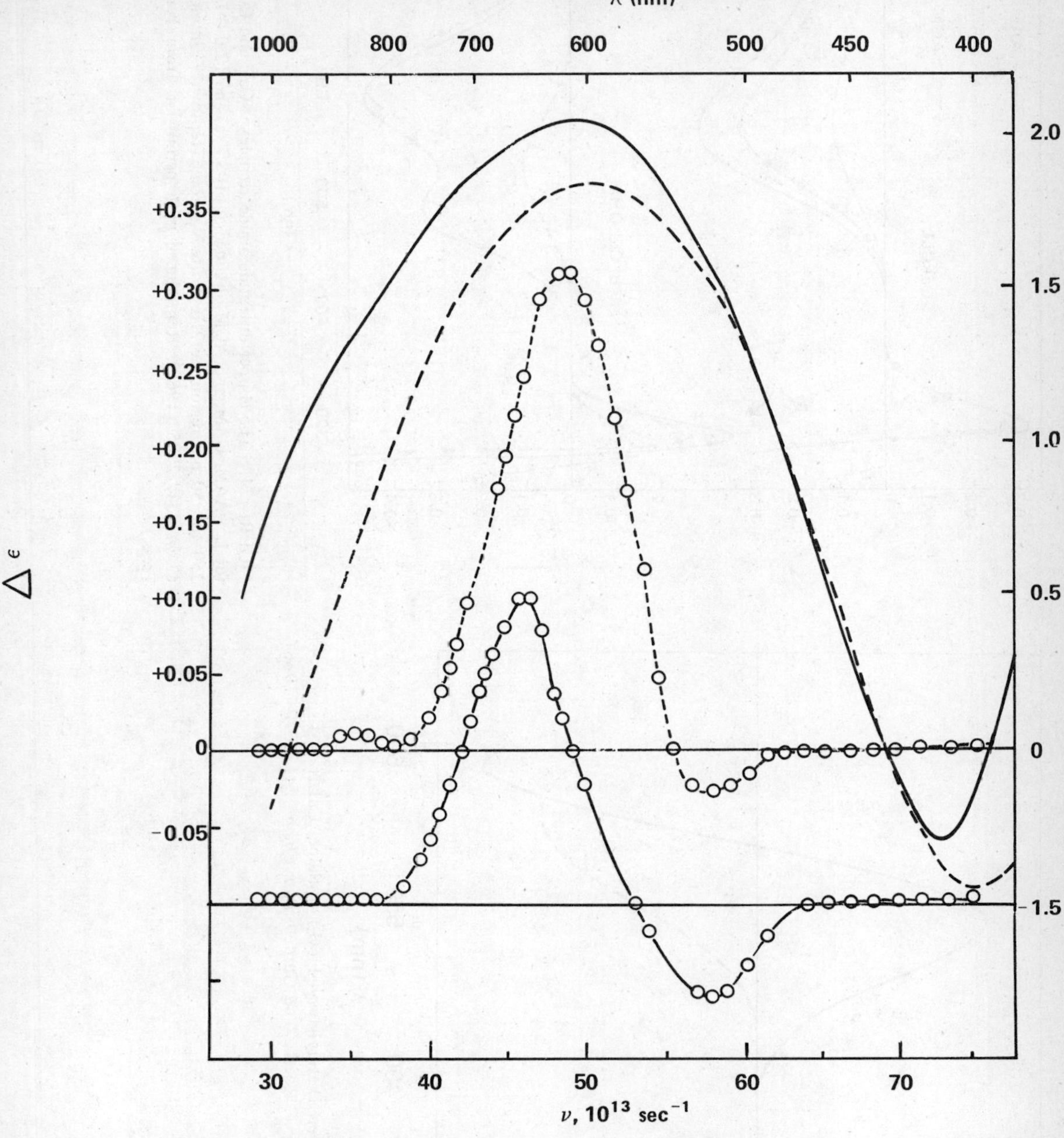

FIGURE 22. Absorption and CD curves of [Cu(L-prol)$_2$] in ethylene glycol (- - -) and in pyridine (–). (From Yasui, *Bull. Chem. Soc. Jap.*, 38, 1748 (1965). With permission.)

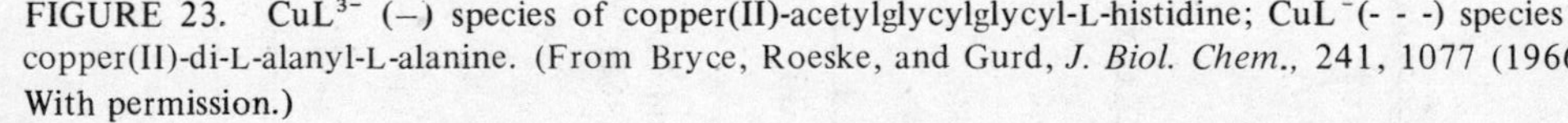

FIGURE 23. CuL^{3-} (–) species of copper(II)-acetylglycylglycyl-L-histidine; CuL^{-}(- - -) species of copper(II)-di-L-alanyl-L-alanine. (From Bryce, Roeske, and Gurd, *J. Biol. Chem.*, 241, 1077 (1966). With permission.)

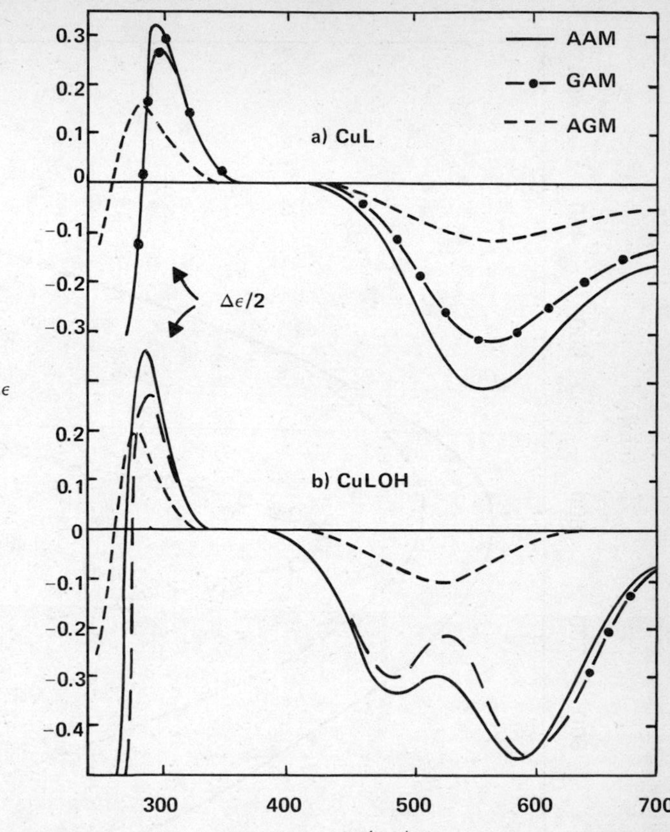

FIGURE 24. The copper dipeptide amide complexes CuL and $CuLOH^{-}$ for L = AAM, GAM, and AGM obtained at pH 8.7 (a) and 12.0 (b). In the 250 to 350 nm region $\Delta\epsilon/2$ is plotted. (From Treptow, *J. Inorg. Nucl. Chem.*, 31, 2985 (1969). Reproduced with permission from Pergamon Press.)

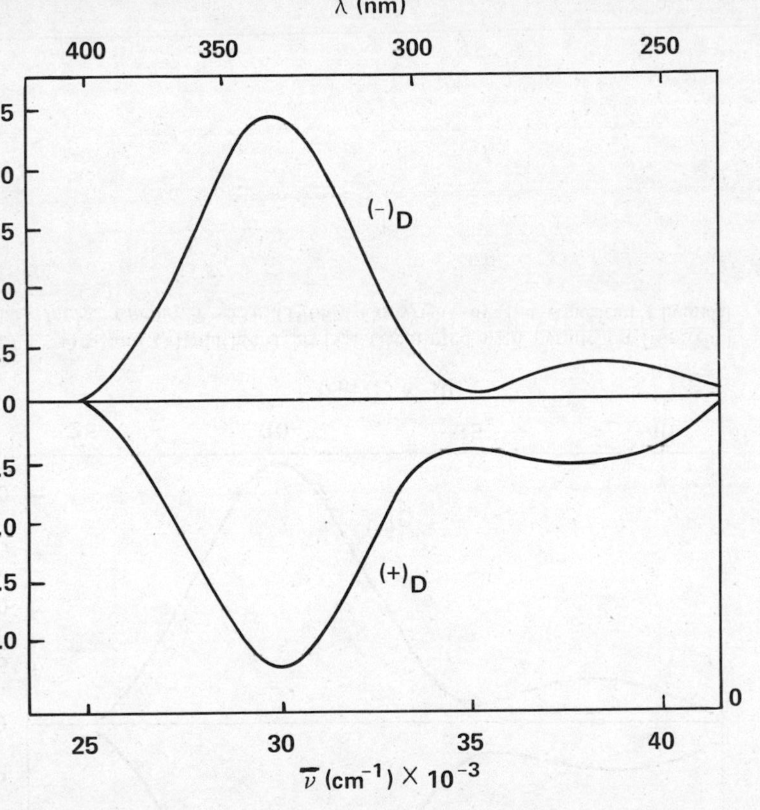

FIGURE 25. (+)D- and (−)D-[Rh(en)$_2$((*S*)-ala)]I$_2$. (Reprinted from Hall and Douglas, *Inorg. Chem.*, 7, 532 (1968). Copyright by the American Chemical Society.)

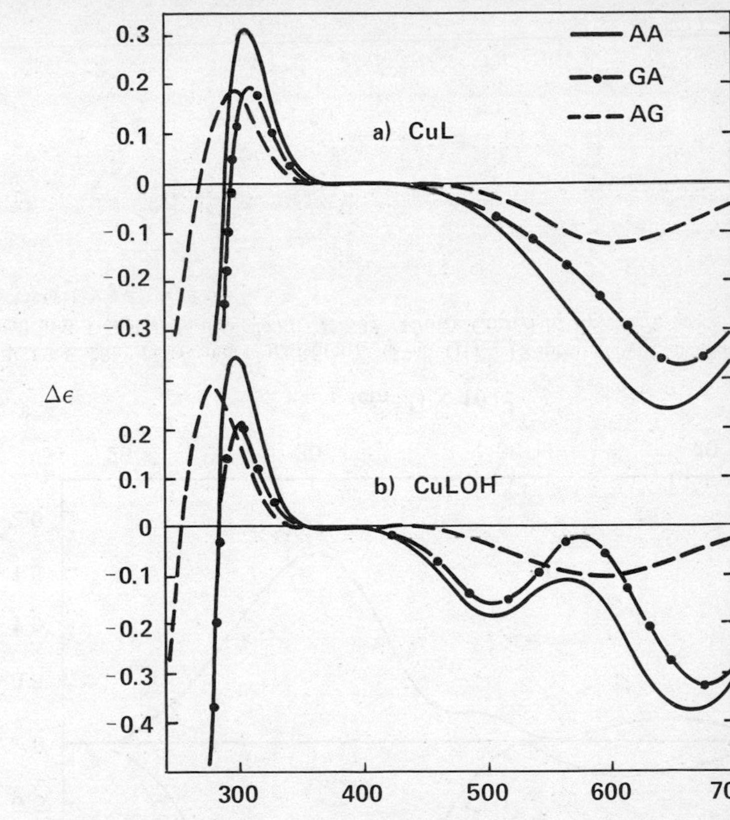

FIGURE 26. The copper dipeptide complexes CuL and CuLOH$^-$ for L = AA, GA, and AG obtained at pH 7.0 (a) and pH 10.8 (b). (From Treptow, *J. Inorg. Nucl. Chem.*, 31, 2985 (1969). Reproduced with permission from Pergamon Press.)

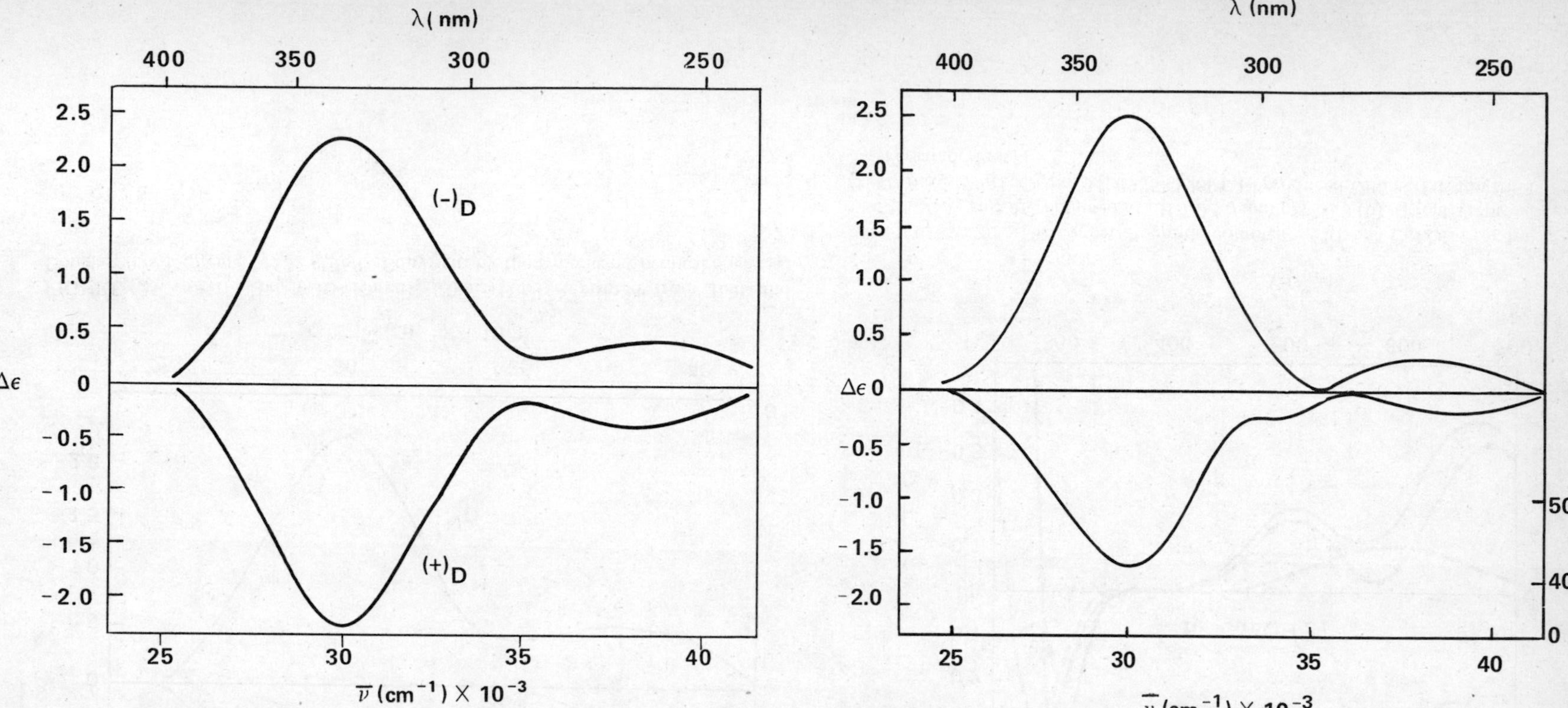

FIGURE 27. (+)D- and (−)D-[Rh(en)$_2$gly]I$_2$. (Reprinted with permission from Hall and Douglas, *Inorg. Chem.*, 7, 531 (1968). Copyright by the American Chemical Society.)

FIGURE 28. (+)D- and (−)D-Rh(en)$_2$((*S*)-met)]I$_2$. (Reprinted with permission from Hall and Douglas, *Inorg. Chem.*, 7, 532 (1968). Copyright by the American Chemical Society.)

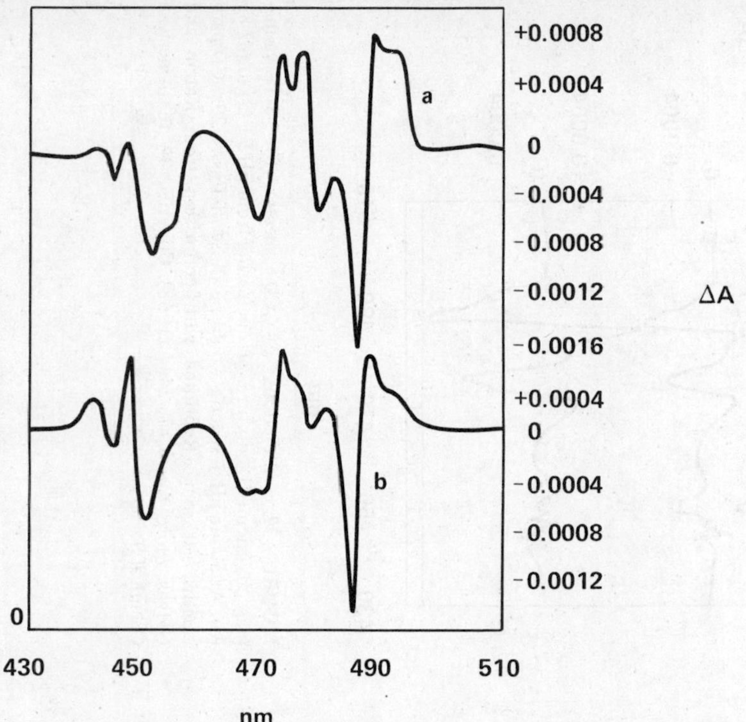

FIGURE 29. Complexes (1:1) of Pr(III) with (a) alanine and (b) asparagine. CD path lengths, 30 mm. [Pr(III)] = ca. O.I.F. (Reprinted with permission from Katzin and Gulyas, *Inorg. Chem.*, 7, 2444 (1968). Copyright by the American Chemical Society.)

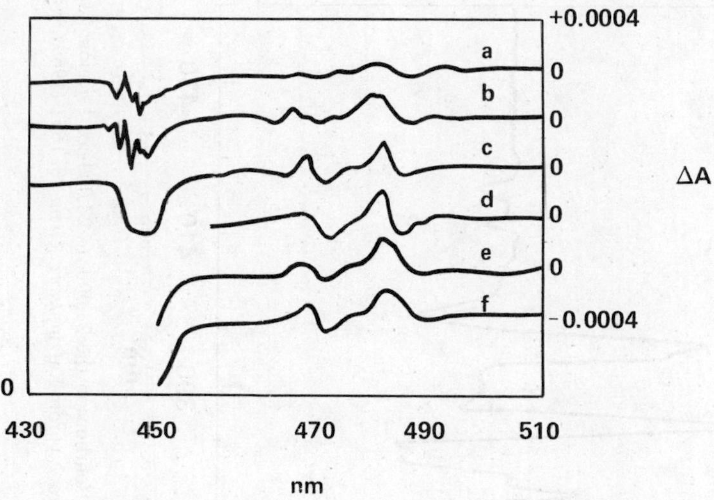

FIGURE 30. Aspartic acid-praseodymium(III) mixtures, with increasing pH; (a) 2:1, pH 6; (b) to (d) 3:1, pH 6, 8, and 9, respectively; (e) and (f) 5:1, pH 8 and 9, respectively. CD path lengths, 30 mm. (Reprinted with permission from Katzin and Gulyas, *Inorg. Chem.*, 7, 2445 (1968). Copyright by the American Chemical Society.)

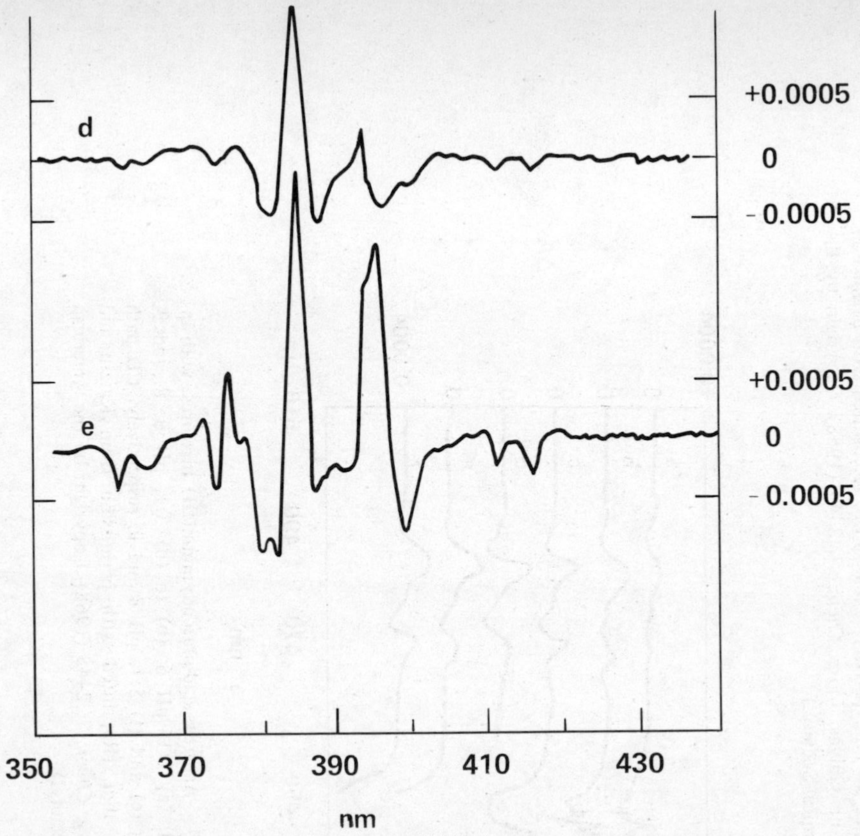

FIGURE 31. Eu(III) solutions in the 23,000- to 28,000-cm^{-1} region: (d) with serine, pH 6; (e) with alanine, pH 6 to 7. (From Katzin, *Inorg. Chem.*, 8, 1651 (1969). With permission.)

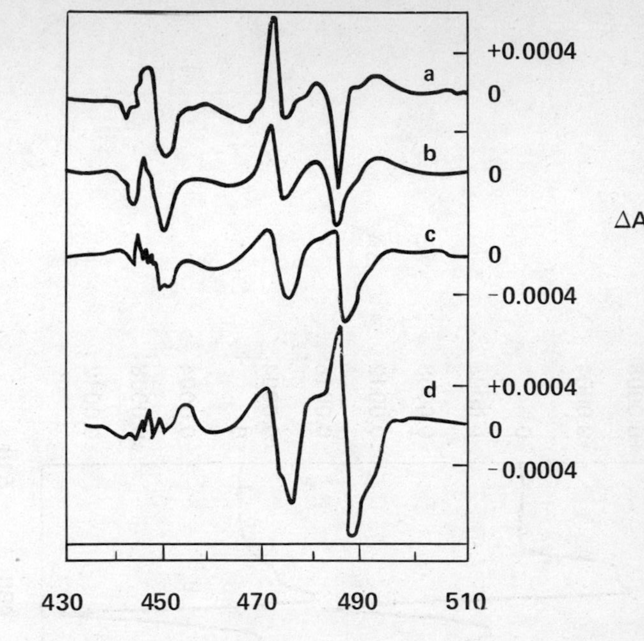

FIGURE 32. Alterations of CD spectra for 3:1 valine-praseodymium(III) mixture, pH increasing from pH 7+ (a) to pH 8+ (d). At about pH 7 spectra would be those of Figure 29a. CD path lengths, 30 mm. (Reprinted with permission from Katzin and Gulyas, *Inorg. Chem.*, 7, 2444 (1968). Copyright by the American Chemical Society.)

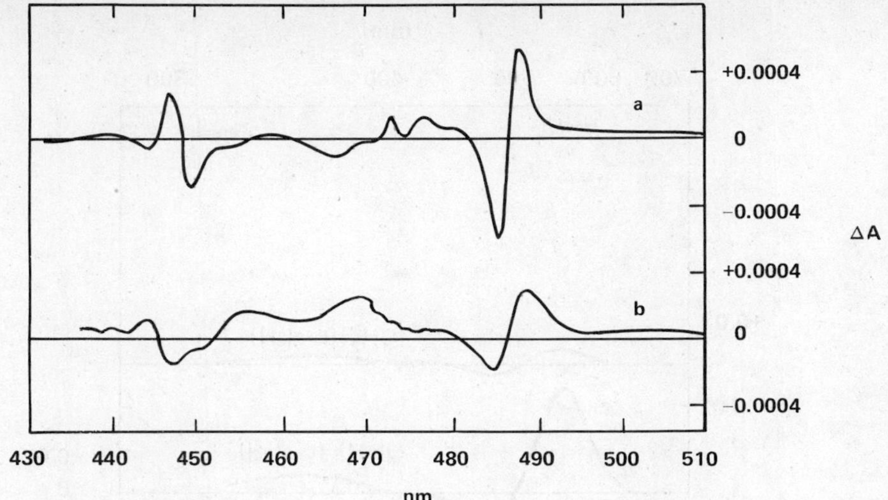

FIGURE 33. Amino acid-praseodymium(III) mixtures (5:1) at pH 9 to 10: (a) asparagine; (b) proline. CD path lengths, 30 mm. (Reprinted with permission from Katzin and Gulyas, *Inorg. Chem.*, 7, 2444 (1968). Copyright by the American Chemical Society.)

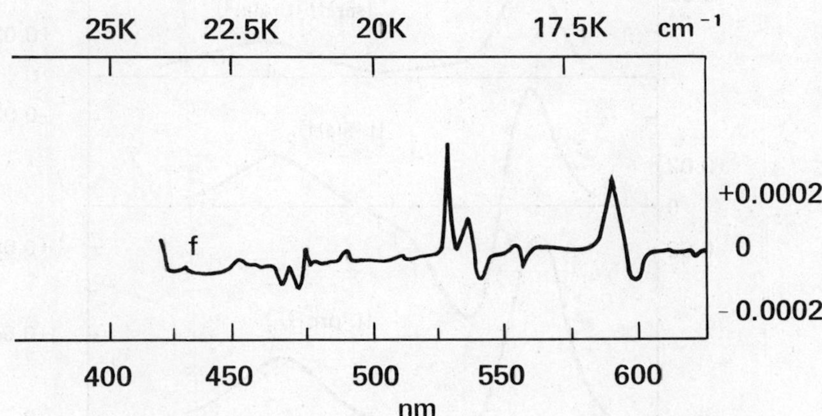

FIGURE 34. Eu(III) solutions with alanine, pH 7. (Reprinted with permission from Katzin, *Inorg. Chem.*, 8, 1651 (1969). Copyright by the American Chemical Society.)

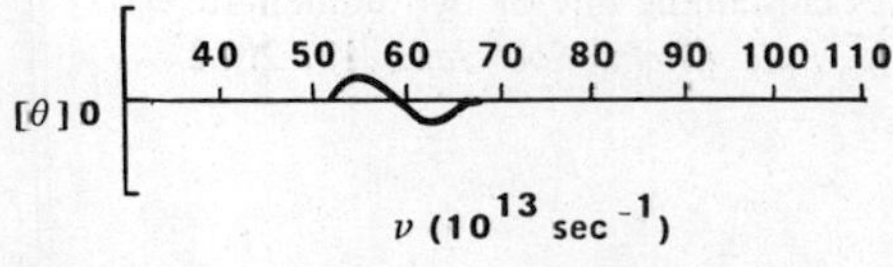

FIGURE 35. $[Co(NH_3)_5(L\text{-}\alpha\text{-alaH})](ClO_4)_3 \cdot 2H_2O$. (From Fujita, Yasui, and Shimura, *Bull. Chem. Soc. Jap.*, 38, 658 (1965). With permission.)

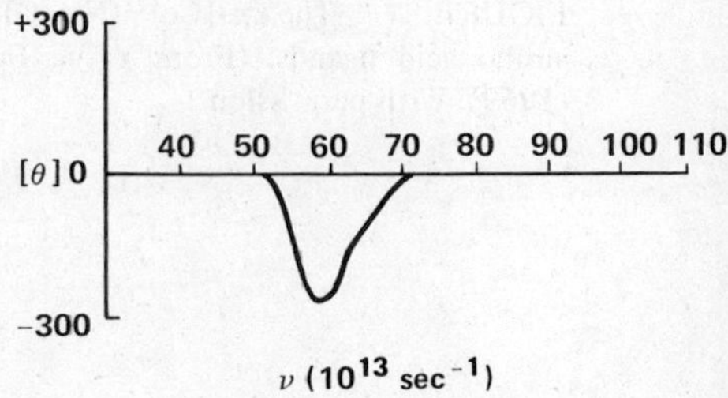

FIGURE 36. $[Co(NH_3)_5](L\text{-phalaH})](ClO_4)_3 \cdot H_2O$. (From Fujita, Yasui, and Shimura, *Bull. Chem. Soc. Jap.*, 38, 658 (1965). With permission.)

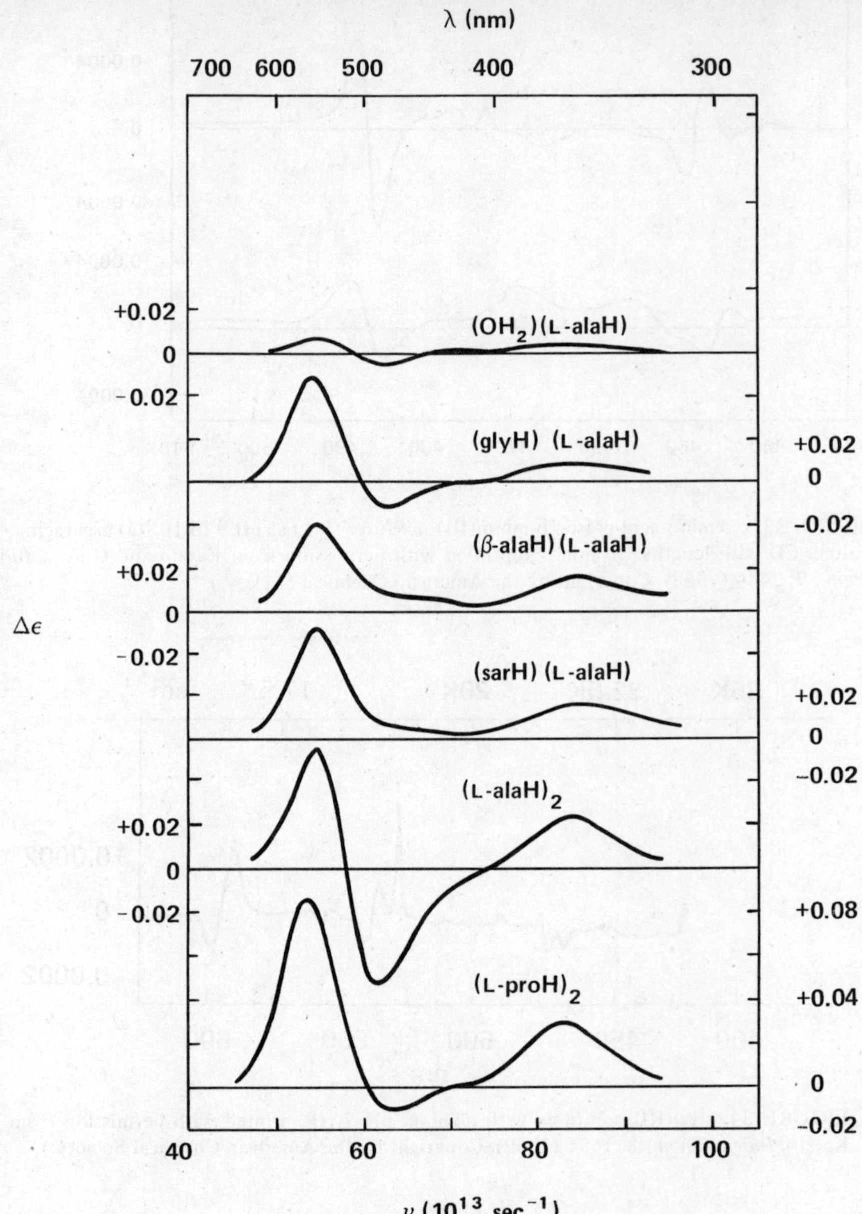

FIGURE 37. The *cis*-[Co^{III}(N)$_4$(O)$_2$] type complexes containing one or two unidentate amino acid ligands. (From Yasui, Fujita, and Shimura, *Bull. Chem. Soc. Jap.*, 42, 2082 (1969). With permission.)

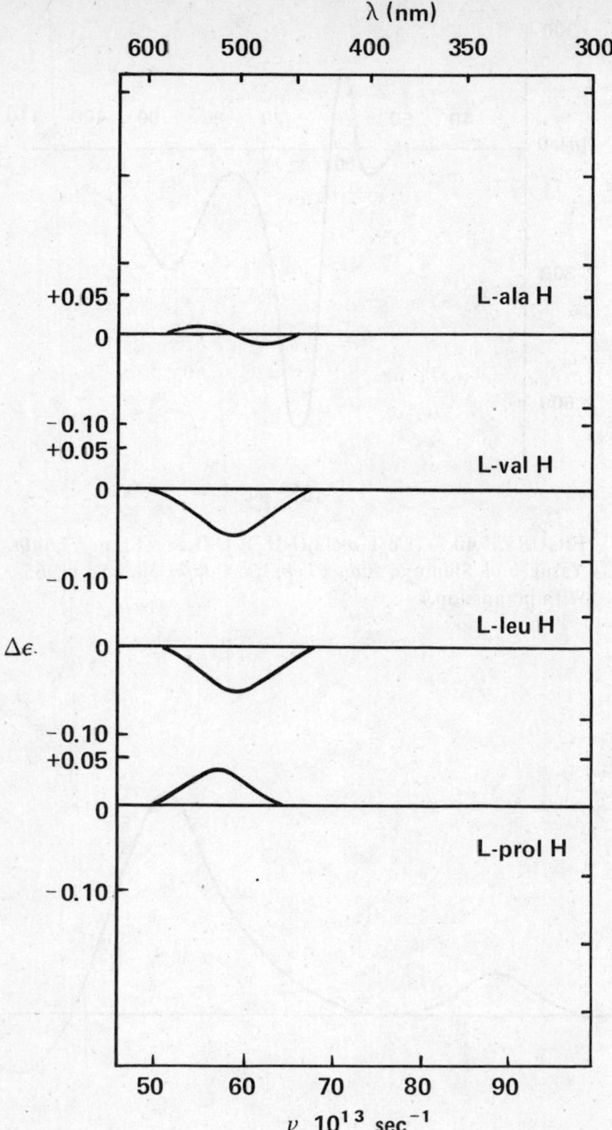

FIGURE 38. $[Co(NH_3)_5(L\text{-}amH)]^{3+}$ ions. (From Yasui, Hidaka, and Shimura, *Bull. Chem. Soc. Jap.*, 39, 2419 (1966). With permission.)

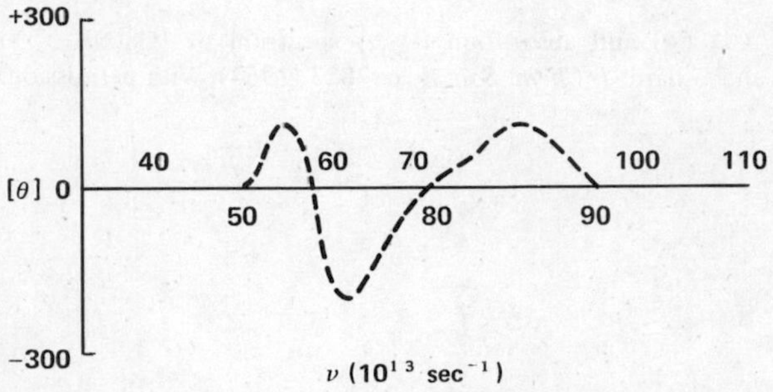

FIGURE 39. *cis*-$[Co(NH_3)_4(L\text{-}\alpha\text{-}alaH)_2](ClO_4)_3 \cdot 2H_2)$. (From Fujita, Yasui, and Shimura, *Bull. Chem. Soc. Jap.*, 38, 659 (1965). With permission.)

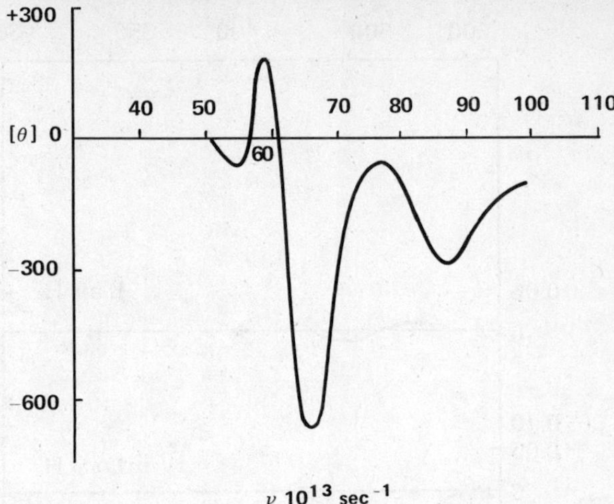

FIGURE 40. $[Co(L\text{-}\alpha\text{-ala})(NH_3)_4]SO_4$. (From Fujita, Yasui, and Shimura, *Bull. Chem. Soc. Jap.*, 38, 658 (1965). With permission.)

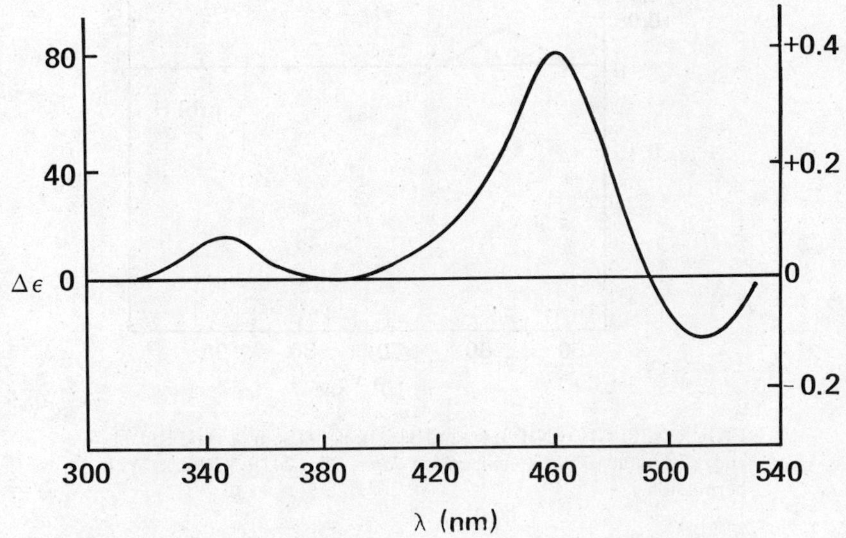

FIGURE 41. CD (–) and absorption (- - -) spectrum of $[Co(NH_3)_4(D\text{-alaninate})]^{2+}$. (From Dunlop and Gillard, *J. Chem. Soc. A*, p. 2823 (1964). With permission.)

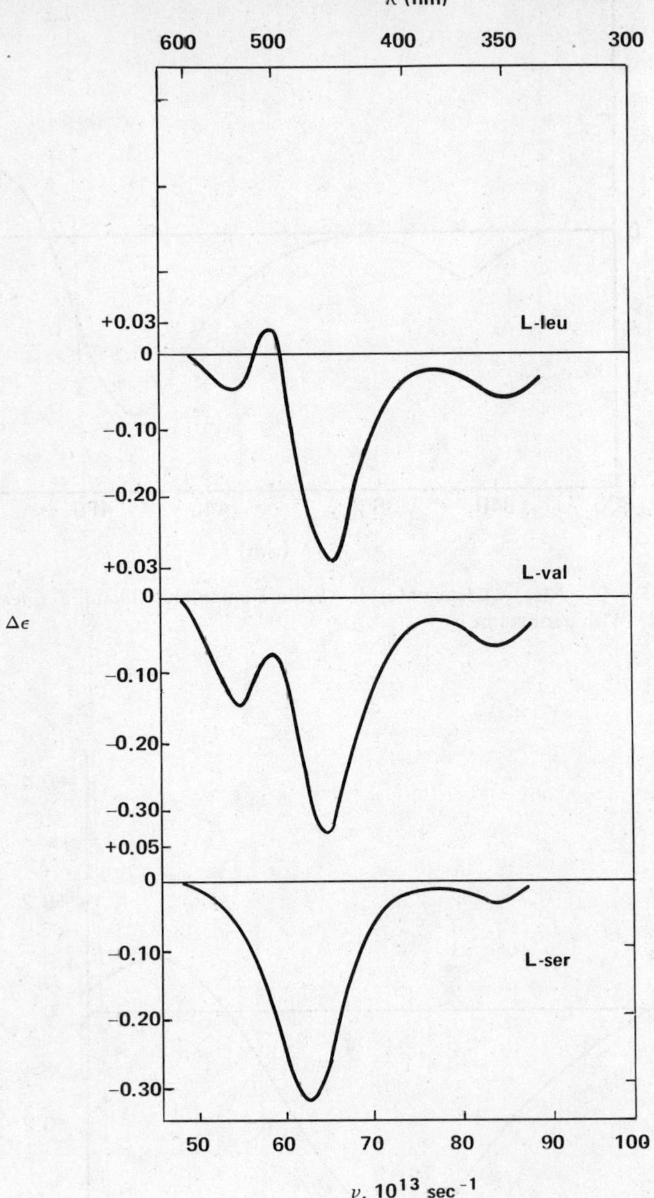

FIGURE 42. $[Co(L\text{-}am)(NH_3)_4]^{2+}$ ions. (From Yasui, Hidaka, and Shimura, *Bull. Chem. Soc. Jap.*, 39, 2421 (1966). With permission.)

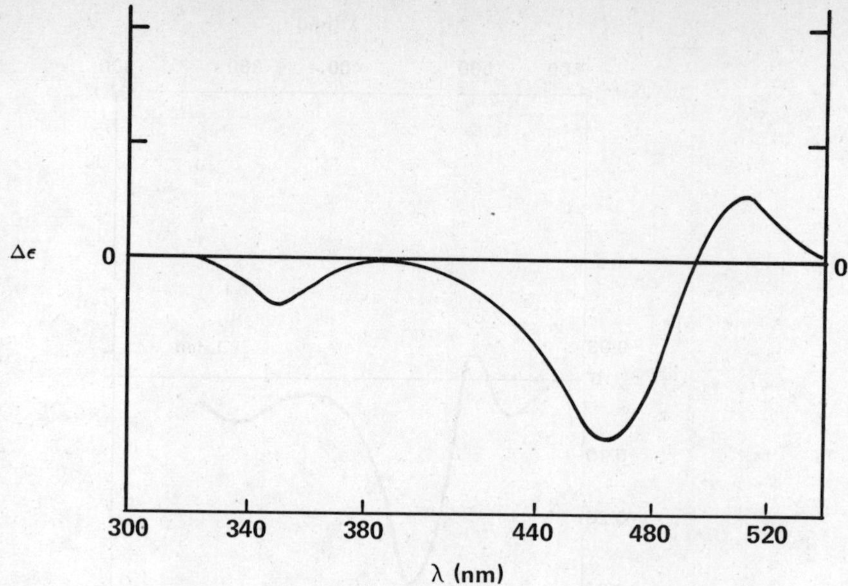

FIGURE 43. $[Co(NH_3)_4(L\text{-leucinate})]^{2+}$. (From Dunlop and Gillard, *J. Chem. Soc. A*, p. 2823 (1964). With permission.)

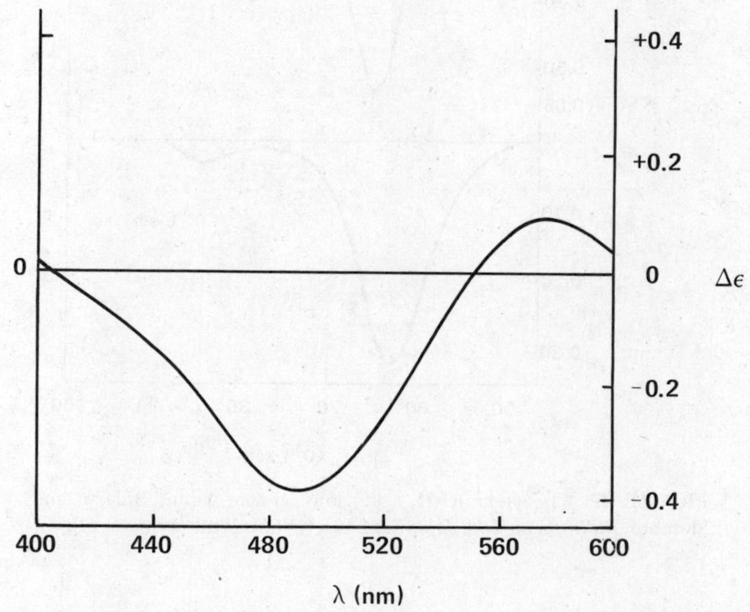

FIGURE 44. $[Co(NH_3)_4(L\text{-histidinate})]^{2+}$. (From Dunlop and Gillard, *J. Chem. Soc. A*, p. 2824 (1964). With permission.)

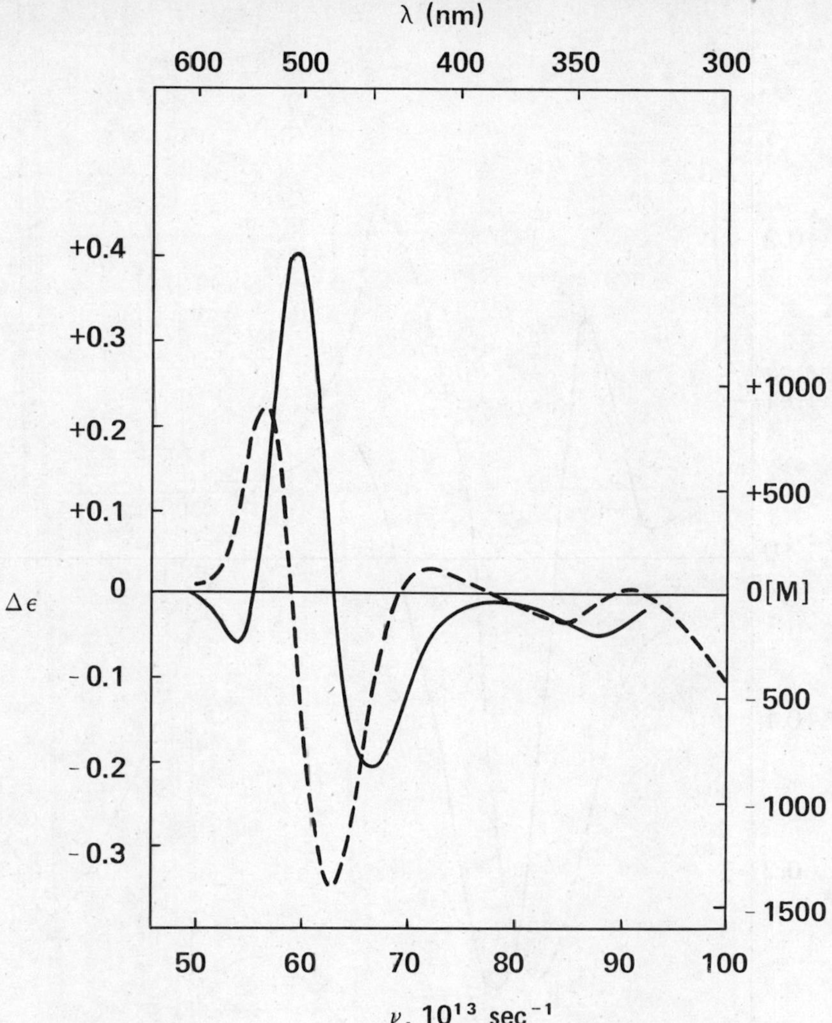

FIGURE 45. CD (–) and RD (- - -) curves of $[Co(L\text{-}prol)(NH_3)_4]^{2+}$ ion. (From Yasui, Hidaka, and Shimura, *Bull. Chem. Soc. Jap.*, 39, 2422 (1966). With permission.)

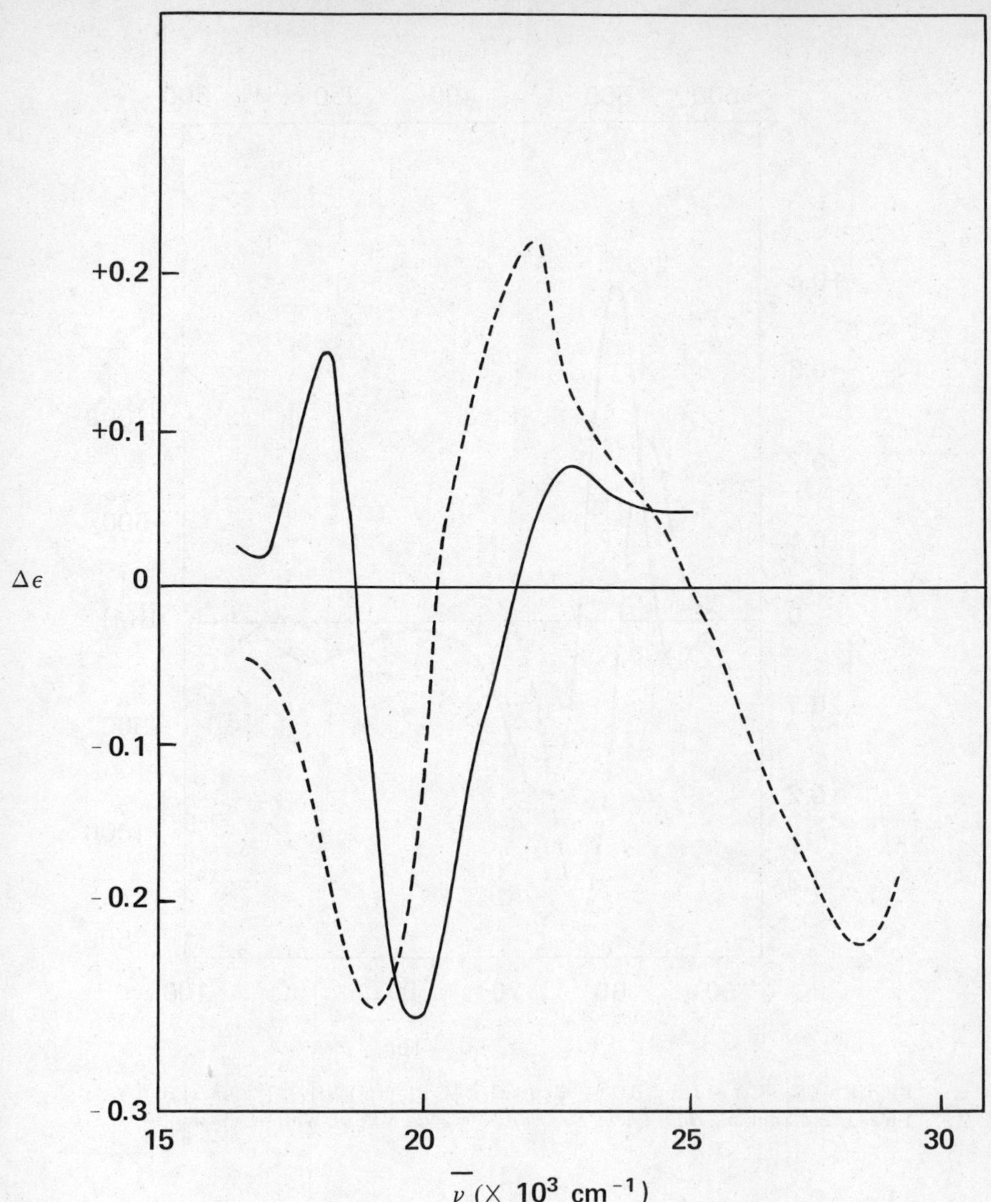

FIGURE 46. $[Co(NH_3)_4(N\text{-}Me\text{-}S\text{-}leu]^{2+}$ (–) and $[Co(NH_3)_4((R)\text{-}pipec)]^{2+}$ (- - -). (Reprinted with permission from Saburi and Yoshikawa, *Inorg. Chem.*, 7, 1894 (1968). Copyright by the American Chemical Society.)

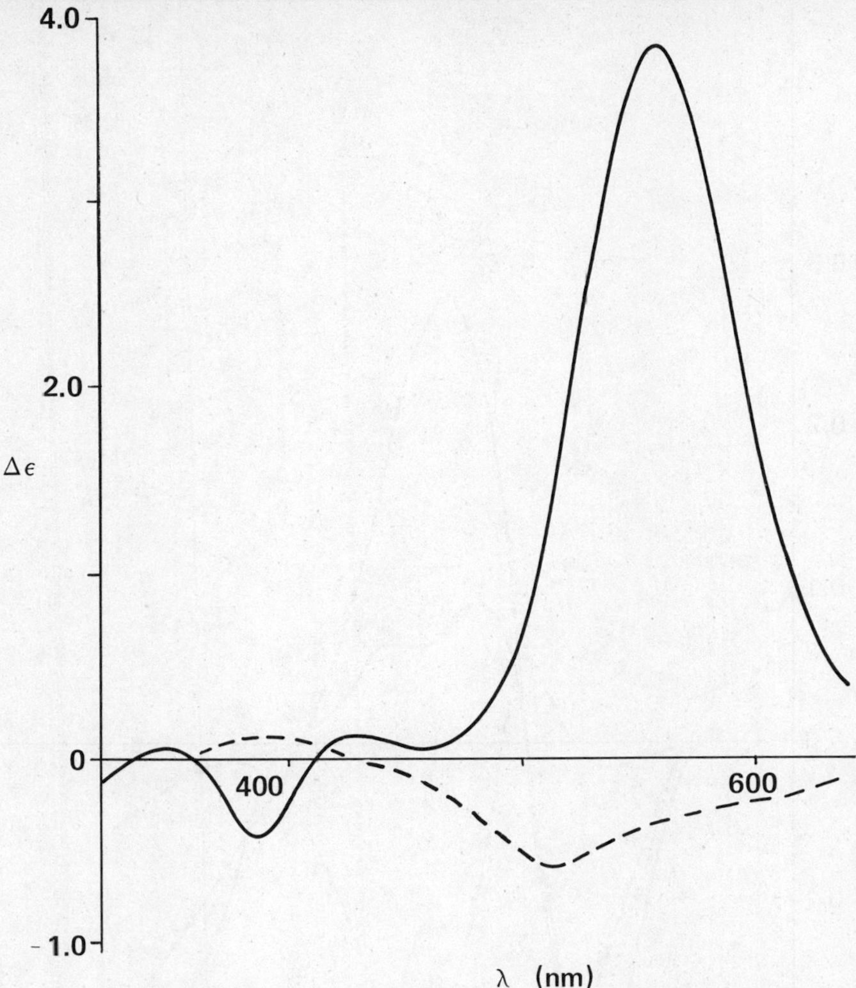

FIGURE 47. (+)-K[Co(L-val)$_2$CO$_3$] in M/5-KHCO$_3$ (–) and (–)-[Co(L-val)$_2$(H$_2$O)$_2$] ClO$_4$ in aqueous solution (- - -). (From Gillard and Price, *J. Chem. Soc. A*, p. 2273 (1971). With permission.)

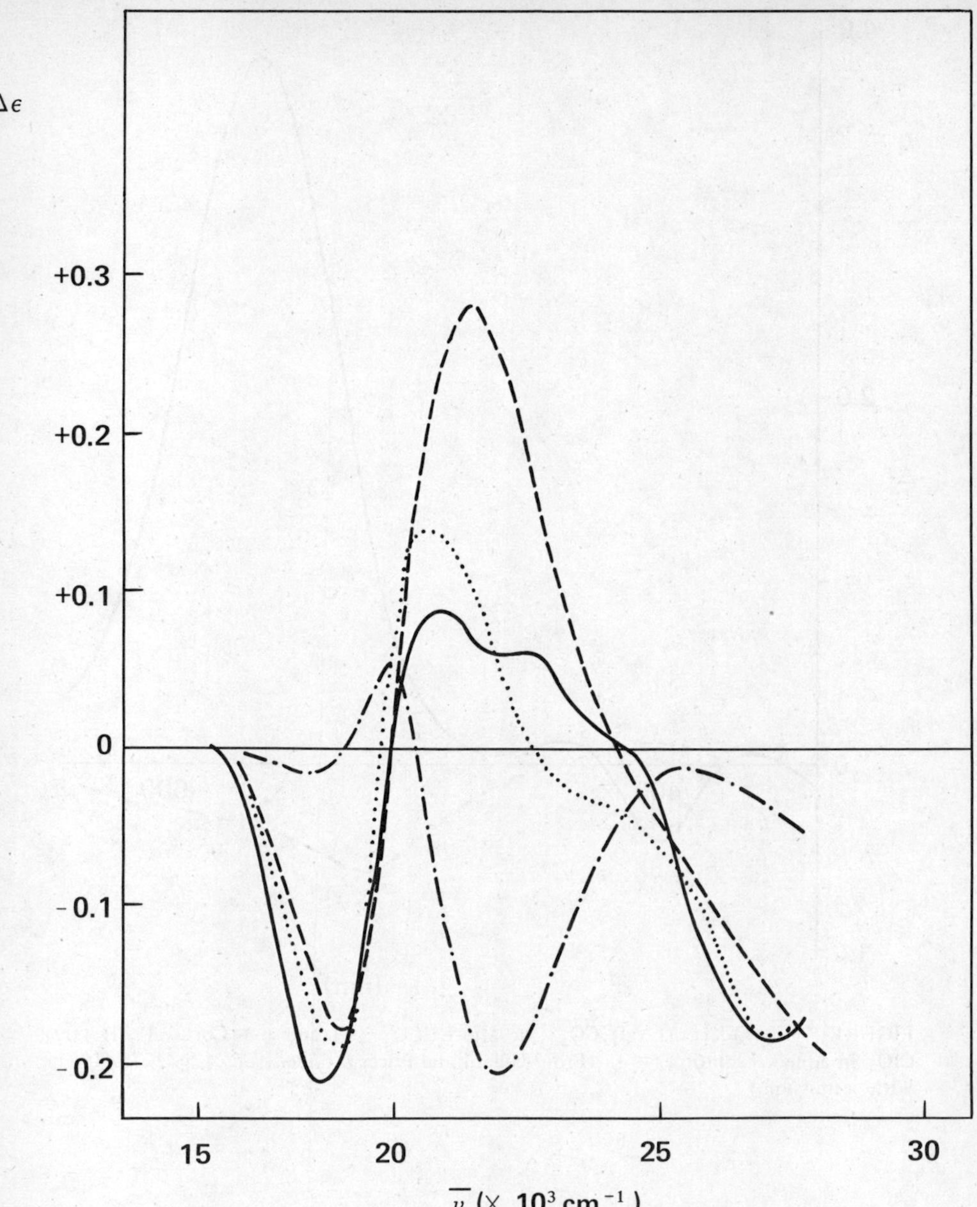

FIGURE 48. $[Co(NH_3)_4(N\text{-Me-(S)-ala}]^{2+}$, $(-)_{436}$-$[Co(NH_3)_4(sar)]^{2+}$, and $[Co(NH_3)_4((S)\text{-ala})]^{2+}$ (—•—), and the composite curve for (-)-$[Co(NH_3)_4(sar)]^{2+}$ +$[Co(NH_3)_4(S)\text{-ala}]^{2+}$ (- - -). (Reprinted with permission from Saburi and Yoshikawa, *Inorg. Chem.*, 7, 1892 (1968). Copyright by the American Chemical Society.)

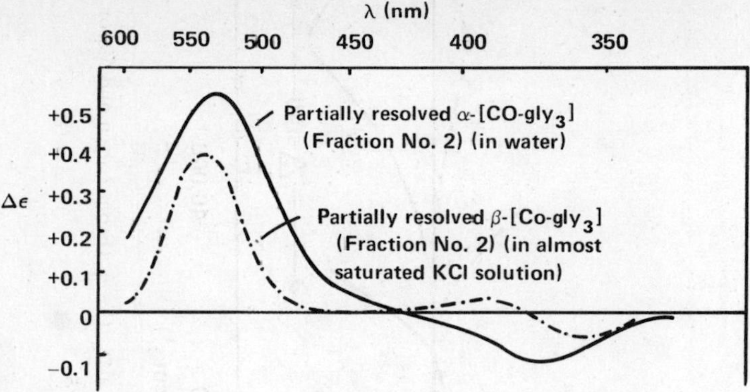

FIGURE 49. α- and β-[Co(gly)$_3$] CD curves for the complexes partially resolved on a starch column. (Reprinted with permission from Douglas and Yamada, *Inorg. Chem.*, 4, 1563 (1965). Copyright by the American Chemical Society.)

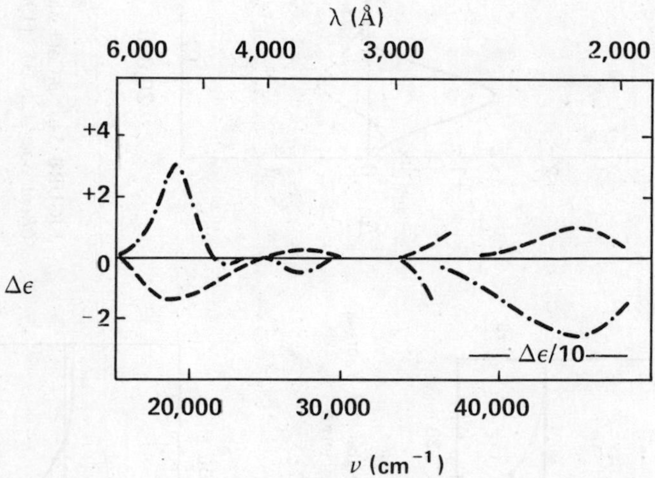

FIGURE 50. α-Co(+ ala)$_3$ (—•—•—) and α′-Co (+ ala)$_3$ (- - -). (From Larsen and Mason, *J. Chem. Soc. A*, p. 314 (1966). With permission.)

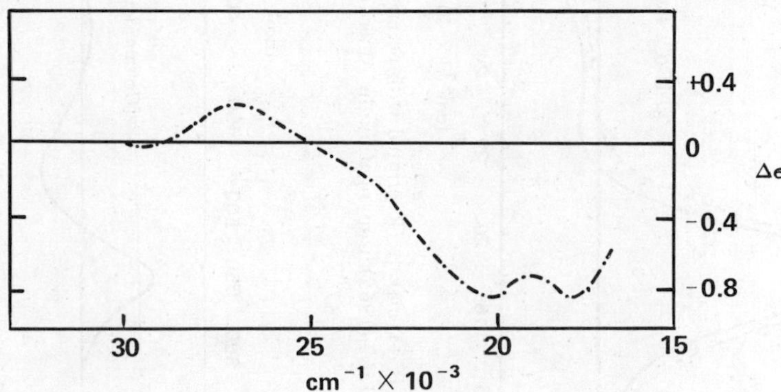

FIGURE 51. (−)α-[Co(L-ala)$_3$]. (From Dunlop and Gillard, *J. Chem. Soc. A*, p. 6534 (1965). With permission.)

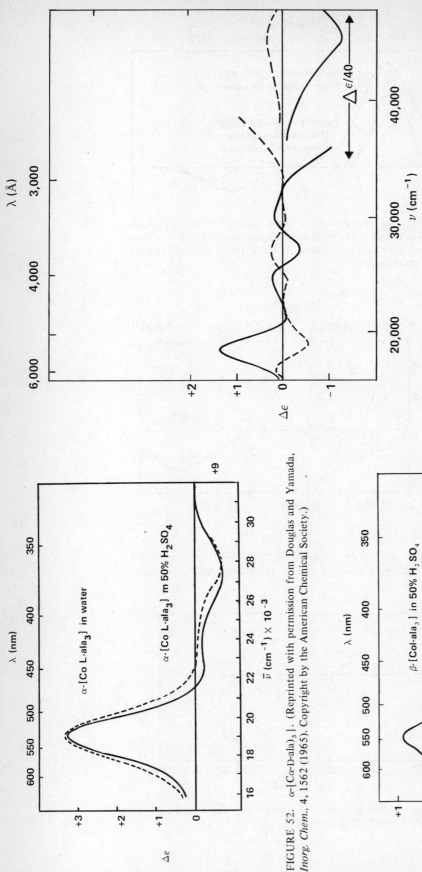

FIGURE 52. α-[Cα-D-ala)$_3$]. (Reprinted with permission from Douglas and Yamada, *Inorg. Chem.*, 4, 1562 (1965). Copyright by the American Chemical Society.)

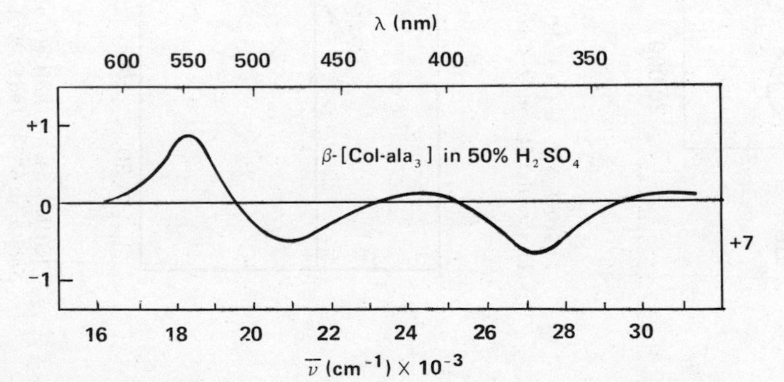

FIGURE 53. β-[Co(L-ala)$_3$]. (Reprinted with permission from Douglas and Yamada, *Inorg. Chem.*, 4, 1563 (1965). Copyright by the American Chemical Society.)

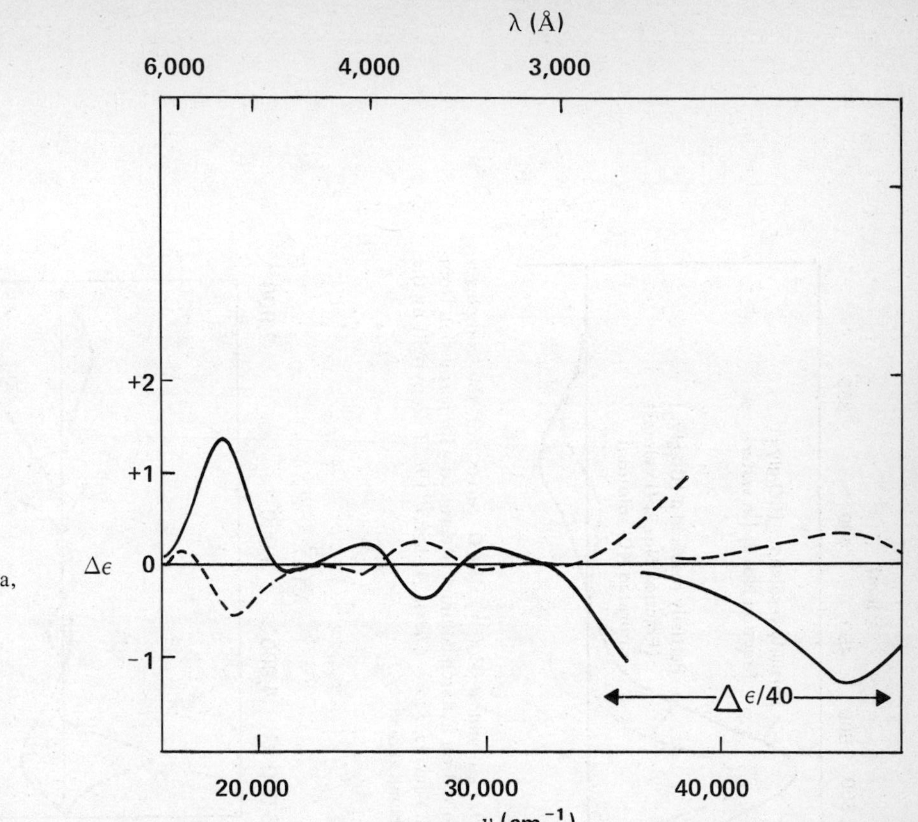

FIGURE 54. β-Co(+ ala)$_3$ (–) and β′-Co(+ ala)$_3$ (- - -). (From Larsen and Mason, *J. Chem. Soc. A*, p. 313 (1966). With permission.)

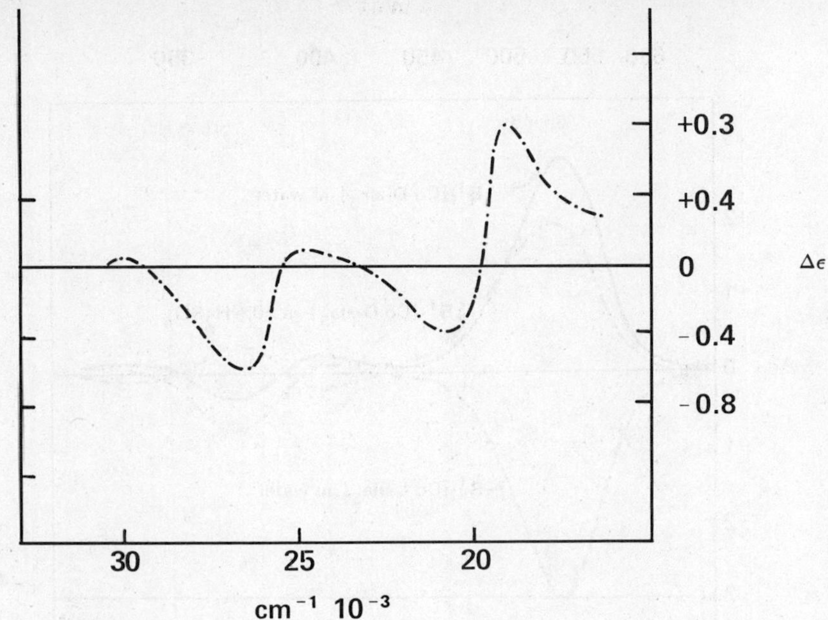

FIGURE 55. (+)β-[Co(L-ala)$_3$]. (From Dunlop and Gillard, *J. Chem. Soc. A*, p.6534 (1965). With permission.)

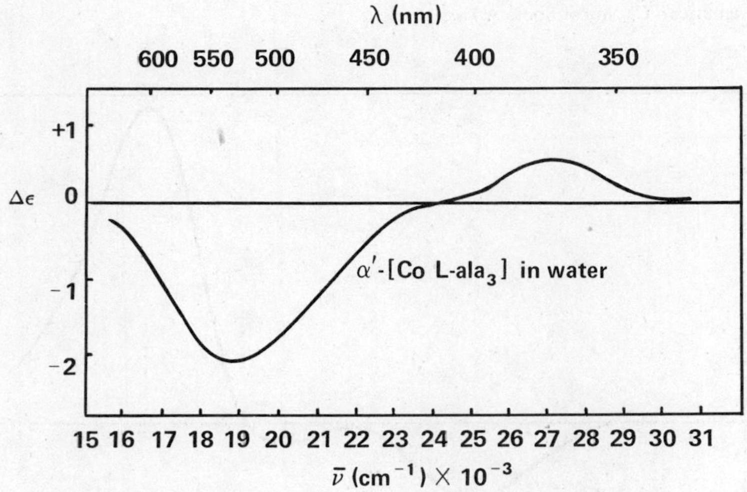

FIGURE 56. α′-[Co(L-ala$_3$)]. (Reprinted with permission from Douglas and Yamada, *Inorg. Chem.*, 4, 1562 (1965). Copyright by the American Chemical Society.)

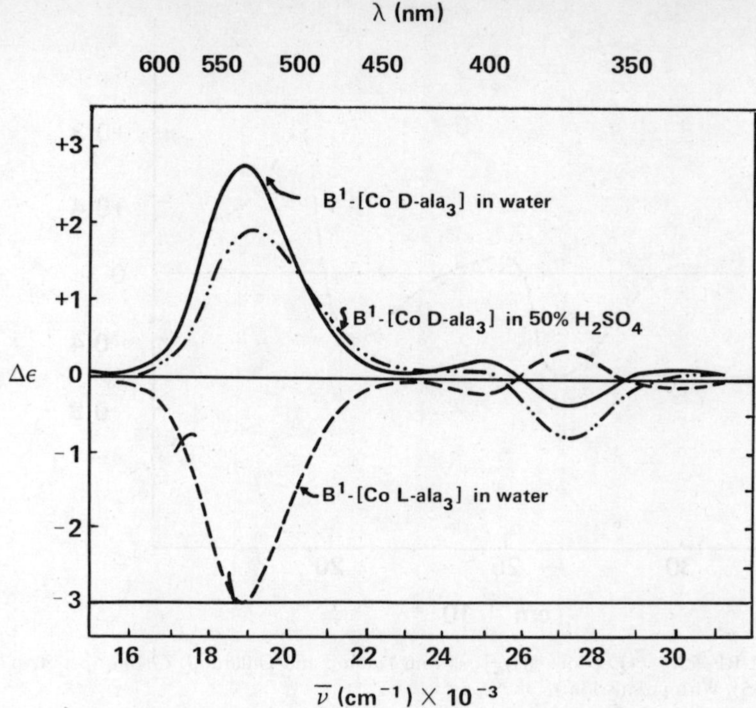

FIGURE 57. β′-[Co(D-ala)$_3$] and β′-[Co(L-ala)$_3$]. (Reprinted with permission from Douglas and Yamada, *Inorg. Chem.*, 4, 1563 (1965). Copyright by the American Chemical Society.)

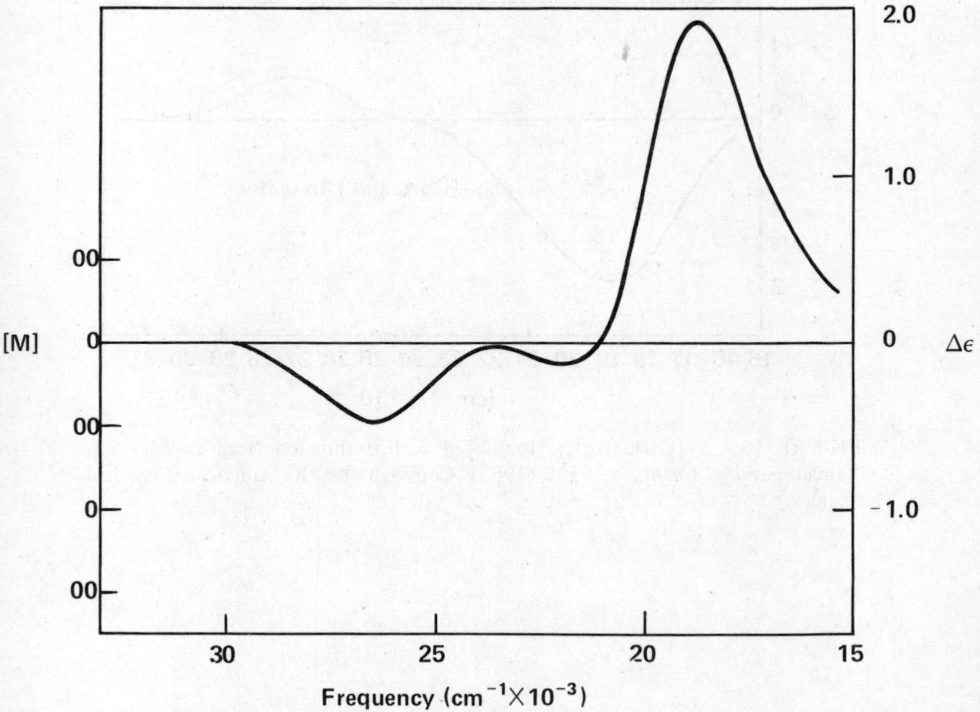

FIGURE 58. 1,2,6-(+)[Co(L-ala)$_3$]. (From Gillard, *Chem. Br.*, 3, 207 (1967). With permission.)

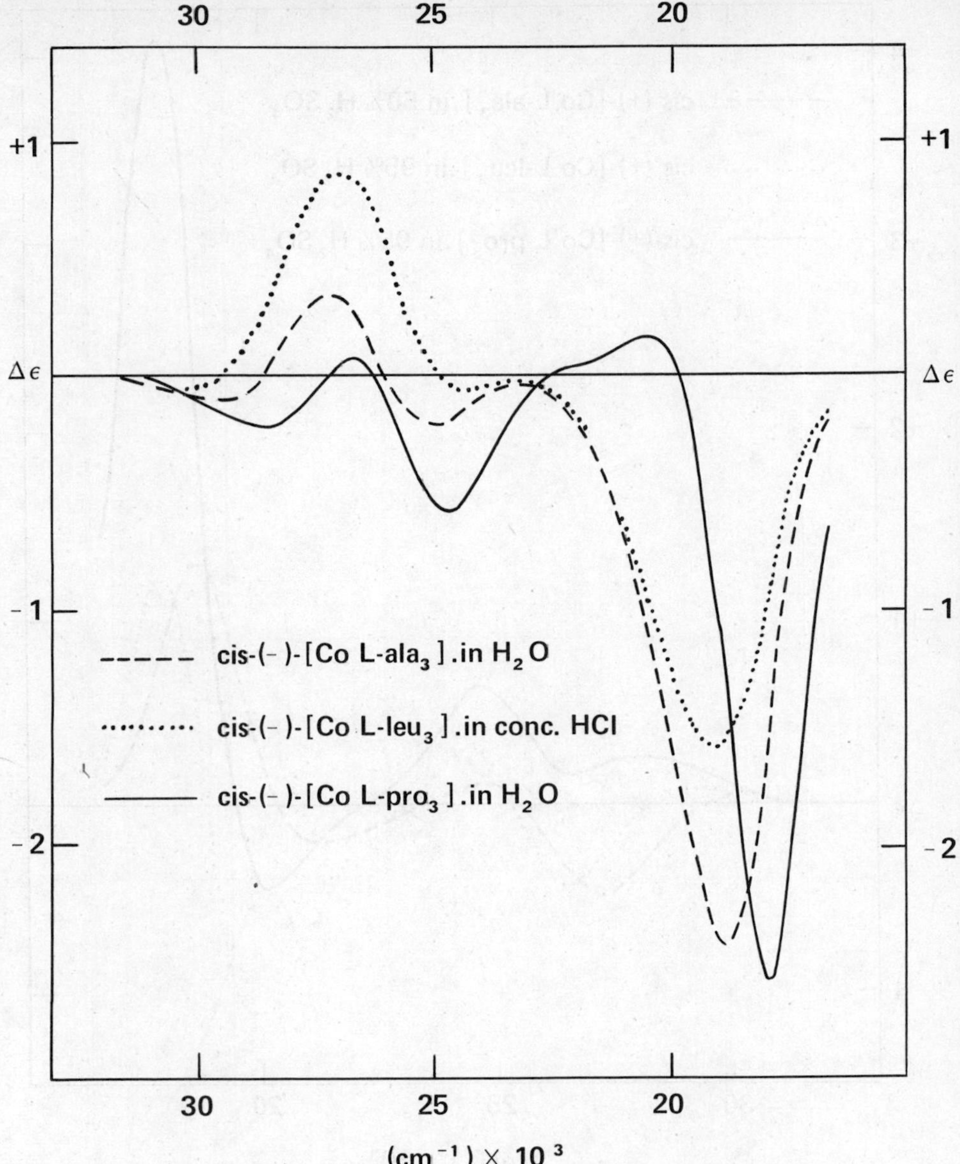

FIGURE 59. *cis*-(−)-[Co(L-amino acid)$_3$] complexes. (Reprinted with permission from Denning and Piper, *Inorg. Chem.*, 5, 1059 (1966). Copyright by the American Chemical Society.)

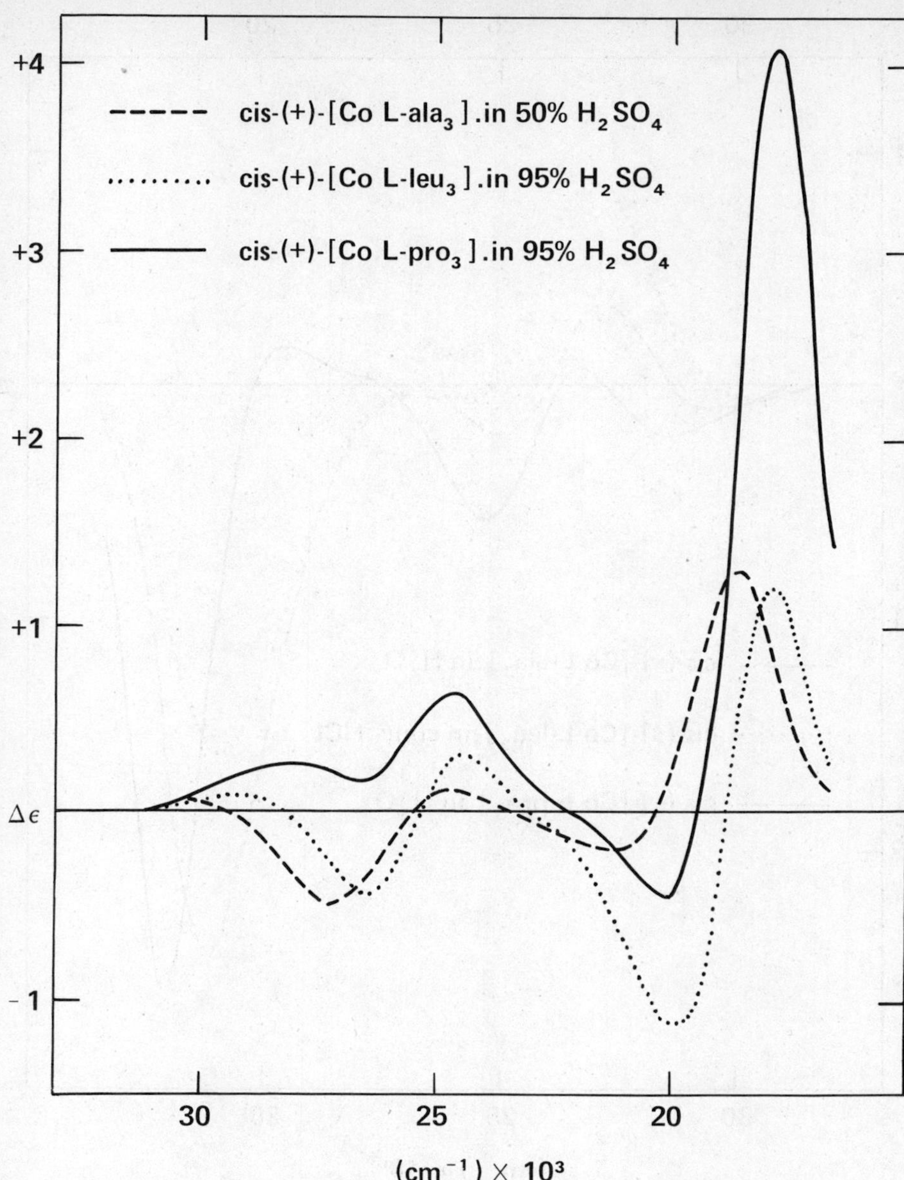

FIGURE 60. *cis*-(+)-[Co(L-amino acid)$_3$] complexes. (Reprinted with permission from Denning and Piper, *Inorg. Chem.*, 5, 1059 (1966). Copyright by the American Chemical Society.)

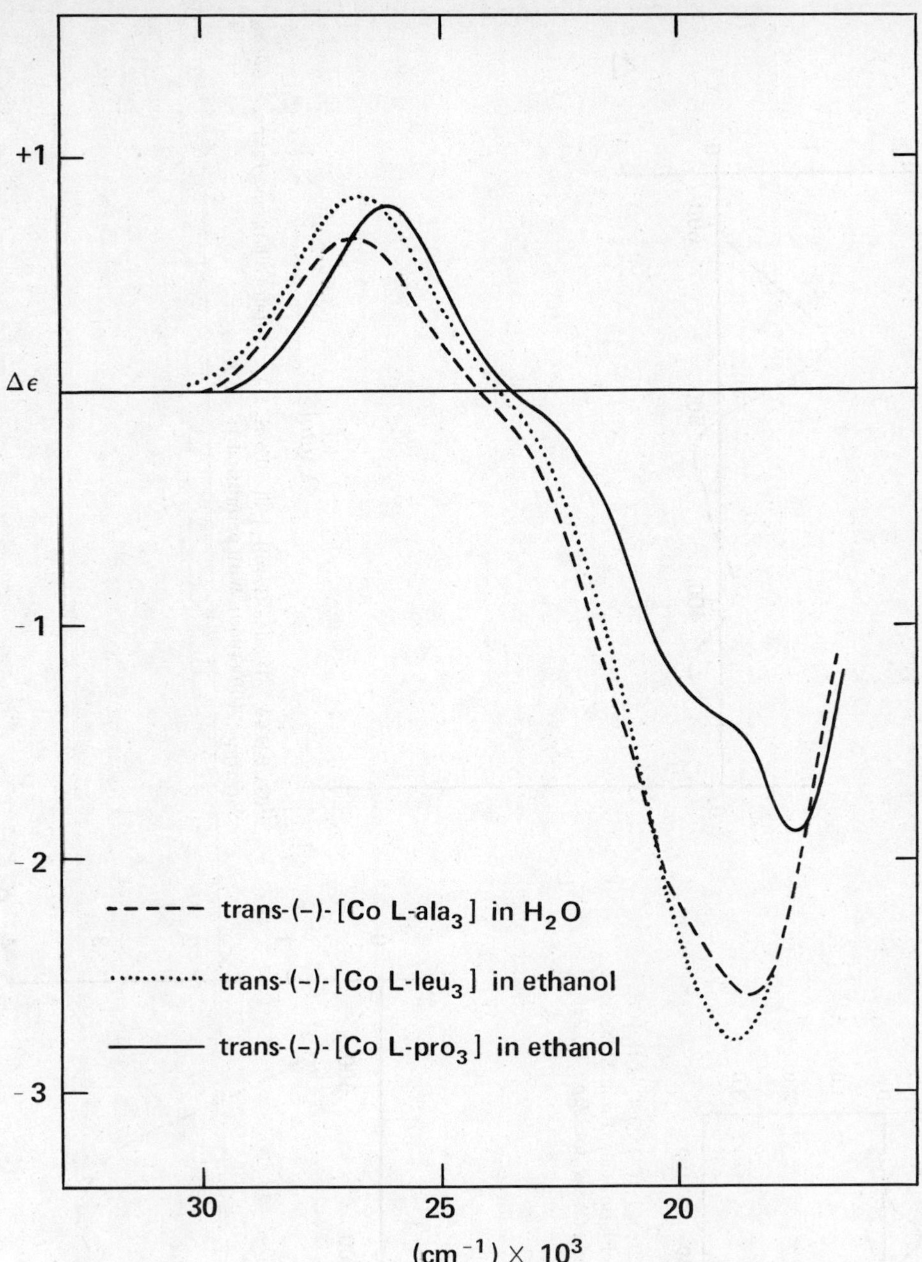

FIGURE 61. *trans*-(−)-[Co(L-amino acid)$_3$] complexes. (Reprinted with permission from Denning and Piper, *Inorg. Chem.*, 5, 1060 (1966). Copyright by the American Chemical Society.)

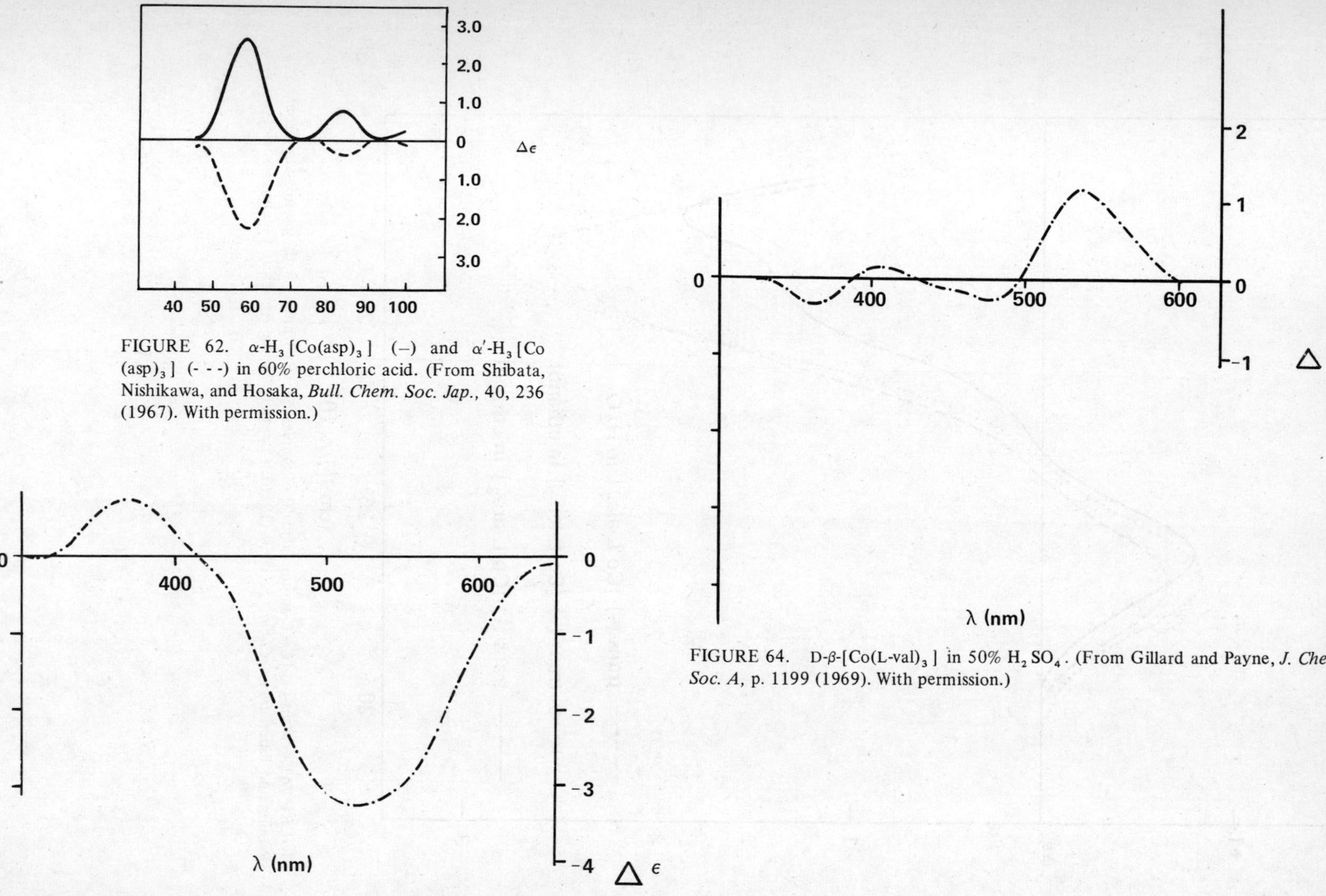

FIGURE 62. α-H_3[Co(asp)$_3$] (–) and α′-H_3[Co(asp)$_3$] (- - -) in 60% perchloric acid. (From Shibata, Nishikawa, and Hosaka, *Bull. Chem. Soc. Jap.*, 40, 236 (1967). With permission.)

FIGURE 63. L-α-[Co(L-val)$_3$]$_3$ in ethanol. (From Gillard and Payne, *J. Chem. Soc. A*, p. 1198 (1969). With permission.)

FIGURE 64. D-β-[Co(L-val)$_3$] in 50% H_2SO_4. (From Gillard and Payne, *J. Chem. Soc. A*, p. 1199 (1969). With permission.)

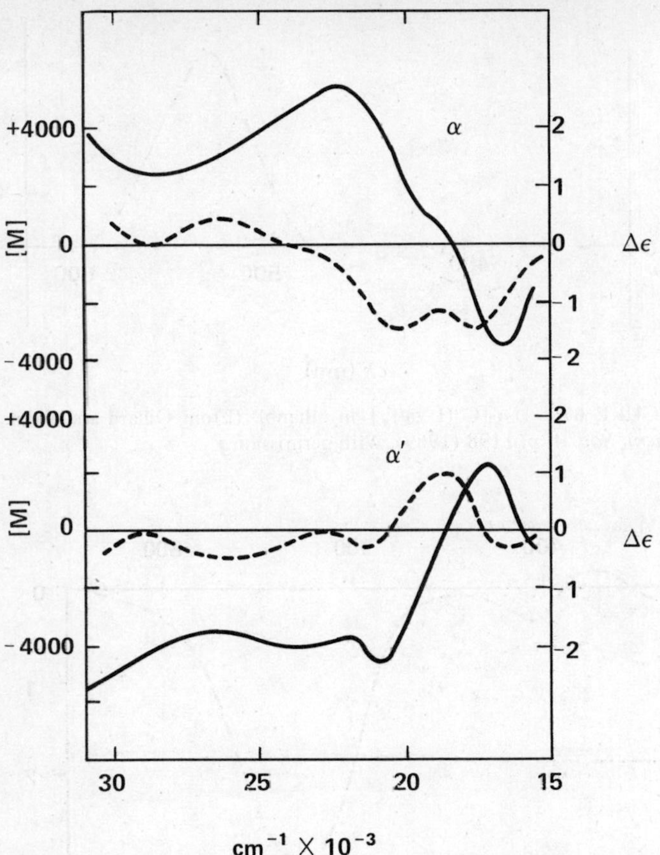

FIGURE 65. RD (–) and CD (- - -) of α- and α′-[Co(L-val)$_3$]. (From Shibata, Nishikawa, and Nishida, *Bull. Chem. Soc. Jap.*, 39, 2310 (1966). With permission.)

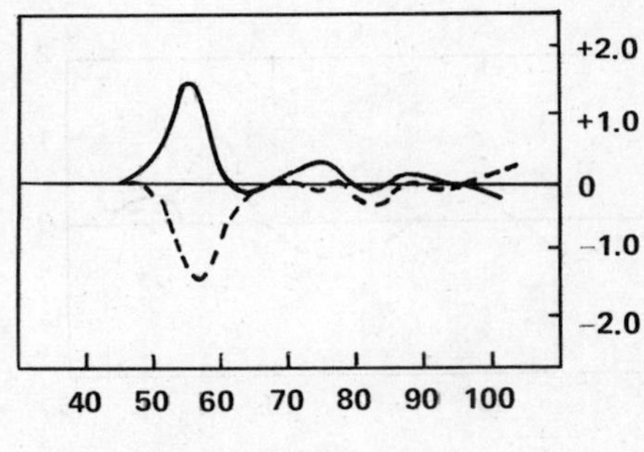

FIGURE 66. β-H_3[Co(asp)$_3$] (–) and β′-H_3[Co(asp)$_3$] (- - -) in 60% perchloric acid. (From Shibata, Nishikawa, and Hosaka, *Bull. Chem. Soc. Jap.*, 40, 236 (1967). With permission.)

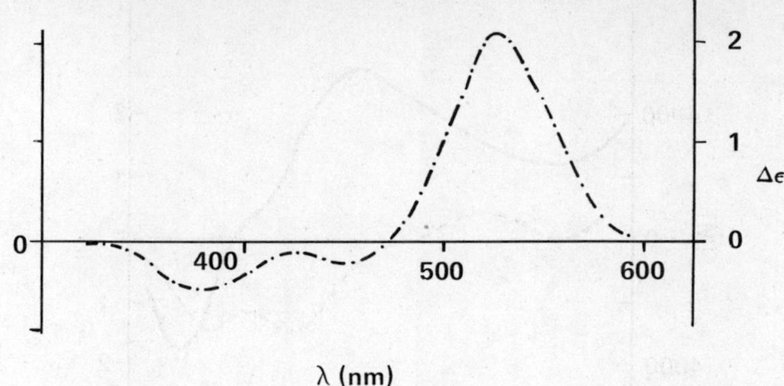

FIGURE 67. D-α-[Co(L-val)$_3$] in ethanol. (From Gillard and Payne, *J. Chem. Soc. A*, p. 1198 (1969). With permission.)

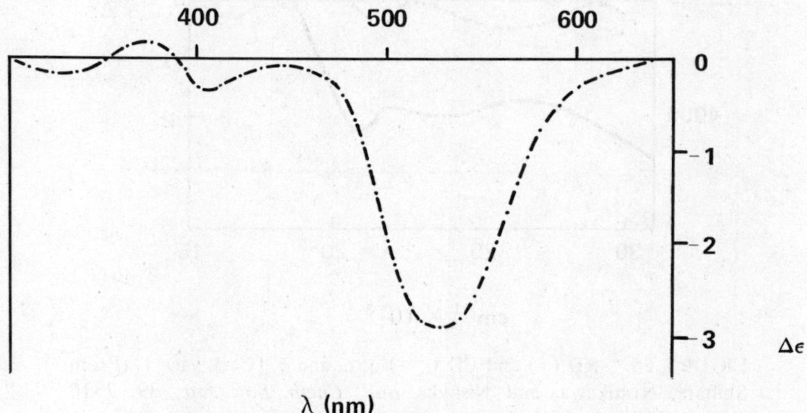

FIGURE 68. L-β-[Co(L-val)$_3$] in water. (From Gillard and Payne, *J. Chem. Soc. A*, p. 1199 (1969). With permission.)

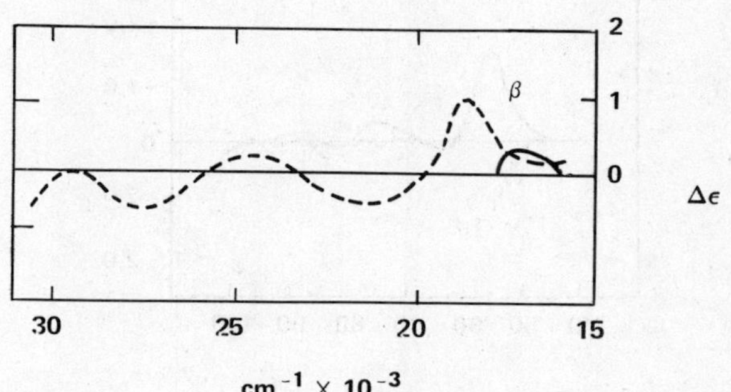

FIGURE 69. β- and β′-[Co(L-val)$_3$]. (From Shibata, Nishikawa, and Nishida, *Bull. Chem. Soc. Jap.*, 39, 2310 (1966). With permission.)

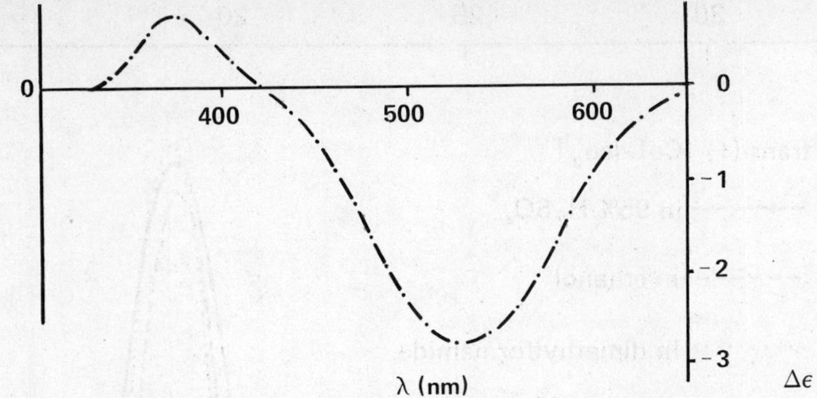

FIGURE 70. L-α-[Co(L-leu)$_3$] in ethanol. (From Gillard and Payne, *J. Chem. Soc. A*, p. 1199 (1969). With permission.)

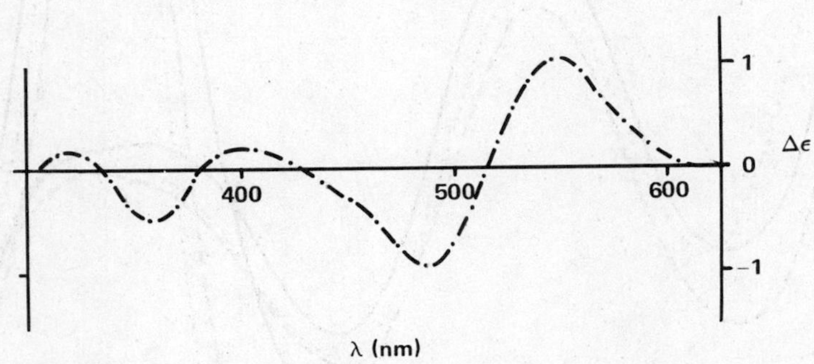

FIGURE 71. D-β-[Co(L-leu)$_3$] in 50% H_2SO_4. (From Gillard and Payne, *J. Chem. Soc. A*, p. 1200 (1969). With permission.)

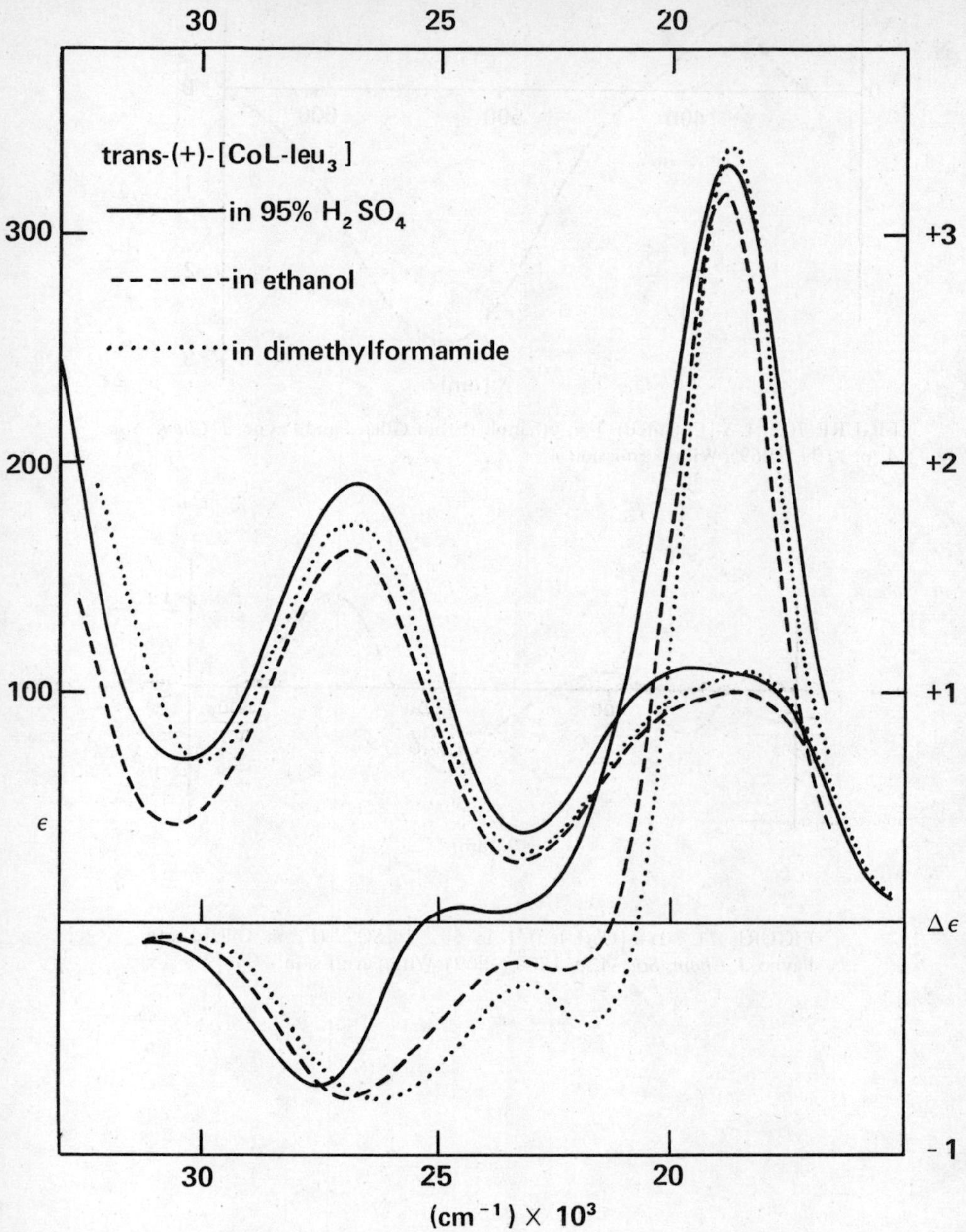

FIGURE 72. Effect of solvent on absorption and CD spectra of a *trans* complex. (Reprinted with permission from Denning and Piper, *Inorg. Chem.*, 5, 1063 (1966). Copyright by the American Chemical Society.)

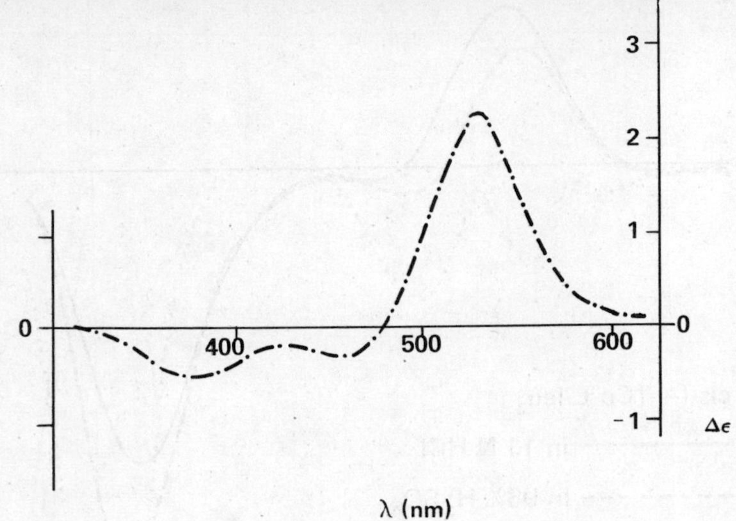

FIGURE 73. D-α-[Co(L-leu)$_3$] in ethanol. (From Gillard and Payne, *J. Chem. Soc. A*, p. 1199 (1969). With permission.)

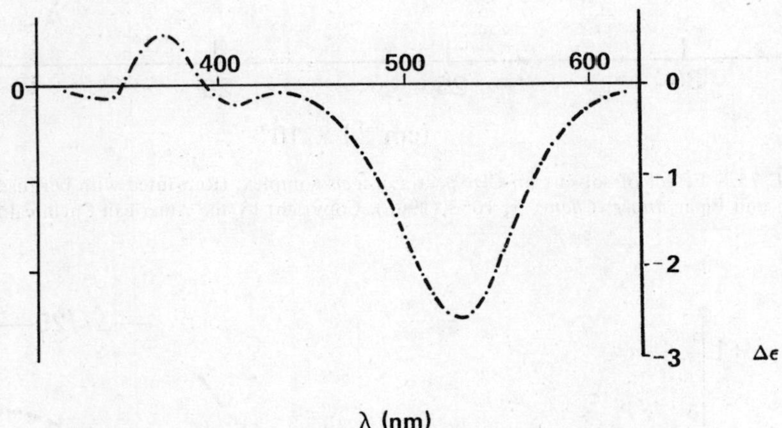

FIGURE 74. L-β-[Co(L-leu)$_3$] in 50% H_2SO_4. (From Gillard and Payne, *J. Chem. Soc. A*, p. 1199 (1969). With permission.)

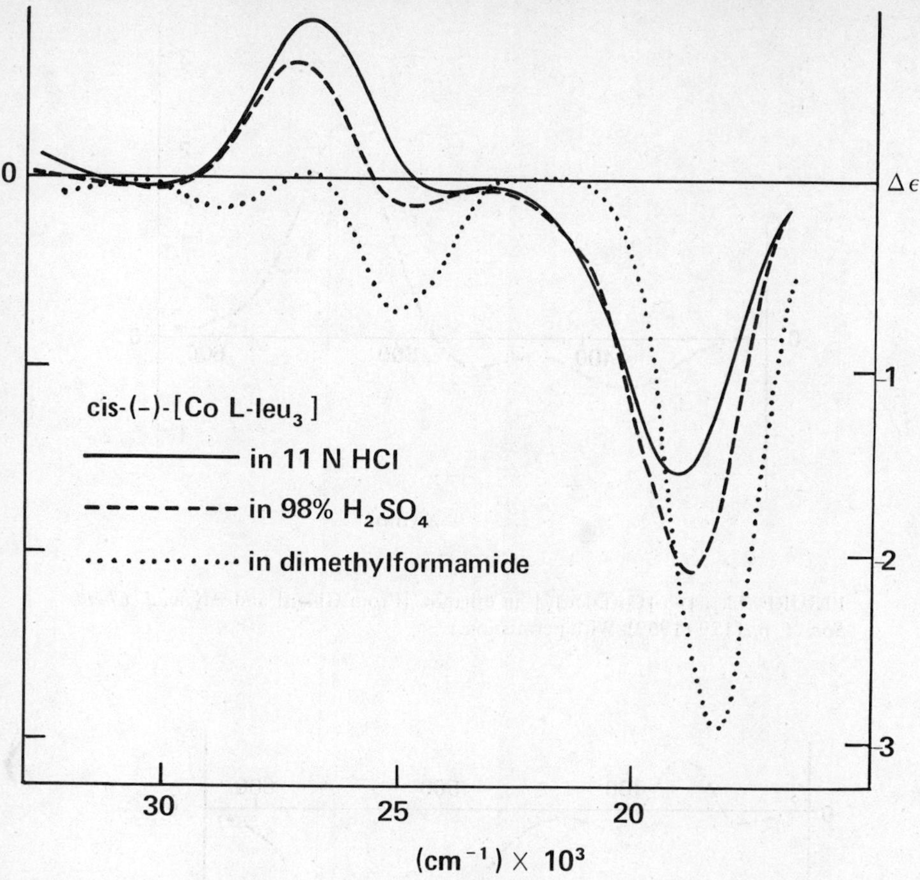

FIGURE 75. Effect of solvent on CD spectra of a *cis* complex. (Reprinted with permission from Denning and Piper, *Inorg. Chem.*, 5, 1063 (1966). Copyright by the American Chemical Society.)

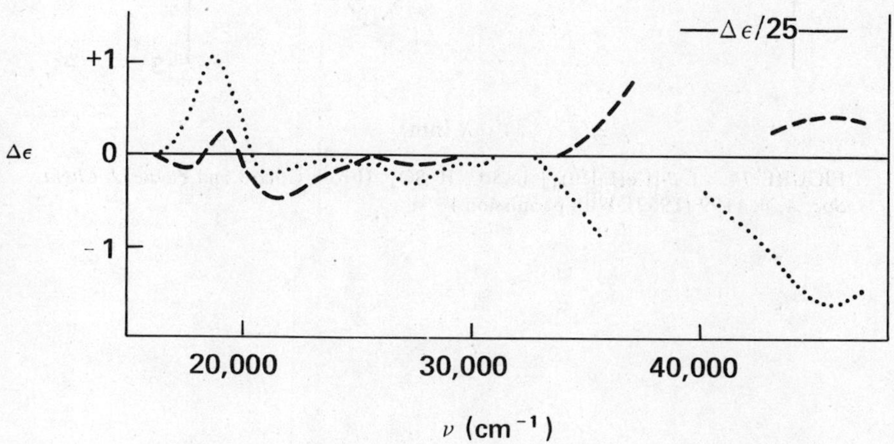

FIGURE 76. α-Co(+ glu)$_3$$^{3-}$ (- - -) and β-Co(+ glu)$_2$$^{3-}$ (···). (From Larsen and Mason, *J. Chem. Soc. A*, p. 314 (1966). With permission.)

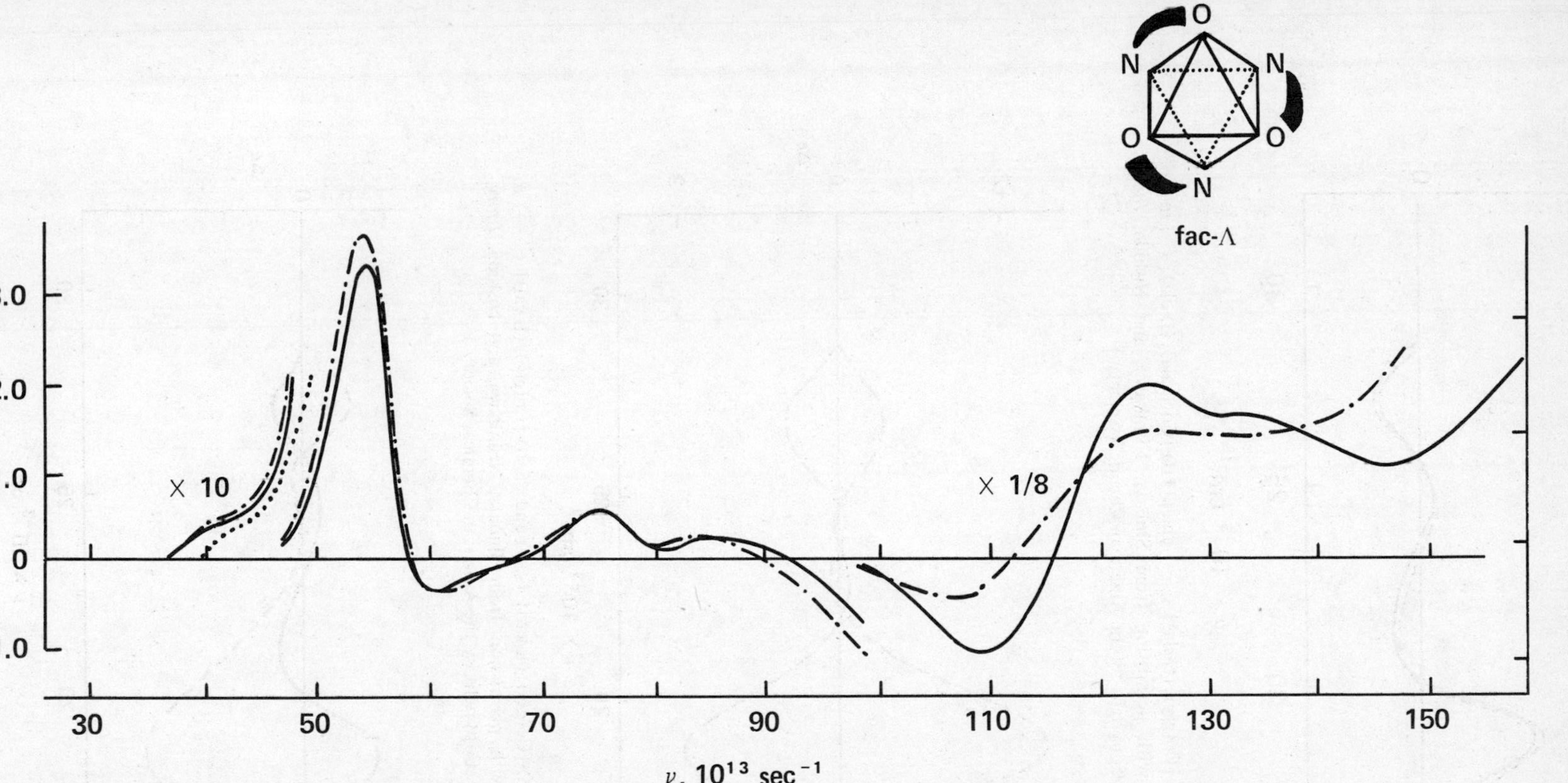

FIGURE 77. $[Co(L\text{-prol})_3]$ (–) and $[Co(L\text{-hydprol})_3]$ (–·–·–) in 70% perchloric acid. (From Yasui, Hidaka, and Shimura, *Bull. Chem. Soc. Jap.*, 38, 2025 (1965). With permission.)

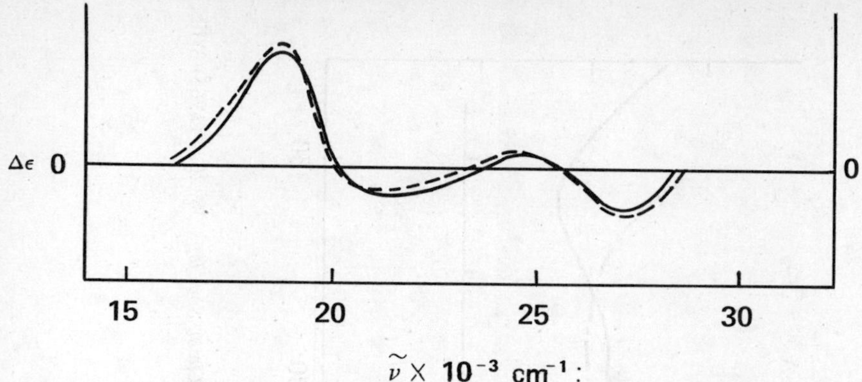

FIGURE 78. β(+)D-[Co(gly)$_2$(L-ala)] (- - -) and β(+)D-[Co(L-ala)$_2$(D-ala)] (–) in 50% H_2SO_4. (Reprinted with permission from Shibata, Nishikawa, and Nishida, *Inorg. Chem.*, 7, 12 (1968). Copyright by the American Chemical Society.)

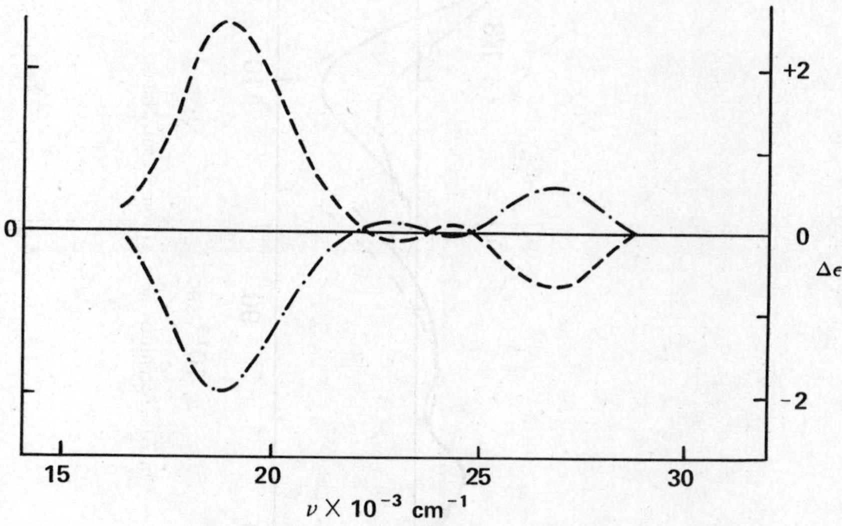

FIGURE 79. α(+)D-[Co(gly)$_2$(L-val)] (- - -) and β(–)D-[Co(gly)$_2$(L-val)] (–•–) in H_2O. (Reprinted with permission from Shibata, Nishikawa, and Nishida, *Inorg. Chem.*, 7, 12 (1968). Copyright by the American Chemical Society.)

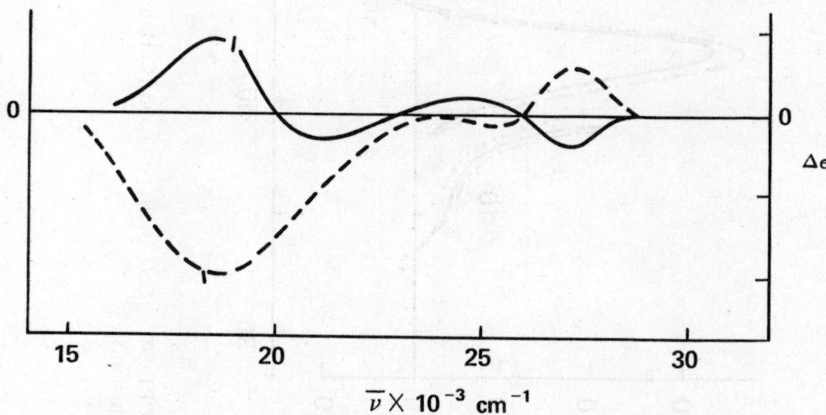

FIGURE 80. β(+)D-[Co(L-val)$_2$(L-ala)] (–) and β(–)D-[Co(L-val)$_2$(D-ala)] (- - -) in 50% H_2SO_4. (Reprinted with permission from Shibata, Nishikawa, and Nishida, *Inorg. Chem.*, 7, 12 (1968). Copyright by the American Chemical Society.)

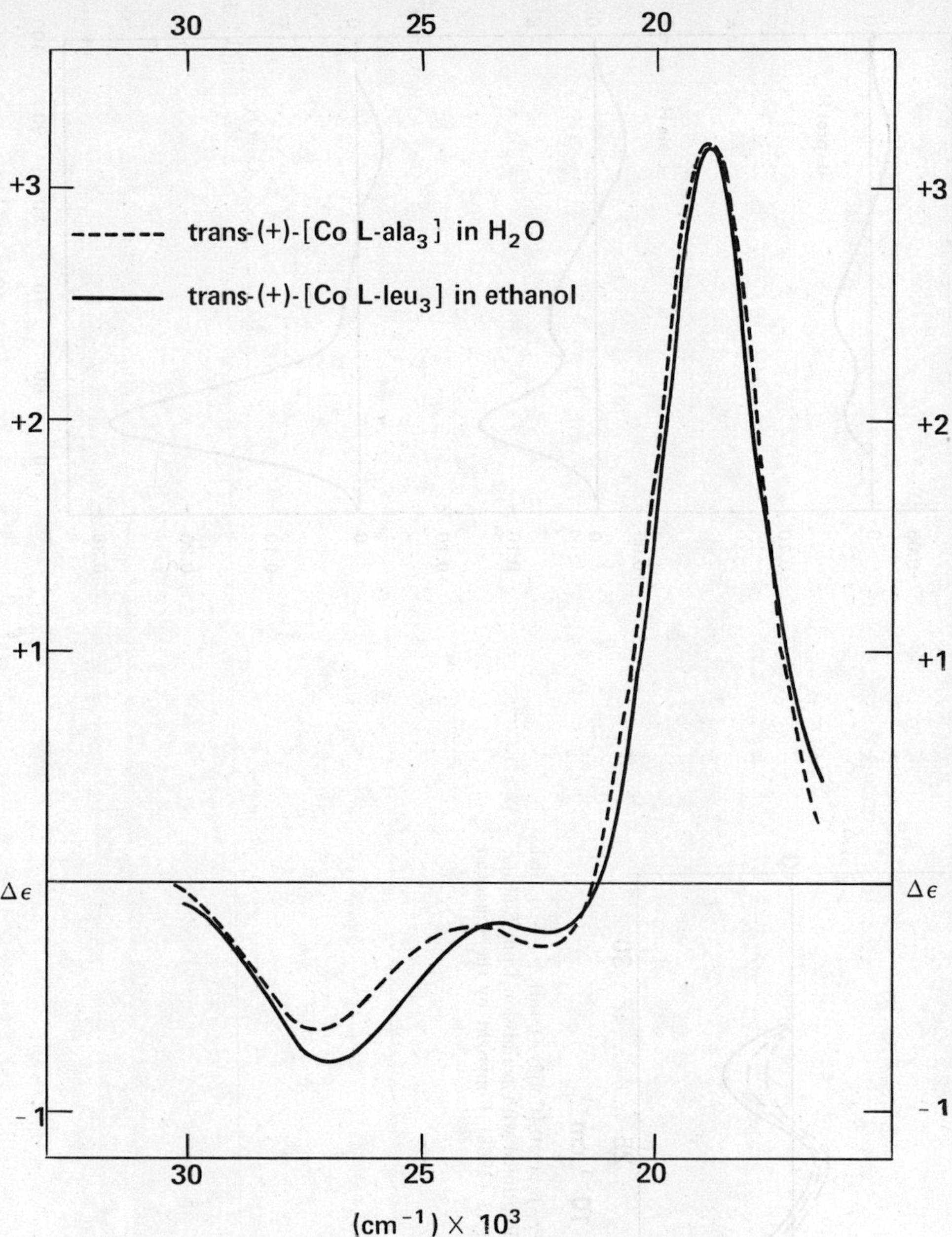

FIGURE 81. *trans*-(+)-[Co(L-amino acid)$_3$] complexes. (Reprinted with permission from Denning and Piper, *Inorg. Chem.*, 5, 1060 (1966). Copyright by the American Chemical Society.)

FIGURE 82. β(+)D-[Co(gly)$_2$(L-val)] (–·–), β(+)D-[Co(gly)(L-val)$_2$] (- - -), and β(+)D-[Co(L-val)$_3$] (–) in 50% H_2SO_4.(Reprinted with permission from Shibata, Nishikawa, and Nishida, *Inorg. Chem.*, 7, 11 (1968). Copyright by the American Chemical Society.)

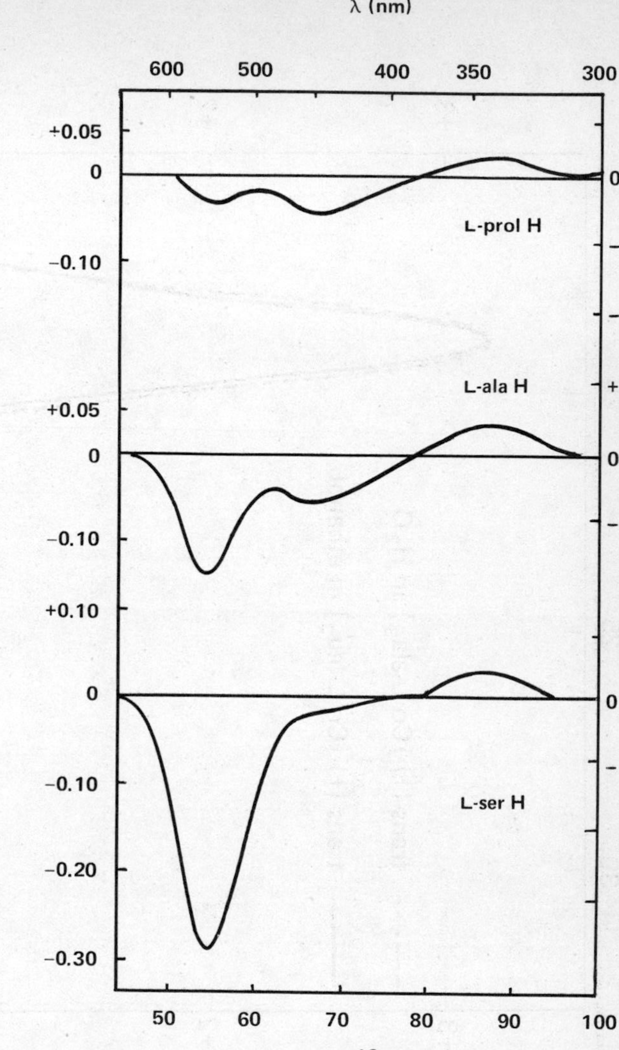

FIGURE 83. *trans*-[Co en$_2$(L-amH)$_2$]$^{3+}$ ions. (From Yasui, Hidaka, and Shimura, *Bull. Chem. Soc. Jap.*, 39, 2422 (1966). With permission.)

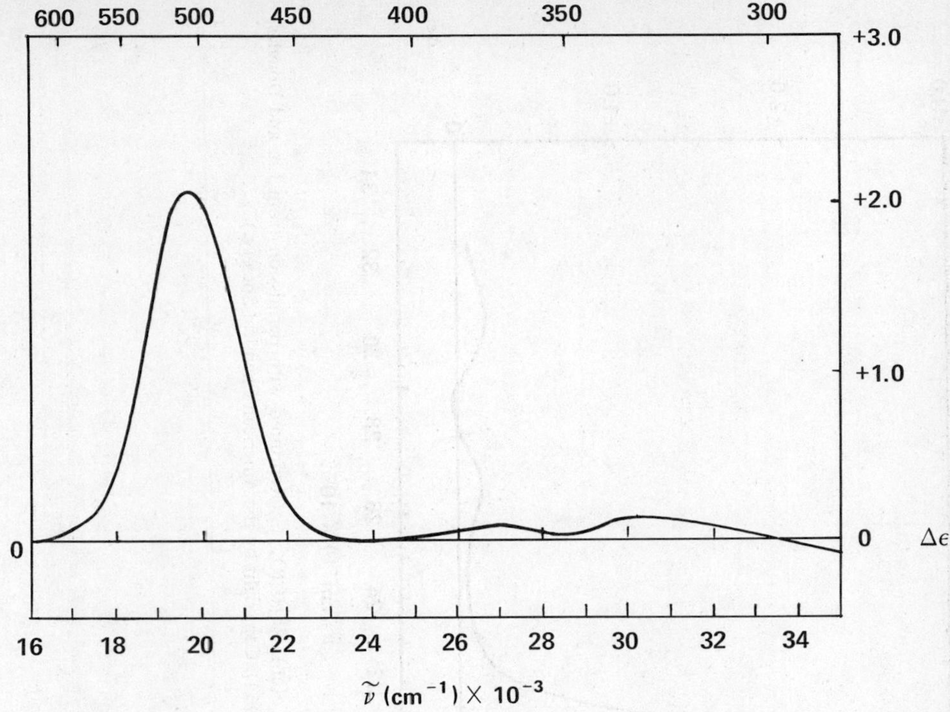

FIGURE 84. $(+)_{546}$-[Co(en)$_2$gly]I$_2$. (Reprinted with permission from Liu and Douglas, *Inorg. Chem.*, 3, 1357 (1964). Copyright by the American Chemical Society.)

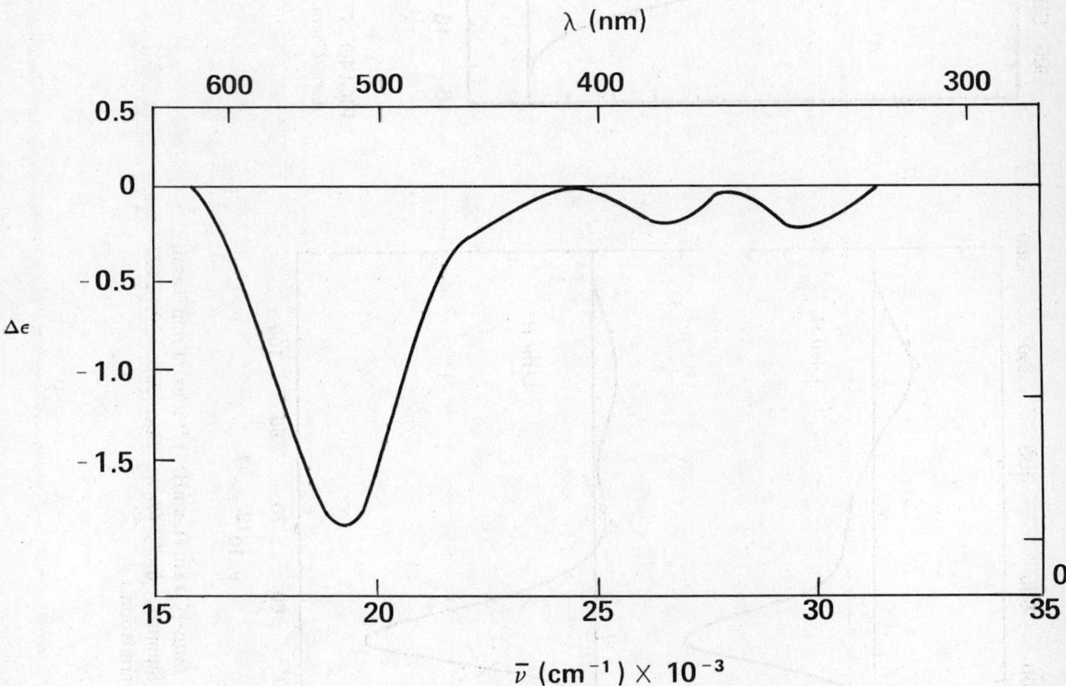

FIGURE 85. (–)D[Co(en)$_2$-(*S*)-pro]I$_2$. (Reprinted with permission from Hall and Douglas, *Inorg. Chem.*, 8, 372 (1969). Copyright by the American Chemical Society.)

FIGURE 86. *trans*-[Co en_2(L-amH)$_2$]$^{3+}$ ions. (From Yasui, Hidaka, and Shimura, *Bull. Chem. Soc. Jap.*, 39, 2423 (1966). With permission.)

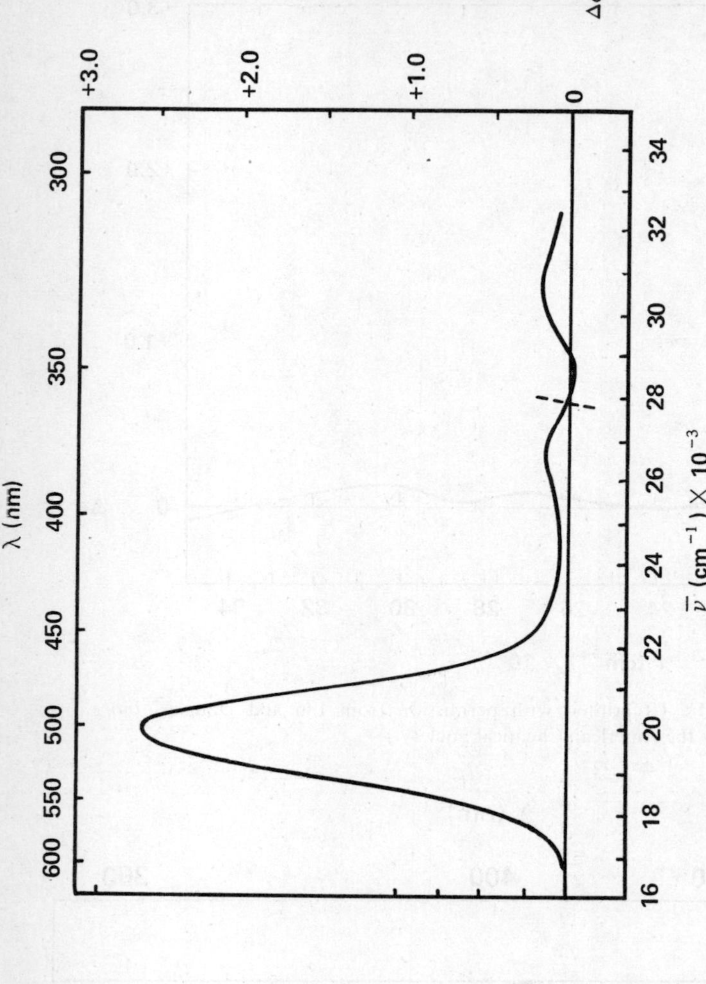

FIGURE 87. $(+)_{546}$-[Co(en)$_2$pic](ClO$_4$)$_2$. (Reprinted with permission from Liu and Douglas, *Inorg. Chem.*, 3, 1360 (1964). Copyright by the American Chemical Society.)

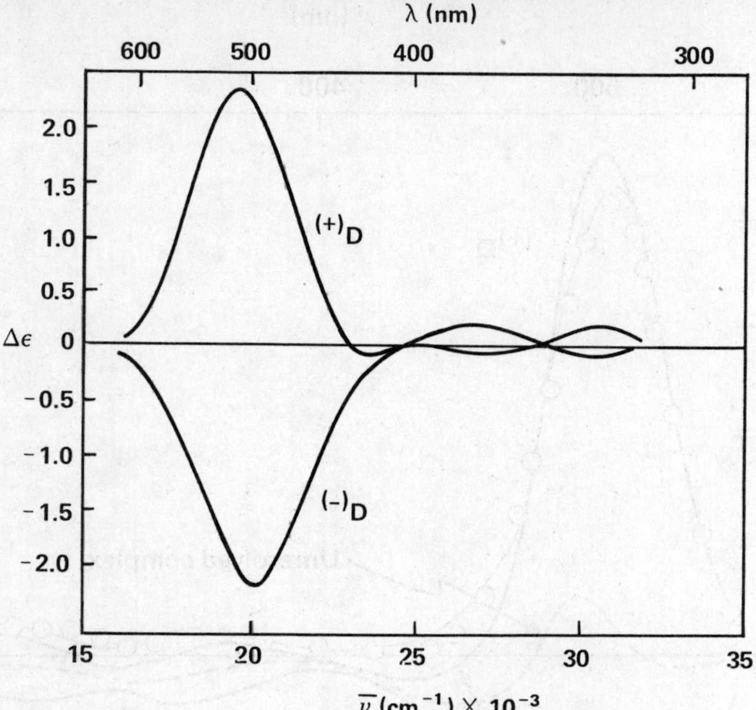

FIGURE 88. Optical isomers of $[Co(en)_2\text{-}(2S,3R)\text{-thr}]I_2$. (Reprinted with permission from Hall and Douglas, *Inorg. Chem.*, 8, 374 (1969). Copyright by the American Chemical Society.)

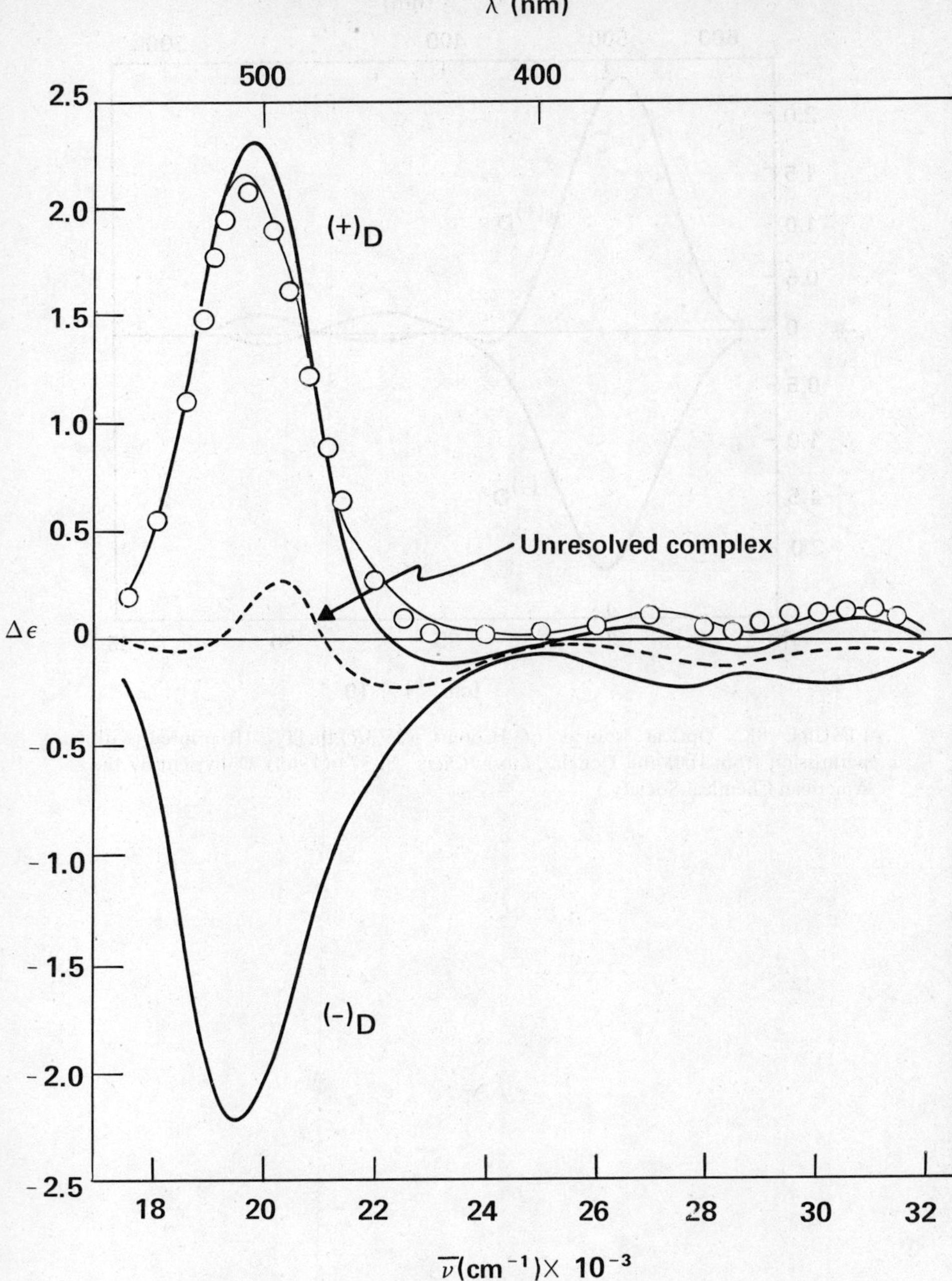

FIGURE 89. CD curves for (+)D, (−)D, and unresolved $[Co(en)_2\text{-}(S)\text{-met}]I_2$. The difference curve for the (+)D isomer and the unresolved complex is shown (- - -) with the experimental points for the CD curve of (+)D-$[Co(en)_2gly]I_2$. (Reprinted with permission from Hall and Douglas, *Inorg. Chem.*, 8, 373 (1969). Copyright by the American Chemical Society.)

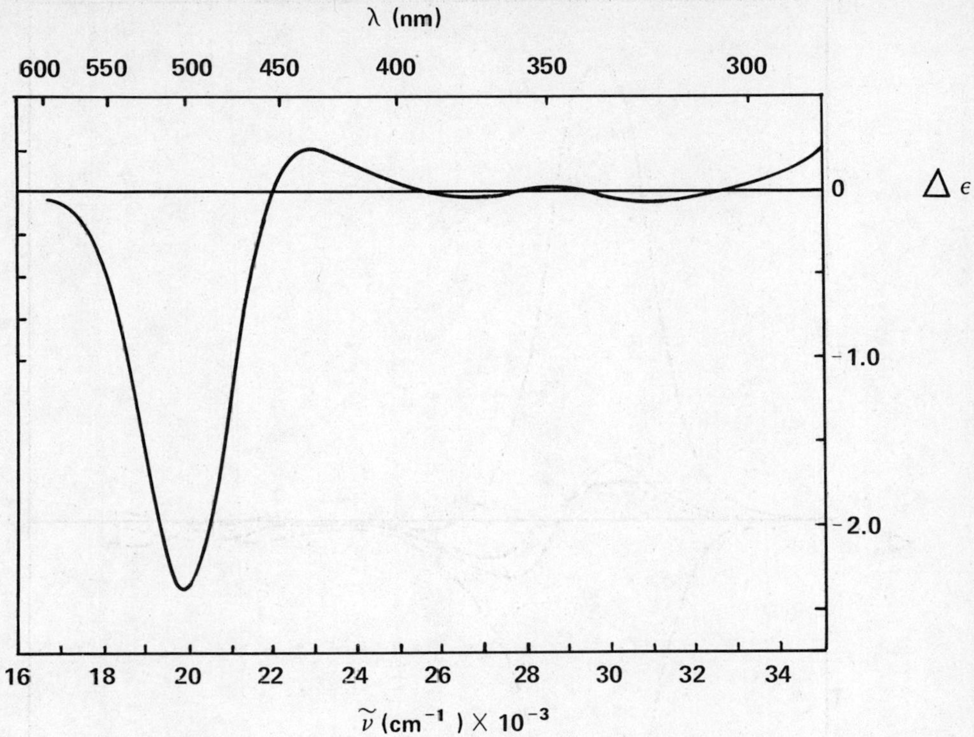

FIGURE 90. $(-)_{546}$-$[Co(en)_2\text{D-leuc}]I_2$. (Reprinted with permission from Liu and Douglas, *Inorg. Chem.*, 3, 1357 (1964). Copyright by the American Chemical Society.)

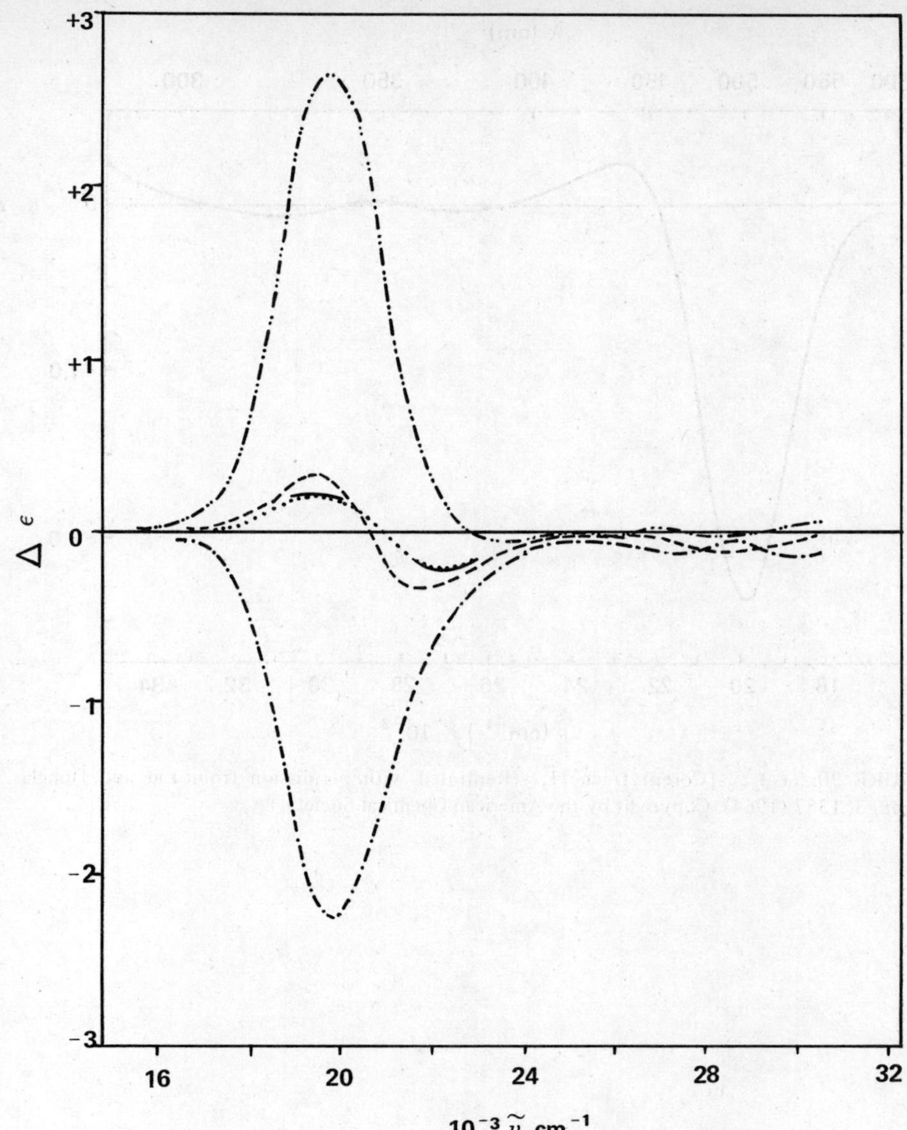

FIGURE 91. Λ-(+)-[Co(L-asp)(en)$_2$]I (E-6) (-·····-), Δ-(–)-[Co(L-asp)(en)$_2$]I (E-7) (-·-·-), the vicinal curve (···), [Co(L-Hasp)(NH$_3$)$_4$](ClO$_4$)$_2$ (- - -), and [Co(L-Hasp)(NH$_3$)$_4$](ClO$_4$)$_2$ at pH ~7 (–). (Reprinted with permission from Kojima and Shibata, *Inorg. Chem.*, 12, 1012 (1973). Copyright by the American Chemical Society.)

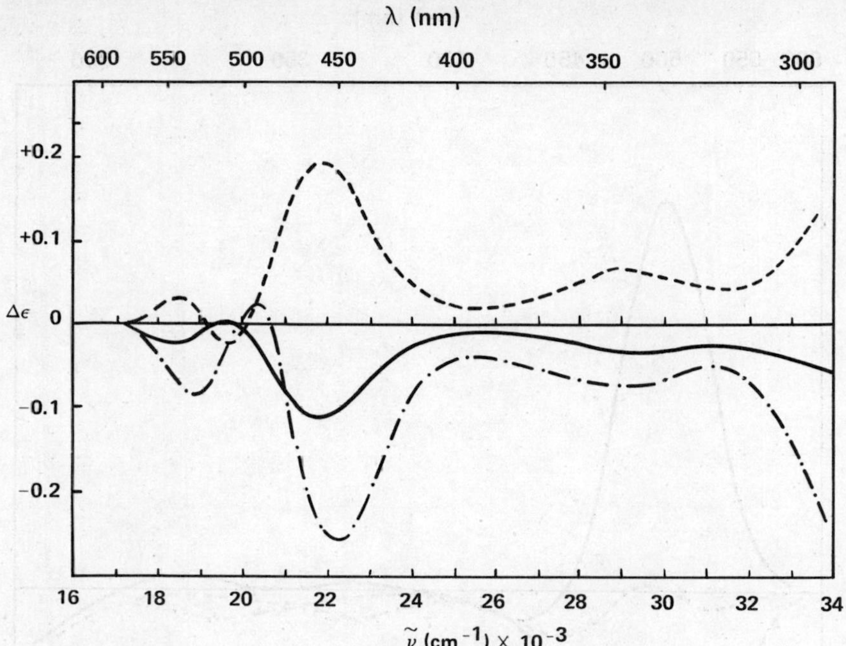

FIGURE 92. $[CO(NH_3)_4$ L-palan$]I_2$ (–), $[Co(NH_3)_4$ D-palan$]I_3 \cdot H_2O$ (-····-), and unresolved $[Co(en)_2$ L-palan$]I_2$ (- - -). (Reprinted with permission from Liu and Douglas, *Inorg. Chem.*, 3, 1360 (1964). Copyright by the American Chemical Society.)

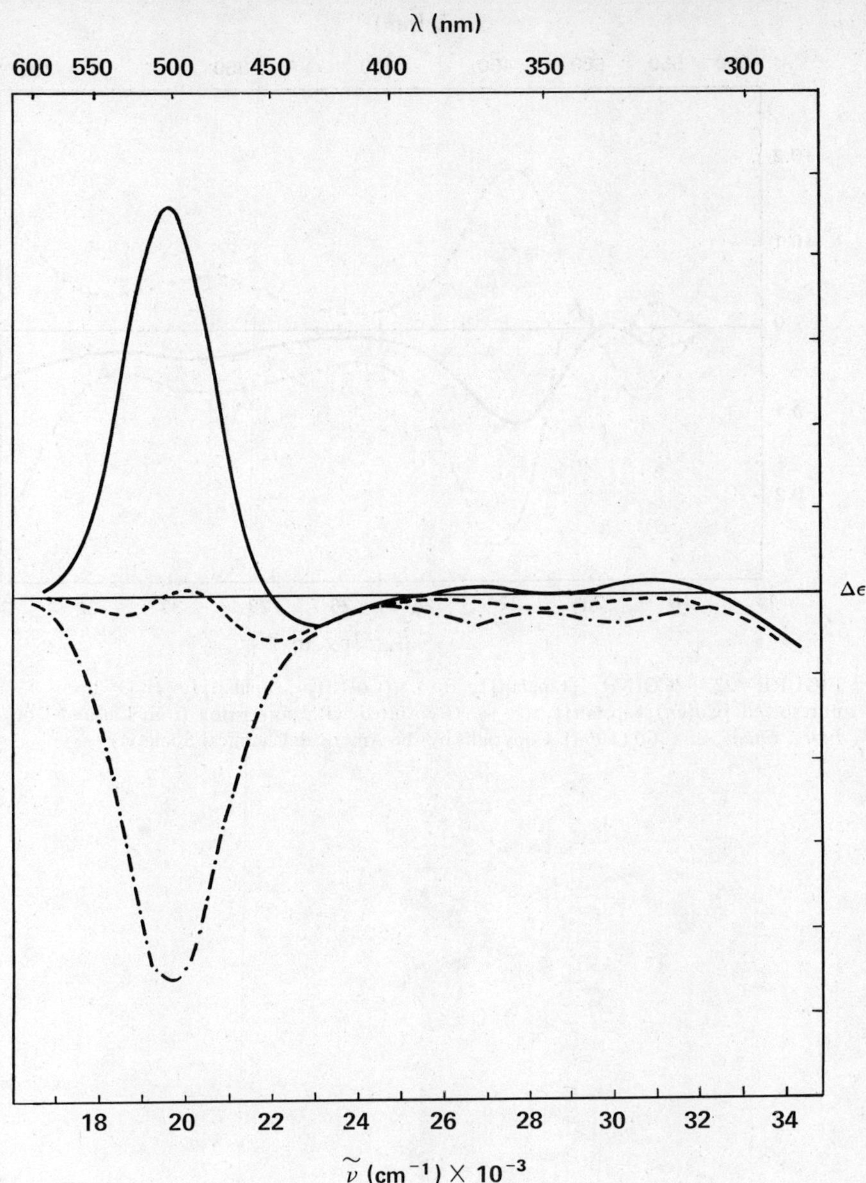

FIGURE 93. [Co(en)$_2$ L-palan]I$_2$; $(+)_{546}$ (–), $(-)_{546}$ (–·–·–) and unresolved (- - -). (Reprinted with permission from Liu and Douglas, *Inorg. Chem.*, 3, 1358 (1964). Copyright by the American Chemical Society.)

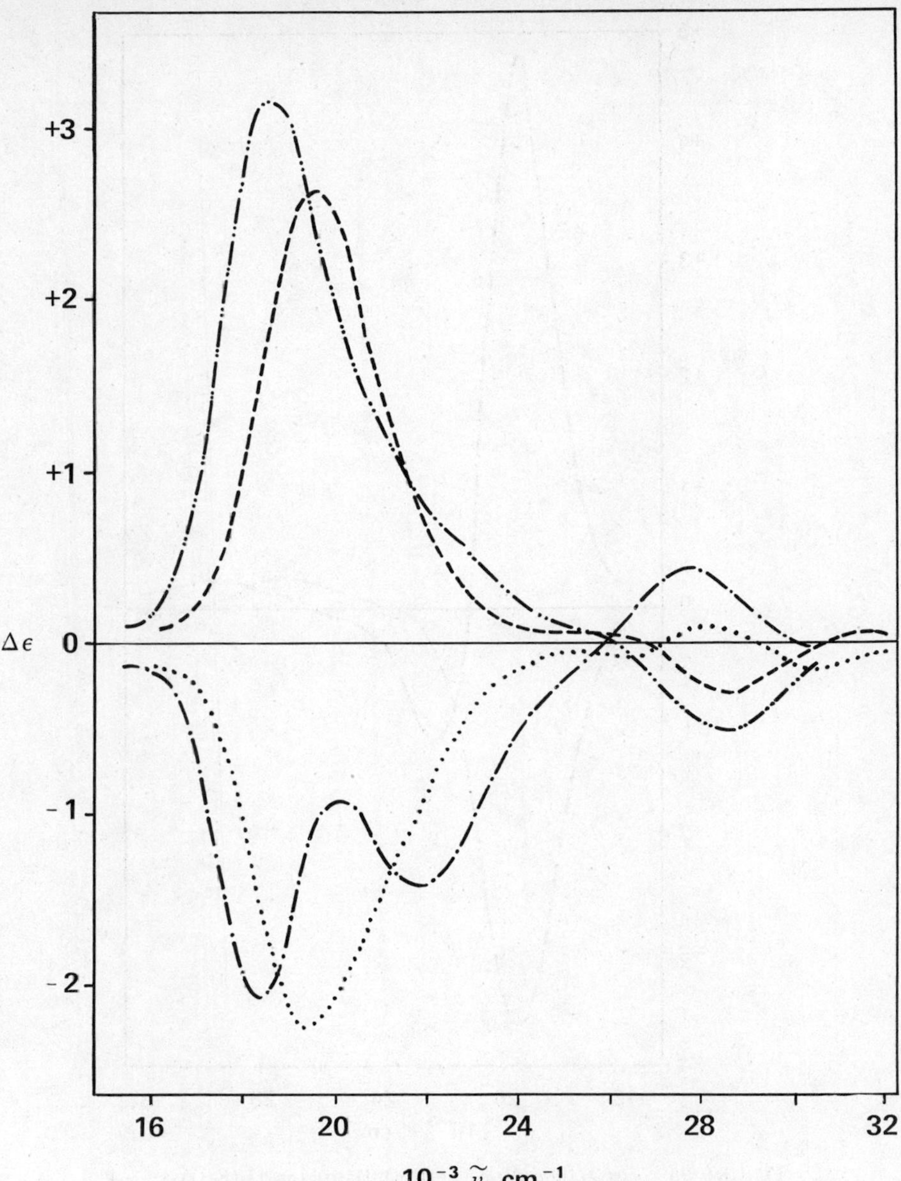

FIGURE 94. *trans(O)*, *cis(N)*-Λ-(+)-[Co(L-Hasp)$_2$(en)ClO$_4$ (E-1) (-·····-), *trans(O)*, *cis(N)*-Δ-(–)-[Co(L-Hasp)$_2$(en)]ClO$_4$ (E-2) (-··-), *cis(O)*, *cis(N)*-Λ-(+)-[Co(L-Hasp)$_2$(en)]Cl (E-4) (- - -), and *cis(O)*, *cis(N)*-Δ-(–)-[Co(L-Hasp)$_2$(en)]I (E-3′) (· · ·). (Reprinted with permission from Kojima and Shibata, *Inorg. Chem.*, 12, 1011 (1973). Copyright by the American Chemical Society.)

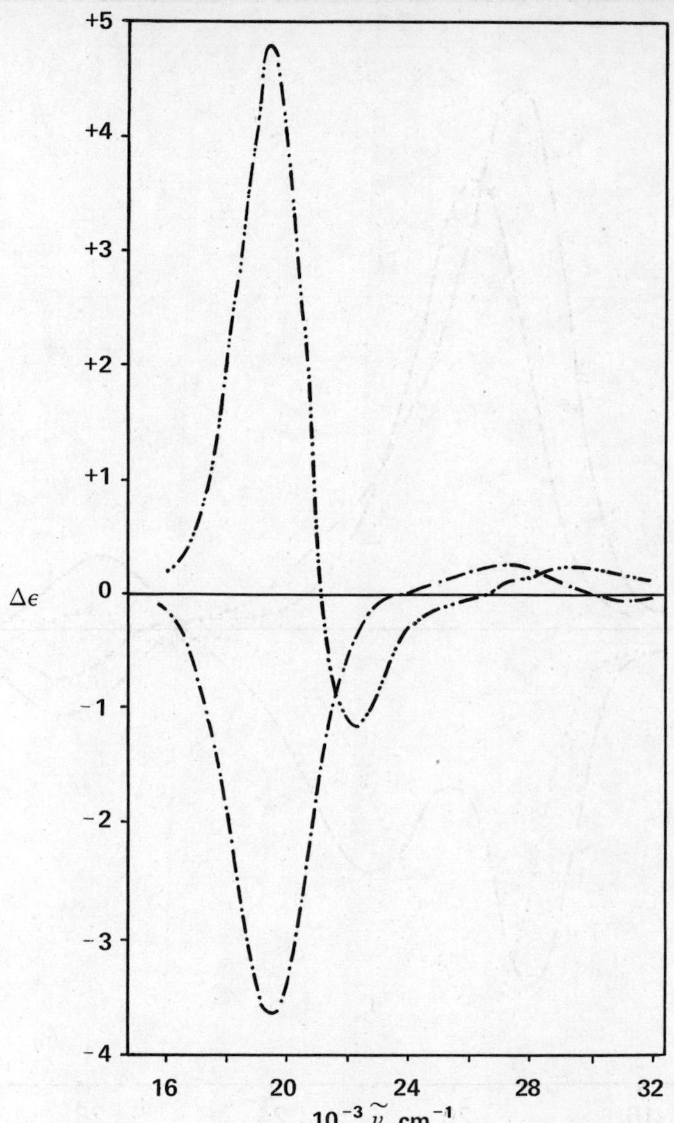

FIGURE 95. *cis(O),trans(N)*-Λ-(+)-[Co(L-Hasp)$_2$(en)]I (E-3) (-·····-) and *cis(O),trans(N)*-Δ-(–)-[Co(L-Hasp)$_2$(en)]ClO$_4$ (E-5) (-···-). (Reprinted with permission from Kojima and Shibata, *Inorg. Chem.*, 12, 1012 (1973). Copyright by the American Chemical Society.)

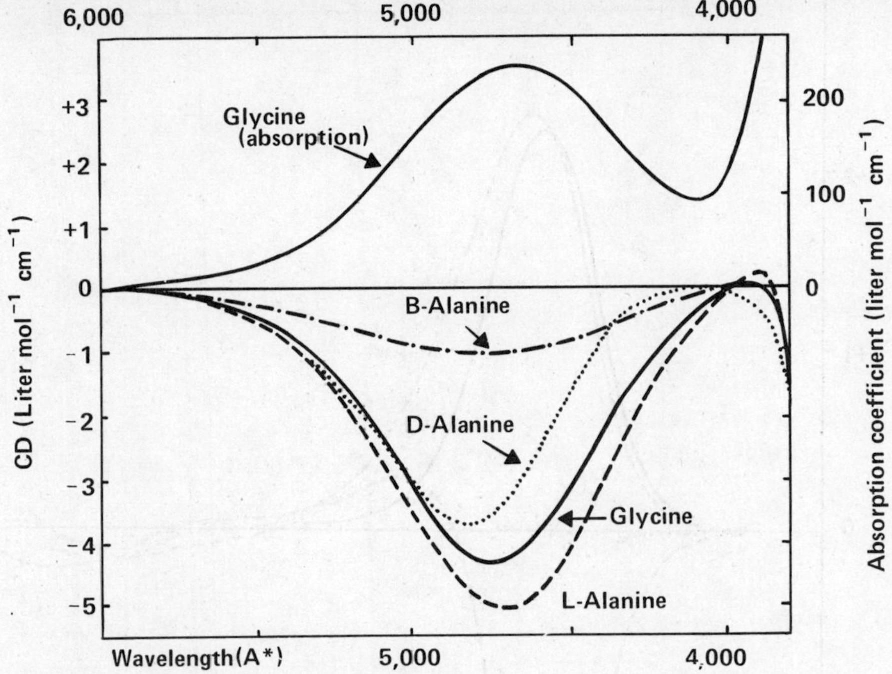

FIGURE 96. *cis*-Dinitrobis(amino acidato)cobalt(III) ions, and the absorption spectrum of a glycine complex ion. (From Denning, Célap, and Radanović, *Inorg. Chim. Acta,* 2, 58 (1968). With permission.)

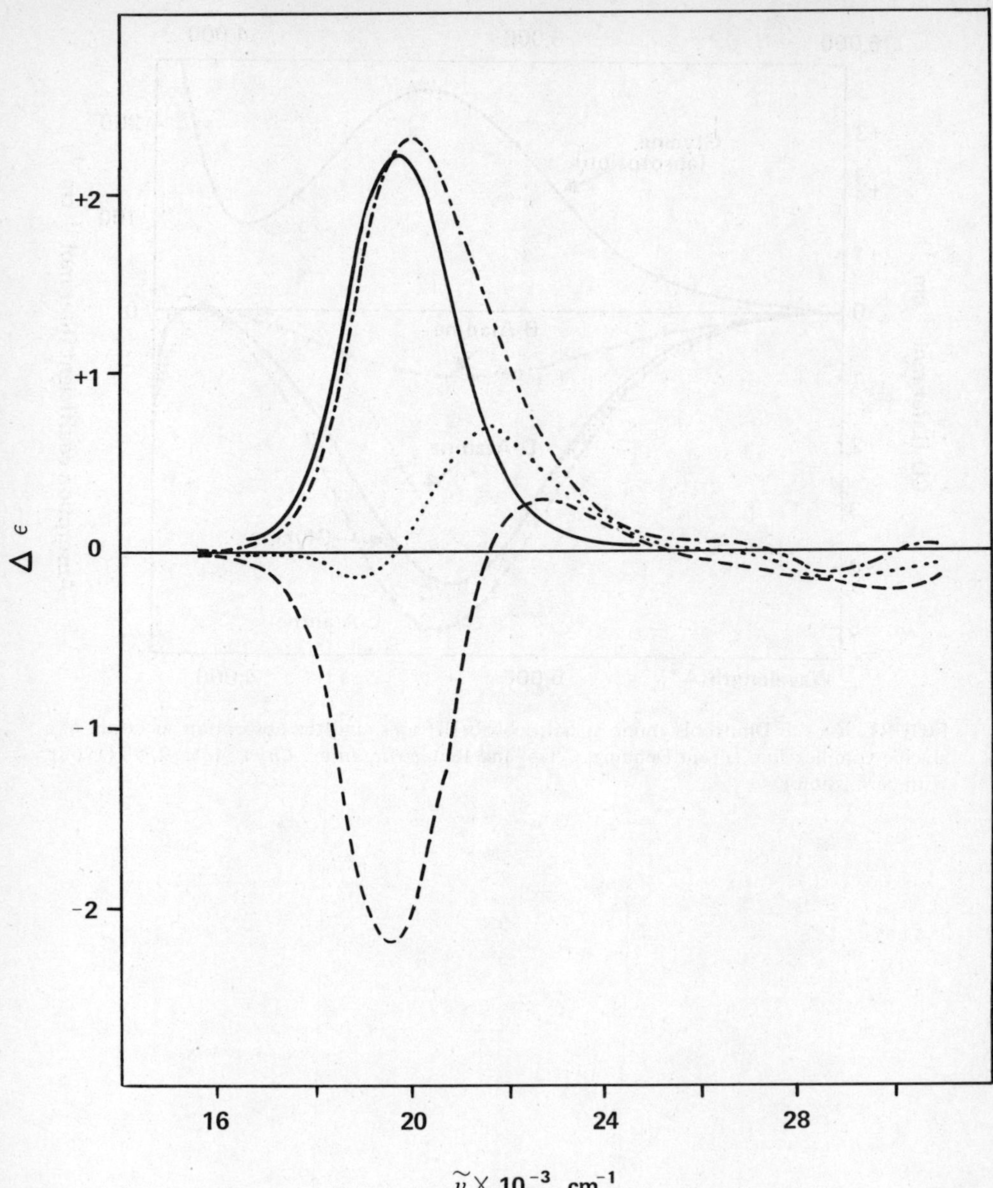

FIGURE 97. (+)[Co(gly)(*l*-pn)$_2$]Cl$_2$ (-·-·-), (–)[Co(gly)(*l*-pn)$_2$]Cl$_2$ (- - -), the vicinal effect (· · ·), and (+)[Co(gly)(*l*-pn)$_2$]Cl$_2$ from which the vicinal effect curve was subtracted (–). (Reprinted with permission from Kojima and Shibata, *Inorg. Chem.*, 9, 245 (1970). Copyright by the American Chemical Society.)

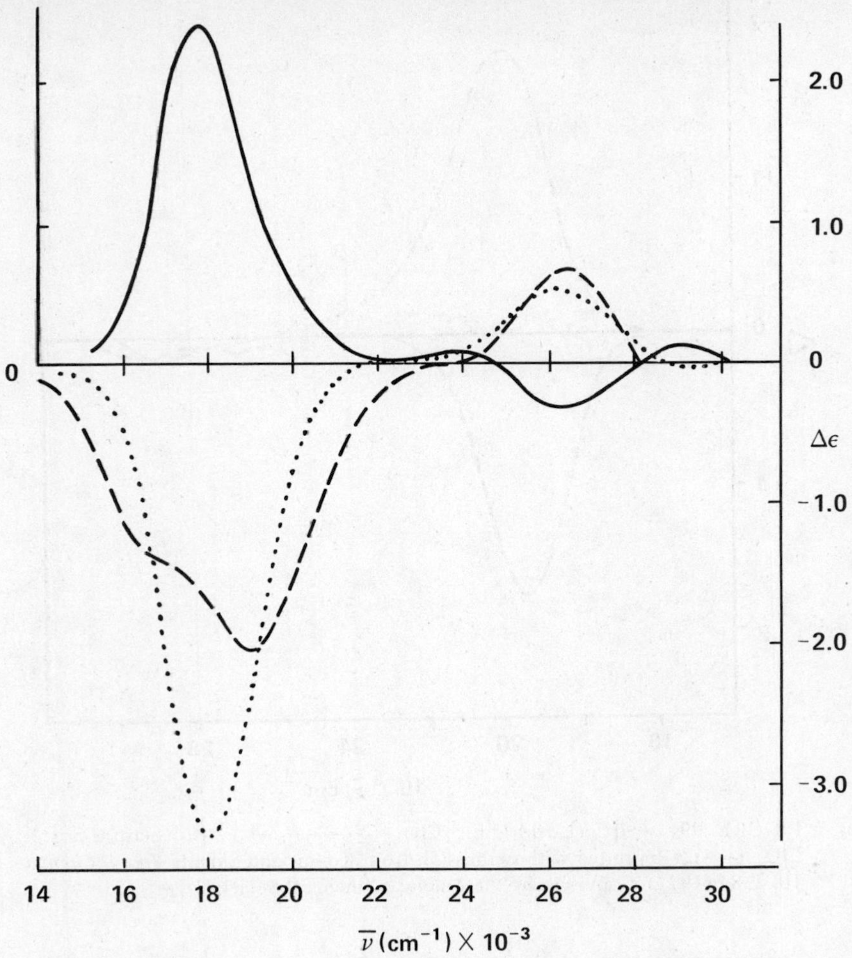

FIGURE 98. Three isomers of $Co(ox)(gly)_2$: (— — —) trans(N) isomer; (- - -) C_2-*cis*(N) isomer; (—) C_1-*cis*(N) isomer. (Reprinted with permission from Matsuoka, Hidaka, and Shimura, *Inorg. Chem.*, 9, 721 (1970). Copyright by the American Chemical Society.)

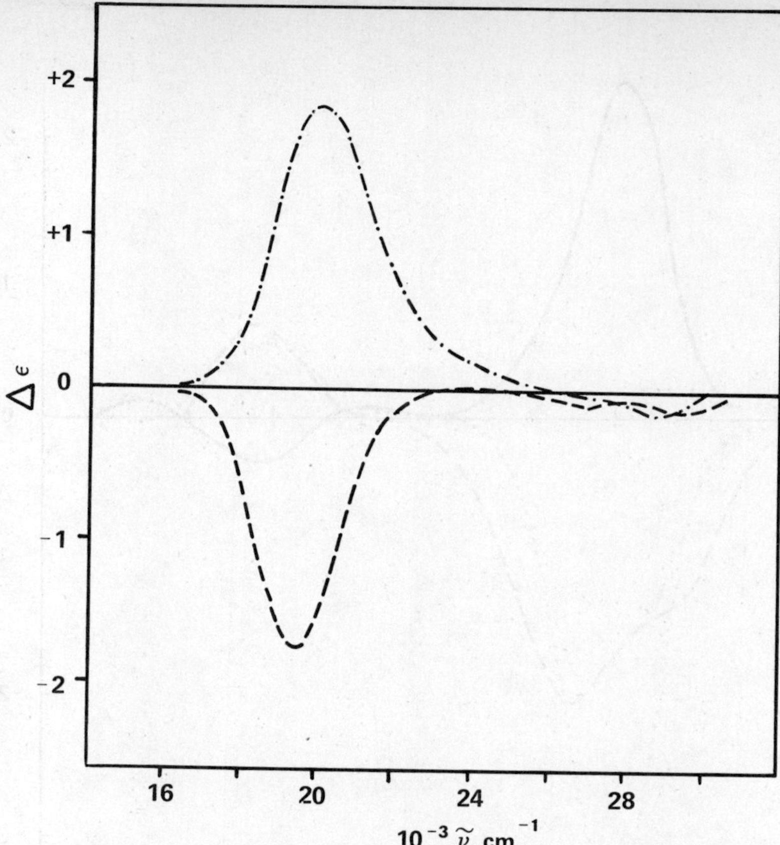

FIGURE 99. (+)[Co(L-asp)(*l*-pn)$_2$]ClO$_4$ (—•—•—) and (–)[Co(L-asp)(*l*-pn)$_2$] ClO$_4$ (- - -). (Reprinted with permission from Kojima and Shibata, *Inorg. Chem.*, 10, 2384 (1971). Copyright by the American Chemical Society.)

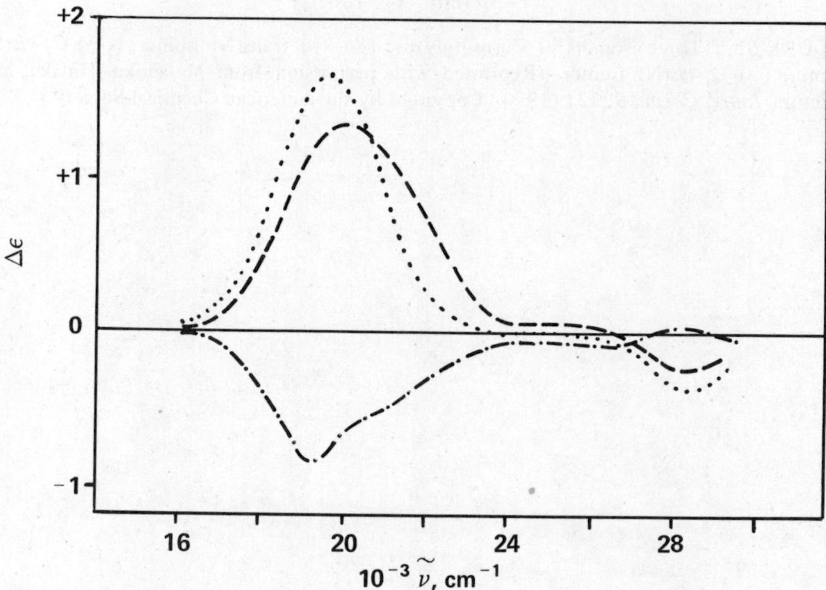

FIGURE 100. *cis*(*O*),*cis*(*N*)-(+) [Co(L-aspH)(*l*-pn)]ClO$_4$ (E-4) (- - -), *cis*(*O*),*cis*(*N*)-(–) [Co(L-aspH)$_2$(*l*-pn)]ClO$_4$ (E-5) (—•—•—), and *cis*(*O*),*trans*(*N*)-(+)[Co(L-aspH)$_2$(*l*-pn)] ClO$_4$ (E-6) (•••). (Reprinted with permission from Kojima and Shibata, *Inorg. Chem.*, 10, 2384 (1971). Copyright by the American Chemical Society.)

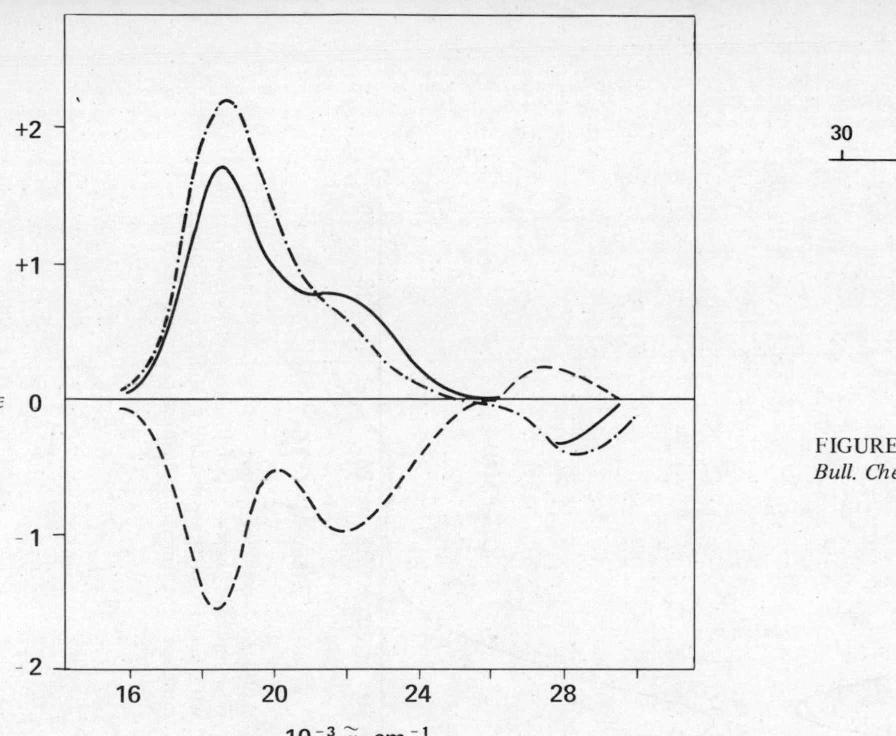

FIGURE 101. *trans(O), cis(N)*-(+)[Co(L-aspH)$_2$(L-pn)]ClO$_4$ (E-1) (–·–·–), *trans(O), cis(N)*-(–)[Co(L-aspH)$_2$(*l*-pn)]ClO$_4$ (E-3) (- - -), and the no. 2 fraction (–). (Reprinted with permission from Kojima and Shibata, *Inorg. Chem.*, 10, 2384 (1971). Copyright by the American Chemical Society.)

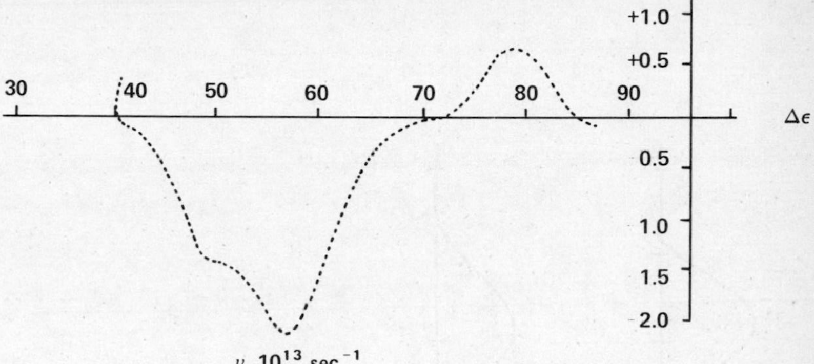

FIGURE 102. $(+)_{546}$-*trans(N)*-Na[Co(ox)(gly)$_2$]. (From Hidaka and Shimura, *Bull. Chem. Soc. Jap.*, 40, 2315 (1967). With permission.)

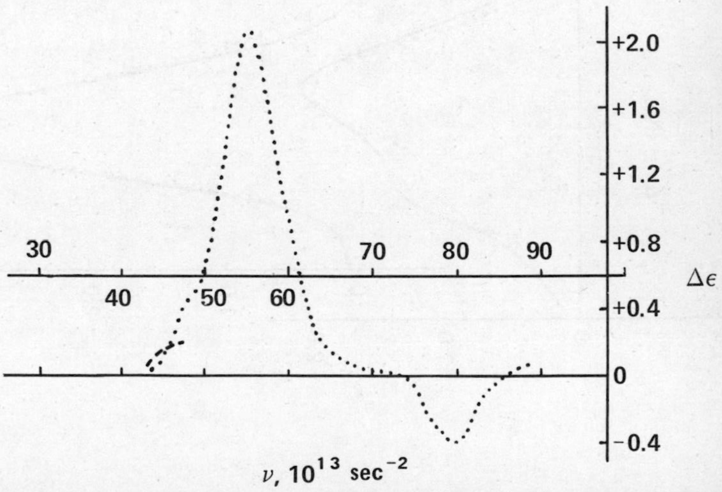

FIGURE 103. $(-)_{546}$-*trans(N)*-K[Co(ox)(L-ala)$_2$]. (From Hidaka and Shimura, *Bull. Chem. Soc. Jap.*, 40, 2315 (1967). With permission.)

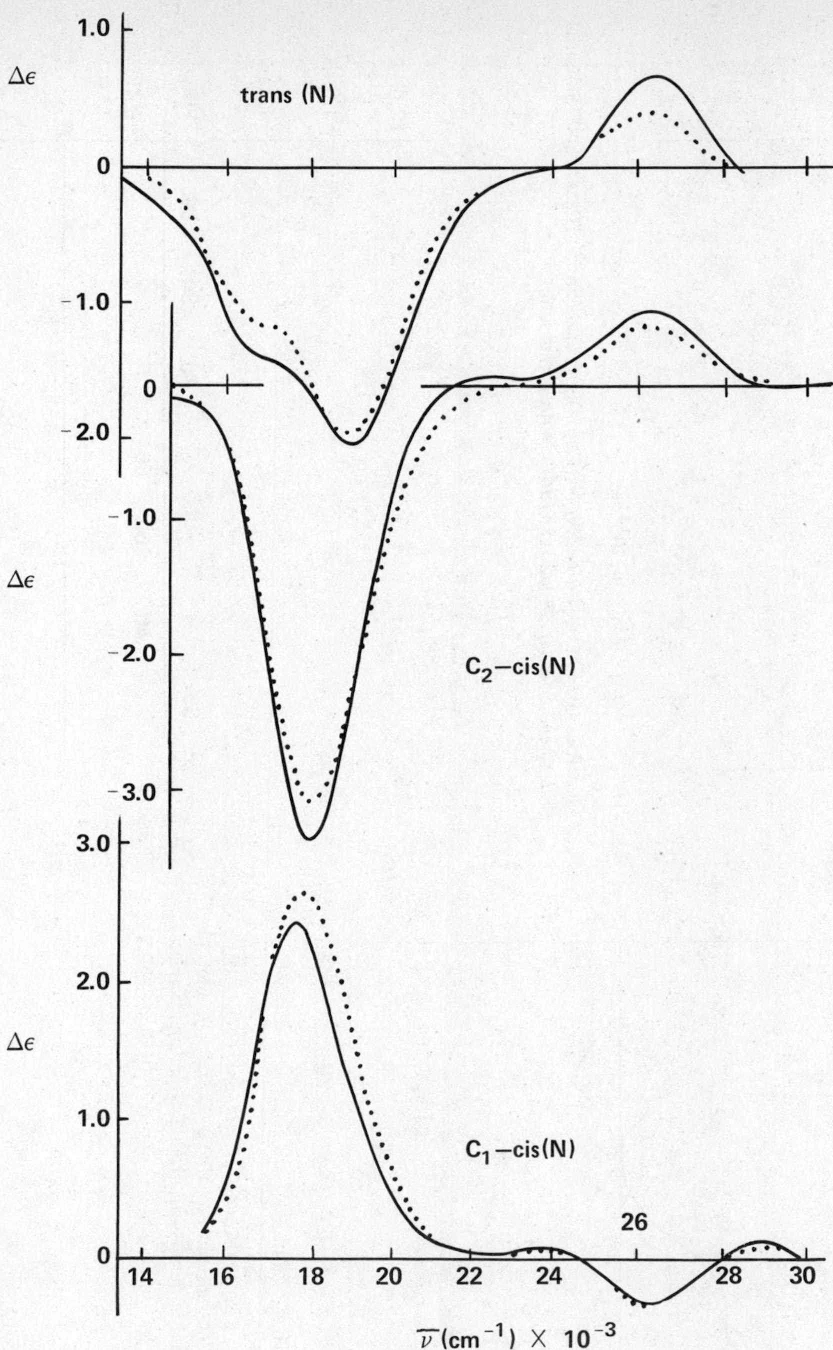

FIGURE 104. Configurational-effect curves of three geometrical isomers of $Co(ox)(L\text{-}ser)_2^-$ (- - -) and observed curves of the corresponding geometrical isomers of $Co(ox)(gly)_2^-$ (—). (Reprinted with permission from Matsuoka, Hidaka, and Shimura, *Inorg. Chem.*, 9, 723 (1970). Copyright by the American Chemical Society.)

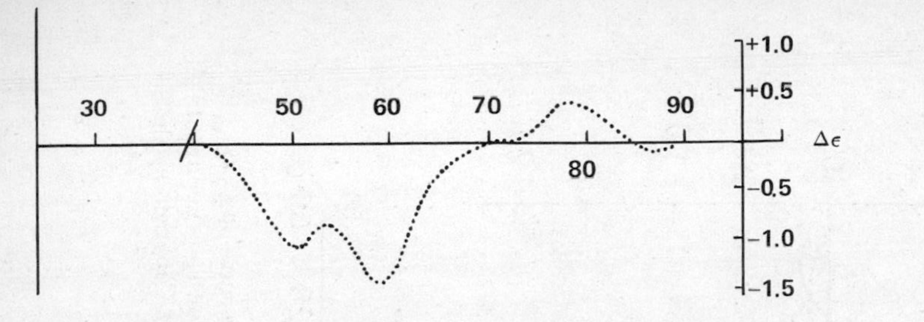

FIGURE 105. $(+)_{546}$-*trans*(N)-K[Co(ox)(L-ala)$_2$]). From Hidaka and Shimura, *Bull. Chem. Soc. Jap.*, 40, 2315 (1967). With permission.)

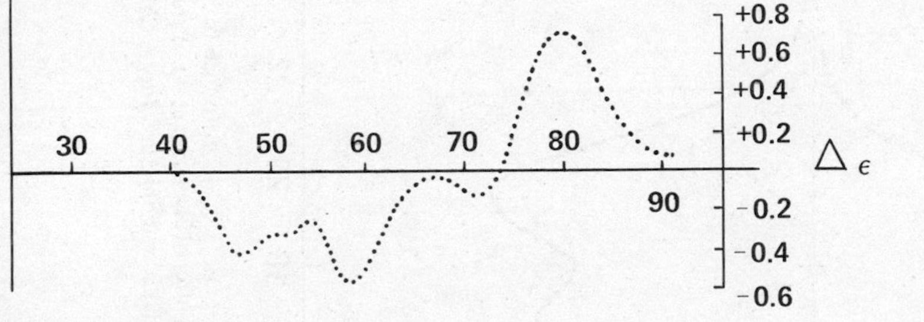

FIGURE 106. $(+)_{546}$-*trans*(N)-K[Co(ox)(β-ala)$_2$]. (From Hidaka and Shimura, *Bull. Chem. Soc. Jap.*, 40, 2316 (1967). With permission.)

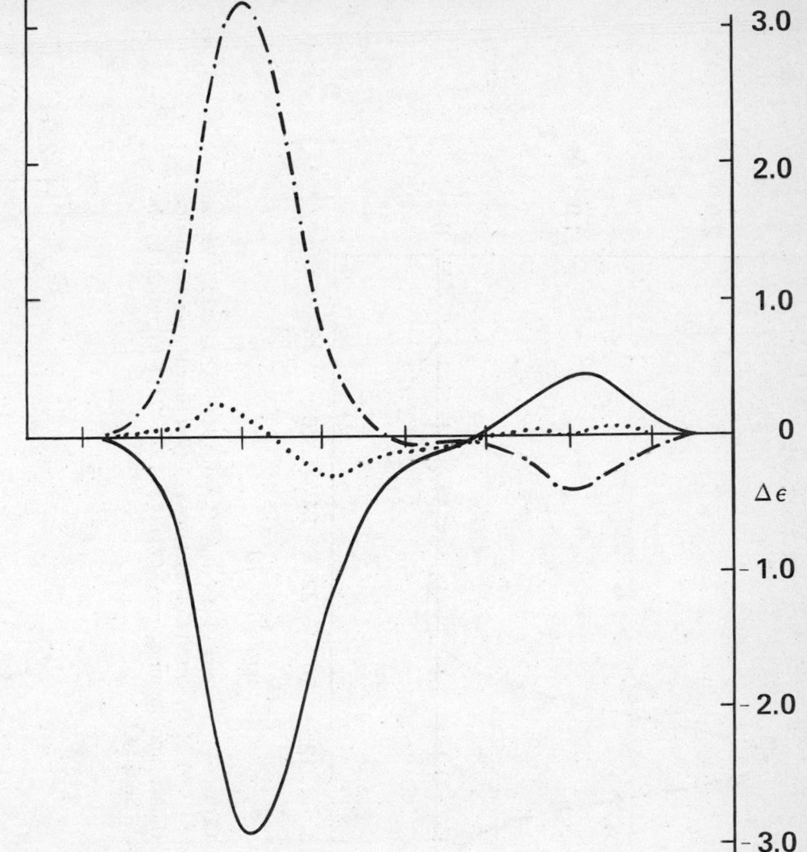

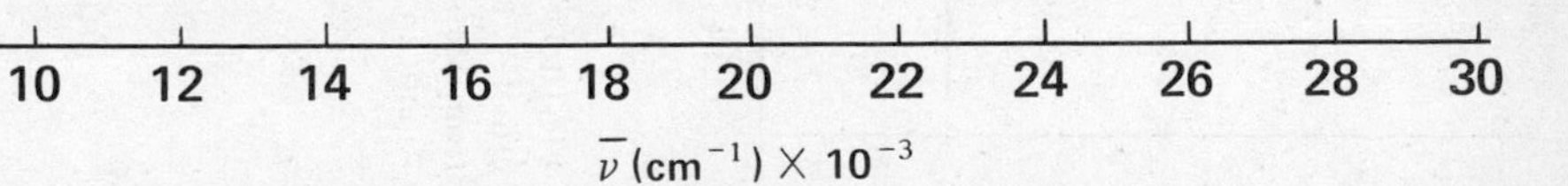

FIGURE 107. CD (—·—·—) curves of Δ-(−)$_{546}$-C_2-*cis* (N)-Co(ox) (L-ser)$_2$$^-$ CD (—) curve of Δ-(+)$_{546}$-C_2-*cis* (N)-Co(ox) (L-ser)$_2$$^-$. Vicinal effect (2L) curve (---) of C_2-*cis* (N)-Co(ox) (L-ser)$_2$$^-$. (Reprinted with permission from Matsuoka, Hidaka, and Shimura, *Inorg. Chem.*, 9, 772 (1970). Copyright by the American Chemical Society.)

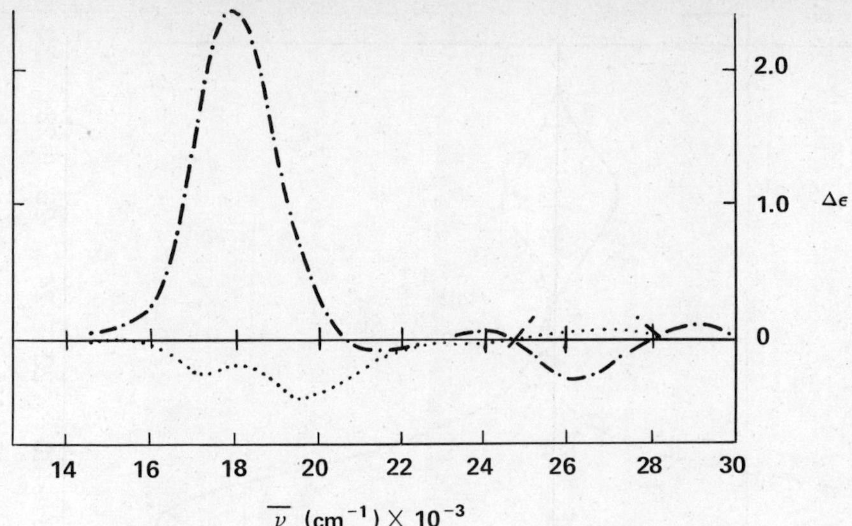

FIGURE 108. CD (—·—·—) curves of Δ-(−)$_{546}$-C_1-*cis*(*N*)-Co(ox)(L-ser)$_2^-$. Vicinal effect (2L) curve (- - -) of C_1-*cis*(*N*)-Co(ox)(L-ser)$_2^-$. (Reprinted with permission from Matsuoka, Hidaka, and Shimura, *Inorg. Chem.*, 9, 722 (1970). Copyright by the American Chemical Society.)

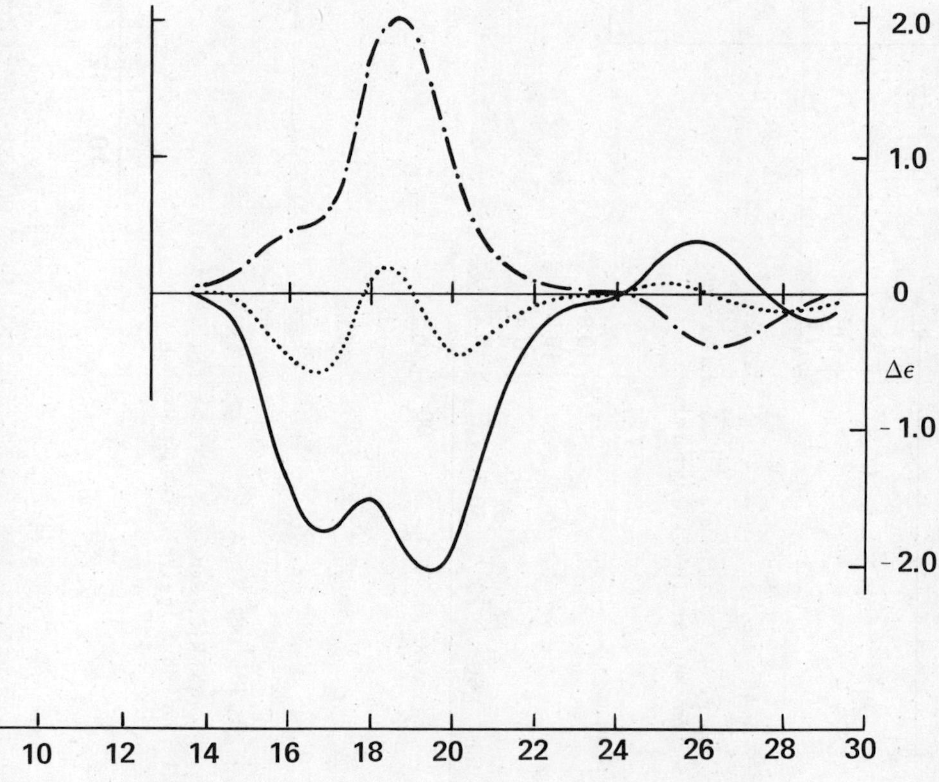

FIGURE 109. CD (—·—·—) curve of Δ-(−)$_{546}$-*trans* (*N*)-Co (ox) (L-ser)$_2^-$. CD (–) curve of Λ-(+)$_{546}$-*trans* (*N*)-Co (ox) (L-ser)$_2^-$. Vicinal effect (2L) curve (- - -) of *trans* (*N*)-Co (ox) (L-ser)$_2^-$. (Reprinted with permission from Matsuoka, Hidaka, and Shimura, *Inorg. Chem.*, 9, 721 (1970). Copyright by the American Chemical Society.)

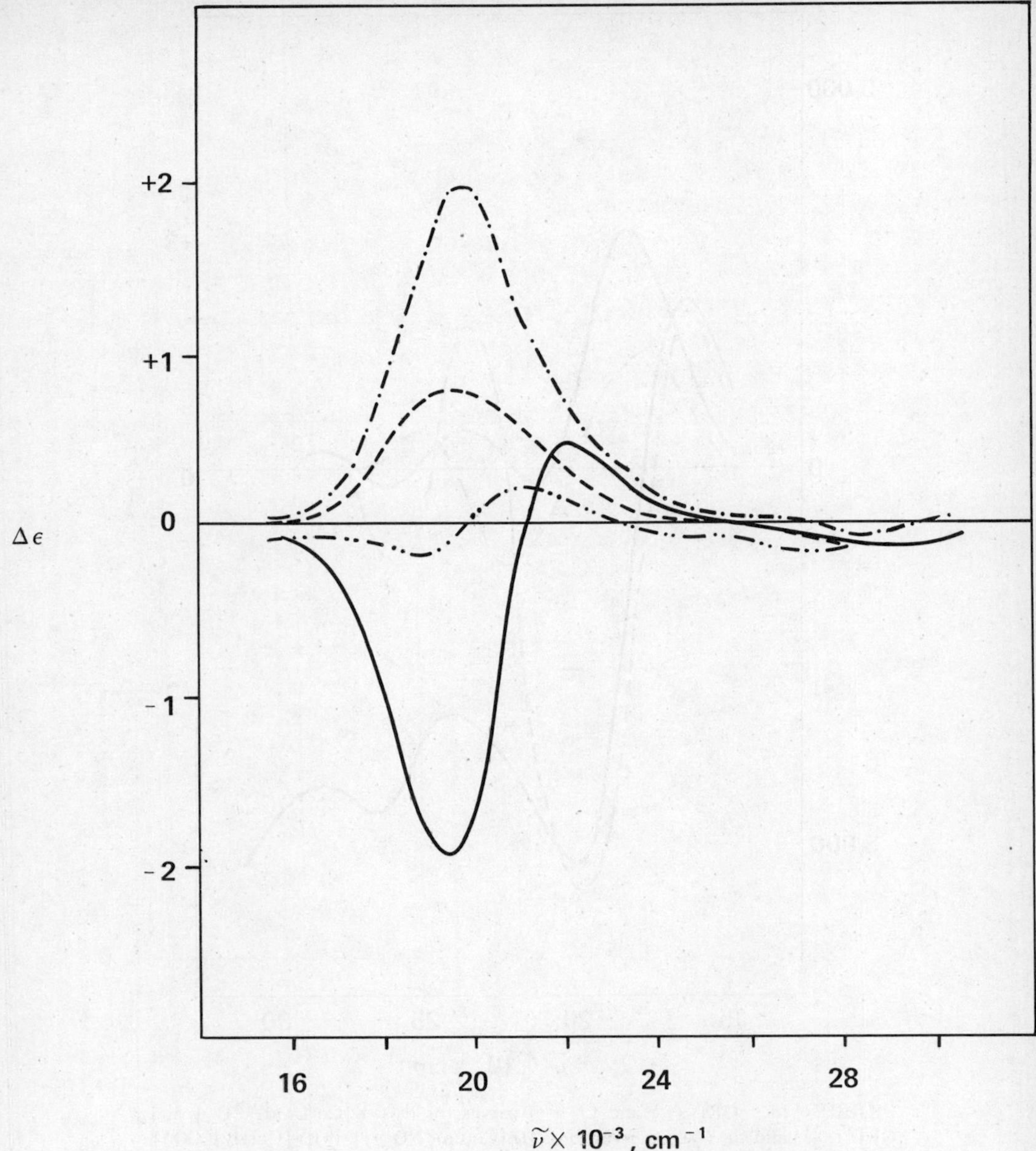

FIGURE 110. *cis (O)*, *cis (N)*-[Co (gly)$_2$ (*l*-pn)] Cl (less soluble (- - -), *cis (O)*, *cis (N)*-[Co (gly)$_2$ (*l*-pn)] Cl (more soluble) (—••—••—), *cis (O)*, *trans (N)*-(+) [Co (gly)$_2$ (*l*-pn)] Cl (more soluble) (—•—•—), and *cis (O)*, *trans (N)*-(–) [Co (gly)$_2$ (*l*-pn)] Cl (less soluble) (–). (Reprinted with permission from Kojima and Shibata, *Inorg. Chem.*, 9, 241 (1970). Copyright by the American Chemical Society.)

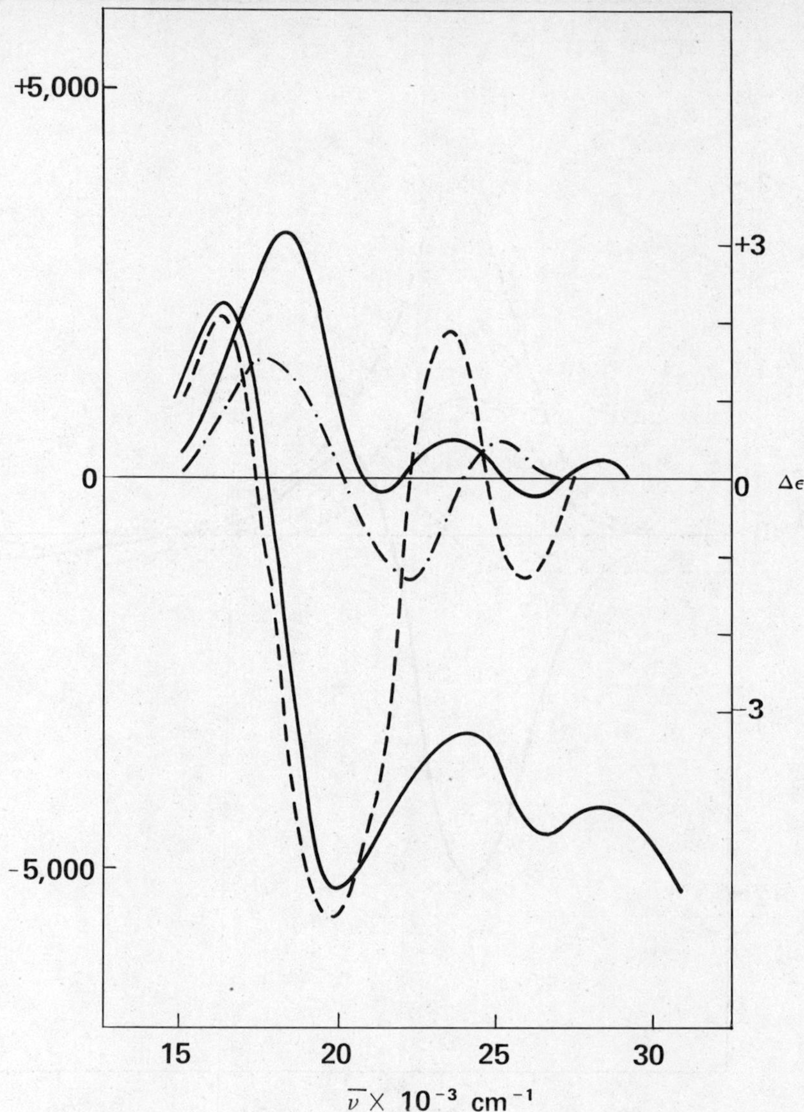

FIGURE 111. ORD (–) and CD (–) curves for (+)D-K[Co(L-val)$_2$CO$_3$] and ORD (- - -) and CD (–•–) curves for (–)D-[Co(en)$_2$NO$_2$)$_2$]•(+)D-[Co(gly)^{2}CO^3] in H_2O. (Reprinted with permission from Shibata, Nishikawa, and Nishida, *Inorg. Chem.*, 7, 11 (1968). Copyright by the American Chemical Society.)

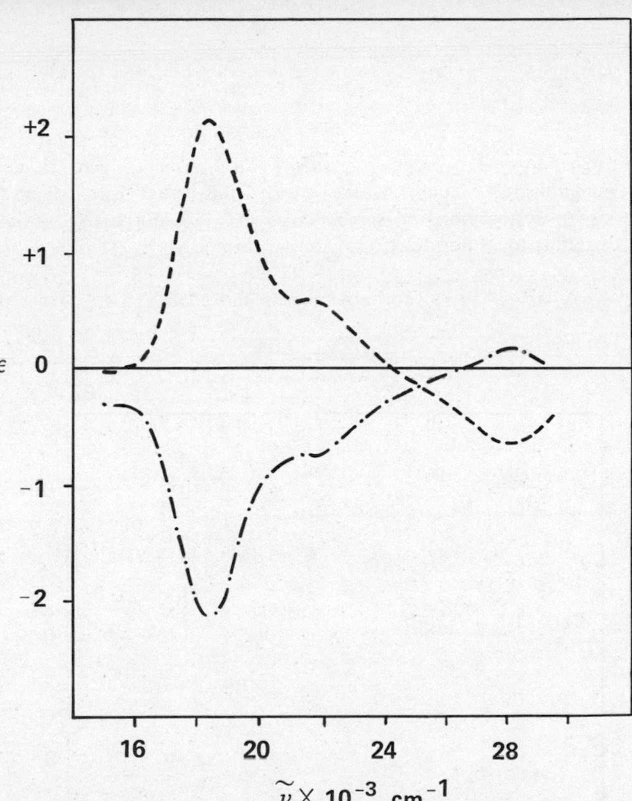

FIGURE 112. *trans(O),cis(N)*-(+)[Co(gly)$_2$(*l*-pn)]Cl (- - -) and *trans(O),cis(N)*-(–)[Co(gly)$_2$(*l*-pn)]Cl (—•—•—). (Reprinted with permission from Kojima and Shibata, *Inorg. Chem.*, 9, 241 (1970). Copyright by the American Chemical Society.)

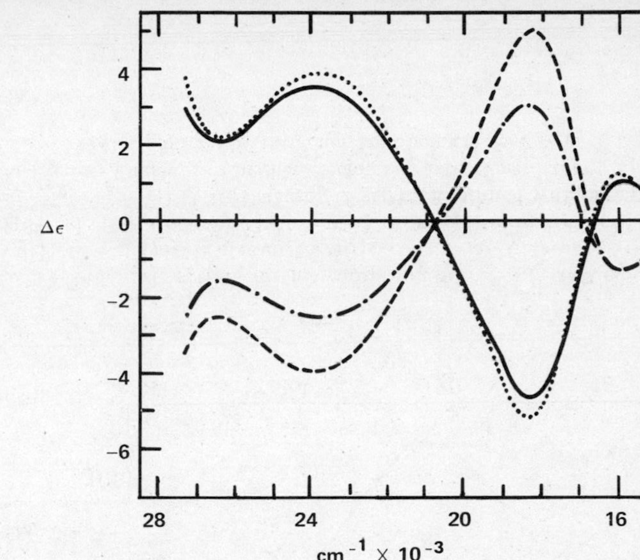

FIGURE 113. CD spectra in chloroform solution·(—) (–)D-[Co(acac)$_2$(*S*-ala)]; (- - -) (+)D-[Co(acac)$_2$(*S*-ala)]; (•••) (–)D-[Co(acac)$_2$(*N*-Me-*S*-ala)]; (-•-•-) (+)D- [Co(acac)$_2$(*N*-Me-*S*-ala)]. (Reprinted with permission from Everett, Finney, Brushmiller, Seematter, and Wingert, *Inorg. Chem.*, 13, 538 (1974). Copyright by the American Chemical Society.)

FIGURE 114. CD spectra in aqueous solutions. (–) (–)D-*cis*-*N*-C_2-[Co(acac)(*S*-val)$_2$]; (- - -) (+)D-*trans*-*N*-C_2-[Co(acac)(*S*-val)$_2$]; (···)(–)D-*trans*-*N*-C_2-[Co(acac)(*S*-val)$_2$]. Absorption spectrum in aqueous solution: (···) (–)D-*trans*-*N*-C_2-[Co(acac)(*S*-val)$_2$]. (Reprinted with permission from Everett, Finney, Brushmiller, Seematter, and Wingert, *Inorg. Chem.*, 13, 538 (1974). Copyright by the American Chemical Society.)

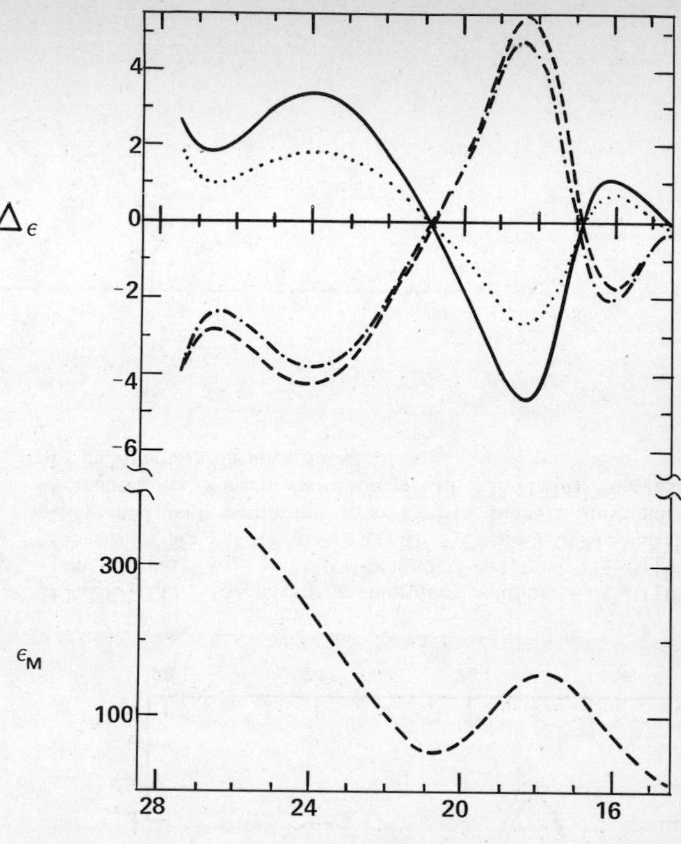

FIGURE 115. CD spectra in chloroform solution: (–) (–)D-[Co(acac)$_2$ (*S*-val)]; (- - -) (+)D-[Co(acac)$_2$ (*S*-val)]; (···) (–)D-[Co(acac)$_2$ (*N*-Me-*S*-val)];(-·-·-) (+)D-[Co(acac)$_2$ (*N*-Me-*S*-val)]. Absorption spectrum in chloroform solution: (- - -) (+)D-[Co(acac)$_2$ (*S*-val)]. (Reprinted with permission from Everett, Finney, Brushmiller, Seematter, and Wingert, *Inorg. Chem.*, 13, 538 (1974). Copyright by the American Chemical Society.)

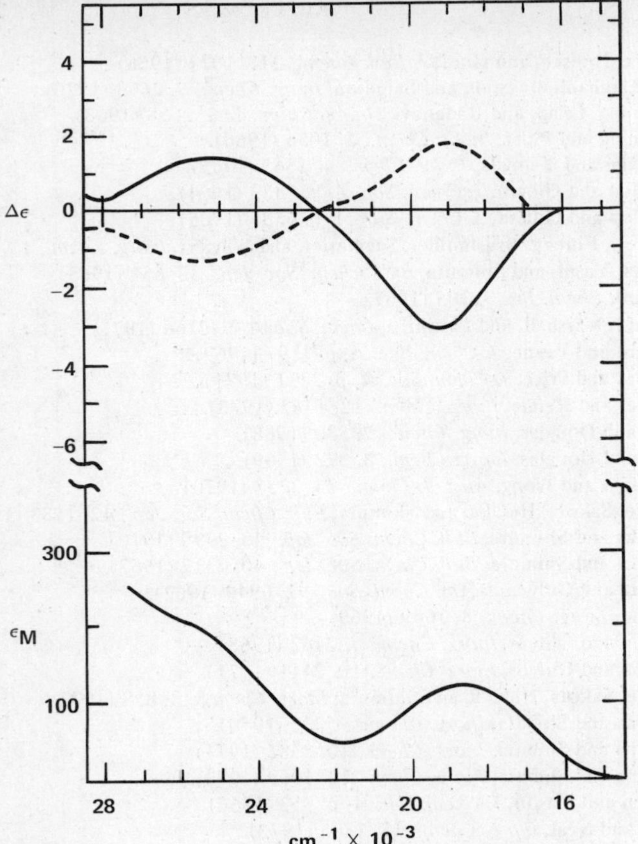

FIGURE 116. CD spectra in aqueous solution: (—) (−)D-*cis*-*N*-C_1-[Co(acac)(*S*-val)$_2$]; (- - -) (+)D-*cis*-*N*-C_1-[Co(acac)(*S*-val)$_2$]. Absorption spectrum in aqueous solution: (—) (−)D-*cis*-*N*-C_1-[Co(acac)(*S*-val)$_2$]. (Reprinted with permission from Everett, Finney, Brushmiller, Seematter, and Wingert, *Inorg. Chem.*, 13, 539 (1974). Copyright by the American Chemical Society.)

REFERENCES

1. Bryce, Roeske, and Gurd, *J. Biol. Chem.*, 241, 1072 (1966).
2. Buckingham, Maxwell, and Sargeson, *Inorg. Chem.*, 9, 2663 (1970).
3. Denning, Célap, and Radanović, *Inorg. Chim. Acta*, 2, 58 (1968).
4. Denning and Piper, *Inorg. Chem.*, 5, 1056 (1966).
5. Douglas and Yamada, *Inorg. Chem.*, 4, 1561 (1965).
6. Dunlop and Gillard, *J. Chem. Soc. A*, p. 2822 (1964).
7. Dunlop and Gillard, *J. Chem. Soc. A*, p. 6531 (1965).
8. Everett, Finney, Brushmiller, Seematter, and Wingert, *Inorg. Chem.*, 13, 536 (1974).
9. Fujita, Yasui, and Shimura, *Bull. Chem. Soc. Jap.*, 38, 654 (1965).
10. Gillard, *Chem. Br.*, 3, 205 (1967).
11. Gillard, Marshall, and Pastini, *J. Chem. Soc. A*, p. 2268 (1971).
12. Gillard and Payne, *J. Chem. Soc. A*, p. 1197 (1969).
13. Gillard and Price, *J. Chem. Soc. A*, p. 2271 (1971).
14. Haines and Reime, *Inorg. Chem.*, 12, 1482 (1973).
15. Hall and Douglas, *Inorg. Chem.*, 7, 530 (1968).
16. Hall and Douglas, *Inorg. Chem.*, 8, 372 (1969).
17. Hawkins and Wong, *Aust. J. Chem.*, 23, 2237 (1970).
18. Koine, Sakota, Hidaka, and Shimura, *Bull. Chem. Soc. Jap.*, 42, 1583 (1969).
19. Hidaka and Shimura, *Bull. Chem. Soc. Jap.*, 43, 2999 (1970).
20. Hidaka and Shimura, *Bull. Chem. Soc. Jap.*, 40, 2312 (1967).
21. Katzin and Gulyes, *J. Am. Chem. Soc.*, 91, 6940 (1969).
22. Katzin, *Inorg. Chem.*, 8, 1649 (1969).
23. Katzin and Gulyas, *Inorg. Chem.*, 7, 2442 (1968).
24. Katzin and Gulyas, *Inorg. Chem.*, 10, 2411 (1971).
25. Koine, Sakota, Hidaka, and Shimura, *Inorg. Chem.*, 12, 859 (1973).
26. Kojima and Shibata, *Inorg. Chem.*, 9, 238 (1970).
27. Kojima and Shibata, *Inorg. Chem.*, 10, 2382 (1971).
28. Kojima and Shibata, *Inorg. Chem.*, 12, 1009 (1973).
29. Larsen and Mason, *J. Chem. Soc. A*, p. 313 (1966).
30. Legg and Neal, *Inorg. Chem.*, 12, 1805 (1973).
31. Liu and Douglas, *Inorg. Nucl. Chem. Lett.*, 4, 15 (1968).
32. Liu and Douglas, *Inorg. Chem.*, 3, 1356 (1964).
33. Marzilli and Buckingham, *Inorg. Chem.*, 6, 1042 (1967).
34. Matsuoka, Hidaka, and Shimura, *Inorg. Chem.*, 9, 719 (1970).
35. Morris and Martin, *Inorg. Chem.*, 10, 964 (1971).
36. Saburi and Yoshikawa, *Inorg. Chem.*, 7, 1890 (1968).
37. Shibata, Nishikawa, and Hosaka, *Bull. Chem. Soc. Jap.*, 40, 236 (1967).
38. Shibata, Nishikawa, and Nishida, *Bull. Chem. Soc. Jap.*, 39, 2310 (1966).
39. Shibata, Nishikawa, and Nishida, *Inorg. Chem.*, 7, 9 (1968).
40. Smith and Douglas, *Inorg. Chem.*, 5, 784 (1966).
41. Treptow, *J. Inorg. Nucl. Chem.*, 31, 2983 (1969).
42. Yasui, *Bull. Chem. Soc. Jap.*, 38, 1746 (1965).
43. Yasui, Fujita, and Shimura, *Bull. Chem. Soc. Jap.*, 42, 2081 (1969).
44. Yasui, Hidaka, and Shimura, *Bull. Chem. Soc. Jap.*, 39, 2417 (1966).
45. Yasui, Hidaka, and Shimura, *Bull. Chem. Soc. Jap.*, 38, 2025 (1965).
46. Yasui, Hidaka, and Shimura, *J. Am. Chem. Soc.*, 87, 2762 (1965).

Table 1
PRIMARY OVERFLOW AMINOACIDURIAS[a]

Disease	Key amino acid(s) affected	Clinical features	Abnormal enzyme (tissue in which defect demonstrated)	Amino acids increased in blood	Amino acids increased in urine	Comments	Treatment	References[b]
Argininosuccinic aciduria	Argininosuccinic acid	Mental retardation, seizures, ataxia, friable hair, reluctance to eat protein; may die in neonatal period	Argininosuccinase (liver, red blood cells and fibroblasts) PD(T)	Argininosuccinic acid; citrulline may be present	Argininosuccinic acid; small amounts of citrulline	Postprandial hyperammonemia; normal BUN; argininosuccinic acid in CSF is 2–3 times normal	Low protein diet	1
Hyperbeta-alaninemia	β-Alanine	Seizures, somnolence	(?) β-Alanine, alpha-ketoglutarate transaminase	β-Alanine, gamma-aminoisobutyric acid (GABA)	β-Alanine, GABA, β-aminoiso-butyric acid	β-Alanine and GABA increased in CSF; β-alanine and carnosine increased in skeletal muscle and brain	(?) Vitamin B_6	2
β-Mercaptolactate cysteinuria	β-Mercaptolactate cysteine disulfide	(?) Mental retardation, susceptibility to infections	Unkown	None	β-Mercaptolactate cysteine disulfide	May be benign; patient was product of sibling mating	Necessity unknown	3
Carnosinemia	Carnosine	Mental retardation, mycolonic seizures, somnolence	Carnosinase (serum)	Carnosine	Carnosine	No anserine, histidine, or 1-methylhistidine in urine	None known	4
Imidazole amino-aciduria	Carnosine	(?) Juvenile amaurotic idiocy. with mental retardation, seizures, retinitis pigmentosa	Unknown	None	Carnosine, anserine, histidine, and 1-methylhistidine	Aminoaciduria in heterozygotes	None known	5

[a]PD(A) = prenatal diagnosis achieved, PD(T) = prenatal diagnosis theoretically possible.
[b]References follow Table 2.

Table 1 (continued)
PRIMARY OVERFLOW AMINOACIDURIAS

Disease	Key amino acid(s) affected	Clinical features	Abnormal enzyme (tissue in which defect demonstrated)	Amino acids increased in blood	Amino acids increased in urine	Comments	Treatment	References[b]
Citrullinemia	Citrulline	Mental retardation, seizures, failure to thrive, vomiting, reluctance to eat protein	Argininosuccinic acid synthetase (liver, fibroblasts) PD(T)	Citrulline	Citrulline	Ammonia intoxication; may have low urea levels in blood and urine	Low protein diet	6
Cystathioninuria	Cystathionine	Probably benign	Cystathionase (liver)	Cystathionine (trace)	Cystathionine, *N*-acetylcystathionine		Large doses of pyridoxine	7
Cystinosis, nephropathic	Cystine	Failure to thrive, vitamin D-resistant rickets, renal tubular defects	Unknown	Essentially normal including cystine and cysteine	Generalized aminoaciduria	Glycosuria, phosphaturia; cystine deposits in many tissues; cystine in leucocytes 80 times normal in patients and 12 times normal in heterozygotes	Symptomatic, low cystine diet and D-penicillamine tried without success; renal transplantation	8
Hyperglycinemia, ketotic (severe infantile)	Glycine	Myoclonic jerks, vomiting, and ketosis; failure to thrive, osteoporosis, neutropenia, frequent infections	Not specific; deficiencies of the following are known: (1) carbamyl PO_4 synthetase, (2) methylmalonyl CoA mutase, and (3) propionyl CoA carboxylase PD(A) for (2) PD(T) for (1) and (3)	Glycine; occasionally lysine	Glycine; occasionally lysine	Ammonia intoxication in (1) and (2)	Low protein diet	9

Table 1 (continued)
PRIMARY OVERFLOW AMINOACIDURIAS

Disease	Key amino acid(s) affected	Clinical features	Abnormal enzyme (tissue in which defect demonstrated)	Amino acids increased in blood	Amino acids increased in urine	Comments	Treatment	References[b]
Hyperglycinemia, nonketotic	Glycine	Mental retardation, myoclonic jerks, spasticity	Probably an enzyme system for formation of NH_3, CO_2, and hydroxymethyltetrahydrofolate from glycine	Glycine	Glycine	Glycine administration accentuates symptoms; oxalate excretion variable	No effective therapy	10
Histidinemia	Histidine	Probably benign	Histidase (skin) PD(T)	Histidine	Histidine imidazole-pyruvic, lactic, and acetic acids; alanine in some	Urocanic acid absent in skin homogenates and sweat	None necessary	6
Homocystinuria	Homocystine	Dislocated lenses, vascular thromboses, Marfan-like syndrome, osteoporosis; mental retardation in some	Cystathionine synthetase (liver) PD(T)	Methionine; small amount of homocysteine	Homocystine, methionine, cysteine-homocysteine disulfide, homolanthionine	Low brain cystathionine levels; two variants: apoenzyme lack and pyridoxine binding defect	Pyridoxine if responds; low methionine diet with cystine supplement if not	11
Maple syrup urine disease (severe infantile)	Leucine, isoleucine, valine	Hypertonicity, seizures, vomiting, coma, early death	Branched-chain keto acid decarboxylase (leucocytes and fibroblasts); absence virtually complete PD(A)	Leucine, isoleucine, valine, and alloisoleucine	Leucine, isoleucine, and valine, together with their three corresponding keto acids	Urine has odor of maple syrup	Partially synthetic diet low in leucine, isoleucine, and valine; peritoneal dialysis during crises	6
Maple syrup urine disease (intermittent form)	Leucine, isoleucine, valine	Episodic vomiting lethargy, coma, usually with infection; normal between episodes	Branched-chain keto acid decarboxylase (leucocytes and fibroblasts); absence incomplete (10–20% of normal) PD(T)	Leucine, isoleucine, valine, and alloisoleucine during acute attacks	Leucine, isoleucine, valine together with their three corresponding keto acids during acute attacks	Odor of maple syrup only when ill; may be fatal during acute attacks	Low protein diet during exacerbations	12

Table 1 (continued)
PRIMARY OVERFLOW AMINOACIDURIAS

Disease	Key amino acid(s) affected	Clinical features	Abnormal enzyme (tissue in which defect demonstrated)	Amino acids increased in blood	Amino acids increased in urine	Comments	Treatment	References[b]
Hyperlysinemia	Lysine	May be benign; mental retardation, seizures, muscular asthenia in some	Lysine-ketoglutarate reductase (fibroblasts and leucocytes) PD(T)	Lysine	Lysine; homoarginine, homocitrulline, and ε-*N*-acetyl lysine in some		Necessity unknown; low protein diet	13
Ornithinemia type I	Ornithine	Mental retardation, myoclonic jerks, ataxia, irritability, refusal to eat protein	Unknown	Ornithine	Homocitrulline; ornithine may be normal	Ammonia intoxication	Low protein diet	14
Ornithinemia type II	Ornithine	Mental retardation prolonged neonatal jaundice, liver cirrhosis, vomiting, osteoporosis	Ornithine keto acid transaminase (liver and fibroblasts) PD(T)	Ornithine	Generalized aminoaciduria, glycosuria		Low protein diet	15
Phenylketonuria, classical	Phenylalanine	Mental retardation, seizures, eczema, fair skin, hair, and eyes	Phenylalanine hydroxylase (liver); absence virtually complete	Phenylalaine	Phenylalanine; phenylpyruvic, lactic, and acetic acids, orthohydroxyphenylacetic acid	May be possible to cease treatment at about age 5 years without intellectual deterioration	Low phenylalanine diet, (?) indefinitely	6
Phenylketonuria, atypical	Phenylalanine	Mental retardation, often less severe than in classical form	Phenylalanine hydroxylase (liver); absence incomplete	Phenylalanine; often levels not as high as in classical form	Phenylalanine; phenylpyruvic, lactic, and acetic acids, orthohydroxyphenylacetic acid	May not need treatment to maintain normal blood phenylalanine after period of most rapid growth	Low phenylalanine diet; phenylalanine given as tolerated	6

Table 1 (continued)
PRIMARY OVERFLOW AMINOACIDURIAS

Disease	Key amino acid(s) affected	Clinical features	Abnormal enzyme (tissue in which defect demonstrated)	Amino acids increased in blood	Amino acids increased in urine	Comments	Treatment	References[b]
Hypophosphatasia	Phospho-ethanolamine	Vitamin D-resistant rickets	Alkaline phosphatase (serum)	Phosphoethanolamine slightly increased	Phosphoethanolamine		None	16
Hyperpipecolic acidemia	Pipecolic acid	Mental retardation, hypotonia, nystagmus, progressive CNS degeneration, vomiting, diarrhea, and hepatomegaly	(?) Pipecolate oxidase	Pipecolic acid	Slight increase in pipecolic acid	Lysine metabolized normally	None	17
Hyperprolinemia type I	Proline	May have mental retardation, seizures, hereditary nephritis, and/or deafness; may be normal	Proline oxidase (liver)	Proline	Proline; hydroxyproline and glycine may be elevated		(?) Low protein, low proline diet	18
Hyperprolinemia type II	Proline	Mental retardation, seizures; may be normal	Δ'-Pyrroline-5-carboxylate dehydrogenase (liver)	Proline	Proline Δ'-pyrroline-5-carboxylate, hydroxyproline, and glycine	No renal disease or deafness	(?) Low protein diet, low proline diet	19
Saccharopinuria	Saccharopine	Mental retardation, short stature	Unknown	Saccharopine, lysine, citrulline, and homocitrulline	Saccharopine, lysine, citrulline, and homocitrulline		None tried	20
Sarcosinemia	Sarcosine	Probably benign; (?) mental retardation, difficulty swallowing	Sarcosine dehydrogenase oxygen oxidoreductase (demethylating)	Sarcosine, ethanolamine	Sarcosine, ethanolamine		Necessity unknown	21

Table 1 (continued)
PRIMARY OVERFLOW AMINOACIDURIAS

Disease	Key amino acid(s) affected	Clinical features	Abnormal enzyme (tissue in which defect demonstrated)	Amino acids increased in blood	Amino acids increased in urine	Comments	Treatment	References[b]
Sulfite oxidase deficiency	S-Sulfo-L-cysteine	Mental retardation, muscular hypertonicity, dislocated lenses	Sulfite oxidase (liver, kidney, brain)	Small amounts of S-sulfo-L-cysteine	S-Sulfo-L-cysteine	Marked decrease in sulfate but increased sulfite and thiosulfate in urine	(?) Low cystine and methionine diet; D-penicillamine, sulfate supplements	22
Tryptophanemia	Tryptophan	Mental retardation, ataxia, short stature, pellagra-like rash	(?) Tryptophan pyrrolase or formylase	Tryptophan	Tryptophan	Decreased excretion of kynurenine after tryptophan load compared to normal	(?) Nicotinamide for rash	23
Tyrosinemia	Tyrosine	Renal tubular defects, failure to thrive, vitamin D-resistant rickets, hepatic cirrhosis	(?) Para-hydroxyphenylpyruvic acid oxidase (liver)	Tyrosine, methionine	Generalized aminoaciduria, especially tyrosine, methionine, para-hydroxyphenylpyruvic, lactic and acetic acids	Glycosuria and phosphaturia	Low tyrosine and phenylalanine diet	24
Hypervalinemia	Valine	Mental retardation	Valine α-keto isovaleric acid transaminase (leucocytes)	Valine	Valine		Low valine diet	25

Compiled by Mary G. Ampola.

Table 2
RENAL (TRANSPORT) AMINOACIDURIAS (PLASMA AMINO ACIDS NORMAL OR LOW)

Disease	Clinical features	Amino acids increased in urine	Amino acids poorly absorbed by intestine	Comments	Treatment	References
Methionine Malabsorption Syndrome	Severe mental retardation, seizures, pungent odor, diarrhea, white hair, hyperpnea, edema, scurvy	Methionine; smaller amounts of valine, leucine, isoleucine, phenylalanine, tyrosine	Methionine (*Escherichia coli* convert some to α-hydroxybutyric acid, which is absorbed and excreted in urine)		Low methionine diet	26
Cystinuria type I	Renal Calculi	Cystine and dibasic amino acids arginine ornithine and lysine; cysteine-homocysteine disulfide	No transport of cystine or dibasic amino acids	Heterozygotes excrete normal amount of cystine and dibasic amino acids but do excrete some disulfide	D-Penicillamine, high fluid intake, alkalinization of urine; low methionine diet	27
Cystinuria type II	Same	Same	Very poor cystine transport; no transport of dibasic amino acids	Heterozygotes excrete moderate excesses of cystine and dibasic amino acids	Same	
Cystinuria type III	Same	Same	Transport of cystine and dibasic amino acids present but subnormal	Heterozygotes excrete mild excesses of cystine and dibasic amino acids, some disulfide	Same	
Cystinuria, isolated	May be benign	Cystine	Unknown	Arginine, ornithine, and lysine in urine normal	None necessary	28
Familial protein intolerance	Liver cirrhosis, neutropenia, vomiting, diarrhea, failure to thrive, refusal to eat protein	Arginine, lysine; ornithine not reported; cystine normal	Probably similar to renal pattern	Usually normal mentality; ammonia elevated in blood after high protein meal; lowered urea production	Low protein diet; arginine supplement to increase conversion of ammonia to urea	29

Table 2 (continued)
RENAL (TRANSPORT) AMINOACIDURIAS (PLASMA AMINO ACIDS NORMAL OR LOW)

Disease	Clinical features	Amino acids increased in urine	Amino acids poorly absorbed by intestine	Comments	Treatment	References
Hartnup Disease	May be benign; periodic ataxia, light-sensitive rash, psychosis, mental retardation in some	Neutral (monamine, monocarboxylic) amino acids; basic amino acids, proline, hydroxyproline, and glycine normal	Neutral amino acids; bacteria convert tryptophan to indoles which are absorbed then excreted in urine		Nicotinamide for rash	30
Lowe's Syndrome	Mental and growth retardation, hypotonia, rickets, lenticular opacities, glaucoma, metabolic acidosis	Severe generalized aminoaciduria	Lysine and arginine defect demonstrated; glycine normal	Urine also contains albumin, glucose phosphate, and is low in acid	Large doses of vitamin D; otherwise none	31

Compiled by Mary G. Ampola.

REFERENCES

1. Moser, Efron, Brown, Diamond, and Neumann, *Am. J. Med.*, 42, 9 (1967).
2. Scriver, Pueschel, and Davies, *N. Engl. J. Med.*, 274, 635 (1966).
3. Ampola, Efron, Bixby, and Meshorer, *Am. J. Dis. Child.*, 117, 66 (1969).
4. Perry, Hansen, Tischler, Bunting, and Berry, *N. Engl. J. Med.*, 217, 1219 (1967).
5. Bessman and Baldwin, *Science*, 135, 789 (1962).
6. Scriver and Rosenberg, *Amino Acid Metabolism and Its Disorders*, W. B. Saunders, Philadelphia, 1973.
7. Perry, Hardwick, Hansen, Love, and Israels, *N. Engl. J. Med.*, 278, 590 (1968).
8. Crawhall, Lietman, Schneider, and Seegmiller, *Am. J. Med.*, 44, 330 (1968).
9. Rampini, Vischer, Curtius, Anders, Tancredi, Frischknecht, and Prader, *Helv. Paediatr. Acta*, 22, 135 (1967).
10. Gerritsen, Nyhan, Rehberg, and Ando, *Pediatr. Res.*, 3, 269 (1967).
11. Carson and Carré, *Arch. Dis. Child.*, 44, 387 (1969).
12. Dancis, Hutzler, and Cox, *Biochem. Med.*, 2, 407 (1969).
13. Dancis, Hutzler, Cox, and Woody, *J. Clin. Invest.*, 48, 1447 (1969).
14. Shih, Efron, and Moser, *Am. J. Dis. Child.*, 117, 83 (1969).
15. Bickel, Feist, Müller, and Quadbeck, *Dtsch. Med. Wochenschr.*, 93, 2247 (1968).
16. Bartter, in *The Metabolic Basis of Inherited Diseases*, Stanbury, Wyngaarden, and Fredrickson, Eds., McGraw-Hill, New York, 1972.
17. Gatfield, Taller, Hinton, Wallace, Abdelnour, and Haust, *Can. Med. Assoc. J.*, 99, 1215 (1968).
18. Efron, *N. Engl. J. Med.*, 272, 1243 (1965).
19. Efron, in *The Metabolic Basis of Inherited Diseases*, Stanbury, Wyngaarden, and Frederickson, Eds., McGraw-Hill, New York, 1966.
20. Carson, Scally, Neill, and Carré, *Nature*, 218, 679 (1969).
21. Gerritsen and Waisman, *N. Engl. J. Med.*, 275, 66 (1966).
22. Irreverre, Mudd, Heizer, and Laster, *Biochem. Med.*, 1, 187 (1967).
23. Tada, Ito, Wada, and Arakawa, *Tohoku J. Exp. Med.*, 80, 118 (1963).
24. Aronsson, Engelson, Jagenburg, and Palmgren, *J. Pediatr.*, 72, 620 (1968).
25. Tada, Wada, and Arakawa, *Am. J. Dis. Child.*, 113, 64 (1967).
26. Hooft, Timmermans, Snoeck, Antener, Oyaert, and van den Hende, *Ann. Paediatr.* (Basel), 205, 73 (1965).
27. Crawhall and Watts, *Am. J. Med.*, 45, 736 (1968).
28. Brodehl, Gallison, and Kowalewski, *Klin. Wochenschr.*, 45, 38 (1967).
29. Kekomaki, Visakorpi, Peerheentupa, and Saxen, *Acta Paediatr. Scand.*, 56, 617 (1967).
30. Pomeroy, Efron, Dayman, and Hoefnagel, *N. Engl. J. Med.*, 278, 1214 (1968).
31. Richards, *Am. J. Dis. Child.*, 109, 185 (1965).

ERRORS OF ORGANIC ACID METABOLISM

Disorder	Clinical features	Enzyme defect	Metabolites accumulated	Treatment
Isovaleric acidemia	Feeding difficulties, recurrent ketoacidosis, mental retardation, "sweaty feet" odor	Isovaleryl-CoA, dehydrogenase (PD)[a]	Isovaleric acid, isovaleryl-glycine	Restricted leucine or protein diet
β-Hydroxyisovaleric acidemia and β-methylcrotonylglycinuria	Neurological abnormalities, mental retardation, feeding difficulties, metabolic ketoacidosis, acrid urine odor	β-Methylcrotonyl-CoA carboxylase (PD)	β-Hydroxyisovaleric acid, β-methylcrotonic acid and their glycine conjugates, tiglylglycine	Restricted protein or leucine diet, biotin
α-Methyl-β-hydroxybutyric and α-methyl acetoacetic acidurias	Vomiting, recurrent ketoacidosis, mental retardation	β-Ketothiolase (PD)	α-Methyl-β-hydroxy butyric acid, α-methylaceto acetic acid, tiglylglycine, hyperglycinemia	Restricted protein diet
Propionic acidemia	Feeding difficulties, vomiting, recurrent ketoacidosis, mental retardation	Propionyl-CoA, carboxylase (PD)	Propionic acid, propionyl glycine, tiglylglycine, hyperglycinemia, hyperammonemia, long chain ketones	Restricted protein diet, biotin
Methylmalonic acidemia	Feeding difficulties, vomiting, seizures, mental retardation, recurrent ketoacidosis	Methylmalonyl-CoA mutase or racemase, B_{12} metabolism (PD)	Methylmalonic acid, hyperglycinemia, hyperammonemia	Vitamin B_{12}, restricted protein diet
Lactic and pyruvic acidosis				
Pyruvate carboxylase deficiency	Severe hypoglycemia of early onset, seizures, mental retardation, Leigh's myeloencephalopathy	Pyruvate carboxylase	Lactate, pyruvate alanine	Restricted carbohydrate diet, biotin
Pyruvate dehydrogenase deficiency	Intermittent ataxia, mental retardation	Pyruvate dehydrogenase complex (PD)	Lactate, pyruvate alanine	Restricted carbohydrate diet, thiamine
Pyroglutamic aciduria	Chronic metabolic acidosis, increased hemolysis	Glutathione synthetase (PD)	Pyroglutamic acid	Bicarbonate
Syndrome of ketoacidosis	Recurrent severe ketoacidosis of early onset	Succinyl-CoA: 3-ketoacid CoA transferase (PD)	Ketones	High glucose and low protein diet
Glutaric acidemia	Neurodegenerative disorder, recurrent ketoacidosis	Glutaryl-CoA dehydrogenase (PD)	Glutaric acid, glutaconic acid, β-hydroxy-glutaric acid	? low lysine and tryptophan diet

[a]PD = prenatal diagnosis either has been made or is potentially possible.

Compiled by Vivian E. Shih.

FREE AMINO ACIDS IN AMNIOTIC FLUID IN EARLY PREGNANCY (13 TO 18 WEEKS) AND AT TERM

	Emery et al.[1]				Levy et al.[2]			
Gestation (week)	13–16		37–40		14–18		39–41	
Number	6		7		15		16	
Amino acid	Mean	Range	Mean	Range	Mean	S.D.	Mean	S.D.
Cysteic acid	9	4–15	11	8–17				
Taurine	270	73–573	139	85–219	43	41	113	49
Hydroxyproline	40	31–51	21	10–37				
Aspartic acid	35	3–76	2	0–4	8	5	26	
Threonine	254	138–333	104	46–167	186	41	139	41
Serine	48	26–91	31	20–52	62	32	119	33
Glutamine	152	34–294	264	118–365	92	64	223	62
Asparagine	24	19–36	5	0–10				
Proline	140	50–229	112	87–172	161	65	103	17
Glutamic acid	299	137–610	31	14–43	196	194	107	126
Citrulline					10	5	11	5
Glycine	206	95–388	152	63–238	160	42	185	75
Alanine	505	326–837	203	104–299	266	67	194	57
α-Aminobutyric acid					14	20	11	4
Valine	263	158–344	46	23–70	133	57	67	18
Cystine	64	38–86	66	40–105	80	11	71	19
Methionine	33	18–58	5	0–9	25	5	12	6
Isoleucine	64	34–88	10	5–16	29	13	20	8
Leucine	143	74–197	19	13–25	76	41	43	26
Tyrosine	76	47–135	14	6–21	62	20	35	10
Phenylalanine	82	49–137	17	8–25	70	25	38	10
Ornithine	62	38–97	24	10–42	49	19	45	16
Lysine	339	182–494	93	50–149	238	102	148	45
Tryptophan	11	0–20	1	0–2				
Histidine	116	71–170	44	28–63	107	27	46	16
Arginine	84	40–143	19	11–26	52	22	28	13

Compiled by Vivian E. Shih.

REFERENCES

1. **Emery et al.,** *Lancet,* 1, 1307 (1970).
2. **Levy et al.,** *Antenatal Diagnosis,* Dorfman, Ed., University of Chicago, Chicago, 1972, 109.

FREE AMINO ACIDS[a] IN BLOOD PLASMA OF NEWBORN INFANTS AND ADULTS

	Newborn[b]			Adult			
				This Lab[c]		Lit.[d]	
Amino acid	Mean	S.D.	Range	Mean	Range	Mean	Range
Taurine	1.76	0.50	0.93–2.70	0.83	0.57–1.73	0.79	0.34–2.10
Urea	16.4						
Hydroxyproline	(0.42)[e]						
Aspartic acid	0.11	0.05	Tr.–0.22	0.22	Tr.–0.72	0.10	0.0–0.32
Threonine	2.59	0.25	1.36–3.99	1.94	1.22–2.93	1.54	0.94–2.30
Serine	1.72[f]	0.36	0.99–2.55[f]	1.18[f]	0.68–2.03[f]	1.21	0.77–1.76
Asparagine	(0.60)[e]			(0.57)[g]		(0.58)	0.54–0.65
Glutamine	11.16	2.04	7.86–14.01	8.30	6.07–10.15	(8.30)	
Glutamic acid	0.76	0.37	0.30–1.57	0.86	0.25–1.73	0.85	0.21–2.82
Proline	2.13	0.38	1.23–3.19	2.71	1.28–5.14	2.12	1.17–3.87
Citrulline	0.28	0.09	0.15–0.50	0.53	0.21–0.97	(0.50)	
Glycine	2.58	0.52	1.68–3.86	1.74	1.08–3.66	1.78	0.90–4.16
Alanine	2.94	0.49	2.10–3.65	3.07	2.22–4.47	2.99	1.87–5.89
α-NH_2-n-butyric acid	0.15	0.06	0.06–0.30	0.17	0.10–0.24	0.21	0.08–0.36
Valine	1.60	0.46	0.94–2.88	1.99	1.36–2.66	2.50	1.65–3.71
Cystine	1.47	0.32	0.85–2.02	1.77	1.15–3.37	1.05	0.20–2.02
Methionine	0.44	0.12	0.13–0.61	0.32	0.23–0.39	0.34	0.09–0.59
Isoleucine	0.52	0.11	0.35–0.69	0.71	0.46–1.15	0.83	0.48–1.28
Leucine	0.95	0.23	0.61–1.43	1.32	0.93–1.78	1.45	0.98–2.30
Tyrosine	1.26	0.30	0.76–1.80	0.91	0.65–1.13	0.94	0.39–1.58
Phenylalanine	1.30	0.23	0.69–1.82	0.95	0.63–1.92	0.88	0.61–1.45
β-Alanine	(0.13)[e]			(0.08)[g]			
Ethanolamine	0.32	0.12	0.16–0.56	0.01	Tr.–0.07		
Ornithine	1.21	0.33	0.65–2.00	0.92	0.43–1.67	0.79	0.39–1.40
Lysine	2.93	0.67	1.67–3.93	2.54	2.11–3.09	2.24	1.21–3.48
Histidine	1.19	0.25	0.76–1.77	1.24	0.97–1.45	1.15	0.49–1.66
Tryptophan	0.65	0.35	Tr.–1.37	0.98	0.51–1.49		
Arginine	0.94	0.30	0.38–1.53	1.43	0.86–2.63	1.30	0.37–2.40

[a]All values expressed in mg per 100 ml plasma.
[b]9 male, 16 female.
[c]3 male, 5 female.
[d]39 male, 37 female: data from 9 laboratories.
[e]Pooled plasma of 10 infants.
[f]Corrected for asparagine.
[g]Single plasma.

Reprinted in modified form from *Pediatrics,* 36, 2 (1965) by Dickinson, Rosenblum and Hamilton. With permission of American Academy of Pediatrics.

Peptides and Polypeptides

A LIST OF SEQUENTIAL POLYPEPTIDES, THEIR METHOD OF PREPARATION AND PRODUCT ANALYSIS

H. J. Goren

Sequential polypeptides are polymers of two or more amino acids in a specific sequence that is repeated throughout the polymer. In recent years these polymers have been used in increasing numbers as models for proteins of biological origin, e.g., structural proteins and enzymes.[1] In addition Lotan, Berger, and Katchalski[2] believe that through the study of the thermodynamics of folding of sequential polypeptides, the understanding of how the secondary and tertiary structure of a protein arises from its primary structure may eventually be realized. The need for such polymers, it appears, will grow.

The synthesis of sequential polypeptides may be broken down into two major steps: The synthesis of the oligopeptide, i.e., the repeating monomer unit, and the polymerization of the oligopeptide. In recent review articles,[1,3] several factors of the polymerization step were put forward as affecting the polymer product.

The table that follows will hopefully not only help the reader to quickly find whether a certain polymer has been prepared, but will also allow him to compare several methods of preparation.

The table contains a column to identify the synthetic sequential polypeptide in the sequence of the monomer unit it was prepared from, columns describing the conditions for polymerization, and columns describing the result of the polymerization. Rather than listing the compounds prepared in alphabetical order, the listing is in a format to assist the reader in carrying out his own comparative analysis. Thus, initially they are grouped according to size of the monomer unit, i.e., dipeptide monomers precede tripeptide monomers, and so on. The order of the synthetic sequential polypeptides follows the bulkiness and functionality of the α-carbon sidechain, from the N-terminal amino acid to the C-terminal amino acid (of the monomer unit). Initially, bulk is the determinant. Glycine, alanine, serine, cysteine, threonine, valine, methionine, leucine, isoleucine, phenylalanine, tyrosine, and tryptophan appear in this order. Basic amino acids are next in the order, i.e., histidine, ornithine, lysine, and arginine; then the acidic amino acids, aspartic and glutamic acid. Each size category of monomers leading to sequential polypeptides is terminated with the imino acids, sarcosine, proline, hydroxyproline, and other analogues of proline.

The synthetic sequential polypeptide column is written in a form to show the final form of the repeating monomer unit. With those polymers containing trifunctional amino acids where the polymer was deprotected after the polymerization step, only the deprotected form is listed in the table. If the sequential polypeptide was never deprotected then this form appears in the table but it appears after the unblocked form, e.g., order of appearance of N-terminal glutamyl tripeptide polymers would be Glu(OH)-, Glu(OMe)-, Glu(OEt)-, Glu(OBzl)-, and Glu(OBu)-.

All amino acids are L unless otherwise designated. An L-amino acid will precede the D-amino acid, which in turn precedes the racemic form. β- and γ- Sequential polypeptides follow their respective α-polypeptides. Some polydepsipeptides are listed in the following table. They occur in the listing after the position of their α-amino acid counterpart, e.g., L-2-hydroxy-4-methyl pentanoic acid, listed as, α-OH-4-Me-pentatoic, would appear after L-leucine.

The procedure for polymerization generally is to dissolve the monomer unit in the active ester salt form (HA · H-AA_1 · · · AA_n-X) or in the free amino and free carboxyl forms (H-AA_1 · · · AA_n-OH) in an appropriate solvent. In Column 2 either -X or the condensing agent is listed for the active ester form of polymerization or for the free peptide form of polymerization, respectively. Column 3 gives the solvent and concentration of the monomer unit. In the active ester form of polymerization, a base is added to neutralize the salt form of the peptide ester. The equivalent amount and the base added are listed in Column 4. The last column of the polymerization procedure gives the length of time and temperature (°C) for the reaction.

The product is described in terms of yield (Column 6), degree of polymerization (Column 7), and degree of racemization (Column 8) of the sequential polypeptide. The degree of polymerization may be determined from the molecular weight of the product which in turn may be determined:

i. from the amount of α-amino groups by the Van Slyke method (V.S.), dinitrophenyl method (Dnp), dansyl method, ninhydrin method (N), and titration with perchloric acid (T-$HClO_4$);

ii. from the amount of α-carboxyl groups by the hydrolysis of *p*-nitrophenol from sequential polypeptides prepared via the *p*-nitrophenyl ester method (ONp), titration with sodium methoxide (T-OMe), and counting the [C^{14}]glycyl alkyl ester (^{14}C);

iii. by sedimentation velocity (Vel.) or sedimentation equilibrium (Eq.) ultracentrifugation methods;

iv. by osmotic pressure (π), light scattering (L.S.), vapor pressure measurements (VP), and viscosity ($[\eta]$);

v. from the elution patterns off gel chromatography, either Sephadex® (G-g) or Bio Gel® (P-p), where g and p are numbers in the table referring to the respective gel size used;

vi. from the elution pattern off CM-cellulose chromatography (CM-Cel);

vii. from the polymer fraction contained by Diaflow membranes (Diaflow). Alternatively, the degree of polymerization may be obtained spectroscopically: Infrared (IR), optical rotation of the sodium D-line ($[m]_D$), and nuclear magnetic resonance (NMR).

The degree of racemization of the product may be determined:

i. by comparing the optical rotation of the sodium D-line of the acid hydrolysate of the sequential polypeptide with an amino acid mixture equal to the polyamino acid (H^+);

ii. enzymatically (E);

iii. by comparing the optical rotatory properties of the synthetic polymer with the same polymer prepared in a racemic free form ([q]); or

iv. by a gas chromatographic procedure (G.C.).

Abbreviations generally follow the IUPAC-IUB Commission on Biochemical Nomenclature, Symbols for Amino-acid Derivatives and Peptides Recommendations (1971), *Eur. J. Biochem.*, 27, 201 (1972), and Abbreviated Nomenclature of Synthetic Polypeptides (Polymerized Amino Acids), Revised Recommendations (1971), *Eur. J. Biochem.*, 26, 301 (1972).

Additional abbreviations which appear follow the philosophy of the IUPAC-IUB recommendations with the exception of those abbreviations that have become standard in the field of synthetic peptide chemistry. These are listed on the next page.

ABBREVIATIONS

Used	For	Recommended (if different)
ONbzl	*p*-Nitrobenzyl ester	
OIbzl	*p*-Iodobenzyl ester	
Azet	Azetidine-2-COOH	
Pipec	Piperidine-2-COOH	
Thz	Thiazolidine-4-COOH	
GcOH	Glycollic acid	
M.A.	Mixed anhydride	
OEtCl$_2$ SalNH$_2$	*N*-Et-3,5-Cl$_2$-Salicylamide ester	
OPhOH	*o*-Hydrophenyl ester	
Et$_4$P$_2$O$_7$	Tetraethylpyrophosphate	
(PhO$_2$P)$_2$O	Bis-*O*-phenylenepyrophosphite	
OPy	3-OH-Pyridine ester	
O(Bu^{i}NO$_2$ MePy)	2-Bu^i-4-NO$_2$-6-Me-pyridine ester	
CMCI	*N*-Cyclohexyl-*N*′-[β-(*N*-methylmorpholinium) ethyl]-carbodiimide *p*-toluenesulfonate	C(NcHx)NMemEt-Tos
HOBztl	l-Hydroxybenzotriazole	
HOBztn	3-OH-4-O-3,4-H$_2$-1,2,3-Benzotriazine	
HCONMe$_2$	Dimethylformamide	
Me$_2$SO	Dimethylsulfoxide	
AcNMe$_2$	Dimethylacetamide	
AcOH	Acetic acid	
H$_4$ furan	Tetrahydrofuran	
OP(NMe$_2$)$_3$	Hexamethylphosphatriamide	
F$_6$ acetone	Hexafluoroacetone	
Et$_2$PO$_3$H	Diethylphosphite	
DCC	Dicyclohexylcarbodiimide	C(NcHx)$_2$
MePdn	1-Me-Pyrrolidone	
Et$_3$N	Triethylamine	
MeMorph	*N*-Methylmopholine	
OMeSO$_2$ Ph	4-(Methylsulfonyl) phenyl ester	
OTcp(2,4,5)	2,4,5-Trichlorophenyl ester	OCl$_3$ ph(2,4,5)
OTcp(2,4,6)	2,4,6-Trichlorophenyl ester	OCl$_3$ ph(2,4,6)
OPcp	Pentachlorophenyl ester	OCl$_5$ ph
ONp	*p*-Nitrophenyl ester	ONph
ODnp(2,4)	2,4-Dinitrophenyl ester	ON$_2$ ph(2,4)
ODnp(2,5)	2,5-Dinitrophenyl ester	ON$_2$ ph(2,5)
SucNBr	*N*-Bromosuccinimide	
SPh	Thiophenyl ester	
EEDQ	*N*-EtOCO-2-EtO-1,2-N$_2$-quinoline	EE-H$_2$ Qnl
Pr^i	Isopropyl	
Dnp	Dinitrophenyl	
F$_3$Ac	Trifluoroacetyl	
OPfp	Pentafluorophenyl ester	OF$_5$ ph
NA	Not applicable	

Preceding the table is an index of the sequential polypeptides listed in alphabetical order with entry numbers corresponding to where the polymers are found in the table. To minimize duplication, the side chain blocking groups and the optical conformation of the amino acids are not included in the index. The table also contains polydepsipeptides and sequential polypeptides where the polyamide backbone is not made up of α-amino α-carboxyl links. These are indexed in alphabetical order at the end of the Index and are appropriately titled. The sequential polypeptides synthesized each year are listed in the annual publication *Amino Acids, Peptides and Proteins* (e.g., Reference 4).

The author wishes to apologize to the research workers in this area for any misrepresentation he has made of their work or failure to list polymers they have prepared. I would welcome receiving corrections and any omissions to the table.

Acknowledgment:

The author is in receipt of support from the Medical Research Council of Canada (MA 4831). I would like to thank the many individuals who have sent me reprints and preprints of their research and Ms. Judy Stearns for the mammoth task of typing the Table and Index.

INDEX TO SEQUENTIAL POLYPEPTIDES

I. α-Amino α-Carboxyl Peptide Backbone

INDEX TO SEQUENTIAL POLYPEPTIDES (continued)

I. α-Amino α-Carboxyl Peptide Backbone

INDEX TO SEQUENTIAL POLYPEPTIDES (continued)

I. α-Amino α-Carboxyl Peptide Backbone

Poly	Entry number
Glu - Ala - Ala -Ala-Ser	526a
- Cys - Cys -Glu-Glu	526b
- Glu -Cys-Glu	527
- Glu - Glu -Cys-Glu	529
-Glu-Glu	537
-Gly	531
-Leu	532–536
-Ser-Glu	528
-Val-Glu	530
Glu - Glu - Glu -Val-Val-Glu	546
- Gly -Glu-Glu-Glu	545
Glu - Ala - Ala -Glu-Ala-Ala-Ala	560
-Lys-Ala-Ala-Ala	559
Gly - Ala	5
- Asp	11
- Glu	12–14
- Gly	3
- Lys	6–10
Gly - Ala - Hyp	127, 128
- Pro	125, 126
- Arg - Arg	148
- Asp - Gly	149, 150
- Azet - Ala	206
- Glu - Gly	151
- Lys	152
- Pro	153
- Gly - Ala	112a
- Gly - Glu	123
- Lys	121, 122
- Phe	113–120
- Pro	124
- His - Gly	138
- Hyp - Gly	198
- Hyp	200–205
- Pro	199
- Lys - Gly	142, 143
- Lys	144–147
- Orn - Gly	139
- Orn	140, 141
-Pipec- Ala	207
- Pro - Ala	171–176
- Glu	187
- Gly	154–170
- Hyp	191–197
- Leu	180–182
- Lys	184–186
- Pro	188–190
- Ser	177–179
- Tyr	183
- Ser - Ala	131
- Gly	129, 130
- Hyp	137
- Pro	132–136
Gly - Gly - Gly -Gly	443
- Hyp -Gly	446, 447
- Lys -Lys	444
- Pro -Gly	445

INDEX TO SEQUENTIAL POLYPEPTIDES (continued)

I. α-Amino α-Carboxyl Peptide Backbone

INDEX TO SEQUENTIAL POLYPEPTIDES (continued)

I. α-Amino α-Carboxyl Peptide Backbone

Poly	Entry number
Pipec- Ala - Gly	442
Pro - Ala	110
- Gly	103–109
Pro - Ala - Gly	392–423
- Glu - Gly	435, 436
- Gly - Ala	386
- Gly	384, 385
- Pro	390, 391
- Ser	387–389
- His - Gly	434
- Hyp - Gly	438, 439
- Leu - Gly	432
- Phe - Gly	433
- Pro - Gly	437
- Ser - Gly	425–431
Pro - Ala - Pro -Ala	522
- Gly - Pro -Gly	521
Pro - Ala - Gly -Pro-Ala-Gly	547, 548
-Pro-Gly	549
- Glu - Gly -Pro-Ala-Gly	551–554
-Glu-Gly	555
- Lys - Gly -Pro-Ala-Gly	550
Sar - Gly	102
Sar - Ala - Gly	383
Ser - Glu	30
- Gly	28, 29
Ser - Gly - Pro	262–265
- Pro - Gly	266–269
Trp - Glu - Ala -Gly	477
Tyr - Glu	41–43
Tyr - Ala - Glu	295, 296
- Gly	294
Tyr - Asp - Ala -Gly	466
- Glu - Ala -Gly	468–470, 472, 473
- Gly -Gly	467
- Tyr -Glu	476
- Val -Gly	475
- Lys - Ala -Gly	465
Val - Ala	35
- Glu	36, 37
- Gly	34
Val - Pro - Gly	275–277
- Val - Ala	274
- Gly	272, 273
Val - Gly - Val -Gly	456

INDEX TO SEQUENTIAL POLYPEPTIDES (continued)

SEQUENTIAL POLYPEPTIDES

		Polymerizing conditions				Product			
Entry No.	Poly	Agent	Concentration (*M*)/solvent	Base/ equivalents	Time/ temperature	Yield (%)	Degree of polymerization/ method	Optical purity (%)	Ref.
1	βAsp	-OPcp	–/$HCONMe_2$	Et_3N/–	–	50[b]	165, 208/Dnp	98.1/H^+	5
2	βAsp	-ONp	–/$HCONMe_2$	Et_3N/–	–	10[b]	185–330/[η], ONp	100/H^+	6
3	Gly-Gly	-OPcp	1/benzene	Et_3N/2.5	–	–	–	–	7
4	Gly-Gc	-ONp	2.7/$HCONMe_2$	Et_3N/1	1–2 days/ room temp.	71	–	NA	8
5	Gly-Ala	-OTcp(2,4,6)	1.65/$HCONMe_2$	Et_3N/1	5 days/ room temp.	48	–	–	9, 10
5a	Gly-Ala	-OTcp(2,4,6)	1.3/$HCONMe_2$	Et_3N/1	–	48	–	–	10a
6	Gly-Lys(HCO)	-OTcp(2,4,5)	–/$HCONMe_2$	Et_3N/–	6 days/ room temp.	–	–	–	11
7	Gly-Lys(Tos)	-OPcp	0.9/Me_2SO	Et_3N/1	6-7 days/20°	–	–	–	12
8	Gly-Lys(Tos)	-OPcp	–	–	–	–	–	–	13
9	Gly-Lys(Tos)	-OTcp(2,4,5)	0.8–1/Me_2So	Et_3N/1	6–7 days/20°	–	–	–	12
10	Gly-Lys(Tos)	-OTcp(2,4,6)	–	–	–	–	–	–	13
11	Gly-Asp(OMe)	-ONp	3/Me_2SO	Et_3N/1.1	5 days/ room temp.	19	–	–	14
12	Gly-Glu	-ONp	>2.5/$HCONMe_2$	Et_3N/1.5	3 days/ room temp.	–	–	–	15
13	Gly-Glu(OBzl)	-OPcp	–	–	–	–	–	–	13
14	Gly-Glu(OBzl)	-OTcp(2,4,6)	–	–	–	–	–	–	13
15	Ala-Gly	-OPcp	1/benzene	Et_3N/2.5	–	60–80	–	–	7
16	Ala-Gly	-OPcp	1.3/benzene[a]	Et_3N/2.5	18 h/ room temp.	50	–	–	16, 17
16a	Ala-Gly	-ONp	1.6/$HCONMe_2$	Et_3N/1	–	50	94/V.S.	–	10a
17	Ala-Gly	-ONp	1.7/$HCONMe_2$	Et_3N/–	5 days/ room temp.	50	94/V.S.	–	9, 10
18	Ala-Gly	-ONp	2.5–5/$HCONMe_2$	Et_3N/1.4	Several days/ room temp.	80	–	–	18

[a]Further solvent was added during the course of the polymerization reaction.
[b]Details of this column or the whole polymerization reaction require further information. See original publication.
[c]Equivalents of DCC added per equivalent of monomer peptide unit.
[d]Yield prior to side-chain deprotection.
[e]For details of molecular weight determination see Mattice, W. L. and Mandelkern, L., *Biochemistry,* 10, 1934 (1971).
[f]Subscripts w and n refer to weight average and number average, respectively

SEQUENTIAL POLYPEPTIDES (continued)

Entry No.	Poly	Polymerizing conditions: Agent	Concentration (*M*)/solvent	Base/ equivalents	Time/ temperature	Product: Yield (%)	Degree of polymerization/ method	Optical purity (%)	Ref.
19	Ala-Gly	-ONp	>2.5/$HCONMe_2$	Et_3N/1.5	3 days/ room temp.	–	94/L.S.	–	19
20	Ala-Gly	-OTcp(2,4,6)	1.7/$HCONMe_2$	Et_3N/1	5 days / room temp.	60	–	–	9, 10
20a	Ala-Gly	-OTcp(2,4,6)	1.4/$HCONMe_2$	Et_3N/1	–	60	–	–	10a
21	Ala-Gly	-OTcp(2,4,6)	1.7/Me_2SO	Et_3N/–	5 days/ room temp.	55	–	–	9, 10
21a	Ala-Gly	-OTcp(2,4,6)	1.4/Me_2SO	Et_3N/10	–	55	–	–	10a
22	Ala-Gly	-OPfp	–/Me_2SO	Et_3N/–	3 days/20°	30–40	94/–	–	20
23	Ala-Gly	Solid phase	NA	NA	NA	NA	5, 10, 20	–	21
24	Ala-Gly	DCC	–	–	–	–	–	–	22
25	Ala-Arg	-OTcp(2,4,5)	0.7/Me_2SO	Et_3N/0.9	5 days/ room temp.	97	5/V.S., 5/IR	–	23
26	DAla-Ala	-ONp	2.5–5/$HCONMe_2$	Et_3N/1.4	Several days/ room temp.	78	–	–	18
27	βAla-Phe	-OPhOH	0.51/Me_2SO	Et_3N/1.5	2.5 days/ room temp.	39	16/Dnp	–	24
28	Ser-Gly	-ONp	>2.5/$HCONMe_2$	Et_3N/1.5	3 days/ room temp.	–	121/L.S.	–	19
29	Ser-Gly	-ONp	–/Me_2SO	Et_3N/–	–	–	12.5/Dnp	100/$[\alpha]_D$	25
30	Ser(Ac)-Glu(OMe)	-ONp	2.1/$HCONMe_2$	Et_3N/1	2–3 days/ room temp.	61	7/ONp	–	26
31	Cys(Bzl)-Gly	-ONp	–/$HCONMe_2$	Et_3N/1	–	65	–	–	27
32	Cys(Bzl)-Cys(Bzl)	-ONp	–/$HCONMe_2$	Et_3N/1	–	44	–	–	28
33	Cys(Bzl)-Glu(OEt)	-ONp	2.5–5/$HCONMe_2$	Et_3N/1.4	Several days/ room temp.	68	–	–	18, 28
34	Val-Gly	-ONSu	0.2/$HCONMe_2$	Et_3N/1.2	5 days/ room temp.	41	–	100/$[\alpha]^b$	29
35	Val-Ala	-ONSu	0.3/$HCONMe_2$	Et_3N/1.2	3 days/ room temp.	92	–	100/$[\alpha]^b$	29
36	Val-Glu(OMe)	-ONp	2.5–5/$HCONMe_2$	Et_3N/1.4	Several days/ room temp.	86	–	–	18, 30
37	Val-Glu(OEt)	-ONp	–/$HCONMe_2$	Et_3N/1	–	75	–	–	27
38	Leu-O-α-OH-4-Me-pentanoyl	-ONSu	2.5/$CHCl_3$	Et_3N/1	6 days/ room temp.	–	Cyclic monomer only (44%)	–	31

SEQUENTIAL POLYPEPTIDES (continued)

		Polymerizing conditions				Product			
Entry No.	Poly	Agent	Concentration (*M*)/solvent	Base/ equivalents	Time/ temperature	Yield (%)	Degree of polymerization/ method	Optical purity (%)	Ref.
39	Leu-O-α-OH-4-Me-pentanoyl	DCC	0.5/acetonitrile	Et_3N/1	12 h/ room temp.	37[b]	Cyclic only	–	31
40	Phe-Gly	-ONp	–	Et_3N/1	–	–	–	–	32
41	Tyr-Glu	-OPcp	2/$AcNMe_2$	Et_3N/–	7 days/ room temp.[b]	77	9/T-$HC1O_4$	–	33
42	Tyr-Glu(OBzl)	-OTcp(2,4,5)	0.9/Me_2SO	Et_3N/2.5	12 days/20°	35	71/–	–	34
42a	Tyr-Glu	-OTcp(2,4,5)	1/Me_2SO	Et_3N/1	–	89	9–10/–	–	34
43	Tyr(Me)-Glu(OEt)	-ONp	–/$HCONMe_2$	Et_3N/1	–	84	–	–	27
44	Lys(Tos)-Gly	-OPcp	–	–	–	–	–	–	13
45	Lys(Tos)-Gly	-OTcp(2,4,6)	–	–	–	–	–	–	13
46	Lys(Cbz)-Gly	-ONp	1/$HCONMe_2$	Et_3N/1.4	24 h/ room temp.	97	–	–	35
46a	Lys-Ala	-OPcp	0.8–0.9/$HCONMe_2$	Et_3N/1	10 days/–	30–60	181/–	–	35a
47	Lys(Cbz)-Glu (OBzl)	-ONp	1/$HCONMe_2$	Et_3N/1.4	24 h/ room temp.	93	–	–	35
47a	Lys-Pro	-OTcp	0.8–1.0/$HCONMe_2$	Et_3N/1	10 days/–	30–60	7, 26/–	–	35a
48	Asp(βAsp-OMe)-OMe	DCC	–/$HCONMe_2$	–	–	–	Low	–	36
49	Asp-Glu	-ONSu	1.3/dioxane[a]	Et_3N/1	6 days/ room temp.	23	57_w and $42^f{}_n$/ G-150	2.1/H^+	37
50	Asp(OBzl)-Glu (OBzl)	-ONp	1/$HCONMe_2$	Et_3N/1.4	24 h/ room temp.	57	–	–	35
51	Glu-Gly	-ONp	0.5/Me_2SO	Et_3N/1.75	20 h/ room temp.	78	457/Dnp, 27/Vel; 258/Dnp, 54/Vel.	98/H^+	38
52	Glu(OMe)-Gly	-OPcp	–	–	–	–	–	–	13
53	Glu(OMe)-Gly	-OTcp(2,4,6)	–	–	–	–	–	–	13
54	Glu(OEt)-Gly	-ONp	>2.5/$HCONMe_2$	Et_3N/1.5	3 days/ room temp.	Low	–	–	19
55	Glu(OBzl)-Gly	-ONp	1/$HCONMe_2$	Et_3N/1.4	24 h/ room temp.	90	–	–	35
56	Glu(OBzl)-Gc	-ONp	1.7/$HCONMe_2$	Et_3N/1	–	–	16% of cyclic monomer	–	8
57	Glu(Gly)-OH	-OPcp	1/$HCONMe_2$	Et_3N/2	3 days/ room temp.	27	15/Eq.	–	39
58	Glu(Gly)-OH	-OPcp	1/$HCONMe_2$	Et_3N/2.55	3 days/ room temp.	32[d]	15/Eq.	100	39

SEQUENTIAL POLYPEPTIDES (continued)

Entry No.	Poly	Agent	Concentration (*M*)/solvent	Base/ equivalents	Time/ temperature	Yield (%)	Degree of polymerization/ method	Optical purity (%)	Ref.
		Polymerizing conditions				Product			
59	Glu(Gly)-OH	-OPcp	1/Me_2SO	Et_3N/2	3 days/ room temp.	47	10/Eq.	–	39
60	Glu(Gly)-OH	-OPcp	1/F_6 acetone	Et_3N/2	3 days/ room temp.	26	5/Eq.	–	39
61	Glu(Gly)-OH	-OPcp	1/F_6 acetone	Et_3N/4	3 days/ room temp.	20	7/Eq.	–	39
62	Glu(Gly)-OH	-OPcp	1/8 *M* urea	Et_3N/1.5	3 days/ room temp.	71	~8/[η]	–	39
63	Glu(Gly)-OH	-OPcp	1/MeOH	Et_3N/1.5	1 day/ room temp.	25	~8/[η]	–	39
64	Glu(Gly)-OH	-OPcp	1/$HCONMe_2$	Et_3N/1.5	3 days/ room temp.	52	10/Eq.	–	39
65	Glu(Gly)-OH	-OPcp	1/$HCONMe_2$	Et_3N/2.5	3 days/ room temp.	35	62/Eq.	–	39
66	Glu(Gly)-OH	-OPcp	1/$HCONMe_2$	Et_3N/3.0	3 days/ room temp.	42	11/Eq.	–	39
67	Glu(Gly)-OH	DCC/1.5	0.18/H_2O/AcOH	N.A.	5 days/ room temp.	33[d]	6/Eq.	100	39
68	Glu(Gly)-OH	M.A.	0.28/$HCONMe_2$	Et_3N/1	4 days/ room temp.	55[d]	6/Eq.	–	39
69	DGlu(Gly)-OH	-OPcp	1/$HCONMe_2$	Et_3N/2.55	3 days/ room temp.	41[d]	62/Eq.	100	39
70	DGlu(Gly)-OH	DCC/1.5	0.18/H_2O/AcOH	NA	5 days/ room temp.	46[d]	7/Eq.	100	39
71	DGlu(Gly)-OH	M.A.	0.28/$HCONMe_2$	Et_3N/1	4 days/ room temp.	51[d]	5/Eq.	–	39
72	Glu(βAla)-OH	-OPcp	1.2/$HCONMe_2$	Et_3N/2.2	Overnight/ room temp.	88	50/Vel.	98/H^+	40
73	DGlu(βAla)-OH	-OPcp	1.2/$HCONMe_2$	Et_3N/2.2	Overnight/ room temp.	71.5	<50/[η]	–	40
74	Glu(γAbu)-OH	-OPcp	1/$HCONMe_2$	Et_3N/2	2 days/ room temp.	45	58/Eq.	100/H^+	41
75	DGlu(γAbu)-OH	-OPcp	1/$HCONMe_2$	Et_3N/2	2 days/ room temp.	41	57/Eq.	100/H^+	41
76	DGlu-Leu	-ONSu	0.3/petroleum[a]	Et_3N/1	4 days/ room temp.[b]	51[b]	–	–	42

SEQUENTIAL POLYPEPTIDES (continued)

		Polymerizing conditions				Product			
Entry No.	Poly	Agent	Concentration (*M*)/solvent	Base/ equivalents	Time/ temperature	Yield (%)	Degree of polymerization/ method	Optical purity (%)	Ref.
77	DGlu-Leu	-ONSu	1.2/benzene	Et_3N/1	4 days/ room temp.	47	–	–	42
78	DGlu-Leu	-ONSu	2.0/H_4 furan	Et_3N/1	8 days/ room temp.[a]	50	40/G-150	97.3/E[b]	42
79	DGlu-Leu	-ONSu	1.2/$HCONMe_2$	Et_3N/1	4 days/ room temp.	41	–	–	42
80	Glu(OBzl)-Tyr(Ac)	-OPcp	1/$HCONMe_2$	Et_3N/1.6	6 days/ room temp.	20	–	–	43
81	Glu(OBzl)-Lys(Cbz)	-ONp	1/$HCONMe_2$	Et_3N/1.4	24 h/ room temp.	78	–	–	35
82	Glu(βAsp-OH)-OH	-OPcp	0.8/$HCONMe_2$	Et_3N/2.5	1 day/ room temp.	43	115/Eq.	–	44
83	Glu(βAsp-OH)-OH	-OPcp	0.8/$HCONMe_2$	Et_3N/3.1	3 days/ room temp.	50	86/Eq.	–	44
84	Glu(βAsp-OH)-OH	-OPcp	0.8/$HCONMe_2$	MeMorph/2.2	3 days/ room temp.	64	29/Eq.	–	44
85	DGlu(DβAsp-OH)-OH	-OPcp	0.8/$HCONMe_2$	Et_3N/2.5	1 day/ room temp.	51	39/Eq.	–	44
86	Glu-Glu	DCC/91[c]	0.5/acetonitrile	NA	24 h/ room temp.	47	114/G-75 or G-100	87/H^+	45
87	Glu-Glu	DCC	5/$HCONMe_2$	NA	–	Low	8/G-75	–	45
88	DGlu-Glu	DCC/91	0.5/acetonitrile	NA	24 h/ room temp.	74	38/G-75	94/H^+	45
89	Glu-DGlu	-ONSu	1.5/$CHCl_3$	Et_3N/1	6 days/ room temp.[a]	42	53/G-150; 48/π	98/[α]	46
90	Glu-DGlu	DCC/91	0.5/acetonitrile	NA	24 h/ room temp.	69	31/G-75	94/H^+	45
91	Glu(OMe)-Glu (OEt)	-ONp	–/$HCONMe_2$	Et_3N/1	–	68	–	–	27
92	Glu(OBzl)-Glu (OBzl)	-ONSu	–/$HCONMe_2$	Et_3N/–	–	–	27/[η]	100/[α][b]	47
93	Glu(OBzl)-Glu (OBzl)	DCC/HONSu	–/$HCONMe_2$	NA	–	~5	7/[η]	100/[α]	47
94	DGlu(OBzl)-Glu (OBzl)	-OPcp	1.8/benzene[a]	Et_3N/1.9	2 days/ room temp.	45–50	36–52/[η]	96–97.5/ [α]	48

SEQUENTIAL POLYPEPTIDES (continued)

	Polymerizing conditions					Product			
Entry No.	Poly	Agent	Concentration (*M*)/solvent	Base/ equivalents	Time/ temperature	Yield (%)	Degree of polymerization/ method	Optical purity (%)	Ref.
95	DGlu(OBzl)-Glu (OBzl)	-ONp	1.8/$HCONMe_2$	Et_3N/2.2	2 days/ room temp.	40–48	27–34/[η]	99/[α]	48
96	DGlu(OBzl)-Glu (OBzl)	-ONp	1.8/Me_2SO	Et_3N/2.2	2 days/ room temp.	10	–	98.2/[α]	48
97	Glu(ONbzl)-Glu (OBzl)	-ONp	1/$HCONMe_2$	Et_3N/1.4	24 h/ room temp.	83	–	–	35
98	Glu(OIbzl)-Glu (OBzl)	-ONp	1/$HCONMe_2$	Et_3N/1.4	24 h/ room temp.	88	–	–	35
99	Glu(γGlu-OH)-OH	-OPcp	1/$HCONMe_2$	Et_3N/2.5	3 days/ room temp.	57	35/Eq., 62/Eq.	100/H^+	49, 50
100	Glu(γGlu-OH)-OH	-SPh	–	Et_3N/–	–	–	–	–	51
101	DGlu(DγGlu-OH) -OH	-OPcp	1/$HCONMe_2$	Et_3N/2.5	3 days/ room temp.	51	30/Eq.	100/H^+	49
102	Sar-Gly	-ONp	1/$HCONMe_2$	Et_3N/1	7 days/ room temp.	67	–	–	52
103	Pro-Gly	-ONp	2/Me_2SO	Et_3N/1.2	3 days/ room temp.	24	65/Vel., 50 and 92/Eq., 20 and 34/VP, 78 and 130/Dnp	–	53
104	Pro-Gly	-ONp	–	–	–	–	104_n,171^f_w/–[e]	–	54
105	Pro-Gly	-ONp	–	–	–	–	–	–	22
106	Pro-Gly	-ODnp(2,4)	–	–	–	–	–	–	22
107	Pro-Gly	-ODnp(2,5)	–	–	–	–	–	–	22
108	Pro-Gly	$Et_4P_2O_7$	–	–	–	–	Low[b]	–	22
109	Pro-Gly	-$NHNH_2$/I_2	–	–	–	–	–	–	22
110	Pro-Ala	-ONp	–	–	–	–	18/–	–	55
111	Hyp-Gly	-ONp	1.4/Me_2SO	Et_3N/1.2	7 days/ room temp.	7	58 and 75/Vel., 49, 55 and 75/Eq., 22, 31, and 27/VP, 47, 82/Dnp	–	53
112	Hyp-Gly	-ONp	–	–	–	–	102_n, 120^f_w/–[e]	–	54
112a	Gly_2-Ala	-OTcp(2,4,5)	1.1/$HCONMe_2$	Et_3N/1–2	7 days/ room temp.	–	59/V.S., IR	–	55a
113	Gly_2-Phe	-OPcp	–/$HCONMe_2$	Et_3N/2.5	–	–	77–96/–	92–95/ $[\alpha]_{223}$	56

SEQUENTIAL POLYPEPTIDES (continued)

Entry No.	Poly	Agent (Polymerizing conditions)	Concentration (*M*)/solvent	Base/ equivalents	Time/ temperature	Yield (%) (Product)	Degree of polymerization/ method	Optical purity (%)	Ref.
114	Gly_2-Phe	-OPcp	1.2/Me_2SO	Et_3N/2	16 h/ room temp.	27b	–	–	57
115	Gly_2-Phe	-ONp	–/Me_2SO	NaONp/–	–	–	–	100/H^+	58
116	Gly_2-Phe	-OPhOH	1/Me_2SO	Et_3N/2	2 days/ room temp.	38	19/Dnp	100/H^+	24
117	Gly_2-Phe	-OPhOH	0.7/Me_2SO	Et_3N/1.5	2 days/ room temp.	45	19/Dnp	100/H^+	24
118	Gly_2-Phe	-$OEtCl_2$ $SalNN_2$	–/$HCONMe_2$	Et_3N/2.5	–	–	54/–	92–95/ $[\alpha]_{223}$	56
119	Gly_2-DPhe	-ONp	–/$HCONMe_2$	Et_3N/2.5	–	–	115/–	95/$[\alpha]_{223}$	56
120	Gly_2-DPhe	-ONp	–/Me_2SO	NaONp/–	–	–	–	96/H^+	56
121	Gly_2-Lys	-OTcp(2,4,5)	–/$HCONMe_2$	Et_3N/–	6 days/ room temp.	–	8–9/–	–	11
122	Gly_2-Lys(Tos)	-ONp	–/Me_2SO	Et_3N/1	5 days/20°	–	31/–	–	59
122a	Gly_2-Lys(Tos)	-OTcp(2,4,5)	0.7/$HCONMe_2$	Et_3N/1–2	7 days/ room temp.	–	101/V.S., IR	–	55a
123	Gly_2-Glu	-ONp	>2.5/$HCONMe_2$	Et_3N/1.5	3 days/ room temp.	–	–	–	15
123a	Gly_2-Glu(OMe)	-OTcp(2,4,5)	0.9/$HCONMe_2$	Et_3N/1–2	7 days/ room temp.	–	12/V.S., IR	–	55a
124	Gly_2-Pro	$Et_4P_2O_7/Et_2PO_3H$(1/1)	1.1	NA	3 h/60–100°b	–	24/V.S., 20/N	–	60
125	Gly-Ala-Pro	$Et_4P_2O_7/Et_2PO_3H$(1/1)	1	NA	3 h/60–100°b	70.2	44/N	–	60
126	Gly-Ala-Pro	$Et_4P_2O_7$	–	–	–	–	–	–	61
127	Gly-Ala-Hyp	-OPcp	0.9/Me_2SO	Et_3N/1.1	Several days/ room temp.	26	83/V.S. and 104/IR	–	62, 63
128	Gly-Ala-Hyp	$Et_4P_2O_2/Et_2PO_3H$(1/1)		NA	3 h/60–100°b	65.4	56/V.S., 28/Vel.	–	60
129	Gly-Ser-Gly	–OPcp	1.2/$HCONMe_2$	Et_3N/1.5	7 days/ room temp.	50	12–15/–	–	10, 64
129a	Gly-Ser-Gly	-OPcp	1.2/$HCONMe_2$	Et_3N/1.6	7 days/ room temp.	100	15 V.S.	–	10a
130	Gly-Ser-Gly	-ONp	–/$HCONMe_2$ Me_2SO	NaONpb/1 Et_3N	–	–	16–23/Dnp	–	25

SEQUENTIAL POLYPEPTIDES (continued)

		Polymerizing conditions				Product			
Entry No.	Poly	Agent	Concentration (*M*)/solvent	Base/ equivalents	Time/ temperature	Yield (%)	Degree of polymerization/ method	Optical purity (%)	Ref.
131	Gly-Ser-Ala	-$NHNH_2/I_2$ [b]	0.3/$AcNMe_2$	Et_3N/7	10 min/ room temp.	50	–	–	65
132	Gly-Ser-Pro	-OPcp	1.7/Me_2SO[b]	Et_3N/–	7 days/ room temp.	Low	10/V.S.	–	66, 67
133	Gly-Ser-Pro	$Et_4P_2O_7$/2	0.1/Et_2PO_3H	NA	72 h/ room temp.	45	12/G-25, G-50	–	68, 69
134	Gly-Ser-Pro	$Et_4P_2O_7$/2	0.26/pyridine[b]	NA	72 h/ room temp.	53	12/G-25, G-50	–	68, 69
135	Gly-Ser-Pro	$Et_4P_2O_7$/2	0.1/pyridine + Et_2PO_3H[b]	NA	72 h/ room temp.	50	12/G-25, G-50	–	68, 69
136	Gly-Ser-Pro	$(PhO_2P)_2O$/5	0.1/Et_2PO_3H[b]	NA	120 h/ room temp.	17	5/G-25, G-50	–	68, 69
137	Gly-Ser-Hyp	-OPcp	1.2/Me_2SO	Et_3N/1.5	5 days/ room temp.	25	22–23/V.S.	–	66, 67
138	Gly-His-Gly	-OTcp(2,4,5)	–/$HCONMe_2$ Me_2SO	Et_3N/–	5–7 days/ room temp.	–	–	–	70
139	Gly-Orn-Gly	DCC	–/Me_2SO	NA	2 days/0°	55	15, 47/Vel.[b]	–	71
140	Gly-Orn-Orn	OTcp(2,4,6)	–/$HCONMe_2$	Et_3N/–[b]	1 day/ room temp.	25	7/Vel.[b]	–	71
141	Gly-Orn-Orn	-$NHNH_2/I_2$	–/$HCONMe_2$	Et_3N/–	2 h/0°	22	5/Vel.[b]	–	71
142	Gly-Lys(Tos)-Gly	-OPcp	–/Me_2SO-$HCONMe_2$	Et_3N/–	–	–	4–13/V.S., 4/IR	–	72
143	Gly-Lys(Tos)-Gly	-OTcp(2,4,5)	–/Me_2SO-$HCONMe_2$	Et_3N/–	–	–	5–28/V.S.	–	72
144	Gly-Lys(Tos)-Lys (Tos)	-OPcp	0.6/Me_2SO	–	5 days/ room temp.	–	34/V.S., 43/IR	–	73
145	Gly-Lys-(Tos)-Lys (Tos)	-ONp	–/Me_2SO	Et_3N/1	5 days/20°C	–	5/–	–	59
146	Gly-Lys(Tos)-Lys (Tos)	-OTcp(2,4,5)	0.7/Me_2SO	–	5 days/ room temp.	–	11/V.S., 9/IR	–	73
147	Gly-Lys(Tos)-Lys (Tos)	-OTcp(2,4,6)	–/Me_2SO	Et_3N/–	4–5 days/ room temp.	–	11–16/–	–	74
148	Gly-Arg-Arg	-OTcp(2,4,5)	1.1/Me_2SO	Et_3N/1	5 days/ room temp.	64	14/V.S., 13/IR	–	23

SEQUENTIAL POLYPEPTIDES (continued)

Entry No.	Poly	Polymerizing conditions: Agent	Concentration (*M*)/solvent	Base/ equivalents	Time/ temperature	Product: Yield (%)	Degree of polymerization/ method	Optical purity (%)	Ref.
149	Gly-Asp(OMe)-Gly[b]	-ONp	1/Me_2SO	Et_3N/2.9	2 days/ room temp.	2[b]	–	–	14
150	Gly-Asp-Gly[b]	-ONp	1/Me_2SO	Et_3N/2.9	2 days/ room temp.	55	24/Dnp	–	14
151	Gly-Glu(OMe)-Gly	-ONp[b]	–/$HCONMe_2$ or Me_2SO	Et_3N/–	144 h/20°	28–35	35/–	–	75
152	Gly-Glu(OBzl)-Lys(Tos)	-OPcp	–/Me_2SO	Et_3N/–	–	–	6/–	–	76
153	Gly-Glu(OMe)-Pro	-OTcp(2,4,5)	1.6/Me_2SO	Et_3N/1.1	7 days/ room temp.	44	7–8/V.S.	–	77
154	Gly-Pro-Gly	-OPcp	1/Me_2SO	Et_3N/1	5 days/ room temp.	–	71/V.S.	–	62, 78
155	Gly-Pro-Gly	-OPcp	1/Me_2SO	Et_3N/1	–	50	71/V.S.	–	63
156	Gly-Pro-Gly	-ONSu	1.3/Me_2SO	Et_3N/1.3	–	70	53/V.S.	–	63, 79
157	Gly-Pro-Gly	-ONp	–/$HCONMe_2$	Et_3N/–	10 days/ room temp.	–	24/V.S.	–	62, 80
158	Gly-Pro-Gly	-ONp	1.2/Me_2SO	Et_3N/1	Overnight/ room temp.	31	28/Vel.	–	81
159	Gly-Pro-Gly	-ONp	1.5/Me_2SO	Et_3N/0.8	–	21	24/V.S.	–	63
160	Gly-Pro-Gly	-ONp	–/Me_2SO	Et_3N, MeMorph/–	–	–	47/[η][b]	–	82
161	Gly-Pro-Gly	-OTcp(2,4,6)	–/$HCONMe_2$	Et_3N/–	5 days/ room temp.	50	4/V.S.	–	62, 80
162	Gly-Pro-Gly	-OTcp(2,4,6)	0.9/Me_2SO	Et_3N/1.1	5 days/20°	42[b]	38/V.S.	–	63
163	Gly-Pro-Gly	-OTcp(2,4,6)	1/Me_2SO	Et_3N/1	5 days/ room temp.	–	38/V.S.	–	62, 78
164	Gly-Pro-Gly	-OTcp(2,4,5)	1/Me_2SO	Et_3N/1	5 days/ room temp.	53	43/V.S.	–	62, 78
165	Gly-Pro-Gly	-OTcp(2,4,5)	1/Me_2SO	Et_3N/1	–	11[b]	43/V.S.	–	63
166	Gly-Pro-Gly	-OPy	1.1/Me_2SO	Et_3N/2.1	–	76	20/V.S.	–	63
167	Gly-Pro-Gly	-O(Bu^iNO_2 MePy Py)	1/Me_2SO	Et_3N/–	5 days/ room temp.	–	14/–	–	83
168	Gly-Pro-Gly	-OQu	–/Me_2SO	Et_2N/2	7 days/20°	–	15/–	–	84, 85
169	Gly-Pro-Gly	-OQu	1/Me_2SO	Et_3N/2.1	–	20	14/V.S.	–	63
170	Gly-Pro-Gly	-OPfp	1.1/Me_2SO	Et_3N/1	–	14	8/V.S.	–	63

SEQUENTIAL POLYPEPTIDES (continued)

Entry No.	Poly	Polymerizing conditions: Agent	Concentration (*M*)/solvent	Base/ equivalents	Time/ temperature	Product: Yield (%)	Degree of polymerization/ method	Optical purity (%)	Ref.
171	Gly-Pro-Ala	-ONSu	1.2/Me_2SO	Et_3N/1	–	31/ 14.8[b]	89/IR/67/IR; 71/V.S.	–	79
172	Gly-Pro-Ala	-ONp	1.1/Me_2SO	Et_3N/1.2	Overnight/ room temp.	–	71/Vel.	–	81
173	Gly-Pro-Ala	-OTcp(2,4,6)	–/$HCONMe_2$	Et_3N/–	5 days/ room temp.	23	19/Dnp	–	80
174	Gly-Pro-Ala	-OPhOH	2/Me_2SO	Et_3N/1.5	2 days/ room temp.	45–66[b]	53/Vel.	–	86
175	Gly-Pro-Ala	$Et_4P_2O_7$	–	–	–	–	–	–	61
176	Gly-Pro-Ala	$Et_4P_2O_7$/Et_2PO_3H(1/1)	1	NA	3 h/60–100°[b]	–	19/V.S.	–	60
177	Gly-Pro-Ser	$Et_4P_2O_7$/2	0.1/Et_2PO_3H	NA	72 h/ room temp.	48	11/G-25	–	68
178	Gly-Pro-Ser	$Et_4P_2O_7$/2	0.2/pyridine; $OP(NMe_2)_3$	NA	72 h/ room temp.	45	12/G-25	–	68
179	Gly-Pro-Ser	$(PhO_2P)_2O$/5	0.1/Et_2PO_3H	NA	120 h/ room temp.	11	3/G-15	–	68
180	Gly-Pro-Leu	-ONSu	1.1/Me_2SO	Et_3N/1	5 days/ room temp.	81[b]	22–15/IR	–	79
181	Gly-Pro-Leu	$Et_4P_2O_7$/Et_2PO_3H	0.9	NA	3 h/60–100°[b]	–	3/V.S., 4/N	–	60
182	Gly-Pro-Leu	$Et_4P_2O_7$/1.2	1.5/Et_2PO_3H	NA	2 h/100°	–	–/NaOMe	–	87
183	Gly-Pro-Tyr	$Et_4P_2O_7$/Et_2PO_3H	0.7	NA	3 h/60–100°[b]	–	11/V.S.	–	60
184	Gly-Pro-Lys	$Et_4P_2O_7$	1/Et_2PO_3H	NA	2 h/100°[b]	–	–	–	60
185	Gly-Pro-Lys (Tos)	-ONp	–/Me_2SO	Et_3N/–	4–5 days/ room temp.	–	16–23/–	–	74
186	Gly-Pro-Lys (Tos)	-OTcp(2,4,6)	–/Me_2SO	Et_3N/–	4–5 days/ room temp.	–	16–23/–	–	74
186a	Gly-Pro-Lys (Tos)	$Et_4P_2O_7$/Et_2PO_3H	0.6	NA	3 h/60–100°[b]	–	23/Vel.	–	60
187	Gly-Pro-Glu (OBzl)	-ONp	1/$HCONMe_2$	Et_3N/1	4 days/20°	30	6/–	–	88

SEQUENTIAL POLYPEPTIDES (continued)

		Polymerizing conditions				Product			
Entry No.	Poly	Agent	Concentration (*M*)/solvent	Base/ equivalents	Time/ temperature	Yield (%)	Degree of polymerization/ method	Optical purity (%)	Ref.
188	Gly-Pro-Pro	-OPcp	1.3/Me_2SO	Et_3N/1	5 days/ room temp.	–	17/Vel. and 38–40/Vel.	–	89
189	Gly-Pro-Pro	-ONp	–/Me_2SO	Et_3N, MeMorph/–	–	–	60/[η][b]	–	82
190	Gly-Pro-Pro	$Et_4P_2O_7$/Et_2PO_3H	0.9	NA	3 h/60–100°[b]	75.3	47/V.S., 20/Vel.	–	60
191	Gly-Pro-Hyp	-OPcp	1/Me_2SO	Et_3N/1	–	57	374/V.S. and >225/IR	–	63
192	Gly-Pro-Hyp	-OPcp	0.9/Me_2SO	Et_3N/1	5 days/ room temp.	57	–	–	62, 90
193	Gly-Pro-Hyp	-ODnp(2,4)	–	–	–	–	–	–	22
194	Gly-Pro-Hyp	-OTcp(2,4,5)	1/Me_2SO	Et_3N/1	5 days/ room temp.	–	75/V.S.	–	62, 78
195	Gly-Pro-Hyp	-OTcp(2,4,5)	1/Me_2SO	Et_3N/1	–	22	75/V.S., 56/IR	–	63
196	Gly-Pro-Hyp	$Et_4P_2O_7$		NA	2 h/100°	–	6–90/Vel.[b]	–	60
196a	Gly-Pro-Hyp	$Et_4P_2O_7$	–	NA	3 h/40–60°[b]	–	28/V.S.	–	60
196b	Gly-Pro-Hyp	$Et_4P_2O_7$	–	NA	3 h/60–80°[b]	–	52/V.S.	–	60
196c	Gly-Pro-Hyp	$Et_4P_2O_7$	–	NA	3 h/60–100°[b]	–	94/V.S.	–	60
196d	Gly-Pro-Hyp	$Et_4P_2O_7$/Et_2PO_3H(2/1)	0.3	NA	4 h/40–80°[b]	–	25/V.S.	–	60
196e	Gly-Pro-Hyp	$Et_4P_2O_7$/Et_2PO_3H(1/1)	0.5	NA	4 h/40–80°[b]	–	52/V.S.	–	60
196f	Gly-Pro-Hyp	$Et_4P_2O_7$/Et_2PO_3H(1/1)	0.9	NA	3 h/60–100°[b]	57	52/V.S., 56/Vel.	–	60
197	Gly-Pro-Hyp	$Et_4P_2O_7$	–	–	–	–	–	–	22
198	Gly-Hyp-Gly	-OPy	1/Me_2SO	Et_3N/–	5 days/ room temp.	39	18/–	–	83
199	Gly-Hyp-Pro	$Et_4P_2O_7$/Et_2PO_3H(1/1)	0.9	NA	3 h/60–100°[b]	84	30/N, 19/Vel.	–	60
200	Gly-Hyp-Hyp	-OPcp	0.9/Me_2SO	Et_3N/1	5 days/20°[b]	10	558/Vel.	–	63, 78
201	Gly-Hyp-Hyp	-OPcp	0.8/Me_2SO	Et_3N/1.1	5 days/ room temp.	89	558/Vel.	–	91, 92
202	Gly-Hyp-Hyp	-OTcp(2,4,5)	1/Me_2SO	Et_3N/1	5 days/20°	43	57/V.S., 141/Vel.	–	62, 63, 78
203	Gly-Hyp-Hyp	-OTcp(2,4,6)	1.1/Me_2SO	Et_3N/1	–	28	25/V.S.	–	63, 78
204	Gly-Hyp-Hyp	-OPfp	1/Me_2SO	Et_3N/1	–	30	14/V.S.	–	63

SEQUENTIAL POLYPEPTIDES (continued)

Entry No.	Poly	Polymerizing conditions: Agent	Concentration (*M*)/solvent	Base/ equivalents	Time/ temperature	Product: Yield (%)	Degree of polymerization/ method	Optical purity (%)	Ref.
205	Gly-Hyp-Hyp	$Et_4P_2O_7/Et_2PO_3H$(1/1)	0.8	NA	3 h/60–100°[b]	22.3	88/V.S., Vel.	–	60
206	Gly-Azet-Ala	-OPhOH	1.1/Me_2SO	MeMorph/3.3	5 days/ room temp.	27	8/P-150	–	93
207	Gly-Pipec-Ala	-OPhOH	1.2/Me_2SO	MeMorph/3	4 days/20°	15	7/P-150	–	93
208	Ala-Gly_2	-ONSu	2.5/Me_2SO	Et_3N/1	48 h/ room temp.	80	62/$[m]_D^{25}$	–	16
209	Ala-Gly_2	-ONSu	2.2/Me_2SO	Et_3N/1	48 h/ room temp.	50	68/$[m]_D^{25}$; 76/ L.S.; 60/Dnp	–	16
210	Ala-Gly_2	-ONSu	5/H_2O	Et_3N/1	48 h/ room temp.	75	11/$[m]_D^{25}$	–	16
211	Ala-Gly_2	-ONSu	5/H_2O	Et_3N/1	48 h/4°	81	10/$[m]_D^{25}$	–	16
212	Ala-Gly_2	-OPcp	2/benzene	Et_3N/2.5	48 h/ room temp.	93	43/$[m]_D^{25}$	–	16, 17
213	Ala-Gly_2	-OPcp	5/H_2O	Et_3N/2.5	48 h/ room temp.	89	7/$[m]_D^{25}$	–	16
214	Ala-Gly_2	-ONp	2.5/$HCONMe_2$	Et_3N/1.5	3 days/ room temp.	–	69/Dnp, 41/ONp	–	94
215	Ala-Gly_2	-ONp	2.2/$HCONMe_2$	Et_3N/1	48 h/ room temp.	98	3/$[m]_D^{25}$	–	16
216	Ala-Gly_2	-ONp	2.2/Me_2SO	Et_3N/1	48 h/ room temp.	80	8/$[m]_D^{25}$	–	16
217	Ala-Gly_2	-ONp	0.8/Me_2SO	Et_3N/1	1 day/ room temp.	–	162/[η]	–	95
218	Ala-Gly_2	-ONp	2.2/$OP(NMe_2)_3$	Et_3N/1	48 h/ room temp.	93	3/$[m]_D^{25}$	–	16
219	Ala-Gly_2	-ONp	4.9/H_2O	Et_3N/1	48 h/ room temp.	77	5/$[m]_D^{25}$	–	16
220	Ala-Gly_2	-ONp	4.9/60% LiBr	Et_3N/1	48 h/ room temp.	94	3/$[m]_D^{25}$	–	16
221	Ala-Gly_2	CMCI	2/H_2O	NA	48 h/ room temp.	92	2/$[m]_D^{25}$	–	16
222	Ala-Gly_2	CMCI/HONSu	2/H_2O	NA	48 h/ room temp.	70	3/$[m]_D^{25}$	–	16

SEQUENTIAL POLYPEPTIDES (continued)

Entry No.	Poly	Polymerizing conditions: Agent	Concentration (*M*)/solvent	Base/ equivalents	Time/ temperature	Product: Yield (%)	Degree of polymerization/ method	Optical purity (%)	Ref.
223	Ala-Gly_2	CMCI/HONSu	2/H_2O	NA	48 h/ room temp.	61	4/$[m]^{25}_D$	–	16
224	DLAla-Gly_2	-ONp	2.5–5.0/$HCONMe_2$	Et_3N/–	–	87	–	NA	18
225	Ala-Gly-Lys	-OPcp	0.9/$HCONMe_2$	Et_3N/2.5	5 days/20°	64	–	–	96
226	Ala-Gly-Sar	-OPcp	–/Me_2SO	Et_3N, MeMorph/–[b]	–	–	Dialyzable[b]	–	82
227	Ala-Gly-Pro	-OPcp	1.4/Me_2SO	MeMorph/2	3 days/ room temp.	70[b]	40_n and $62^f{}_w$/ P-150	–	97
228	Ala-Gly-Pro	$Et_4P_2O_7$/Et_2PO_3H(1/1)	1	NA	1 h/60° + 2 h/100°	–	55/V.S.	–	60
229	Ala-Gly-Pro	$Et_4P_2O_7$	0.5/Et_2PO_3H	NA	1 h/65° + 1 h/100°	–	53/–	–	80
230	DAla-Gly-Pro	-OPcp	1.4/Me_2SO	MeMorph/2	3 days/ room temp.	30	11_n and $19^f{}_w$/ P-150	–	97
231	DAla-Gly-DPro	-OPcp	1.4/Me_2SO	MeMorph/2	3 days/ room temp.	53	40_n and $56^f{}_w$/ P-150	–	97
232	Ala-Gly-DPro	-OPcp	1.4/Me_2SO	MeMorph/2	4 days/ room temp.	41.5	10_n and $16^f{}_w$/ P-150	–	97
233	Ala-Gly-Thz	-OPcp	1.2/Me_2SO	MeMorph/2	4 days/ room temp.	61.5[b]	35/[η] –	–	93
234	Ala-Gly-Pipec	-OPcp	–/Me_2SO	–	–	–	No polymerization	–	93
235	βAla-Gly-Pro	-ONp	–/Me_2SO	Et_3N, MeMorph/–	–	–	22/[η][b]	–	82
236	Ala_2-Gly	-ONSu	0.3/Me_2SO	Et_3N/2	7 days/ room temp.[b]	92	23 and 11/Vel.	–	98
237	Ala_2-Gly	-OPcp	0.6/benzene[a]	Et_3N/2 5	48 h/ room temp.[b]	81	–	–	16
238	Ala_2-Gly	-ONp	0.8/Me_2SO	Et_3N/1	1 day/ room temp.	–	176/[η]	–	95
239	Ala_2-Gly	–		–	–	–	–	–	99
240	Ala_2-Gc	-ONp	1.6/$HCONMe_2$	Et_3N/1.5	–	68	–	–	8
241	Ala_2-Lys	–	–	–	–	–	–	–	99
242	Ala_2-Lys(Cbz)	-OTcp(2,4,5)	1.06/$HCONMe_2$	Et_3N/1	10 days/ room temp.	57	20/V.S.	–	100

SEQUENTIAL POLYPEPTIDES (continued)

		Polymerizing conditions				Product			
Entry No.	Poly	Agent	Concentration (*M*)/solvent	Base/equivalents	Time/temperature	Yield (%)	Degree of polymerization/method	Optical purity (%)	Ref.
243	Ala_2-Lys(Cbz)	-OTcp(2,4,5)	0.96/$HCONMe_2$	Et_3N/2.5	10 days/room temp.	46	11/V.S.	–	100
244	Ala_2-Arg[b]	-OTcp(2,4,5)	1/Me_2SO	Et_3N/3	5 days/room temp.	83	23/V.S., 21/IR	–	23
245	Ala-Val-Ala	-ONSu	0.5/$HCONMe_2$	Et_3N/1.2	2 days/room temp.	100	–	100/[α][b]	29
246	Ala-Val-Ala	-ONSu	0.3/Me_2SO	Et_3N/1.2	4 days/room temp.	98	–	100/[α][b]	29
247	Ala-Val-Ala	-ONSu	1.0/Me_2SO	Et_3N/1.2	4 days/room temp.	100	–	100/[α][b]	29
248	Ala-Met-Ala	-ONSu	1.3/$HCONMe_2$	Et_3N/1.2	1 day/room temp.	100	–	100/[α][b]	29
249	Ala-Phe-Gly	-OPcp	0.8/$HCONMe_2$	Et_3N/2.2	48 h/room temp.	84	16/Eq.	–	101
249a	Ala-Orn-Gly	-OTcp(2,4,5)	0.6/$HCONMe_2$	Et_3N/–	–	–	91/–	–	101a
249b	Ala-Orn-Ala	-OTcp(2,4,5)	0.6/$HCONMe_2$	Et_3N/2.8	6 days/room temp.	56	101/–	–	101a
249c	Ala-Orn-Glu	-OTcp(2,4,5)	–/$HCONMe_2$	Et_3N/–	–	–	61/–	–	101a
250	Ala-Lys(Cbz)-Ala	-OTcp(2,4,5)	1/$HCONMe_2$	Et_3N/1	12 days/room temp.	47.5	8/V.S.	–	100, 102
251	Ala-Lys(Cbz)-Ala	-OTcp(2,4,5)	1/$HCONMe_2$	Et_3N/2.5	12 days/room temp.	50	10/V.S.	–	100, 102
252	Ala-Arg_2	-OTcp(2,4,5)	0.9/$HCONMe_2$	Et_3N/3	5 days/room temp.	87	11/V.S., 13/IR	–	23
253	Ala-Glu(OEt)-Gly	-ONp	2.5/$HCONMe_2$	Et_3N/1.5	3 days/room temp.	–	55/L.S.; 29/ONp	–	94
254	Ala-Pro-Gly	-ONSu	0.5/Me_2SO[a]	Et_3N/2[b]	5 days/room temp.	81	–	–	103
255	Ala-Pro-Gly	-ONSu	1/Me_2SO	Et_3N/1	3 days/room temp.	62	48/Eq[b]	–	104
256	Ala-Pro-Gly	-ONp	0.5/Me_2SO[a]	Et_3N/2[b]	5 days/room temp.	60	62 and 24/Vel.	–	103
257	Ala-Pro-Gly	-ONp	0.8/Me_2SO[a]	Et_3N/1.2	4 days/room temp.	41	11/Eq.	–	53
258	Ala-Pro-Gly	M.A.	NA	NA	NA	NA	6	NA	103

SEQUENTIAL POLYPEPTIDES (continued)

		Polymerizing conditions				Product			
Entry No.	Poly	Agent	Concentration (*M*)/solvent	Base/ equivalents	Time/ temperature	Yield (%)	Degree of polymerization/ method	Optical purity (%)	Ref.
259	Ala-Pro-Gly	$Et_4P_2O_7$	–	–	–	–	–	–	22
260	Ala-Pro_2	$Et_4P_2O_7/Et_2PO_3H$(1/1)	0.9	NA	3h/60–100°[b]	48.4	36/N	–	60
261	Ala-Hyp_2	$Et_4P_2O_7/Et_2PO_3H$(1/1)	0.8	NA	3 h/60–100°[b]	–	58/N 220/Vel.	–	60
262	Ser-Gly-Pro	$Et_4P_2O_7$/2	0.1/Et_2PO_3H	NA	12 h/ room temp.	31	8/G-50	–	68, 69
263	Ser-Gly-Pro	$Et_4P_2O_7$/2	0.1/pyridine	NA	36 h/ room temp.	37	10/G-50	–	68, 69
264	Ser-Gly-Pro	$Et_4P_2O_7$/2	0.2/Et_2PO_3H: pyridine (1:1)	NA	72 h/ room temp.	40	10/G-50	–	68, 69
265	Ser-Gly-Pro	$(PhO_2P)_2O$/5	0.1/$OP(NMe_2)_3$	NA	240 h/ room temp.	11	4/G-50	–	68, 69
266	Ser-Pro-Gly	-ONp	1.1/Me_2SO	Et_3N/1.4	3 days/ room temp.	61	46/Vel.; 41/Eq.	–	53
267	Ser-Pro-Gly	$Et_4P_2O_7$/2	–/Et_2PO_3H	NA	24 h/ room temp.	49	12/G-50	–	68, 69
268	Ser-Pro-Gly	$Et_4P_2O_7$/2	–/pyridine	NA	72 h/ room temp.	48	16/G-50	–	68, 69
269	Ser-Pro-Gly	$Et_4P_2O_7$/2	–/$OP(NMe_2)_3$: Et_2PO_3H	NA	100 h/ room temp.	54	16/G-50	–	68, 69
270	Cys(Bzl)-Gly_2	-ONp	2.5–5/$HCONMe_2$	Et_3N/1.4	Several days/ room temp.	40	–	–	18
271	Cys(Bzl)-[Glu (OEt)]$_2$	-ONp	–/$HCONMe_2$	Et_3N/1	–	92	90/ONp	–	27, 28
272	Val_2-Gly	-ONSu	0.5/$HCONMe_2$	Et_3N/1.2	5 days/ room temp.	100	–	100/[α][b]	29
273	Val_2-Gly	-ONSu	0.2/$HCONMe_2$	Et_3N/1.2	5 days/ room temp.	91	–	100/[α][b]	29
274	Val-Val-Ala	-ONSu	0.9/Me_2SO	Et_3N/1.2	3 days/ room temp.	100	–	100/[α][b]	29
275	Val-Pro-Gly	-ODnp(2,4)	–	–	–	–	–	–	22
276	Val-Pro-Gly	-ODnp(2,5)	–	–	–	–	431/L.S.	–	22
277	Val-Pro-Gly	$-NHNH_2/I_2$	–	–	–	–	–	–	22

SEQUENTIAL POLYPEPTIDES (continued)

		Polymerizing conditions				Product			
Entry No.	Poly	Agent	Concentration (*M*)/solvent	Base/ equivalents	Time/ temperature	Yield (%)	Degree of polymerization/ method	Optical purity (%)	Ref.
278	Met_2-Ala	-ONp	2.8/$CHCl_3$	Et_3N/1	48 h/ room temp.	38	6/Vel.	–	105
279	Met_2-Ala	-ONp	2.3/$CHCl_3$	Et_3N/1.5	48 h/ room temp.	58	9/[η]	–	106
280	DL-Leu-Gly_2	-Cl	–	–	7 h/135°[b]	–	5/V.S.	–	107
281	Leu-Gly-Gly	-ONp	–/Me_2SO	Et_3N/–	–	–	–	–	108
282	Leu-Gly-Pro	$Et_4P_2O_7$/Et_2PO_3H(1/1)	0.9	NA	3 h/60–100°[b]	73.5	33/V.S., 34/N	–	60
283	Leu-Leu-Gly	-ONp	–/Me_2SO	Et_3N/–	–	–	–	–	108
284	Leu-Leu-O-α-OH-4-Me-pentanoyl	-ONSu	2.5/various solvents	Et_3N/1	–	–	No polymer	–	31
285	Leu-Leu-O-α-OH-4-Me-pentanoyl	-Cl	3.5/benzene	Et_3N/2	24 h/ room temp.	45	96/dansyl	100/[b]	31
286	Leu_2-Asp(OBzl)	-ONp	1.1/Me_2SO	Et_3N/1.3	4 days/ room temp.	85	–	93/H^+	109
287	Leu_2-Asp(OBzl)	-ONp	1.1/$CHCl_3$	Et_3N/1.3	4 days/ room temp.	–	–	–	109
288	Leu-Orn(F_3Ac)-Leu	-ONSu	1.1/$HCONMe_2$	Et_3N/1	5 days/ room temp.	100[b]	9/Eq. and 23/Eq.	–	110
289	Leu-Orn(F_3Ac)-Leu	$Et_4P_2O_7$/1	0.3/pyridine	NA	5 days/ room temp.	83	6/Eq.	–	110
290	Leu-Arg-Leu[b]	NA	NA	NA	NA	NA	NA	NA	110
290a	Nva-Gly-Pro	-ONp	–/Me_2SO[a]	Et_3N/3[b]	4 days/ room temp.	52.5[b]	39/[η]	–	110a
290b	Nva-Gly-Pro	M.A.	NA	NA	NA	NA	61/–	–	110a
291	Phe-Gly_2	-OPcp	–/$HCONMe_2$	Et_3N/2.5	–	–	48/–	100/ $[\alpha]_{223}$	56
292	DPhe-Gly_2	-OPcp	–/$HCONMe_2$	Et_3N/2.5	–	–	92/–	100/ $[\alpha]_{223}$	56
293	Phe-Pro-Gly	-ONp	1.1/Me_2SO	Et_3N/2	87 h/ room temp.	70	–	–	111
294	Tyr(Me)-Ala-Gly	-ONp	–/$HCONMe_2$	Et_3N/1	–	82	–	–	27
295	Tyr-Ala-Glu	-ONSu	0.25/$HCONMe_2$[a,b]	Et_3N/1.2	6 days/ room temp.	100	173/Eq.[b]	–	112
296	Tyr-Ala-Glu	-ONSu[b]	NA	NA	NA	NA	2,3,4	–	113

SEQUENTIAL POLYPEPTIDES (continued)

Entry No.	Poly	Polymerizing conditions: Agent	Concentration (*M*)/solvent	Base/ equivalents	Time/ temperature	Product: Yield (%)	Degree of polymerization/ method	Optical purity (%)	Ref.
297	His-Gly_2	-ONp	–	$Et_3N/1$	–	–	–	–	32
298	His(Cbz)-Gly_2	-OTcp(2,4,5)	–/$HCONMe_2$: Me_2SO	Et_3N/–	5–7 days/	–	–	–	70
299	His-Ala-Glu[b]	-ONSu	1/$HCONMe_2$ a	$Et_3N/2$	7 days/ room temp.	88	30/Vel., Eq.γ	100/H^+	114
300	Lys-Gly-Ala	-OPcp	0.9/$HCONMe_2$	$Et_3N/2.5$	5 days/20°	55	–	–	96
301a	Lys-Ala-Gly	-ONSu	1/$HCONMe_2$[a]	$(Pr^i)_2EtN/1$	4 days/ room temp.	72	143/G-25, 119/π	99.3[b]/H^+ 100/H^+	115a 115b
301b	Lys-Ala-Gly	-OPcp	10^{-2}/$HCONMe_2$[a,b]	$Et_3N/4$	28 h/ room temp.[b]	57	51/π		
302	Lys-Ala_2	-ONSu	0.4/$HCONMe_2$	$Et_3N/1.3$	7 days/ room temp.	88.5	48/CM–Cell.	>99.5 ala/G.C.[b]	116
303	Lys-Ala_2	-OPhOH	0.9/Me_2SO	$Et_3N/2$	4 days/ room temp.	40	25_n and $41^f{}_w$/ P-100	–	117, 118
304	Lys(Cbz)-Ala_2	-OTcp(2,4,5)	1/$HCONMe_2$	$Et_3N/1$	10 days/ room temp.	81	22/V.S.	–	100, 102
305	Lys(Cbz)-Ala_2	-OTcp(2,4,5)	1/$HCONMe_2$	$Et_3N/2.5$	10 days/ room temp.	70	8/V.S.	–	100, 102
306	Lys-Ala-Glu	-ONSu	0.4/$HCONMe_2$	$Et_3N/1.25$	6 days/ room temp.	100	Mostly low/G_{50}	–	119
307	Lys-Ala-Glu	-ONSu	0.1/benzene	$Et_3N/1.25$	6 days/ room temp.	100	Mostly low/G_{50}	–	119
308	$[Lys(Cbz)]_2$-Gly	-OTcp(2,4,5)	0.9/$HCONMe_2$	$Et_3N/1$	12 days/ room temp.	52	10/V.S.	–	100, 102
309	$[Lys(Cbz)]_2$-Gly	-OTcp(2,4,5)	0.7/$HCONMe_2$	$Et_3N/2.5$	12 days/ room temp.	46	9/V.S.	–	100, 102
310	$[Lys(Cbz)]_2$-Ala	-OTcp(2,4,5)	0.8/$HCONMe_2$	$Et_3N/1$	10 days/ room temp.	33	10/V.S.	–	100, 102
310a	Lys_2-Pro	-OPcp	0.5–0.6/$HCONMe_2$	$Et_3N/1$	10 days/–	30–60	14/–	–	35a
310b	Lys_2-Pro	-OTcp(2,4,5)	0.6–0.7/$HCONMe_2$	$Et_3N/1$	10 days/–	30–60	3/–	–	35a
311	Lys-Arg-Gly	-OTcp(2,4,5)	1/$HCONMe_2$	$Et_3N/2.5$	5 days/20°	95	21/V.S. and 19/IR	–	120
312	Lys-Arg-Ala	-OTcp(2,4,5)	0.9/$HCONMe_2$	$Et_3N/2.5$	5 days/20°	66	15/V.S. and IR	–	120
313	Lys-Glu-Ala	-$NHNH_2$/Bu^t-NO_2	0.03/benzene: $HCONMe_2$ (6:1)	MeMorph/1.1	2 weeks/ room temp.	37	9/[η]	–	121

SEQUENTIAL POLYPEPTIDES (continued)

Entry No.	Poly	Agent	Concentration (*M*)/solvent	Base/ equivalents	Time/ temperature	Yield (%)	Degree of polymerization/ method	Optical purity (%)	Ref.
	Polymerizing conditions					Product			
313a	Lys-Pro_2	-OPcp	0.6–0.7/$HCONMe_2$	Et_3N/1	10 days/–	30–60	14/–	–	35a
313b	Lys-Pro_2	-OTcp(2,4,5)	0.7–0.9/$HCONMe_2$	Et_3N/1	10 days/–	30–60	61/–	–	35a
314	Asp(OMe)-Gly_2	-ONp	2.4/Me_2SO	Et_3N/1	Overnight/ room temp.	–	30/[η]	100/H^+	14
315	Asp(OMe)-Gly_2	-ONp	1.8/Me_2SO	Et_3N/1	Overnight/ room temp.	65	23/[η]	100/H^+	14
316	Asp(OMe)-Gly_2	-ONp	1.55/Me_2SO	Et_3N/1	Overnight/ room temp.	70	26/[η]	100/H^+	14
317	Asp(OMe)-Gly_2	-ONp	0.9/Me_2SO	Et_3N/1	Overnight/ room temp.	–	40/Dnp	100/H^+	14
318	Asp(OMe)-Gly_2	-ONp	0.83/Me_2SO	Et_3N/1	Overnight/ room temp.	45	33/[η]	100/H^+	14
319	Asp(OMe)-Gly_2	-ONp	0.66/Me_2SO	Et_3N/1	Overnight/ room temp.	40	21/[η]	100/H^+	14
320	Asp(OMe)-Gly_2	-ONp	–/Me_2SO	Et_3N/1	–	–	15/N	–	32
321	Asp(OMe)-Gly_2	-ONp	1.1/$HCONMe_2$	Et_3N/1	Overnight/ room temp.	60	14/[η]	100/H^+	14
322	Asp(OMe)-Gly_2	-ONp	0.26/$HCONMe_2$	Et_3N/1	Overnight/ room temp.	30	8/[η]	100/H^+	14
323	Asp(OMe)-Gly_2	-ONp	1.78/MePdn	Et_3N/1	Overnight/ room temp.	65	33/[η]	100/H^+	14
324	Asp(OMe)-Gly_2	-ONp	1.78/$OP(NMe_2)_3$	Et_3N/1	Overnight/ room temp.	50	17/[η]	100/H^+	14
325	Asp-Gly-Gly[b]	-ONp	0.9/Me_2SO	Et_3N/1.5	3 days	100	24/Dnp	–	14
326	Asp-Ser(Ac)-Gly	-ONp	–/Me_2SO	Et_3N/–	–	–	18/Vel.	–	25
327	Asp(OMe)-Ser-Gly	-ONp	1/Me_2SO	NaONp/1	50 h/ room temp.	75	33/Dnp; 37/Vel.	100/[α]	25
328	Asp(OMe)-Ser-Gly	-ONp	1.2–0.6/Me_2SO[a]	Et_3N/1.25	4 days/ room temp.	84	21/Dnp; 18/Vel.	100/[α]	25
329	Asp(OMe)-Ser-(Ac)-Gly	DCC/1–2	0.12–0.18/$HCONMe_2$	NA	–	–	–	–	25
330	Asp-Ser-Gly	-ONp[b]	–	–	–	–	13/Dnp; 21/Vel.	<100/[α]	25
331	Asp-Cys-Gly	-OPcp	1/$HCONMe_2$	Et_3N/2	48 h/ room temp.	39[b]	22/Eq.[b]	–	122
332	Asp(Cys-Gly)-OH	-OPcp	1/$HCONMe_2$	Et_3N/2	48 h/ room temp.	73[b]	25/Eq.[b]	–	122

SEQUENTIAL POLYPEPTIDES (continued)

		Polymerizing conditions				Product			
Entry No.	Poly	Agent	Concentration (M)/solvent	Base/ equivalents	Time/ temperature	Yield (%)	Degree of polymerization/ method	Optical purity (%)	Ref.
333	Asp(OMe)-Phe-Gly	-ONp	–	–	–	–	–	–	32
334	Asp_2-Glu	-ONSu	0.6/benzene[a]	Et_3N/1	6 days/ room temp.[b]	53	39_w and $30^f{}_n$/ G-150	97.4/H^+	37
335	Asp_2-DGlu	-ONSu	0.6/H_4 furan[a]	Et_3N/1	6 days/ room temp.[b]	28	30_w and $25^f{}_n$/ G-150	98.1/H^+	37
336	Asp-DGlu-Glu	-ONSu	0.8/dioxane[a]	Et_3N/1	6 days/ room temp.[b]	25	29_w and $24^f{}_n$/ G-150	97.5/H^+	37
337	Glu(OEt)-Gly_2	-ONp	2.5–5/$HCONMe_2$	Et_3N/1.4	–	72	–	–	18
338	Glu(OEt)-Gly_2	-ONp	2.5/$HCONMe_2$	Et_3N/1.5	3 days/ room temp.	–	25/Dnp; 17/ONp	–	94, 123
339	Glu-Gly-Ala	-OPcp	–	–	–	–	26/V.S.[b]	–	124
340	Glu-Gly-Ala	-ONp	–	–	–	–	34/V.S.[b]	–	124
341	Glu(OBzl)-Gly-Ala	-ONp[b]	–/$HCONMe_2$ or Me_2SO	Et_3N/–	240 h/20°	28–35	23/–	–	75
342	Glu(OMe)-Gly-Lys (Tos)	-OTcp(2,4,6)	0.7/Me_2SO	Et_3N/1	6–7 days/20°	–	–	–	123, 12
343	Glu(OBzl)-Gly-Lys (Tos)	-OPcp	0.6/Me_2SO	Et_3N/1	6–7 days/20°	–	–	–	123, 12
344	Glu(OBzl)-Gly-Lys (Tos)	-ONp	0.7/Me_2SO	Et_3N/1	6–7 days/20°	–	–	–	123, 12
345	Glu-Ala-Gly[b]	-OPcp	0.36/$HCONMe_2$: Me_2SO	Et_3N/1.25[b]	172 h/ room temp.	83[b]	26/Dnp, 39/Vel.	100/H^+	125
346	Glu(Ala-Gly)-OH	-OPcp	0.1/Me_2SO	Et_3N/1.7	108 h/ room temp.	55[b]	31/Dnp, 52/Vel.	100/H^+	125
347	Glu-Ala_2	-OPhOH	0.75/Me_2SO	Et_3N/2	4 days/ room temp.	65	9–22/[η]	–	117
348	Ala-Glu-Ala	-OPhOH	0.23/Me_2SO	Et_3N/1	Several weeks/ room temp.	29	<74/G-75	–	126
349	Glu-Ala-Glu	-OPcp	1/Me_2SO	Et_3N/2	16 h/ room temp.	59[b]	44/Dnp, 76/Vel.	100/H^+	57
350	Glu-Ala-Glu	-ONp	0.7/$HCONMe_2$	Et_3N/1	–	40	61/Dnp	98/H^+	127
351	Glu-Ser-Gly	-ONp	0.5/Me_2SO[a]	Et_3N/1[b]	22 h/ room temp.	45	15–38/Dnp, Vel. and Eq.	90–95/H^+	38
352	Glu(OMe)-Ser(Ac)-Glu(OMe)	-ONp	1.2/$HCONMe_2$	Et_3N/1.3	–	100	15–20/ONp	–	26

SEQUENTIAL POLYPEPTIDES (continued)

Entry No.	Poly	Agent	Concentration (*M*)/solvent	Base/ equivalents	Time/ temperature	Yield (%)	Degree of polymerization/ method	Optical purity (%)	Ref.
	Polymerizing conditions					Product			
353	Glu-Cys-Gly[b]	-OPcp	1/$HCONMe_2$	Et_3N/2	48 h/ room temp.	54	55/Eq.	–	122
354	Glu(Cys-Gly)-OH[b]	-OPcp	1/$HCONMe_2$	Et_3N/2	48 h/ room temp.	84	31/Eq.	–	122
355	Glu(OEt)-Cys(Bzl) -Glu(OEt)	-ONp	2.5–5/$HCONMe_2$	Et_3N/1.3	–	92	–	–	18, 28
356	Glu(OMe)-Val-Glu (OMe)	-ONp	2.5–5/$HCONMe_2$	Et_3N/1.3	–	92	–	–	18, 30
357	Glu-Tyr-Glu	-OPhOH	1.5/$HCONMe_2$[a,b]	Et_3N/2	96 h/ room temp.	68	25/π	–	128
358	Glu(OBzl)-Tyr(Bzl) -Glu(OBzl)	-OPhOH	1/$HCONMe_2$	Et_3N/2	7 days/ room temp.	77.5	28/[η]	–	128
359	Glu(OBzl)-Tyr(Bzl) -Glu(OBzl)	-OPhOH	0.1/$HCONMe_2$	Et_3N/2	7 days/ room temp.	34.4	11/[η]	–	128
360	[Glu(OEt)]$_2$-Gly	-ONp	–/$HCONMe_2$	Et_3N/1	–	95	30–50/ONp	–	123, 129
361	DGlu$_2$-Leu	-ONSu	0.05/petroleum[a]	Et_3N/1	4 days/ room temp.[b]	29	–/[b]	–	42
362	DGlu$_2$-Leu	-ONSu	0.15/petroleum[a]	Et_3N/1	4 days/ room temp.[b]	36	–	–	42
363	DGlu$_2$-Leu	-ONSu	0.33/petroleum[a]	Et_3N/1	4 days/ room temp.[b]	34	40 and 23/G-150	97.6/E	42
364	DGlu$_2$-Leu	-ONSu	0.4/petroleum[a]	Et_3N/1	4 days/ room temp.[b]	40	–	–	42
365	DGlu$_2$-Leu	-ONSu	1.5/petroleum[a]	Et_3N/1	4 days/ room temp.[b]	29	–	–	42
366	DGlu$_2$-Leu	-ONSu	0.3/benzene	Et_3N/1	4 days/ room temp.[b]	29	–	–	42
367	DGlu$_2$-Leu	-ONSu	1.0/benzene	Et_3N/1	4 days/ room temp.[b]	28	–	–	42
368	DGlu$_2$-Leu	-ONSu	2.4/benzene	Et_3N/1	4 days/ room temp.[b]	38	–	–	42
369	DGlu$_2$-Leu	-ONSu	3.0/benzene	Et_3N/1	4 days/ room temp.[b]	32	–	–	42
370	DGlu$_2$-Leu	-ONSu	0.12/$HCONMe_2$	Et_3N/1	4 days/ room temp.[b]	0	–	–	42

SEQUENTIAL POLYPEPTIDES (continued)

		Polymerizing conditions				Product			
Entry No.	Poly	Agent	Concentration (*M*)/solvent	Base/ equivalents	Time/ temperature	Yield (%)	Degree of polymerization/ method	Optical purity (%)	Ref.
371	DGlu$_2$-Leu	-ONSu	0.3/HCONMe$_2$	Et$_3$N/1	4 days/ room temp.[b]	0	–	–	42
372	DGlu$_2$-Leu	-ONSu	1.0/HCONMe$_2$	Et$_3$N/1	4 days/ room temp.[b]	16	–	–	42
373	DGlu$_2$-Leu	-ONSu	3.0/HCONMe$_2$	Et$_3$N/1	4 days/ room temp.[b]	27	–	–	42
374	DGlu$_2$-Leu	-ONSu	0.3/Me$_2$SO	Et$_3$N/1	4 days/ room temp.[b]	17	–	–	42
375	DGlu$_2$-Leu	-ONSu	1.0/Me$_2$SO	Et$_3$N/1	4 days/ room temp.[b]	28	–	–	42
376	DGlu$_2$-Leu	-ONSu	3.0/Me$_2$SO	Et$_3$N/1	4 days/ room temp.[b]	21	–	–	42
377	DGlu$_2$-Leu	-ONSu	2.7/H$_4$ furan	Et$_3$N/1	4 days/ room temp.[b]	21	–	–	42
378	Glu$_2$-DGlu	DCC/91	0.5/acetonitrile	NA	24 h/ room temp.	44	12/G-75	–	45
379	DGlu-Glu-DGlu	-ONSu	0.3/CHCl$_3$[a]	Et$_3$N/1	2 days/ room temp.	54	57_w and $43^f{}_n$/ G-150, 69/π	–	46
380	[Glu(OBzl)]$_3$	-OPcp	–/HCONMe$_2$	–	–	–	38/[η]	<10/ $[\alpha]_{546}$	47
381	[Glu(OBzl)]$_3$	-OPhOH	–/HCONMe$_2$	Et$_3$N/–	–	50	17/[η]	100/ $[\alpha]_{546}$	47
382	Glu(OBzl)-DGlu (OBzl)-Glu(OBzl)	-OPhOH	2/benzene	Et$_3$N/2	4 days/ room temp.	–	8/[η][b]	–	130
383	Sar-Ala-Gly	-OPcp	–/Me$_2$SO	Et$_3$N, MeMorph/ –[b]	–	–	Dialyzable[b]	–	82
384	Pro-Gly$_2$	-ONp	2.5–5/HCONMe$_2$	Et$_3$N/1.4	–	39	–	–	18
385	Pro-Gly$_2$	-NHNH$_2$/Suc NBr	–/AcNMe$_2$	–	–	–	6/Dnp	–	131
386	Pro-Gly-Ala	Et$_4$P$_2$O$_7$	–	–	–	–	–	–	61
387	Pro-Gly-Ser	Et$_4$P$_2$O$_7$/2	0.09/Et$_2$PO$_3$H	NA	72 h/ room temp.	52	10/G-50	–	68
388	Pro-Gly-Ser	Et$_4$P$_2$O$_7$/2	0.18/pyridine	NA	72 h/ room temp.	50	12/G-50	–	68

SEQUENTIAL POLYPEPTIDES (continued)

		Polymerizing conditions				Product			
Entry No.	Poly	Agent	Concentration (*M*)/solvent	Base/ equivalents	Time/ temperature	Yield (%)	Degree of polymerization/ method	Optical purity (%)	Ref.
389	Pro-Gly-Ser	$(PhO_2P)_2O$/5	0.1/Et_2PO_3H	NA	120 h/ room temp.	13	4/G-50	–	68
390	Pro-Gly-Pro	DCC/HCBztn	0.04/$HCONMe_2$	NA	66 h/20°	55[b]	20/G-50	–	132
391	Pro-Gly-Pro	$Et_4P_2O_7$/1	1.7/pyridine	NA	5 days/ room temp.[b]	64	27/Eq., 26/Vel.[b]	–	133
392	Pro-Ala-Gly	-ONSu	1/Me_2SO	MeMorph/4.5	4 days/ room temp.	16	–	–	97
393	Pro-Ala-Gly	-OPcp	0.9/Me_2SO	MeMorph/2	6 days/ room temp.	34	–	–	97
394	Pro-Ala-Gly	-ONp	0.5/Me_2SO[a]	Et_3N/3[b]	20 h/room temp. + 21 h/50°[b]	81	55/Vel.; 84 and 32/G-50	100/[b]	134
395	Pro-Ala-Gly	-ONp	0.9/Me_2SO	MeMorph/4.5	4 days/ room temp.	22	–	–	97
396	Pro-Ala-Gly	DCC/HONSu	0.08/$HCONMe_2$	NA	232 h/0°	3.0	16/G-50	–	132
397	Pro-Ala-Gly	DCC/HONSu	0.08/$HCONMe_2$	NA	24 h/–10°	1.1[b]	9/G-50	–	132
398	Pro-Ala-Gly	DCC/HONSu	0.08/$HCONMe_2$	NA	66 h/–10°	0.1	19/G-50	–	132
399	Pro-Ala-Gly	DCC/HONSu	0.08/$HCONMe_2$	NA	232h/0°	0.1	18/G-50	–	132
400	Pro-Ala-Gly	DCC/HONSu	0.08/$HCONMe_2$	NA	24 h/20°	0.9[b]	16/G-50	–	132
401	Pro-Ala-Gly	DCC/HONSu	0.08/$HCONMe_2$	NA	66 h/20°	0.1	20/G-50	–	132
402	Pro-Ala-Gly	DCC/HONSu	0.08/$HCONMe_2$	NA	232 h/20°	0.2	22/G-50	–	132
403	Pro-Ala-Gly	DCC/HONSu	0.08/pyridine	NA	24 h/–10°	1.6[b]	9/G-50	–	132
404	Pro-Ala-Gly	DCC/HONSu	0.08/pyridine	NA	24 h/20°	0.1[b]	20/G-50	–	132
405	Pro-Ala-Gly	DCC/HONSu	0.08/Et_2PO_3H	NA	66 h/20°	1.5[b]	16/G-50	–	132
406	Pro-Ala-Gly	DCC/HOBztn	0.08/$HCONMe_2$	NA	232 h/–20°	0.2	20/G-50	–	132
407	Pro-Ala-Gly	DCC/HOBztn	0.08/$HCONMe_2$	NA	24 h/–10°	0.3[b]	29/G-50	–	132
408	Pro-Ala-Gly	DCC/HOBztn	0.08/$HCONMe_2$	NA	66 h/–10°	0.3	22/G-50	–	132
409	Pro-Ala-Gly	DCC/HOBztn	0.08/$HCONMe_2$	NA	232 h/0°	0.5	27/G-50	–	132
410	Pro-Ala-Gly	DCC/HOBztn	0.08/$HCONMe_2$	NA	24 h/20°	0.2[b]	36/G-50	–	132
411	Pro-Ala-Gly	DCC/HOBztn	0.08/$HCONMe_2$	NA	66 h/20°	0.5	29/G-50	–	132
412	Pro-Ala-Gly	DCC/HOBztn	0.08/$HCONMe_2$	NA	232 h/20°	0.4	38/G-50	–	132
413	Pro-Ala-Gly	DCC/HOBztn	0.44/$HCONMe_2$	NA	24 h/20°	51	9/G-50[b]	–	132
414	Pro-Ala-Gly	DCC/HOBztn	0.44/$HCONMe_2$	NA	66 h/20°	57	16/G-50[b]	–	132
415	Pro-Ala-Gly	DCC/HOBztn	0.08/pyridine	NA	24 h/–10°	0.2[b]	22/G-50	–	132
416	Pro-Ala-Gly	DCC/HOBztn	0.08/pyridine	NA	24 h/20°	0.4[b]	29/G-50	–	132
417	Pro-Ala-Gly	DCC/HOBztn	0.08/Et_2Pd_3H	NA	66 h/20°	0.2[b]	22/G-50	–	132

SEQUENTIAL POLYPEPTIDES (continued)

		Polymerizing conditions				Product			
Entry No.	Poly	Agent	Concentration (*M*)/solvent	Base/ equivalents	Time/ temperature	Yield (%)	Degree of polymerization/ method	Optical purity (%)	Ref.
418	Pro-Ala-Gly	DCC/HOBztn	0.44/CH_2Cl_2	NA	66 h/20°	38	22/G-50[b]	–	132
419	Pro-Ala-Gly	DCC/HOBztl	0.08/$HCONMe_2$	NA	24 h/–10°	0.1	18/G-50	–	132
420	Pro-Ala-Gly	DCC/HOBztl	0.08/pyridine	NA	24 h/–10°	0.6[b]	11/G-50	–	132
421	Pro-Ala-Gly	$Et_4P_2O_7$	–	NA	–/10°–20°	2	49/G-75	–	135
422	Pro-Ala-Gly	$(PhO_2P)_2O$	–	–	–	–	–	–	22
423	Pro-Ala-Gly	Solid phase	NA	NA	NA	NA	3,4	–	136
424	Pro-Ala-Gc	-ONp	1.5/$HCONMe_2$	Et_3N/1.5	–	90	6/ONp	–	8
425	Pro-Ser-Gly	-OPcp	0.9/Me_2SO	–	6 days/40–50°	100	14/V.S.	–	137
426	Pro-Ser-Gly	-ONp	1.1/$HCONMe_2$	Et_3N/1.4	24 h/ room temp.	65	66–87/Vel.	100/E	134
427	Pro-Ser-Gly	2/DCC/HONSu/4	0.1/$HCONMe_2$	NA	60 h/ room temp.	32	21/G-25	–	68
428	Pro-Ser-Gly	DCC/HOBztn	0.04/$HCONMe_2$	NA	60 h/20°	40	13/G-50	–	132
429	Pro-Ser-Gly	$Et_4P_2O_7$/2	0.2/pyridine	NA	72 h/ room temp.	56	28/G-25	–	68
430	Pro-Ser-Gly	$Et_4P_2O_7$/2	0.2/Et_2PO_3H	NA	72 h/ room temp.	60	23/G-25	–	68
431	Pro-Ser-Gly	$(PhO_2P)_2O$/5	0.1/Et_2PO_3H	NA	100 h/ room temp.	21	7/G-25	–	68
432	Pro-Leu-Gly	$Et_4P_2O_7$/1.4	1.4/Et_2PO_3H	NA	2 h/100°	–	22/T-OMe	–	87
433	Pro-Phe-Gly	-ONp	1.2/Me_2SO	Et_3N/1.4	5 days/ room temp.	72	–	–	111
434	Pro-His(Cbz)-Gly	-OTcp(2,4,5)	0.7/Me_2SO	Et_3N/4.4	7 days/ room temp.	73	6/V.S.	–	77
435	Pro-Glu(OBu^t)-Gly	-OPcp	0.8/Me_2SO	MeMorph/2	3 days/ room temp.	70	47/–[b]	–	136
436	Pro-Gly(OBu^t)-Gly	DCC/HOBztl	0.8/$HCONMe_2$	MeMorph/1	3 days/ room temp.	30	15/–	–	136
437	Pro-Pro-Gly	Solid phase	NA	NA	NA	NA	10 and 20	–	138
438	Pro-Hyp-Gly	-ONp	0.4/Me_2SO[a]	Et_3N/1.3	2 days/ room temp.	59	169/Vel., 157/ Eq., 56/Dnp	–	53
439	Pro-Hyp-Gly	Solid phase	NA	NA	NA	NA	5 and 10	–	139
440	Hyp-Ser-Gly	-OTcp(2,4,6)	1/Me_2SO	Et_3N/1.4	6 days/ room temp.	10	22/V.S.	–	140

SEQUENTIAL POLYPEPTIDES (continued)

		Polymerizing conditions				Product			
Entry No.	Poly	Agent	Concentration (*M*)/solvent	Base/ equivalents	Time/ temperature	Yield (%)	Degree of polymerization/ method	Optical purity (%)	Ref.
441	Hyp-Glu-Gly	-OTcp(2,4,6)	1.6/Me_2SO	Et_3N/1.4	10 days/ room temp.	90[b]	22/V.S., 26/N[b]	–	141
442	Pipec-Ala-Gly	-OPcp	2/Me_2SO	MeMorph/ 2.25	6 days/ room temp.	16	Low	–	93
443	Gly-Gly-Gly-Gly	-OPcp	1/benzene	Et_3N/2.5	–	–	–	–	7
444	Gly_2-[Lys(Tos)]$_2$	-ONp	–/Me_2SO	Et_3N/1	5 days/20°	–	15/–	–	59
445	Gly_2-Pro-Gly	-ONp	–	–	–	–	73_n, 100_w/–[e, f]	–	54
446	Gly_2-Hyp-Gly	-ONp	–	–	–	–	132_n, 229_w/–[e, f]	–	54
447	Gly_2-Hyp-Gly	-ONp	0.9/Me_2SO	Et_3N/1.4	3 days/ room temp.	46	60/Eq, 18/π, 106/Dnp	–	53
448	Ala-Gly_3	-OPcp	1/benzene[a]	Et_3N/2.5	–	–	–	–	7
449	Ala-Gly_3	-OPcp	0.7/benzene	Et_3N/2.5	18 h/ room temp.	60	–	–	16, 17
450	Ala_2-Gly_2	-OPcp	0.5/benzene[a]	Et_3N/2.5	24 h/ room temp.	81	–	–	16
451	Ala_3-Gly	-ONp	0.6/Me_2SO	Et_3N/1	1 day/ room temp.	–	74/[η]	–	95
452	Ala_3-Lys(Cbz)	-OTcp(2,4,5)	1/$HCONMe_2$	Et_3N/1	12 days/ room temp.	30	7/V.S.	–	100
453	Ala_3-Lys(Cbz)	-OTcp(2,4,5)	1/$HCONMe_2$	Et_3N/2.5	10 days/ room temp.	20	8/V.S.	–	100
454	Ala-Val_2-Ala	-ONSu	0.1/$HCONMe_2$	Et_3N/1.2	3 days/ room temp.	100	–	1 100/[α][b]	29
455	Ala-Glu-Ala-Gly	-OPcp	0.11/Me_2SO[b]	Et_3N/3.5	7 days/ room temp.	30.3[b]	305/G-100	–	141
456	Val-Gly-Val-Gly	-ONSu	0.3/$HCONMe_2$	Et_3N/1.2	5 days/ room temp.	98	–	100/[α][b]	29
457	Leu-Gly-Leu-Gly	-ONp	–/Me_2SO	Et_3N/–	–	–	–	–	108
458	Leu_3-Gly	-ONp	–/Me_2SO	Et_3N/–	–	–	–	–	108
459	Leu-Leu-Leu-O-α-OH-4-Me-pentanoyl	-Cl[b]	3.5/benzene	Et_3N/2	24 h/ room temp.	42	102/dansyl	100/–	31
460	Leu-O-α OH-4-Me-pentanoyl-Leu-O-α-OH-4-Me-pentanoyl	-Cl	3.5/benzene	Et_3N/2	24 h/ room temp.	38	54/dansyl	100/–	31
461	Phe-Asp-Val-Gly	-OPcp	0.1/Me_2SO[b]	Et_3N/3	7 days/ room temp.	6	120/Diaflow[b]	–	142

SEQUENTIAL POLYPEPTIDES (continued)

Entry No.	Poly	Polymerizing conditions: Agent	Concentration (*M*)/solvent	Base/ equivalents	Time/ temperature	Product: Yield (%)	Degree of polymerization/ method	Optical purity (%)	Ref.
462	Phe-Glu-Ala-Gly[b]	-OPcp	0.1/Me_2SO[b]	Et_3N/3.5	7 days/ room temp.	–	248/G-100	–	143
463	Phe-Glu-Ala-Gly[b]	-OPcp	0.1/Me_2SO	Et_3N/3.5	3 days/ room temp.	41	25/G-50	–	144
464	Phe-Glu-Val-Gly	-OPcp	0.1/Me_2SO[b]	Et_3N/3	7 days/ room temp.	21	120/Diaflow[b]	–	142
465	Tyr-Lys(αGlu) -Ala-Gly[b]	-OPcp[b]	0.1/Me_2SO	Et_3N/3	7 days/ room temp.	51[b]	36/G-50	–	145
466	Tyr-Asp-Ala-Gly	-OPcp	0.1/Me_2SO	Et_3N/3.5	7 days/ room temp.	10[b]	123/Diaflow[b]	–	146
467	Tyr-Glu-Gly_2	-OPcp	0.09/Me_2SO	Et_3N/2.3	6 days/ room temp.	42.5[b]	246/G-100	–	147
468	Tyr-Glu-Ala-Gly	-OPcp[b]	0.1/Me_2SO	Et_3N/2.5	12 h/ room temp.	53.3[b]	310/ ^{14}C; 245/π	–	148
469	Tyr-Glu-Ala-Gly	-OPcp	0.2/benzene	Et_3N/2.2	1 week/4°	71	126/[η]	–	121
470	Tyr-Glu-Ala-Gly[b]	-OPcp	0.07/Me_2SO	Et_3N/3.5	7 days/ room temp.	85	238/G-100	–	149
471	Tyr-Glu(Ala-Gly) -OH	-OPcp	0.1/Me_2SO	Et_2N/3.2	7 days/ room temp.	36[b]	43/G-50	–	150
472	Tyr(Me)-Glu-Ala-Gly	-OPcp	0.1/Me_2SO	Et_3N/5	6 days/ room temp.	24[b]	230/G-100	–	151
473	Tyr-Glu(Gly-OH)- Ala-Gly	-OPcp	0.1/Me_2SO	Et_3N/3.4	7 days/ room temp.	27	105/Diaflow[b]	–	152
474	Tyr-Glu-Ala-βAla	-OPcp	0.1/Me_2SO	Et_3N/3.5	1 week/ room temp.	30	115/Diaflow[b]	–	153
475	Tyr-Glu-Val-Gly	-OPcp	0.1/Me_2SO	Et_3N/2.5	5 days/ room temp.	61[b]	346/ ^{14}C, 210/G-100	–	154
476	Tyr-Glu-Tyr-Glu	-OPhOH	1.5/$HCONMe_2$[a]	Et_3N/2	96 h/ room temp.	66	27/π[b]	–	128
477	Trp-Glu-Ala-Gly	-OPcp[b]	0.1/Me_2SO	Et_3N/2.5	7 days/ room temp.	47	68/G-50	–	155
478	$[Orn(Cbz)]_2$-Gly_2	-OTcp(2,4,5)	0.8/$HCONMe_2$	Et_3N/1	10 days/ room temp.	50	3/V.S.	–	100
479	$[Orn(Cbz)]_2$-Ala_2	-OTcp(2,4,5)	0.7/$HCONMe_2$	Et_3N/1	10 days/ room temp.	35	6/V.S.	–	100

SEQUENTIAL POLYPEPTIDES (continued)

		Polymerizing conditions				Product			
Entry No.	Poly	Agent	Concentration (*M*)/solvent	Base/ equivalents	Time/ temperature	Yield (%)	Degree of polymerization/ method	Optical purity (%)	Ref.
479a	Orn_3-Gly	-OPcp	0.2/$HCONMe_2$	Et_3N/1	7 days/20°	30–40	28/V.S.	–	155a
480	Lys-Gly-Lys-Gly	-$OMeSO_2Ph$	0.1/Me_2SO	Et_3N/2.5	6 days/ room temp.	92^b	21/G-100	–	156
481	Lys-Ala_3	-OPhOH	0.3/Me_2SO	$Et_3N/2^b$	1 week/ room temp.	16	94/G-75^b	100/H^+	157
482	$[Lys(Cbz)]_2$-Gly_2	-OTcp(2,4,5)	0.6/$HCONMe_2$	Et_3N/1	10 days/ room temp.	28	5/V.S.	–	100
483	$[Lys(Cbz)]_2$-Ala_2	-OTcp(2,4,5)	0.6/$HCONMe_2$	Et_3N/1	10 days/ room temp.	40	4/V.S.	–	100
483a	Lys_3-Gly	-OPcp	0.2/$HCONMe_2$	Et_3N/1	7 days/20°	30–40	57/V.S.	–	155a
484	$[Lys(Cbz)]_3$-Ala	-OTcp(2,4,5)	0.6/$HCONMe_2$	Et_3N/1	10 days/ room temp.	20	6/V.S.	–	100
485	$[Lys(Cbz)]_3$-Ala	-OTcp(2,4,5)	0.6/$HCONMe_2$	Et_3N/2.5	12 days/ room temp.	20	3/V.S.	–	100
485a	Lys_3-Pro	-OPcp	0.4–0.5/$HCONMe_2$	Et_3N/1	10 days/–	30–60	29/–	–	35a
485b	Lys_3-Pro	-OTcp(2,4,5)	0.4–0.5/$HCONMe_2$	Et_3N/1	10 days/–	30–60	10/–	–	35a
486	Glu-Ala-Glu-Ala	-OPcp	0.8/$HCONMe_2$	Et_3N/2	2 days/ room temp.	53^b	11/Eq.	100/H^+	158
487	Glu(OMe)-Ala-Glu(OMe)-Ala	-OPhOH	1.1/Me_2SO	Et_3N/2.5	14 days/ room temp.	69	12/[η]	–	159
488	Glu(NH_2)-Ala-Glu(NH_2)-Ala	-OPhOH	0.8/Me_2SO	Et_3N/2.3	6 days/ room temp.	50^b	13–25/[η]	–	159
489	Glu(OEt)-Cys(Bzl)-$[Glu(OEt)]_2$	-ONp	–/$HCONMe_2$	Et_3N/1	–	82	60/ONp	–	27
490	Glu-Tyr-Ala-Gly	-OPcp	0.1/Me_2SO	Et_3N/5	6 days/ room temp.	49.5^b	238/G-100	–	160
491	Glu-His-Lys-Tyr	DCC/1.2	0.3/$HCONMe_2$	NA	6 days/ room temp.	95	34/Dnp	–	161
492	Glu-Lys-Ala-Gly	-OPcp	0.054/$HCONMe_2$	Et_3N/1.1	14 days/ room temp.	88	8/$[\eta]^b$	–	162
493	Glu-Lys-Ala-Gly	-OPcp	0.054/benzene	Et_3N/1.1	14 days/ room temp.	90	32/Eq., 29/$[\eta]^b$	–	162
494	Glu_2-Ala-Gly	-OPcp	0.1/Me_2SO	Et_3N/2.9	Overnight/ room temp.	87	674/G-75; 596/π	100/H^+	163

SEQUENTIAL POLYPEPTIDES (continued)

Entry No.	Polymerizing conditions					Product			
	Poly	Agent	Concentration (*M*)/solvent	Base/ equivalents	Time/ temperature	Yield (%)	Degree of polymerization/ method	Optical purity (%)	Ref.
495	[Glu(OMe)]$_2$-Ser(Ac)-Glu(OMe)	-ONp	1/HCONMe$_2$	Et$_3$N/1.2	–	100	15–20/ONp	–	26
496	[Glu(OEt)]$_2$-Cys (Bzl)-Glu(OEt)	-ONp	–/HCONMe$_2$	Et$_3$N/1	–	92	48/ONp	–	27
497	[Glu(OMe)]$_2$-Val-Glu(OMe)	-ONp	2.5–5/HCONMe$_2$	Et$_3$N/1.4	–	93	32/ONp	–	18, 30
498	Glu$_2$-Tyr-Glu	-OPhOH	1.5/HCONMe$_2$	Et$_3$N/2	96 h/ room temp.	72[b]	15/π	–	128
499	[Glu(OEt)]$_3$-Gly	-ONp	–/HCONMe$_2$	Et$_3$N/1	4 days/ room temp.	95	30–50/ONp	–	129
500	DGlu$_3$-Leu	-ONSu	0.3/H$_4$ furan	Et$_3$N/1	4 days/ room temp.	66	–	–	42
501	DGlu$_3$-Leu	-ONSu	1.0/H$_4$ furan	Et$_3$N/1	4 days/ room temp.	76	–	–	42
502	DGlu$_3$-Leu	-ONSu	1.2/H$_4$ furan	Et$_3$N/1	4 days/ room temp.	56	15, 26, and 32/ G-150	97/E	42
503	DGlu$_3$-Leu	-ONSu	2.0/H$_4$ furan	Et$_3$N/1	4 days/ room temp.	70	–	–	42
504	DGlu$_3$-Leu	-ONSu	0.3/dioxane	Et$_3$N/1	4 days/ room temp.	53	–	–	42
505	DGlu$_3$-Leu	-ONSu	1.0/dioxane	Et$_3$N/1	4 days/ room temp.	56	–	–	42
506	DGlu$_3$-Leu	-ONSu	2.0/dioxane	Et$_3$N/1	4 days/ room temp.	65	–	–	42
507	DGlu$_3$-Leu	-ONSu	0.3/benzene	Et$_3$N/1	4 days/ room temp.	51	–	–	42
508	DGlu$_3$-Leu	-ONSu	1.0/benzene	Et$_3$N/1	4 days/ room temp.	52	–	–	42
509	DGlu$_3$-Leu	-ONSu	3.3/benzene	Et$_3$N/1	4 days/ room temp.	60	–	–	42
510	DGlu$_3$-Leu	-ONSu	0.1/petroleum	Et$_3$N/1	4 days/ room temp.	50	–	–	42
511	DGlu$_3$-Leu	-ONSu	0.3/petroleum	Et$_3$N/1	4 days/ room temp.	49	–	–	42

SEQUENTIAL POLYPEPTIDES (continued)

	Polymerizing conditions					Product			
Entry No.	Poly	Agent	Concentration (*M*)/solvent	Base/ equivalents	Time/ temperature	Yield (%)	Degree of polymerization/ method	Optical purity (%)	Ref.
512	DGlu$_3$-Leu	-ONSu	0.3/HCONMe$_2$	Et$_3$N/1	4 days/ room temp.	8[b]	–	–	42
513	DGlu$_3$-Leu	-ONSu	2.1/Me$_2$SO	Et$_3$N/1	4 days/ room temp.	50	–	–	42
514	DGlu$_3$-Leu	-ONSu	No solvent	Et$_3$N/1	4 days/ room temp.	53	–	–	42
515	Glu$_3$-DGlu	DCC/91	0.5/acetonitrile	NA	24 h/ room temp.	74	19/G-75	90/H$^+$	45
516	Glu$_3$-DGlu	DCC/91	5/HCONMe$_2$	NA	24 h/ room temp.	0	–	–	45
517	DGlu$_2$-Glu-DGlu	-ONSu	0.4/CHCl$_3$ [a]	Et$_3$N/1	6 days/ room temp	26	21_w, 18^f_n/G-150, 17/π	–	46
518	Glu(OBzl)-DGlu (OBzl)-Glu(OBzl)-DGlu(OBzl)	-OPhOH	2/benzene	Et$_3$N/2	4 days/ room temp.	–	35/L.S., 8/[η] [b]	100/[α] [b]	130
519	DGlu(OBzl)-Glu (OBzl)-DGlu (OBzl)-Glu(OBzl)	-OPhOH	2/benzene	Et$_3$N/2	4 days/ room temp.	–	8/[η] [b]	100/[α] [b]	130
520	DGlu(OBzl)-Glu (OBzl)-DGlu (OBzl)-Glu(OBzl)	-OPhOH	–/benzene	–	–	–	9/[η], 23/[η]	–	164, 165
521	Pro-Gly-Pro-Gly	-NHNH$_2$/I$_2$	–	–	–	–	–	–	22
522	Pro-Ala-Pro-Ala	-ONp	–	–	–	–	15_n, 18_w/–[e, f]	–	54
523a	Ala$_2$-Glu-Ala-Ser	-OPcp	0.5/Me$_2$SO	Et$_3$N/1	6 days/ room temp.	36	–	–	166a
523b	Ala$_2$-Gly$_3$	-OPcp	0.6/benzene	Et$_3$N/2.5	18 h/ room temp.	52	–	–	16
524	Ala-DGlu-Lys-DAla-Gly	-OPcp	0.056/HCONMe$_2$	Et$_3$N/2.2	14 days/ room temp.	91[b]	34/[η]	–	162
525	Ala-DGlu-Lys-DAla-Gly	-OPcp	0.064/benzene	Et$_3$N/2.5	14 days/ room temp.	100[b]	11/[η]	–	162
526a	Glu-Ala$_3$-Ser	-OPcp	0.5/Me$_2$SO	Et$_3$N/1	6 days/ room temp.	36	–	–	166a
526b	Glu(OEt)-[Cys (Bzl)]$_2$-[Glu (OEt)]$_2$	-ONp	–/HCONMe$_2$	Et$_3$N/1	–	100	–	–	27, 28

SEQUENTIAL POLYPEPTIDES (continued)

	Polymerizing conditions					Product			
Entry no.	Poly	Agent	Concentration (*M*)/solvent	Base/ equivalents	Time/ temperature	Yield (%)	Degree of polymerization/ method	Optical purity (%)	Ref.
527	Glu(OEt)-Cys(Bzl)-Glu(OEt)-Cys(Bzl)-Glu(OEt)	-ONp	–/$HCONMe_2$	Et_3N/1	–	94	–	–	27, 28
528	[Glu(OMe)]$_3$-Ser(Ac)-Glu(OMe)	-ONp	0.8/$HCONMe_2$	Et_3N/1	2–3 days/ room temp.	100	15–20/ONp	–	26
529	[Glu(OEt)]$_3$-Cys(Bzl)-Glu(OEt)	-ONp	–/$HCONMe_2$	Et_3N/1	–	100	–	–	27, 28
530	[Glu(OMe)]$_3$-Val–Glu(OMe)	-ONp	2.5–5.0/$HCONMe_2$	Et_3N/1.2	–	94	70–80/ONp	–	18, 30
531	[Glu(OEt)]$_4$-Gly	-ONp	–/$HCONMe_2$	Et_3N/1	–	96	30–50/ONp	–	120
532	DGlu$_4$-Leu	-ONSu	0.62/benzene	Et_3N/1	4 days/ room temp.	57	33/G-150	96.9/E	42
533	DGlu$_4$-Leu	-ONSu	1/benzene	Et_3N/1	4 days/ room temp.	58	–	–	42
534	DGlu$_4$-Leu	-ONSu	1/petroleum	Et_3N/1	4 days/ room temp.	40	–	–	42
535	DGlu$_4$-Leu	-ONSu	1/H_4 furan	Et_3N/1	4 days/ room temp.	48	–	–	42
536	DGlu$_4$-Leu	-ONSu	1/dioxane	Et_3N/1	4 days/ room temp.	51	–	–	42
537	DGlu$_3$-Glu-DGlu	-ONSu	0.4/$CHCl_3$[a]	Et_3N/1	6 days/ room temp.	58	21_w, $17^f{}_n$/G-150; 12/π	–	46
538	Ala-Gly-Ala-Gly-Ala-Ser	-ONp	0.9/$HCONMe_2$	Et_3N/1.3	2–3 days/ room temp.	96	40/ONp	–	166b
539	Ala$_2$-Gly-Pro-Ala-Gly	DCC/HOBztn	0.2/$HCONMe_2$	NA	3 days/ room temp.	–	9/P-4[b]	–	136
540	Ala$_2$-Gly-Pro$_2$-Gly	-ONSu	0.6/Me_2SO	Et_3N/1	Several days/ room temp.	–	13/Eq.	–	167
541	Ala$_3$-Gly$_3$	-OPcp	0.5/benzene[a]	Et_3N/2.5	26 h/ room temp.	50	–	–	16
542	Ala-Pro-Gly-Pro-Ala-Gly	-ONSu	0.6/Me_2SO	Et_3N/1	Several days/ room temp.	–	18/Eq.	–	167
543	Ala-Pro-Gly-Pro-Pro-Gly	-ONSu	0.6/Me_2SO	Et_3N/1	Several days/ room temp.	–	24/Eq.	–	167
544	Cys(Bzl)-Gly$_2$-Glu(OMe)-Val-Glu	-ONp	2.5–5.0/$HCONMe_2$	Et_3N/1.4	–	65	–	–	18

SEQUENTIAL POLYPEPTIDES (continued)

	Polymerizing conditions					Product			
Entry no.	Poly	Agent	Concentration (*M*)/solvent	Base/ equivalents	Time/ temperature	Yield (%)	Degree of polymerization/ method	Optical purity (%)	Ref.
544a	Lys-Gly-Pro_2-Gly-Pro	-ONp	0.66/$HCONMe_2$	Et_3N/2	48 h/ room temp.	64	16/Dnp	–	167a
545	$[Glu(OEt)]_2$-Gly-$[Glu(OEt)]_3$	-ONp	–/$HCONMe_2$	Et_3N/1	–	98	100/ONp	–	129
546	$[Glu(OMe)]_3$-Val_2 Glu(OMe)	-ONp	–/$HCONMe_2$	Et_3N/1	–	89	70–80/ONp	–	27, 30
547	Pro-Ala-Gly-Pro-Ala-Gly	DCC/HOBztn	0.2/$HCONMe_2$	NA	60 h/ room temp.	–	9/P-2[b]	–	136
548	Pro-Ala-Gly-Pro-Ala-Gly	DCC/HOBztn	0.4/$HCONMe_2$	NA	15 min/ room temp.	84	1/P-2[b]	–	136
549	Pro-Ala-Gly-Pro-Pro-Gly	-ONSu	0.6/Me_2SO	Et_3N/1	Several days/ room temp.	–	24/Eq.	–	167
550	Pro-Lys-Gly-Pro-Ala-Gly	-OPcp	–	–	–	–	<10/–	–	136
551	Pro-Glu(OBu^t)-Gly-Pro-Ala-Gly	-ONSu	0.4–0.7/$HCONMe_2$	MeMorph/2	3 days/ room temp.	35	18/–	–	136
552	Pro-Glu(OBu^t)-Gly-Pro-Ala-Gly	-OPcp	0.6/Me_2SO	MeMorph/2	3 days/ room temp.	43	25/–	–	136
553	Pro-Glu(OBu^t)-Gly-Pro-Ala-Gly	EEDQ	0.5/$HCONMe_2$	MeMorph/1.1	3 days/ room temp.	20	18/–	–	136
554	Pro-Glu(OBu^t)-Gly Pro-Ala-Gly	DCC/HOBztl	0.5/$HCONMe_2$	MeMorph/1.1	3 days/ room temp.	10	9/–	–	136
555	Pro-Glu(OBu^t)-Gly-Pro-Glu(OBu^t)-Gly	EEDQ	–	–	–	20–25	15/–	–	136
556	Ala_2-Glu(OBzl)-Ala_3-Glu(OBzl)	–	–	–	–	–	–	–	168
557	Leu-Glu-Lys-Ala-Glu-Ala-Gly	-ONp	0.5/Me_2SO	Et_3N/2	5 days/ room temp.	37	4_n, $9^f{}_w$/P-150	–	169
558	Leu-Glu-Lys-Ala-Glu-Ser-Gly	-ONp	0.3/Me_2SO	Et_3N/2	5 days/ room temp.	62	8_n, $15^f{}_w$/P-150	–	169
559	Glu-Ala_2-Lys-Ala_3	-OPhOH	0.3/Me_2SO	Et_3N/2[b]	1 week/ room temp.	25	8/G-25[b]	–	170
560	Glu-Ala_2-Glu-Ala_3	-OPhOH	0.3/Me_2SO	Et_3N/2[b]	1 week/ room temp.	60	13, 26/G-50[b]	–	170
561	Ala_2-Gly-Ala_2-Gly-Pro-Ala-Gly	DCC/	0.8/$HCONMe_2$	NA	5 days/ room temp.	–	8/P-4[b]	–	136

SEQUENTIAL POLYPEPTIDES (continued)

REFERENCES

1. Johnson, *J. Pharm. Sci.*, 63, 313 (1974).
2. Lotan, Berger, and Katchalski, *Ann. Rev. Biochem.*, 41, 869 (1972).
3. Goren, *CRC Crit. Rev. Biochem.*, 2, 197 (1974).
4. Jones, in *Amino Acids, Peptides and Proteins,* Vol. 3, Young, Ed., The Chemical Society, London, 1971, chapt. 3.
5. Kovacs, Ballina, Rodin, Balasubramanian, and Applequist, *J. Am. Chem. Soc.*, 87, 119 (1965).
6. Kovacs, Ballina, and Rodin, *Chem. Ind.*, p. 1955 (1963).
7. Brack and Spach, *C. R. Acad. Sci. C,* 271, 916 (1970).
8. Stewart, *Aust. J. Chem.*, 22, 1291 (1969).
9. Poroshin, Chuvaeva, and Shibnev, *Dokl. Akad. Nauk Tadzh. SSR,* 12(7), 28 (1969).
10. Shibnev, Chuvaeva, Khalikov, and Poroshin, *Izv. Akad. Nauk SSSR Ser. Khim.*, p. 2409 (1970).

10a. Shibnev, Chuvaeva, Khalikov, and Poroshin, *Izv. Akad. Nauk SSSR Ser. Khim.*, p. 2566 (1970).

11. Burichenko, Ovchinnikova, Dakhte, and Poroshin, *Dokl. Akad. Nauk Tadzh. SSR,* 14(3), 30 (1971).
12. Yusupov, Zegelman, Sharifova, and Poroshin, *Dokl. Akad. Nauk Tadzh. SSR,* 12(9), 32 (1969).
13. Zegelman, Yusupov, Demyanik, Sharifova, and Poroshin, *Dokl. Akad. Nauk Tadzh. SSR,* 14(4), 31 (1971).
14. De Tar, Gauge, Honsberg, and Honsberg, *J. Am. Chem. Soc.*, 89, 988 (1967).
15. Rippon, Lowbridge, and Walton, *Abstr. Biophys. Soc.*, 13, 287a (1973).
16. Brack and Spach, *Biopolymers,* 11, 563 (1972).
17. Brack, Caille, and Spach, *Monatsh. Chem.*, 103, 1604 (1972).
18. Stewart, *Aust. J. Chem.*, 18, 887 (1965).
19. Anderson, Chen, Rippon, and Walton, *J. Mol. Biol.*, 67, 459 (1972).
20. Shibnev and Chuvaeva, *Izv. Akad. Nauk SSSR Ser. Khim.*, p. 954 (1967).
21. Ebihara and Kishida, *Nippon Kagaka Zasshi,* 90, 819 (1969).
22. Huggins, Ohtsuka, and Morimota, *J. Polym. Sci. C.*, 23, 343 (1968).
23. Burichenko, Poroshin, Kasimova, and Shibnev, *Izv. Akad. Nauk SSSR Ser. Khim.*, p. 2596 (1972).
24. Cowell and Jones, *J. Chem. Soc. Sect. C,* p. 1082 (1971).
25. De Tar, Rogers, and Bach, *J. Am. Chem. Soc.*, 89, 3039 (1967).
26. Stewart, *Aust. J. Chem.*, 21, 1935 (1968).
27. Stewart, *Aust. J. Chem.*, 19, 1503 (1966).
28. Fraser, Harrap, MacRae, Stewart, and Suzuki, *J. Mol. Biol.*, 14, 423 (1965).
29. Katakai, Oya, Toda, Uno, and Iwakura, *Macromolecules,* 6, 827 (1973).
30. Fraser, Harrap, MacRae, Stewart, and Suzuki, *J. Mol. Biol.*, 12, 482 (1965).
31. Ridge, Rydon, and Snell, *J. Chem. Soc. Perkin Trans. I,* p. 2041 (1972).
32. De Tar, Honsberg, Honsberg, Wieland, Gouge, Bach, Tahara, Brinigar, and Rogers, *J. Am. Chem. Soc.*, 85, 2873 (1963).
33. Yamamoto and Noguchi, *J. Biochem.*, 67, 103 (1970).
34. Ismailov and Shibnev, *Izv. Akad. Nauk Tadzh. SSSR Ser. Khim.*, p. 1872 (1973).
35. Ledger and Stewart, *Aust. J. Chem.*, 20, 2509 (1967).

35a. Maryash and Shibnev, *Khim. Prir. Soedin.*, p. 213 (1974).

36. Hardy, Haylock, and Rydon, *J. Chem. Soc. Perkin Trans. I,* p. 605 (1972).
37. Ali, Hardy, and Rydon, *J. Chem. Soc. Perkin Trans. I,* p. 1070 (1972).
38. De Tar and Vajda, *J. Am. Chem. Soc.*, 89, 998 (1967).
39. Kovacs, Schmit, and Ghatak, *Biopolymers,* 6, 817 (1968).
40. Kovacs and Johnson, *J. Chem. Soc.* (Lond.), p. 6777 (1965).
41. Kovacs, Kapoor, Ghatak, Mayers, Giannasio, Giannotti, Senyk, Nitecki, and Goodman, *Biochemistry,* 11, 1953 (1972).
42. Hardy, Rydon, and Thompson, *J. Chem. Soc. Perkin Trans. I,* p. 5 (1972).
43. Scatherin, Tamburro, and Rocchi, *Gazz. Chim. Ital.*, 98, 1247 (1968).
44. Rodin and Kovacs, *Bioorg. Chem.*, 2, 65 (1972).
45. Marlborough and Rydon, *J. Chem. Soc. Perkin Trans. I,* p. 1 (1972).
46. Hardy, Rydon, and Storey, *J. Chem. Soc. Perkin Trans. I,* p. 1523 (1972).
47. Trudelle, *Chem. Commun.*, p. 639 (1971).
48. Heitz and Spach, *Macromolecules,* 4, 429 (1971).
49. Kovacs, Schmit, and Johnson, *Can. J. Chem.*, 47, 3690 (1969).
50. Balasubramanian, Kalita, and Kovacs, *Biopolymers,* 12, 1089 (1973).
51. Bruckner, Szekerke, and Kovacs, *Naturwissenschaften,* 43, 107 (1956).
52. Stewart, *Aust. J. Chem.*, 22, 2451 (1969).
53. De Tar, Albers, and Gilmore, *J. Org. Chem.*, 37, 4377 (1972).

SEQUENTIAL POLYPEPTIDES (continued)

54. **Mattice and Mandelkern,** *Biochemistry,* 10, 1926 (1971).
55. **Mattice and Mandelkern,** *J. Am. Chem. Soc.,* 92, 5285 (1970).
55a. **Berbiera, Burichenko, Zegelman, and Shibnev,** *Khim. Prir. Soedin.,* p. 206 (1974).
56. **Kovacs, Mayers, Johnson, Giannotti, Cortegiano and Roberts,** in *Progress in Peptide Research,* Lande, Ed., Gordon and Breach, New York, 1972, 185.
57. **Kovacs, Giannotti, and Kapoor,** *J. Am. Chem. Soc.,* 88, 2282 (1966).
58. **De Tar and Estrin,** *Tetrahedron Lett.,* p. 5985 (1966).
59. **Poroshin and Burichenko,** *Dokl. Akad. Nauk Tadzh. SSR,* 11(11), 28 (1968).
60. **Shibnev and Lazareva,** *Izv. Akad. Nauk SSSR Ser. Khim.,* p. 398 (1969).
61. **Heidemann and Bernhardt,** *Nature,* 216, 263 (1967).
62. **Shibnev, Chuvaeva, and Poroshin,** *Izv. Akad. Nauk SSSR Ser. Khim.,* p. 2527 (1969).
63. **Shibnev, Chuvaeva, and Poroshin,** *Izv. Akad. Nauk SSSR Ser. Khim.,* p. 121 (1970).
64. **Khalikov, Poroshin, and Shibnev,** *Dokl. Akad. Nauk Tadzh. SSR,* 11(7), 29 (1968).
65. **Wolman, Gallop, Patchornik, and Berger,** *J. Am. Chem. Soc.,* 84, 1889 (1962).
66. **Shibnev, Khalikov, Finogenova, and Poroshin,** *Izv. Akad. Nauk SSSR Ser. Khim.,* p. 399 (1970).
67. **Poroshin, Khalikov, and Shibnev,** *Dokl, Akad. Nauk Tadzh. SSR,* 12(3), 26 (1969).
68. **Heidemann and Nil,** *Z. Naturforsch.,* 24, 843 (1969).
69. **Heidemann and Nil,** *Z. Naturforsch.,* 24, 837 (1969).
70. **Poroshin, Khalikov, Ismailov, Maryash, Dakhte, and Shibnev,** *Dokl. Akad. Nauk Tadzh. SSR,* 14(6), 34 (1971).
71. **Davydov and Debabov,** *Zh. Obshch. Khim.,* 37, 1000 (1967).
72. **Burichenko, Korycekina, and Poroshin,** *Dokl. Akad. Nauk Tadzh. SSR,* 14(5), 16 (1971).
73. **Poroshin, Burichenko, Maryash, and Shibnev,** *Izv. Akad. Nauk SSSR Ser. Khim.,* p. 1276 (1971).
74. **Shibnev, Grechishko, and Poroshin,** *Izv. Akad. Nauk SSSR Ser. Khim.,* p. 2327 (1967).
75. **Zegelman, Yusupov, Salitra, and Poroshin,** *Dokl. Akad. Nauk Tadzh. SSR,* 15(5), 33 (1972).
76. **Burichenko, Morozova, and Poroshin,** *Dokl. Akad. Nauk Tadzh. SSR,* 13, 26 (1970).
77. **Shibnev, Khalikov, Ismailov, and Poroshin,** *Izv. Akad. Nauk SSSR Ser. Khim.,* p. 1874 (1973).
78. **Shibnev, Chuvaeva, Martynova, and Poroshin,** *Izv. Akad. Nauk SSSR Ser. Khim.,* p. 637 (1969).
79. **Shibnev, Chuvaeva, Martynova, and Poroshin,** *Izv. Akad. Nauk SSSR Ser. Khim.,* p. 2532 (1969).
80. **Shibnev, Lisovenko, Chuvaeva, and Poroshin,** *Izv. Akad. Nauk SSSR Ser. Khim.,* p. 2564 (1968).
81. **Bloom, Dasgupta, Patel, and Blout,** *J. Am. Chem. Soc.,* 88, 2035 (1966).
82. **Fairweather and Jones,** *Immunology,* 25, 241 (1973).
83. **Shibnev, Dyermaev, Chuvaeva, Smirnov, and Poroshin,** *Izv, Akad. Nauk SSSR Ser. Khim.,* p. 1634 (1967).
84. **Shibnev, Poroshin, Chuvaeva, and Martynova,** *Izv. Akad. Nauk SSSR Ser. Khim.,* p. 1144 (1968).
85. **Poroshin, Chuvaeva, and Shibnev,** *Dokl. Akad. Nauk Tadzh. SSR,* 12(12), 21 (1969).
86. **Jones,** *Chem. Commun.,* p. 1436 (1969).
87. **Kitaoka, Sakakibara, and Tani,** *Bull. Chem. Soc. Jap.,* 31, 802 (1958).
88. **Poroshin, Chuvaeva, and Shibnev,** *Dokl. Akad. Nauk Tadzh. SSR,* 12(11), 15 (1969).
89. **Shibnev, Finogenova, and Poroshin,** *Dokl. Akad. Nauk SSSR,* 198(4), 862 (1971).
90. **Poroshin, Chuvaeva, and Shibnev,** *Dokl. Akad. Nauk Tadzh. SSR,* 12(8), 21 (1969).
91. **Shibnev, Chuvaeva, and Poroshin,** *Izv. Akad. Nauk SSSR Ser. Khim.,* p. 225 (1968).
92. **Shibnev, Chuvaeva, and Poroshin,** *Izv. Akad. Nauk SSSR Ser. Khim.,* p. 1825 (1968).
93. **Fairweather and Jones,** *J. Chem. Soc. Perkin Trans. I,* p. 2475 (1972).
94. **Anderson, Rippon, and Walton,** *Biochem. Biophys. Res. Commun.,* 39, 802 (1970).
95. **Takahashi,** *Bull. Chem. Soc. Jap.,* 42, 521 (1969).
96. **Chuvaeva, Morozova, Shibnev, and Poroshin,** *Dokl. Akad. Nauk Tadzh. SSR,* 13(9), 28 (1970).
97. **Fairweather and Jones,** *J. Chem. Soc. Perkin Trans. I,* p. 1908 (1972).
98. **Doyle, Traub, Lorenzi, Brown, and Blout,** *J. Mol. Biol.,* 51, 47 (1970).
99. **Parrish,** Ph.D. thesis, Harvard University, 1969.
100. **Shibnev and Maryash,** *Izv. Akad. Nauk SSSR Ser. Khim.,* p. 435 (1973).
101. **Scatturin, Tamburro, Rocchi, and Coletta,** *Gazz. Chim. Ital.,* 96, 1393 (1966).
101a. **Burichenko, Kamuroza, and Shibnev,** *Khim. Prir. Soedin.,* p. 201 (1974).
102. **Shibnev, Maryash, and Poroshin,** *Dokl. Akad. Nauk SSSR,* 207(3), 625 (1972).
103. **Lorenzi, Doyle, and Blout,** *Biochemistry,* 10, 3046 (1971).
104. **Segal and Traub,** *J. Mol. Biol.,* 43, 487 (1969).
105. **Dellacherie and Néel,** *C.R. Acad. Sci.,* 266, 527 (1968).
106. **Dellacherie and Néel,** *Bull. Soc. Chim. Fr.,* p. 1218 (1969).
107. **Liwschitz, Zilkha, Borensztain, and Frankel,** *J. Org. Chem.,* 21, 1531 (1956).
108. **Iio and Takahashi,** personal communication.
109. **D'Alagni, Bemparad, and Garofolo,** *Polymers,* 13, 419 (1972).

SEQUENTIAL POLYPEPTIDES (continued)

110. **Fridkin, Frenkel, and Ariely,** *Biopolymers,* 8, 661 (1969).
111. **Tamburro, Scatturin, and Marchiori,** *Gazz. Chim. Ital.*, 98, 638 (1968).
112. **Ramachandran, Berger, and Katchalski,** *Biopolymers,* 10, 1829 (1971).
113. **Schechter, Schechter, Ramachandran, Conway-Jacobs, and Sela,** *Eur. J. Biochem.*, 20, 301 (1971).
114. **Goren, Katchalski, and Fridkin,** *Int. Cong. Biochem.*, 89 (1973), in preparation.
115a. **Cernosek, Malin, Wells, and Fasman,** *Biochemistry,* 13, 1252 (1974).
115b. **Johnson,** *J. Chem. Soc. Sect. C,* p. 3008 (1968).
116. **Yaron, Tal (Turkeltaub), and Berger,** *Biopolymers,* 11, 2461 (1972).
117. **Cowell and Jones,** *J. Chem. Soc. Perkin Trans. I,* p. 2236 (1972).
118. **Cowell and Jones,** *Chem. Commun.*, p. 1009 (1971).
119. **Fletcher (Weighardt) and Goren,** in preparation.
120. **Burichenko, Poroshin, Kasimova, Mirzoev, and Shibnev,** *Izv. Akad. Nauk SSSR Ser. Khim.*, p. 2597 (1972).
121. **Zeiger,** personal communication.
122. **Kovacs, Kalita, and Ghatak,** *J. Org. Chem.*, 37, 30 (1972).
123. **Rippon, Chen, Anderson, and Walton,** *Biopolymers,* 11, 1411 (1972).
124. **Zegelman, Yusupov, and Poroshin,** *Dokl. Akad. Nauk SSSR,* 13(2), 25 (1970).
125. **Johnson,** *J. Chem. Soc. Sect. C,* p. 2638 (1967).
126. **Gruetzmacher, Treiber, and Walton,** *Biopolymers,* (1974), in press.
127. **Kovacs and Kapoor,** *J. Am. Chem. Soc.*, 87, 118 (1965).
128. **Trudelle,** *J. Chem. Soc. Perkin Trans. I,* p. 1001 (1973).
129. **Stewart,** *Aust. J. Chem.*, 19, 2373 (1966).
130. **Caille, Heitz, and Spach,** *J. Chem. Soc. Perkin Trans. I,* (1974), in press.
131. **Wolman, Gallop, and Patchornik,** *J. Am. Chem. Soc.*, 83, 1263 (1961).
132. **Heidemann and Meisel,** *Makromol. Chem.*, 166, 1 (1973).
133. **Engel, Kurtz, Katchalski, and Berger,** *J. Mol. Biol.*, 17, 255 (1966).
134. **Brown, Di Cortao, Lorenzi, and Blout,** *J. Mol. Biol.*, 63, 85 (1972).
135. **Heidemann and Bernhardt,** *Nature,* 220, 1326 (1968).
136. **Heidemann,** personal communication.
137. **Shibnev, Khalikov, Finogenova, and Poroshin,** *Izv. Akad. Nauk SSSR Ser. Khim.*, p. 2822 (1970).
138. **Sakakibara, Kishida, Kikuchi, Sakai and Kakiuchi,** *Bull. Chem. Soc. Jap.*, 41, 1273 (1968).
139. **Sakakibara, Inouye, Shudo, Kishida, Kobayashi, and Prockop,** *Biochim. Biophys. Acta,* 303, 198 (1973).
140. **Shibnev, Khalikov, Finogenova, and Poroshin,** *Izv. Akad. Nauk SSSR Ser. Khim.*, p. 880 (1970).
141. **Johnson, Cheng, and Tsang,** *J. Med. Chem.*, 15, 95 (1972).
142. **Johnson,** *J. Pharm. Sci.*, 62, 1564 (1973).
143. **Johnson and Chen,** *J. Pharm. Sci.*, 60, 330 (1971).
144. **Johnson and Cheng,** *J. Med. Chem.*, 14, 1238 (1971).
145. **Johnson,** *J. Pharm. Sci.*, 61, 1654 (1972).
146. **Johnson and Cheng,** *J. Med. Chem.*, 16, 415 (1973).
147. **Johnson and Trask,** *J. Med. Chem.*, 14, 251 (1971).
148. **Johnson and Trask,** *J. Chem. Soc. Sect. C,* p. 2644 (1969).
149. **Johnson,** *J. Pharm. Sci.*, 59, 1849 (1970).
150. **Johnson,** *J. Med. Chem.*, 15, 423 (1972).
151. **Johnson and Chen,** *J. Med. Chem.*, 14, 640 (1971).
152. **Johnson and Cheng,** *J. Pharm. Sci.*, 62, 917 (1973).
153. **Johnson and Cheng,** *J. Med. Chem.*, 17, 320 (1974).
154. **Johnson and Trask,** *J. Chem. Soc. Sect. C,* 2247 (1970).
155. **Johnson and Tsang,** *J. Med. Chem.*, 15, 488 (1972).
155a. **Maryash and Shibnev,** *Khim. Prir. Soedin.*, p. 213 (1974).
156. **Johnson and Rea,** *Can. J. Chem.*, 48, 2509 (1970).
157. **Wender, Bensusan, Treiber, and Walton,** *Biopolymers,* 13, 1929 (1974).
158. **Kovacs, Cervi, Kapoor, and Kalita,** personal communication.
159. **Jones and Walker,** *J. Chem. Soc. Perkin Trans. I,* p. 2923 (1972).
160. **Johnson,** *J. Med. Chem.*, 14, 488 (1971).
161. **Vajda,** *Acta Chim. Acad. Sci. Hung.*, 46, 221 (1965).
162. **Zeiger, Lange, and Maurer,** *Biopolymers,* 12, 2135 (1973).
163. **Johnson,** *J. Chem. Soc. Sect. C,* p. 1412 (1969).
164. **Lotz, Heitz, and Spach,** *C. R. Acad. Sci. Ser. C,* 276, 1373 (1973).
165. **Spach and Heitz,** *C. R. Acad. Sci. Ser. C,* 276, 1373 (1973).
166a. **Chakravarty, Mathur, and Dhar,** *Indian J. Biochem. Biophys.*, 10, 233 (1973).
166b. **Stewart,** *Aust. J. Chem.*, 19, 489 (1966).

SEQUENTIAL POLYPEPTIDES (continued)

167. **Segal,** *J. Mol. Biol.*, 43, 497 (1969).
167a. **Sigler and Walton,** personal communication.
168. **Walton,** personal communication.
169. **Cowell and Jones,** *J. Chem. Soc. Perkin Trans. I*, p. 1814 (1972).
170. **Treiber and Walton,** *Biopolymers,* (1974), in press.

PEPTIDES PREPARED BY SOLID PHASE PEPTIDE SYNTHESIS[a]

G. R. Marshall and R. B. Merrifield

The following table contains peptides prepared by the Solid Phase method of peptide synthesis, or by closely related methods, that were published before November, 1969. The method involves a stepwise elongation of the peptide chain while it is covalently bound to an insoluble solid support. Intermediates are purified by filtration and washing, but are not isolated. After the desired sequence is assembled the peptide is cleaved from the solid and is isolated. Some of the peptides were prepared by an automated apparatus that utilizes this principle.[21]

The nomenclature and symbols for naming synthetic modifications of natural peptides follow the Tentative Rules of the IUPAC-IUB Commission on Biochemical Nomenclature. Disulfide linkages have been omitted to conserve space.

[a]Amino acid residues represent the natural (L) form. Polyfunctional amino acids are joined by normal α-peptide linkage unless indicated otherwise.

Abbreviations: Aoc, adamantyloxycarbonyl; Bip(BMU), benzoylisopropenyl; BME, 2-mercaptoethanol; Boc, *t*-butyloxycarbonyl; CAA, chloroacetic acid; Cap, ε-aminocaproic acid; DCC, *N,N'*-dicyclohexylcarbodiimide; Dim, N-(5,5'-dimethyl-3-oxo-1-cyclohexen-1-yl); DMF, *N,N*-dimethylformamide; Dnp, 2,4-dinitrophenyl; $HOCH_2$-S-DVB, hydroxymethylated copolymer of styrene and divinylbenzene; Hyphlac, *p*-hydroxyphenyllactic acid; Hyv, hydroxyisovaleric acid; Lac, lactic acid; MCP, mercaptopropionyl; NCA, *N*-carboxyanhydride; Nps, *o*-nitrophenylsulfenyl; NEPIS, *N*-ethyl-5-phenylisoxazolium-3'-sulfonate; Oct, octanoyl; OH-AER, hydroxide form anion-exchange resin; ONp, *p*-nitrophenyl ester; OPcp, pentachlorophenyl ester; OSu, *N*-hydroxysuccinimide ester; Phlac, phenyllactic acid; pMZ, *p*-methoxybenzyloxycarbonyl; PCA, pyrrolidone carboxylic acid (pyroglutamic acid); S-DVB-2%, chloromethylated styrene-2%-divinylbenzene; STP, sodium thiophenol; TFA, trifluoroacetic acid; The, β-(2-thienyl)-alanine; Tos, toluenesulfonyl; Tyr(Me), tyrosine *o*-methyl ether; Und, ω-aminoundecanoic acid; Z, benzyloxycarbonyl.

PEPTIDES PREPARED BY SOLID PHASE PEPTIDE SYNTHESIS

Entry number	Number of residues	Trivial name or source	Sequence
1	2	**Leucylglycine**	H-Leu-Gly-OH
2	2	**Leucylglycine**	H-Leu-Gly-OH
3	2	**Valylvaline**	H-Val-Val-OH
4	2	**Dimedonyl-valylalanine**	Dim-Val-Ala-OH
5	2	**TMV protein—(111–112)**	H-Thr-Arg-OH
6	2	**Alanylphenylalanine**	H-Ala-Phe-OH
7	2	**di[1,6—diaminohexane—sebacic acid amide]**	H-[NH$(CH_2)_6$NH-CO$(CH_2)_8$CO$]_2$OH
8	2	**Octanoyl TMV protein—(111–112)**	Oct-Thr-Arg-OH
9	2	**Valylalaline,** protected	Dim-Val-Ala-OMe
10	2	**Tyrosylglycine,** protected	Boc-Tyr(Bzl)-Gly-OMe
11	2	**Asparaginylalanine,** protected	Boc-Asn-Ala-OMe
12	2	**Lysylglycine,** protected	Boc-Lys(Z)-Gyl-OMe
13	2	**Histidylphenylalanine,** protected	Boc-His(Bzl)-Phe-OMe
14	2	**Aspartylalanine,** protected	Boc-Asp(OMe)-Ala-OMe
15	2	**Tryptophanylphenylalanine**	H-Try-Phe-OH
16	2	**Glutamyllysine**	H-Glu-Lys-OH
17	3	**Prolylleucylglycine amide,** protected	Z-Pro-Leu-Gly-NH_2
18	3	**Aspartylalanylvaline**	H-Asp-Ala-Val-OH
19	3	**Phenylalanyllysylglycylhydrazide,** protected	Z-Phe-Lys(Boc)-Gly-$NHNH_2$
20	3	**Phenylalanylglycylglycine**	H-Phe-Gly-Gly-OH
21	3	**Leucylphenylalanylvaline**	H-Leu-Phe-Val-OH
22	3	**Dimedonyl-valylleucylalanine**	Dim-Val-Leu-Ala-OH
23	3	**Prolyllysylglycinamide,** protected	Z-Pro-Lys(Tos)-Gly-NH_2
24	3	**TMV protein—(110–112)**	H-Ala-Thr-Arg-OH
25	3	**Cysteinylcysteinylalanine**	H-Cys-Cys-Ala-OH
26	3	**Tri[1,6—diaminohexane—sebacic acid amide]**	H[NH$(CH_2)_6$NHCO$(CH_2)_8$CO$]_3$OH
27	3	**Prolylleucylglycinamide,** protected	Z-Pro-Leu-Gly-NH_2
28	3	**Leucyltyrosylalanylhydrazide,** protected	Boc-Leu-Tyr-Ala-$NHNH_2$
29	3	**Glutathione,** protected	Z-Glu-OH-Cys(Bzl)-Gly-OH
30	3	**Glycylphenylalanylalanine**	H-Gly-Phe-Ala-OH
31	3	**Octanoyl TMV protein—(110–112)**	Oct-Ala-Thr-Arg-OH
32	3	**Leucylcysteinylalanine,** protected	Boc-Leu-Cys(Bzl)-Ala-OMe
33	3	**Alanylleucylphenylalanine,** protected	Boc-Ala-Leu-Phe-OMe
34	3	**Alanylalanylalanine**	H-Ala-Ala-Ala-OH
35	3	**Phenylalanylalanylserine,** protected	Boc-Phe-Ala-Ser(Bzl)-OMe
36	3	**Alanylglutamylalanine,** protected	Boc-Ala-Glu(OMe)-Ala-OMe
37	3	**Valylglycylaspartic acid,** protected	Boc-Val-Gly-Asp(OMe)-OMe
38	3	**Glycylvalylthreonine,** protected	Z-Gly-Val-Thr(Bzl)-OMe
39	3	**Glycyllysylglycine,** protected	Ac-Gly-Lys-Gly-OH
40	3	**Phenylalanyllysylglycine,** protected	Z-Phe-Lys(Boc)-Gly-$NHNH_2$
41	3	**Methionylglutamyllysine**	H-Met-Glu-Lys-OH
42	3	**Alanylserylphenylalanine**	H-Ala-Ser-Phe-OH
43	3	**Alanylthreonylphenylalanine**	H-Ala-Thr-Phe-OH
44	4	**Leucylalanylglycylvaline**	H-Leu-Ala-Gly-Val-OH
45	4	**Leucylalanylglycylvaline**	H-Leu-Ala-Gly-Val-OH
46	4	**Glycylglycylleucylglycine**	H-Gly-Gly-Leu-Gly-OH
47	4	**Leucylaspartylalanylvaline**	H-Leu-Asp-Ala-Val-OH
48	4	**Glycylaspartylserylglycine**	H-Gly-Asp-Ser-Gly-OH
49	4	**Phenylalanylglycylglycylglycine**	H-Phe-Gly-Gly-Gly-OH
50	4	**Pyroglutamylphenylalanylalanylarginine**	PCA-Phe-Ala-Arg-OH
51	4	**Dimedonyl-leucylvalylalanylphenylalanine**	Dim-Leu-Val-Ala-Phe-OH
52	4	**Phenylalanylphenylalanylphenylalanyl-glycylhydrazide,** protected	Aoc-Phe-Phe-Phe-Gly-$NHNH_2$
53	4	**TMV protein—(109–112)**	H-Asp-Ala-Thr-Arg-OH
54	4	**Seryltryptophanyllysylarginine**	H-Ser-Trp-Lys-Arg-OH
55	4	**Leucylphenylalanylalanylalanine,** protected	Z-Leu-Phe-Ala-Ala-OH
56	4	**Octanoyl TMV protein—(109–112)**	Oct-Asp-Ala-Thr-Arg-OH
57	4	**Valylglycolylhydroxylvalerylphenyllactic acid**	H-Val-Glyc-Hyv-Phlac-OH
58	4	**Leucylalanylglycylphenylalanine**	H-DL-Leu-Ala-Gly-Phe-OH

PEPTIDES PREPARED BY SOLID PHASE PEPTIDE SYNTHESIS (continued)

Entry number	Reference	Solid support	Type of bond	Monomer protection	Coupling reagent	Deprotection	Cleavage
1	1	S-DVB-4%	Bzl	Z	ONp	HBr-HOAc	HBr-HOAc
2	2, 3	Popcorn polymer	Z	OEt	Mixed anh.	NaOH	15% HBr-HOAc
3	4	S-DVB-2%	Bzl	pMZ	DCC	TFA	HBr-TFA
4	5	S-DVB-2%	Bzl	Boc	DCC	HCl-HOAc	NaOH-EtOH
5	38	S-DVB-2%	Bzl	Boc	DCC	HCl-HOAc	HBr-TFA
6	48	S-DVB-2%	Alkyl	Z	Mixed anh.	HBr-HOAc	NaOH-Diox-EtOH
7	36	S-DVB-2%	Bzl	Boc	DCC	TFA	HBr-TFA
8	111	S-DVB-2%	Bzl	Boc	DCC	HCl-HOAc	HBr-TFA
9	98	S-DVB-2%	Bzl	Boc	DCC	HCl-HOAc	OH-AER/MeOH
10	98	S-DVB-2%	Bzl	Boc	DCC	HCl-HOAc	OH-AER/MeOH
11	98	S-DVB-2%	Bzl	Boc	ONp	HCl-HOAc	OH-AER/MeOH
12	98	S-DVB-2%	Bzl	Boc	DCC	HCl-HOAc	OH-AER/MeOH
13	98	S-DVB-2%	Bzl	Boc	DCC	HCl-HOAc	OH-AER/MeOH
14	98	S-DVB-2%	Bzl	Boc	DCC	HCl-HOAc	OH-AER/MeOH
15	60	S-DVB-2%	Bzl	Boc	DCC	HCl-HOAc	HF,HBr-TFA
16	109	S-DVB-2%	Bzl	Boc	DCC	HCl-HOAc	HBr-TFA
17	6	NO_2-S-DVB-2%	NO_2-Bzl	Z	ONp	HBr-HOAc	NH_3
18	7	S-DVB-2%	Bzl	Boc	DCC	HCl-HOAc	HBr-TFA
19	8	S-DVB-2%	Bzl	Boc,Nps,Z	DCC	CH_3C-SNH_2	N_2H_4
20	9	S-DVB-2%	Bzl	Boc	DCC	HCl-HOAc	HBr-TFA
21	4	S-DVB-2%	Bzl	pMZ	DCC-HOSu	TFA	HBr-TFA
22	5	S-DVB-2%	Bzl	Boc	DCC	HCl-HOAc	NaOH
23	23	S-DVB-2%	Bzl	Boc	DCC,ONp	HCl-HOAc	NH_3-EtOH
24	38	S-DVB-2%	Bzl	Boc	DCC	HCl-HOAc	HBr-TFA
25	111	S-DVB-2%	Bzl	Boc	DCC	HCl-HOAc	HBr-TFA
26	36	S-DVB-2%	Bzl	Boc	DCC	TFA	HBr-TFA
27	37	S-DVB-2%	Bzl	Boc	ONp,OSu-triazole	HCl-HOAc	NH_3
28	134	S-DVB-2%	Bzl	Boc	ONp,OSu	HCl-Diox.	NH_2NH_2-DMF
29	70	Phenol-trioxane	Phenyl	Z	DCC,ONp	HBr-HOAc	NaOH-MeOH
30	39	S-DVB-2%	Bzl	Boc	Mixed anh.	HCl-HOAc	NaOH-EtOH
31	110	S-DVB-2%	Bzl	Boc	DCC	HCl-HOAc	HBr-TFA
32	98	S-DVB-2%	Bzl	Boc	DCC	HCl-HOAc	OH-AER/MeOH
33	98	S-DVB-2%	Bzl	Boc	DCC	HCl-HOAc	OH-AER/MeOH
34	62	S-DVB-2%	Bzl	Boc	DCC,OSu	HCl-HOAc	HBr-TFA
35	98	S-DVB-2%	Bzl	Boc	DCC	HCl-HOAc	OH-AER/MeOH
36	98	S-DVB-2%	Bzl	Boc	DCC	HCl-HOAc	OH-AER/MeOH
37	98	S-DVB-2%	Bzl	Boc	DCC	HCl-HOAc	OH-AER/MeOH
38	98	S-DVB-2%	Bzl	Boc	DCC	HCl-HOAc	OH-AER/MeOH
39	108	S-DVB-2%	Bzl	Boc	DCC	HCl-HOAc	HBr-TFA
40	53	S-DVB-2%	Bzl	Bpoc	DCC	CAA	NH_2NH_2-EtOH
41	109	S-DVB-2%	Bzl	Boc	DCC	HCl-HOAc	HBr-TFA
42	68, 109	S-DVB-2%	Bzl	Boc	DCC	HCl-HOAc	HBr-TFA
43	68, 109	S-DVB-2%	Bzl	Boc	DCC	HCl-HOAc	HBr-TFA
44	10	Br-S-DVB-2%	BrBzl	Z	DCC	HBr-HOAc	HBr-HOAc
45	11	NO_2-S-DVB-2%	NO_2Bzl	Z	DCC	HBr-HOAc	NaOH
46	12	Polystyrene	Bzl	Boc	OSu	HCl-DMF	HBr-TFA
47	7	S-DVB-2%	Bzl	Boc	DCC	HCl-HOAc	HBr-TFA
48	13	S-DVB-2%	Bzl	Boc	DCC	HCl-HOAc	HBr-TFA
49	9	S-DVB-2%	Bzl	Boc	DCC	HCl-HOAc	HBr-TFA
50	14	S-DVB-2%	Bzl	Boc	DCC	HCl-HOAc	HF
51	5	S-DVB-2%	Bzl	Boc	DCC	HCl-HOAc	NaOH
52	70	Phenyl-trioxane	Phenyl	Aoc	DCC	HCl-MeOH	NH_2NH_2-DMF
53	38	S-DVB-2%	Bzl	Boc	DCC	HCl-HOAc	HBr-TFA
54	93	S-DVB-2%	Bzl	Boc	DCC	HCl-Diox.	HF
55	64	S-DVB-2%	Phenacyl	Boc	OSu	TFA	Sodium Thiophenol (STP)
56	110	S-DVB-2%	Bzl	Boc	DCC	HCl-HOAc	HBr-TFA
57	41	S-DVB-2%	Bzl	t-butyl	Mixed anh.	TFA	HBr-TFA
58	45	S-DVB-2%	Bzl	Boc	DCC	HCl-HOAc	HBr-TFA

PEPTIDES PREPARED BY SOLID PHASE PEPTIDE SYNTHESIS (continued)

Entry number	Number of residues	Trivial name or source	Sequence
59	4	**Leucylleucylglycylphenylalanine**	H-DL-Leu-Leu-Gly-Phe-OH
60	4	**Tryptophanylphenylalanylprolylalanine,** protected	Boc-Try-Phe-Pro-Ala-OMe
61	4	**Alanylalanylalanylglycine**	H-Ala-Ala-Ala-Gly-OH
62	4	**Leucylalanylglycylvaline**	H-Leu-Ala-Gly-Val-OH
63	4	**Leucyllalanylglycylvaline**	H-Leu-Ala-Gly-Val-OH
64	4	**Phenylalanylvalylalanylleucine,** protected	Z-Phe-Val-Ala-Leu-$NHNH_2$
65	4	**TMV protein (108–112)**	H-Leu-Asp-Ala-Thr-OH
66	4	**Alanylthreonylarginylarginine**	H-Ala-Thr-Arg-Arg-OH
67	4	**Tritiated tetralysine**	H-(^{3}H)-Lys-Lys-Lys-Lys-OH
68	4	**Gastrin tetrapeptide**	Boc-Try-Met-Asp-Phe-NH_2
69	4	**Angiotensin-(1–4),** protected	Z-Asp(Bzl)-Arg(NO_2)-Val-Tyr(Bzl)-$NHNH_2$
70	4	**Angiotensin-(4–8),** protected	H-Val-His(Bzl)-Pro-Phe-OH
71	4	**Leucylleucylleucyltyrosine,** protected	Boc-Leu-Leu-Leu-Tyr(Bzl)-OH
72	4	**Threonyltryptophanyllysylglycine,** protected	Ac-Thr-Try-Lys-Gly-OH
73	4	**Threonylglycyllysylglycine,** protected	Ac-Thr-Gly-Lys-Gly-OH
74	4	**Staphylococcal Nuclease—(136–139),** protected	H-Lys(TFA)-Leu-Asn-Ile-OH
75	4	**Prolyltyrosyllysylmethionine,** protected	H-Pro-Tyr-Lys(Boc)-Met-$NHNH_2$
76	5	**Phenylalanylglycylglycylglycylglycine**	H-Phe-Gly-Gly-Gly-Gly-OH
77	5	**β-(2-Thienyl)-alanylglycylglycylglycylglycine**	H-The-Gly-Gly-Gly-Gly-OH
78	5	**Dimedonyl-valylvalylphenylalanylleucyl-alanine**	Dim-Val-Val-Phe-Leu-Ala-OH
79	5	**[Ala^5, Val^8]-eledoisin—(5–9),** protected	Boc-Ala-Ala-Phe-Val-Gly-OH
80	5	**TMV protein—(108–112)**	H-Leu-Asp-Ala-Thr-Arg-OH
81	5	**Prolylalanylphenylalanylphenylalanyl-leucine**	H-Pro-Ala-Phe-Phe-Leu-OH
82	5	**Dansylprolylleucylglycylprolylarginine**	Dansyl-Pro-Leu-Gly-Pro-Arg-OH
83	5	**TMV protein—(108–112)**	H-Thr-Val-Val-Gln-Arg-OH
84	5	**Threonylvalylglutaminylvalylarginine**	H-Thr-Val-Gln-Val-Arg-OH
85	5	**Physalaemin—(7–11),** protected	Aoc-Phe-Tyr-Gly-Leu-Met-NH_2
86	5	**[Ala^{108}]TMV protein (108–112)**	H-Ala-Asp-Ala-Thr-Arg-OH
87	5	**[Ile^{108}]TMV protein (108–112)**	H-Ile-Asp-Ala-Thr-Arg-OH
88	5	**[Tyr^{108}]TMV protein (108–112)**	H-Tyr-Asp-Ala-Thr-Arg-OH
89	5	**[D-Leu^{108}]TMV protein (108–112)**	H-D-Leu-Asp-Ala-Thr-Arg-OH
90	5	**[Glu^{109}]TMV protein (108–112)**	H-Leu-Glu-Ala-Thr-Arg-OH
91	5	**[Asn^{109}]TMV protein (108–112)**	H-Leu-Asn-Ala-Thr-Arg-OH
92	5	**[Leu^{110}]TMV protein (108–112)**	H-Leu-Asp-Leu-Thr-Arg-OH
93	5	**[Gly^{110}]TMV protein (108–112)**	H-Leu-Asp-Gly-Thr-Arg-OH
94	5	**[Leu^{111}]TMV protein (108–112)**	H-Leu-Asp-Ala-Leu-Arg-OH
95	5	**[Gly^{111}]TMV protein (108–112)**	H-Leu-Asp-Ala-Gly-Arg-OH
96	5	**[Ser^{111}]TMV protein (108–112)**	H-Leu-Asp-Ala-Ser-Arg-OH
97	5	**Retro-TMV protein (108–112)**	H-Arg-Thr-Ala-Asp-Leu-OH
98	5	**Threonylleucylaspartylalanylthreonine**	H-Thr-Leu-Asp-Ala-Thr-OH
99	5	**Aspartylalanylthreonylarginylarginine**	H-Asp-Ala-Thr-Arg-Arg-OH
100	5	**Bradykinin-potentiating Factor**	PCA-Lys-Trp-Ala-Pro-OH
101	5	**Methionylphenylalanylglycylleucyl-alanine,** protected	Boc-Met-Phe-Gly-Leu-Ala-OMe
102	5	**Phenylalanylleucylleucylglycylphenyl-alanine**	H-DL-Phe-Leu-Leu-Gly-Phe-OH
103	5	**TMV protein—(120–124),** protected	H-Ala-Ile-Arg(NO_2)-Ser-Ala-OH
104	5	**Leucylalanylvalylphenylalanyl glycine,** protected	Z-Leu-Ala-Val-Phe-Gly-OH
105	5	**Glucagon—(25–29)**	H-Try-Leu-Met-Asn-Thr-OH
106	5	**Histidylphenylalanylarginyltryptophanyl-glycine**	H-His-Phe-Arg-Try-Gly-OH
107	5	**Serylprolylprolylprolylglycine**	H-Ser-Pro-Pro-Pro-Gly-OH
108	6	**Secretin—(22–27)-hexapeptide amide**	H-Leu-Leu-Glu-Gly-Leu-Val-NH_2

PEPTIDES PREPARED BY SOLID PHASE PEPTIDE SYNTHESIS (continued)

Entry number	Reference	Solid support	Type of bond	Monomer protection	Coupling reagent	Deprotection	Cleavage
59	45	S-DVB-2%	Bzl	Boc	DCC	HCl-HOAc	HBr-TFA
60	98	S-DVB-2%	Bzl	Boc	DCC	HCl-HOAc	OH-AER/MeOH
61	62	S-DVB-2%	Bzl	Boc	OSu	HCl-HOAc	HBr-TFA
62	123	S-DVB-2%	Z	Boc-$NHNH_2$	Azide	HCl-Diox.	HBr-TFA
63	113, 127	S-DVB-2%	Benzhydryl	Bip(BMV)	DCC	0.1N HCl/THF, TSA/THF	5% $TFA/CHCl_3$,0.5N $HCl/HOAc-CHCl_3$
64	115	S-DVB-2%	Boc-hydrazide	Boc	DCC	0.5%TFA/ CH_2Cl_2	50%TFA/CH_2Cl_2
65	135	S-DVB-2%	Bzl	Boc	DCC	HCl-HOAc	HBr-TFA
66	135	S-DVB-2%	Bzl	Boc	DCC	HCl-HOAc	HBr-TFA
67	96	S-DVB-2%	Bzl	Boc	DCC	HCl-Diox.	HBr-TFA
68	61	S-DVB-2%	Bzl	Boc,Nps	DCC,OSu	HCl	NH_3-MeOH
69	104	S-DVB-2%	Boc-hydrazide	Bpoc	DCC	0.5%TFA/ CH_2Cl_2	50%TFA/CH_2Cl_2
70	104	S-DVB-2%	t-Butyl	Phth,Bpoc	DCC	NH_2NH_2,0.5% TFA/CH_2Cl_2	50%TFA/CH_2Cl_2
71	105	S-DVB-2%	Bzl	Boc	DCC	HCl-HOAc	NaOH-Diox.
72	108	S-DVB-2%	Bzl	Boc	DCC	HCl-HOAc	HBr-TFA
73	108	S-DVB-2%	Bzl	Boc	DCC	HCl-HOAc	HBr-TFA
74	128	S-DVB-2%	Bzl	Boc	DCC, ONp	HCl-Diox.	HF
75	53	S-DVB-2%	Bzl	Bpoc	DCC	CAA	NH_2NH_2-EtOH
76	9	S-DVB-2%	Bzl	Boc	DCC	HCl-HOAc	HBr-TFA
77	9	S-DVB-2%	Bzl	Boc	DCC	HCl-HOAc	HBr-TFA
78	5	S-DVB-2%	Bzl	Boc	DCC	HCl-HOAc	NaOH
79	16	S-DVB-2%	Bzl	Boc	DCC	HCl-HOAc	NaOH-EtOH
80	38	S-DVB-2%	Bzl	Boc	DCC	HCl-HOAc	HBr-TFA
81	37	S-DVB-2%	Bzl	Boc	ONp,OSu-triazole	HCl-HOAc	HBr-TFA
82	82	S-DVB-2%	Bzl	Boc	DCC	HCl-HOAc	HBr-TFA
83	49	S-DVB-2%	Bzl	Boc	DCC,ONp	HCl-HOAc	HBr-TFA
84	49	S-DVB-2%	Bzl	Boc	DCC,ONp	HCl-HOAc	HBr-TFA
85	70	Phenol-trioxane	Phenyl	Aoc	DCC,ONp	HCl-MeOH	NH_3-MeOH
86	135	S-DVB-2%	Bzl	Boc	DCC	HCl-HOAc	HBr-TFA
87	135	S-DVB-2%	Bzl	Boc	DCC	HCl-HOAc	HBr-TFA
88	135	S-DVB-2%	Bzl	Boc	DCC	HCl-HOAc	HBr-TFA
89	135	S-DVB-2%	Bzl	Boc	DCC	HCl-HOAc	HBr-TFA
90	135	S-DVB-2%	Bzl	Boc	DCC	HCl-HOAc	HBr-TFA
91	135	S-DVB-2%	Bzl	Boc	DCC,ONp	HCl-HOAc	HBr-TFA
92	135	S-DVB-2%	Bzl	Boc	DCC	HCl-HOAc	HBr-TFA
93	135	S-DVB-2%	Bzl	Boc	DCC	HCl-HOAc	HBr-TFA
94	135	S-DVB-2%	Bzl	Boc	DCC	HCl-HOAc	HBr-TFA
95	135	S-DVB-2%	Bzl	Boc	DCC	HCl-HOAc	HBr-TFA
96	135	S-DVB-2%	Bzl	Boc	DCC	HCl-HOAc	HBr-TFA
97	135	S-DVB-2%	Bzl	Boc	DCC	HCl-HOAc	HBr-TFA
98	135	S-DVB-2%	Bzl	Boc	DCC	HCl-HOAc	HBr-TFA
99	135	S-DVB-2%	Bzl	Boc	DCC	HCl-HOAc	HBr-TFA
100	95	S-DVB-2%	Bzl	Boc	DCC	HCl-Diox.	HF
101	98	S-DVB-2%	Bzl	Boc	DCC	HCl-HOAc	OH-AER/MeOH
102	46	Phenol-CH_2O	Bzl	Boc	DCC	HCl-HOAc	HBr-TFA
103	67, 118	Styrene	Bzl	Boc	Mixed anh.	HCl-THF	HBr-benzene
104	129	Sephadex LH20	Alkyl	Boc	OSu,ONp	TFA	NaOH
105	60	S-DVB-2%	Bzl	Boc	DCC,ONp	HCl-HOAc	HF,HBr-TFA
106	90	S-DVB-2%	Bzl	Boc	DCC	HCl-HOAc	NaOEt/EtOH
107	109	S-DVB-2%	Bzl	Boc	DCC	HCl-HOAc	HBr-TFA
108	15	$HOCH_2$-S-DVB	Bzl	Boc	ONp	HCl-HOAc	NH_3-MeOH

PEPTIDES PREPARED BY SOLID PHASE PEPTIDE SYNTHESIS (continued)

Entry number	Number of residues	Trivial name or source	Sequence
109	6	**Dimedonyl-valylglycylvalylphenylalanyl-leucylalanine**	Dim-Val-Gly-Val-Phe-Leu-Ala-OH
110	6	**Glycyllysylasparaginyllysylglycylarginine**	H-Gly-Lys-Asn-Lys-Gly-Arg-OH
111	6	**Secretin—(22–27)**, protected	Boc-Leu-Leu-Gln-Gly-Leu-Val-OMe
112	6	**TMV protein—(107–112)**	H-Thr-Leu-Asp-Ala-Thr-Arg-OH
113	6	**Chymotrypsinogen—(52–57)**	H-Val-Val-Thr-Ala-Ala-His-OH
114	6	**Lysylcysteinylalanylvalylcysteinyl-arginine**	H-Lys-Cys-Ala-Val-Cys-Arg-OH
115	6	**Staphylococcal nuclease—(100–105)-**, protected	Boc-Leu-Ala-Tyr-Ile-Tyr-Ala-$NHNH_2$
116	6	**Physalaemin—(6–11)**, protected	Aoc-Lys(Aoc)-Phe-Tyr-Gly-Leu-Met-NH_2
117	6	**[Tyr(Me)^8]-physalaemin—(6.11) -**, protected	Aoc-Lys(Aoc)-Phe-Tyr(Me)-Gly-Leu-Met-NH_2
118	6	**[Phe^6, Tyr(Me)^8]-physalaemin—(6.11)**, protected	Aoc-Phe-Phe-Tyr(Me)-Gly-Leu-Met-NH_2
119	6	**[Phe^6, Phe^8]-physalaemin—(6.11)**, protected	Aoc-Phe-Phe-Phe-Gly-Leu-Met-NH_2
120	6	**Valylglutamyllysylalanylalanylalanine**	H-Val-Glu-Lys-Ala-Ala-Ala-OH
121	6	**Prolylprolylphenylalanylphenylalanylalanyl-alanine**, protected	Boc-Pro-Pro-Phe-Phe-Ala-Ala-OH
122	6	**Alanylarginylthreonylalanylaspartylleucine**	H-Ala-Arg-Thr-Ala-Asp-Leu-OH
123	6	**Leucylaspartylalanylthreonylarginyl-arginine**	H-Leu-Asp-Ala-Thr-Arg-Arg-OH
124	6	**Alanylleucyltyrosylleucylvalyl-cysteinylamide**, protected	Z-Ala-Leu-Tyr-Leu-Val-Cys(Bzl)-NH_2
125	6	**Glycyllysylphenylalanylglycylleucyl-alanine**, protected	Boc-Gly-Lys(Z)-Phe-Gly-Leu-Ala-OMe
126	6	**Glutamyllysyllysylserylleucylproline**	H-Glu-Lys-Lys-Ser-Leu-Pro-OH
127	6	**Alanylalanylalanylalanylalanylalanine**	H-Ala-Ala-Ala-Ala-Ala-Ala-OH
128	6	**Ribonuclease— (3–8)**, protected	Boc-Thr-Ala-Ala-Ala-Lys(Z)-Phe-$NHNH_2$
129	6	**Aspartylisoleucylalanylmethionylglutamylly-sine**	H-Asp-Ile-Ala-Met-Glu-Lys-OH
130	7	**Eledoisin analog**	H-Gly-Ala-Phe-Val-Gly-Leu-Met-NH_2
131	7	**Myoglobin—(147–153)**	H-Lys-Glu-Leu-Gly-Tyr-Gln-Gly-OH
132	7	**TMV protein—(106–112)**	H-Glu-Thr-Leu-Asp-Ala-Thr-Arg-OH
133	7	**ACTH—(4–10)**	H-Met-Glu-His-Phe-Arg-Trp-Gly-OH
134	7	**Angiotensin—(2–8)**, protected	H-Arg(NO_2)-Val-Tyr-Ile-His(Bzl)-Pro-Phe-OH
135	7	**Glycylphenylalanylleucylphenylalanyl-leucylglycylphenylalanine**	H-DL-Gly-Phe-Leu-Phe-Leu-Gly-Phe-OH
136	7	**Hemoglobin A—chain (1–7)** (Human)	H-Val-Leu-Ser-Pro-Ala-Asp-Lys-OH
137	7	**Parathyroid hormone (18–24)**	H-Glu-Arg-Val-Glu-Trp-Leu-Arg-OH
138	7	**cyclo-[glycyl-*p*-aminobenzoylglycyl-histidylglycyl-*p*-aminobenzoyl-ε-aminocaproyl]**	cyclo-[Gly-Pab-Gly-His-Gly-Pab-Acap-]
139	7	**Glutamylaspartylisoleucylalanyl-methionyl glutamyl lysine**	H-Glu-Asp-Ile-Ala-Met-Glu-Lys-OH
140	8	**[Ile^5]-angiotensin II**	H-Asp-Arg-Val-Tyr-Ile-His-Pro-Phe-OH
141	8	**[β-Asp^1,Ile^5]-angiotensin II**	H-β-Asp-Arg-Val-Tyr-Ile-His-Pro-Phe-OH
142	8	**[Asn^1,Ile^5]-angiotensin II**	H-Asn-Arg-Val-Tyr-Ile-His-Pro-Phe-OH
143	8	**[Ala^3,Ile^5]-angiotensin II**	H-Asp-Arg-Ala-Tyr-Ile-His-Pro-Phe-OH
144	8	**Des-Arg^1-[Thr^6]-bradykinin**	H-Pro-Pro-Gly-Phe-Thr-Pro-Phe-Arg-OH
145	8	**Des-Arg^1-[Phe^6]-bradykinin**	H-Pro-Pro-Gly-Phe-Phe-Pro-Phe-Arg-OH
146	8	**Des-Arg^1-[Lys(Tos)^6]-bradykinin**	H-Pro-Pro-Gly-Phe-Lys(Tos)-Pro-Phe-Arg-OH
147	8	**Des-Arg^1-[D-Pro^2]-bradykinin**	H-D-Pro-Pro-Gly-Phe-Ser-Pro-Phe-Arg-OH
148	8	**Des-Arg^1-[D-Pro^3]-bradykinin**	H-Pro-D-Pro-Gly-Phe-Ser-Pro-Phe-Arg-OH
149	8	**Des-Arg^1-[D-Pro^7]-bradykinin**	H-Pro-Pro-Gly-Phe-Ser-D-Pro-Phe-Arg-OH
150	8	**Des-Arg^1-[D-Arg^9]-bradykinin**	H-Pro-Pro-Gly-Phe-Ser-Pro-Phe-D-Arg-OH
151	8	**Des-Arg^1-[Tyr(Me)^{5,8}]-bradykinin**	H-Pro-Pro-Gly-Tyr(Me)-Ser-Pro-Tyr(Me)-Arg-OH
152	8	**Des-Arg^1-[Tyr(Me)^5,Thr^6,Leu^8]-bradykinin**	H-Pro-Pro-Gly-Tyr(Me)-Thr-Pro-Leu-Arg-OH
153	8	**Des-Arg^1-[Tyr(Me)^{5,8},Gly^6]-bradykinin**	H-Pro-Pro-Gly-Tyr(Me)-Gly-Pro-Tyr(Me)-Arg-OH
154	8	**Des-Arg^1-[D-Pro^{2,3,7},Tyr(Me)^{5,8}]-bradykinin**	H-D-Pro-D-Pro-Gly-Tyr(Me)-Gly-D-Pro-Tyr(Me)-Arg-OH

PEPTIDES PREPARED BY SOLID PHASE PEPTIDE SYNTHESIS (continued)

Entry number	Reference	Solid support	Type of bond	Monomer protection	Coupling reagent	Deprotection	Cleavage
109	5	S-DVB-2%	Bzl	Boc	DCC	HCl-HOAc	NaOH
110	14	S-DVB-2%	Bzl	Boc	DCC	HCl-HOAc	HF
111	15	$HOCH_2$-S-DVB	Bzl	Boc	ONp	HCl-HOAc	NH_3-MeOH
112	38	S-DVB-2%	Bzl	Boc	DCC	HCl-HOAc	HBr-TFA
113	88	S-DVB-2%	Bzl	Boc	DCC	HCl-Diox.	HBr-TFA
114	93	S-DVB-2%	Bzl	Boc	DCC	HCl-Diox.	HF
115	134	S-DVB-2%	Bzl	Boc	ONp,OSu	HCl-Diox.	NH_2NH_2-DMF
116	69	Phenol-trioxane	Phenyl	Aoc	DCC,ONp	HCl-MeOH	NH_3-MeOH
117	69	Phenol-trioxane	Phenyl	Aoc	DCC,ONp	HCl-MeOH	NH_3-MeOH
118	69	Phenol-trioxane	Phenyl	Aoc	DCC,ONp	HCl-MeOH	NH_3-MeOH
119	69	Phenol-trioxane	Phenyl	Aoc	DCC,ONp	HCl-MeOH	NH_3-MeOH
120	93	S-DVB-2%	Bzl	Boc	DCC	HCl-HOAc	HBr-TFA
121	64	S-DVB-2%	Phenacyl	Boc	OSu	TFA	STP
122	135	S-DVB-2%	Bzl	Boc	DCC	HCl-HOAc	HBr-TFA
123	135	S-DVB-2%	Bzl	Boc	DCC	HCl-HOAc	HBr-TFA
124	94	S-DVB-2%	Bzl	Boc	DCC	HCl-HOAc	NH_3
125	98	S-DVB-2%	Bzl	Boc	DCC	HCl-HOAc	OH-AER-MeOH
126	69	S-DVB-2%	Bzl	Boc	NEPIS, DCC-HoSu	HCl-Diox.	HBr-TFA
127	62	S-DVB-2%	Bzl	Boc	OSu	HCl-HOAc	HBr-TFA
128	85	S-DVB-2%	Bzl	Boc	DCC	HCl-HOAc	NH_2NH_2-DMF
129	109	S-DVB-2%	Bzl	Boc	DCC	HCl-HOAc	HBr-TFA
130	16	S-DVB-2%	Bzl	Boc	DCC	HCl-HOAc	HBr-TFA
131	89	S-DVB-2%	Bzl	Boc	DCC,ONp	HCl-HOAc	HBr-TFA
132	38	S-DVB-2%	Bzl	Boc	DCC	HCl-HOAc	HBr-TFA
133	57	S-DVB-2%	Bzl	Boc	DCC	HCl-Diox.	HBr-TFA
134	7	S-DVB-2%	Bzl	Boc	DCC	HCl-HOAc	HBr-TFA
135	46	Phenol-CH_2O	Bzl	Boc	DCC	HCl-HOAc	HBr-TFA
136	99	S-DVB-2%	Bzl	Boc	DCC	HCl-HOAc	HBr-TFA
137	121	S-DVB-2%	Bzl	Boc,Bpoc	DCC	HCl-Diox.,15% TFA/CH_2Cl_2	HF
138	117	S-DVB-2%	Bzl	Boc	DCC	HCl-HOAc	HBr-TFA
139	109	S-DVB-2%	Bzl	Boc	DCC	HCl-HOAc	HBr-TFA
140	7	S-DVB-2%	Bzl	Boc	DCC	HCl-HOAc	HBr-TFA
141	7	S-DVB-2%	Bzl	Boc	DCC	HCl-HOAc	HBr-TFA
142	7	S-DVB-2%	Bzl	Boc	DCC	HCl-HOAc	HBr-TFA
143	17	S-DVB-2%	Bzl	Boc	DCC	HCl-HOAc	HBr-TFA
144	18	S-DVB-2%	Bzl	Boc	DCC	HCl-Diox.	HBr-TFA
145	18	S-DVB-2%	Bzl	Boc	DCC	HCl-Diox.	HBr-TFA
146	18	S-DVB-2%	Bzl	Boc	DCC	HCl-Diox.	HBr-TFA
147	18	S-DVB-2%	Bzl	Boc	DCC	HCl-Diox.	HBr-TFA
148	18	S-DVB-2%	Bzl	Boc	DCC	HCl-Diox.	HBr-TFA
149	18	S-DVB-2%	Bzl	Boc	DCC	HCl-Diox.	HBr-TFA
150	18	S-DVB-2%	Bzl	Boc	DCC	HCl-Diox.	HBr-TFA
151	18	S-DVB-2%	Bzl	Boc	DCC	HCl-Diox.	HBr-TFA
152	18	S-DVB-2%	Bzl	Boc	DCC	HCl-Diox.	HBr-TFA
153	18	S-DVB-2%	Bzl	Boc	DCC	HCl-Diox.	HBr-TFA
154	18	S-DVB-2%	Bzl	Boc	DCC	HCl-Diox.	HBr-TFA

PEPTIDES PREPARED BY SOLID PHASE PEPTIDE SYNTHESIS (continued)

Entry number	Number of residues	Trivial name or source	Sequence
155	8	**Des-Arg^{1}-[Leu^{5,8},Thr^{6}]-bradykinin**	H-Pro-Pro-Gly-Leu-Thr-Pro-Leu-Arg-OH
156	8	**Des-Arg^{1}-[Leu^{5,8},Gly^{6}]-bradykinin**	H-Pro-Pro-Gly-Leu-Gly-Pro-Leu-Arg-OH
157	8	**Des-Arg^{1}-[Tyr(Me)^{5,8},Gly^{6}]-bradykinin**	H-Pro-Pro-Gly-Tyr(Me)-Gly-Pro-Tyr(Me)-Arg-OH
158	8	**Des-Arg^{1}-[Tyr(Me)^{5,8},Asn^{6}]-bradykinin**	H-Pro-Pro-Gly-Thr(Me)-Asn-Pro-Tyr(Me)-Arg-OH
159	8	**Des-Arg^{1}-[Tyr(Me)^{5,8},Thr^{6}]-bradykinin**	H-Pro-Pro-Gly-Tyr(Me)-Thr-Pro-Tyr(Me)-Arg-OH
160	8	**Nylon 6 oligomers**	H-[Cap]$_n$-OH; n = 2–8
161	8	**TMV protein—(105–112)**	H-Ala-Glu-Thr-Leu-Asp-Ala-Thr-Arg-OH
162	8	**Histidylphenylalanylvalylglutamyllysylalanylalanylalanine**	H-His-Phe-Val-Glu-Lys-Ala-Ala-Ala-OH
163	8	**Staphylococcal nuclease—(132–139)** protected	Boc-Glu(OBzl)-Ala-Gln-Ala-Lys(TFA)-Leu-Ile-OH
164	8	**[Ile^{1},Ile^{5}]-angiotensin II**	H-Ile-Arg-Val-Tyr-Ile-His-Pro-Phe-OH
165	8	**[Ala^{3},Ile^{5}]-angiotensin II**	H-Asp-Arg-Ala-Tyr-Ile-His-Pro-Phe-OH
166	8	**[Ile^{5},Ala^{8}]-angiotensin II**	H-Asp-Arg-Val-Tyr-Ile-His-Pro-Ala-OH
167	8	**[Ala^{5}]-angiotensin II**	H-Asp-Arg-Val-Tyr-Ala-His-Pro-Phe-OH
168	8	**[Val^{5},Phlac^{8}]-angiotensin II**	H-Asp-Arg-Val-Tyr-Val-His-Pro-Phlac-OH
169	8	**Glycylprolylglycylprolylprolylglycylalanyllysine**	H-Gly-Pro-Gly-Pro-Pro-Gly-Ala-Lys-OH
170	8	**Phenylalanylphenylalanylglycylleucylleucylleucylglycylphenylalanine**	H-DL-Phe-Phe-Gly-Leu-Leu-Leu-Gly-Phe-OH
171	8	**[D-Val^{5}]-angiotensin II**	H-Asp-Arg-Val-Tyr-D-Val-His-Pro-Phe-OH
172	8	**[Val^{5},D-Pro^{7}]-angiotensin II**	H-Asp-Arg-Val-Tyr-Val-His-D-Pro-Phe-OH
173	8	**[Deuterated-Val^{5}] angiotensin II**	H-Asp-Arg-Val-Tyr-(^{2}H)-Val-His-Pro-Phe-OH
174	8	**Lysozyme-(64–71)**	H-Cys-Asn-Asp-Gly-Arg-Thr-Pro-Gly-OH
175	8	**[Asn^{1},HyV^{5}]-angiotensin II**	H-Asn-Arg-Val-Tyr-HyV-His-Pro-Phe-OH
176	8	**[Asn^{1},HyPhlac^{4},Val^{5}]-angiotensin II**	H-Asn-Arg-Val-HyPhlac-Val-His-Pro-Phe-OH
177	8	**Leucylalanyltyrosyllysyllysyllysyllysyllysine,** protected	H-Leu-Ala-Tyr-[Lys(TFA)]$_5$-OH
178	8	**[Pro^{3},Ile^{5}]-angiotensin II**	H-Asp-Arg-Pro-Tyr-Ile-His-Pro-Phe-OH
179	8	**Polyglutamyl folic acid**	Pteroyl-(γ-Glu)$_n$-Glu-OH, n = 0–6
200	8	**Prolylglycylthreonyllysylmethionylisoleucylphenylalanylalanine**	H-Pro-Gly-Thr-Lys-Met-Ile-Phe-Ala-OH
201	8	**[Gly^{1},Gly^{2},Ile^{5}]-angiotensin II**	H-Gly-Gly-Val-Tyr-Ile-His-Pro-Phe-OH
202	8	**Glucagon (22–29)**	H-Phe-Val-Gln-Try-Leu-Met-Asn-Thr-OH
203	9	**Bradykinin**	H-Arg-Pro-Pro-Gly-Phe-Ser-Pro-Phe-Arg-OH
204	9	**Bradykinin**	H-Arg-Pro-Pro-Gly-Phe-Ser-Pro-Phe-Arg-OH
205	9	**D-Bradykinin**	H-D-Arg-D-Pro-D-Pro-Gly-D-Phe-D-Ser-D-Pro-D-Phe-D-Arg-OH
206	9	**Acetyl-bradykinin**	Ac-Arg-Pro-Pro-Gly-Phe-Ser-Pro-Phe-Arg-OH
207	9	**[Thr^{6}]-bradykinin**	H-Arg-Pro-Pro-Gly-Phe-Thr-Pro-Phe-Arg-OH
208	9	**[Phe^{6}]-bradykinin**	H-Arg-Pro-Pro-Gly-Phe-Phe-Pro-Phe-Arg-OH
209	9	**[Lys(Tos)^{6}]-bradykinin**	H-Arg-Pro-Pro-Gly-Phe-Lys(Tos)-Pro-Phe-Arg-OH
210	9	**[D-Pro^{2}]-bradykinin**	H-Arg-D-Pro-Pro-Gly-Phe-Ser-Pro-Phe-Arg-OH
211	9	**[D-Pro^{3}]-bradykinin**	H-Arg-Pro-D-Pro-Gly-Phe-Ser-Pro-Phe-Arg-OH
212	9	**[D-Pro^{7}]-bradykinin**	H-Arg-Pro-Pro-Gly-Phe-Ser-D-Pro-Phe-Arg-OH
213	9	**Acetyl-[D-Pro^{7}]-bradykinin**	Ac-Arg-Pro-Pro-Gly-Phe-Ser-D-Pro-Phe-Arg-OH
214	9	**[D-Arg^{9}]-bradykinin**	H-Arg-Pro-Pro-Gly-Phe-Ser-Pro-Phe-D-Arg-OH
215	9	**[Tyr(Me)^{5,8}]-bradykinin**	H-Arg-Pro-Pro-Gly-Tyr(Me)-Ser-Pro-Tyr(Me)-Arg-OH
216	9	**Acetyl-[Tyr(Me)^{5,8}]-bradykinin**	Ac-Arg-Pro-Pro-Gly-Tyr(Me)-Ser-Pro-Tyr(Me)-Arg-OH
217	9	**[Thr^{6},Leu^{8}]-bradykinin**	H-Arg-Pro-Pro-Gly-Phe-Thr-Pro-Leu-Arg-OH
218	9	**[Gly^{6},Tyr(Me)^{8}]-bradykinin**	H-Arg-Pro-Pro-Gly-Phe-Gly-Pro-Tyr(Me)-Arg-OH
219	9	**Acetyl-[Gly^{6},Tyr(Me)^{8}]-bradykinin**	Ac-Arg-Pro-Pro-Gly-Phe-Gly-Pro-Tyr(Me)-Arg-OH
220	9	**[D-Pro^{2,3,7}]-bradykinin**	H-Arg-D-Pro-D-Pro-Gly-Phe-Ser-D-Pro-Phe-Arg-OH
221	9	**[Leu^{5,8},Thr^{6}]-bradykinin**	H-Arg-Pro-Pro-Gly-Leu-Thr-Pro-Leu-Arg-OH
222	9	**[Leu^{5,8},Gly^{6}]-bradykinin**	H-Arg-Pro-Pro-Gly-Leu-Gly-Pro-Leu-Arg-OH
223	9	**[Tyr(Me)^{5,8},Gly^{6}]-bradykinin**	H-Arg-Pro-Pro-Gly-Tyr(Me)-Gly-Pro-Tyr(Me)-Arg-OH
224	9	**[Tyr(Me)^{5,8},Asn^{6}]-bradykinin**	H-Arg-Pro-Pro-Gly-Tyr(Me)-Asn-Pro-Tyr(Me)-Arg-OH
225	9	**[D-Arg^{1,9},D-Phe^{5,8},D-Ser^{6}]-bradykinin**	H-D-Arg-Pro-Pro-Gly-D-Phe-D-Ser-Pro-D-Phe-D-Arg-OH
226	9	**Des-Arg^{1}-endo-Phe^{8a}-bradykinin**	H-Pro-Pro-Gly-Phe-Ser-Pro-Phe-Phe-Arg-OH

PEPTIDES PREPARED BY SOLID PHASE PEPTIDE SYNTHESIS (continued)

Entry number	Reference	Solid support	Type of bond	Monomer protection	Coupling reagent	Deprotection	Cleavage
155	18	S-DVB-2%	Bzl	Boc	DCC	HCl-Diox.	HBr-TFA
156	18	S-DVB-2%	Bzl	Boc	DCC	HCl-Diox.	HBr-TFA
157	18	S-DVB-2%	Bzl	Boc	DCC	HCl-Diox.	HBr-TFA
158	18	S-DVB-2%	Bzl	Boc	DCC	HCl-Diox.	HBr-TFA
159	18	S-DVB-2%	Bzl	Boc	DCC	HCl-Diox.	HBr-TFA
160	19	S-DVB-2%	Bzl	Boc	DCC	TFA	HBr-HOAc
161	38	S-DVB-2%	Bzl	Boc	DCC	HCl-HOAc	HBr-TFA
162	93	S-DVB-2%	Bzl	Boc	DCC	HCl-HOAc	HBr-TFA
163	132	S-DVB-2%	Bzl	Boc	OSu,ONp	HCl-Diox.	HBr-TFA
164	71	S-DVB-2%	Bzl	Boc	DCC	HCl-HOAc	HBr-TFA
165	17	S-DVB-2%	Bzl	Boc	DCC	HCl-HOAc	HBr-TFA
166	91	S-DVB-2%	Bzl	Boc	DCC	HCl-HOAc	HBr-TFA
167	71	S-DVB-2%	Bzl	Boc	NEPIS	HCl-HOAc	HBr-TFA
168	40	S-DVB-2%	Bzl	Boc	DCC, Mixed anh.	HCl-HOAc	HBr-TFA
169	59	S-DVB-2%	Bzl	Boc	DCC	HCl-HOAc	HBr-TFA
170	45	S-DVB-2%	Bzl	Boc	DCC	HCl-HOAc	HBr-TFA
171	92	S-DVB-2%	Bzl	Boc	DCC	HCl-HOAc	HBr-TFA
172	92	S-DVB-2%	Bzl	Boc	DCC	HCl-HOAc	HBr-TFA
173	92	S-DVB-2%	Bzl	Boc	DCC	HCl-HOAc	HBr-TFA
174	107	S-DVB-2%	Bzl	Boc	DCC,ONp	HCl-HOAc	HBr-TFA
175	97	S-DVB-2%	Bzl	Boc	DCC	HCl-HOAc	HBr-TFA
176	97	S-DVB-2%	Bzl	Boc	DCC	HCl-HOAc	HBr-TFA
177	69	S-DVB-2%	Bzl	Boc	NEPIS, Azide	HCl-Diox.	HBr-TFA
178	71	S-DVB-2%	Bzl	Boc	DCC	HCl-HOAc	HBr-TFA
179	103	S-DVB-2%	Bzl	Boc	Mixed anh.	TFA-CH_2Cl_2	HBr-TFA
200	108	S-DVB-2%	Bzl	Boc	DCC	HCl-HOAc	HBr-TFA
201	125	S-DVB-2%	Bzl	Boc	DCC	HCl-HOAc	HBr-TFA
202	60	S-DVB-2%	Bzl	Boc	DCC,ONp	HCl-HOAc-BME	HF
203	20, 21	S-DVB-2%	Bzl	Boc	DCC	HCl-HOAc	HBr-TFA
204	14	S-DVB-2%	Bzl	Boc	DCC	HCl-HOAc	HF
205	22	S-DVB-2%	Bzl	Boc	DCC	HCl-Diox.	HBr-TFA
206	18	S-DVB-2%	Bzl	Boc	DCC	HCl-Diox.	HBr-TFA
207	18	S-DVB-2%	Bzl	Boc	DCC	HCl-Diox.	HBr-TFA
208	18	S-DVB-2%	Bzl	Boc	DCC	HCl-Diox.	HBr-TFA
209	18	S-DVB-2%	Bzl	Boc	DCC	HCl-Diox.	HBr-TFA
210	18	S-DVB-2%	Bzl	Boc	DCC	HCl-Diox.	HBr-TFA
211	18	S-DVB-2%	Bzl	Boc	DCC	HCl-Diox.	HBr-TFA
212	18	S-DVB-2%	Bzl	Boc	DCC	HCl-Diox.	HBr-TFA
213	18	S-DVB-2%	Bzl	Boc	DCC	HCl-Diox.	HBr-TFA
214	18	S-DVB-2%	Bzl	Boc	DCC	HCl-Diox.	HBr-TFA
215	18	S-DVB-2%	Bzl	Boc	DCC	HCl-Diox.	HBr-TFA
216	18	S-DVB-2%	Bzl	Boc	DCC	HCl-Diox.	HBr-TFA
217	18	S-DVB-2%	Bzl	Boc	DCC	HCl-Diox.	HBr-TFA
218	18	S-DVB-2%	Bzl	Boc	DCC	HCl-Diox.	HBr-TFA
219	18	S-DVB-2%	Bzl	Boc	DCC	HCl-Diox.	HBr-TFA
220	18	S-DVB-2%	Bzl	Boc	DCC	HCl-Diox.	HBr-TFA
221	18	S-DVB-2%	Bzl	Boc	DCC	HCl-Diox.	HBr-TFA
222	18	S-DVB-2%	Bzl	Boc	DCC	HCl-Diox.	HBr-TFA
223	18	S-DVB-2%	Bzl	Boc	DCC	HCl-Diox.	HBr-TFA
224	18	S-DVB-2%	Bzl	Boc	DCC	HCl-Diox.	HBr-TFA
225	18	S-DVB-2%	Bzl	Boc	DCC	HCl-Diox.	HBr-TFA
226	18	S-DVB-2%	Bzl	Boc	DCC	HCl-Diox.	HBr-TFA

PEPTIDES PREPARED BY SOLID PHASE PEPTIDE SYNTHESIS (continued)

Entry number	Number of residues	Trivial name or source	Sequence
227	9	**Des-Arg[1]-bradykininyl-glycine**	H-Pro-Pro-Gly-Phe-Ser-Pro-Phe-Arg-Gly-OH
228	9	**Des-Arg[1]-bradykininyl-arginine**	H-Pro-Pro-Gly-Phe-Ser-Pro-Phe-Arg-Arg-OH
229	9	**Des-Arg[1]-[Tyr(Me)[5,8]]-bradykininyl-glycine**	H-Pro-Pro-Gly-Tyr(Me)-Ser-Pro-Tyr(Me)-Arg-Gly-OH
230	9	**D-Retrobradykinin**	D-Arg-D-Phe-D-Pro-D-Ser-D-Phe-Gly-D-Pro-D-Pro-D-Arg-OH
231	9	**Lysine vasopressin**, protected	Tos-Cys(Bzl)-Tyr-Phe-Gln-Asn-Cys(Bzl)-Pro-Lys(Tos)-Gly-NH_2
232	9	**Chymotrypsin (193–201) nonapeptide**	H-Gly-Asp-Ser-Gly-Gly-Pro-Leu-Val-Cys(Bzl)-OH
233	9	**[Tyr(Me)[5],Thr[6],Tyr(Me)[8]]-bradykinin**	H-Arg-Pro-Pro-Gly-Tyr(Me)-Thr-Pro-Tyr(Me)-Arg-OH
234	9	**Oxytocin**	H-Cys-Tyr-Ile-Gln-Asn-Cys-Pro-Leu-Gly-NH_2
235	9	**Oxytocin**	H-Cys-Tyr-Ile-Gln-Asn-Cys-Pro-Leu-Gly-NH_2
236	9	**Oxytocin**	H-Cys-Tyr-Ile-Gln-Asn-Cys-Pro-Leu-Gly-NH_2
237	9	**Oxytocin**, protected	H-Cys(Bzl)-Tyr-Ile-Gln-Asn-Cys(Bzl)-Pro-Leu-Gly-NH_2
238	9	**Oxytocin**, protected	Z-Cys(Bzl)-Tyr(Bzl)-Ile-Gln-Asn-Cys(Bzl)-Pro-Leu-Gly-NH
239	9	**Oxytocin**, protected	Aoc-Cys-Tyr-Ile-Gln-Asn-Cys-Pro-Leu-Gly-NH_2
240	9	**Deamino-oxytocin**	MCP-Tyr-Ile-Gln-Asn-Cys-Pro-Leu-Gly-NH_2
241	9	**Deamino-oxytocin**, protected	MCP(Bzl)-Tyr(Bzl)-Ile-Gln-Asp-Cys(Bzl)-Pro-Leu-Gly-NH_2
242	9	**[Lys[8]]-vasopressin**	H-Cys-Tyr-Phe-Gln-Asn-Cys-Pro-Lys-Gly-NH_2
243	9	**[Lys[8]]-vasopressin**	H-Cys-Tyr-Phe-Gln-Asn-Cys-Pro-Lys-Gly-NH_2
244	9	**[Lys[8]]-vasopressin**, protected	Tos-Cys(Bzl)-Tyr-Phe-Gln-Asn-Cys(Bzl)-Pro-Lys(Tos)-Gly-NH_2
245	9	**[Lys[8]]-vasopressin**, protected	Z-Cys(Bzl)-Tyr(Bzl)-Phe-Gln-Asn-Cys(Bzl)-Pro-Lys(Z)-Gly-NH_2
246	9	**[Ser[4],Gln[8]]-oxytocin**	H-Cys-Tyr-Ile-Ser-Asn-Cys-Pro-Gln-Gly-NH_2
247	9	**[Ser[4],Gln[8]]-oxytocin**, protected	Z-Cys(Bzl)-Tyr(Bzl)-Ile-Ser-(Bzl)-Asn-Cys(Bzl)-Pro-Gln-Gly-NH_2
248	9	**Oxytocinoic acid**, protected	H-Cys(Bzl)-Tyr-Ile-Gln-Asn-Cys(Bzl)-Pro-Leu-Gly-OH
249	9	**TMV protein—(104–112)**	H-Thr-Ala-Glu-Thr-Leu-Asp-Ala-Thr-Arg-OH
250	9	**DNP-nonalysine**	Dnp-Lys-Lys-Lys-Lys-D-Lys-Lys-Lys-Lys-Lys-OH
251	9	**[Phe[8]]-oxytocin**	H-Cys-Tyr-Ile-Gln-Asn-Cys-Pro-Phe-Gly-NH_2
252	9	**[Pro[4],Ile[8]]-oxytocin**	H-Cys-Tyr-Ile-Pro-Asn-Pro-Ile-Gly-NH_2
253	9	**Oxytocinoic acid methyl ester**, protected	H-Cys(Bzl)-Tyr-Ile-Gln-Asn-Cys(Bzl)-Pro-Leu-Gly-OMe
254	9	**[Pro[4],Gln[8]]-oxytocin**	H-Cys-Tyr-Ile-Pro-Asn-Cys-Pro-Gln-Gly-NH_2
255	9	**[Pro[4],Leu[8]]-oxytocin**	H-Cys-Tyr-Ile-Pro-Asn-Cys-Pro-Leu-Gly-NH_2
256	9	**Mellitin—(1–9)**	H-Gly-Ile-Gly-Ala-Val-Leu-Lys-Val-Leu-OH
257	9	**Oxytocin methylamide**, protected	Z-Cys(Bzl)-Tyr-Ile-Gln-Asn-Cys(Bzl)-Pro-Leu-Gly-$NHCH_3$
258	9	**Oxytocin hydrazide**, protected	Z-Cys(Bzl)-Tyr-Ile-Gln-Asn-Cys(Bzl)-Pro-Leu-Gly-$NHNH_2$
259	9	**Oxytocin hydroxyamide**, protected	Z-Cys(Bzl)-Tyr-Ile-Gln-Asn-Cys(Bzl)-Pro-Leu-Gly-NHNHOH
260	9	**Ribonuclease—(1–8)**, protected	Boc-Lys(Boc)-Glu(OBu)-Thr-Ala-Ala-Ala-Lys(Z)-Phe-$NHNH_2$
261	10	**TMV Protein (103–112) decapeptide**	H-Thr-Thr-Ala-Glu-Thr-Leu-Asp-Ala-Thr-Arg-OH
262	10	**Nylon 11 oligomers**	H-$[Und]_n$-OH, n = 2–10
263	10	**Nylon 6.6 oligomers**	H-$[NH(CH_2)_6NHCO(CH_2)_4CO]_n$-OH, n = 2–10
264	10	**Bradykininyl-glycine**	H-Arg-Pro-Pro-Gly-Phe-Ser-Pro-Phe-Arg-Gly-OH
265	10	**Endo-Phe[8a]-bradykinin**	H-Arg-Pro-Pro-Gly-Phe-Ser-Pro-Phe-Phe-Arg-OH
266	10	**Bradykininyl-arginine**	H-Arg-Pro-Pro-Gly-Phe-Ser-Pro-Phe-Arg-Arg-OH
267	10	**[Tyr(Me)[5,8]]-bradykininyl-phenylalanine**	H-Arg-Pro-Pro-Gly-Tyr(Me)-Ser-Pro-Tyr(Me)-Arg-Phe-OH
268	10	**[Tyr(Me)[5,8]]-bradykininyl-glycine**	H-Arg-Pro-Pro-Gly-Tyr(Me)-Ser-Pro-Tyr(Me)-Arg-Gly-OH
269	10	**Gly-[Gly[6],Tyr(Me)[8]]-bradykinin**	H-Gly-Arg-Pro-Pro-Gly-Phe-Gly-Pro-Tyr(Me)-Arg-OH
270	10	**TMV protein—(103–112)**	H-Thr-Thr-Ala-Glu-Thr-Leu-Asp-Ala-Thr-Arg-OH
271	10	**Staphylococcal nuclease—(108–117)**, protected	Boc-Leu-Ala-Lys(TFA)-Val-Ala-Tyr-Val-Tyr-Lys(TFA)-Pro-OH

PEPTIDES PREPARED BY SOLID PHASE PEPTIDE SYNTHESIS (continued)

Entry number	Reference	Solid support	Type of bond	Monomer protection	Coupling reagent	Deprotection	Cleavage
227	18	S-DVB-2%	Bzl	Boc	DCC	HCl-Diox.	HBr-TFA
228	18	S-DVB-2%	Bzl	Boc	DCC	HCl-Diox.	HBr-TFA
229	18	S-DVB-2%	Bzl	Boc	DCC	HCl-Diox.	HBr-TFA
230	18	S-DVB-2%	Bzl	Boc	DCC	HCl-Diox.	HBr-TFA
231	23	S-DVB-2%	Bzl	Boc	DCC	HCl-HOAc	NH_3
232	24	S-DVB-2%	Bzl	Boc	DCC	HCl-Diox.	HBr-TFA
233	88	S-DVB-2%	Bzl	Boc	DCC	HCl-Diox.	HBr-TFA
234	42, 50	S-DVB-2%	Bzl	Boc	DCC,ONp	HCl-HOAc,TFA	NH_3-MeOH
235	51	S-DVB-2%	Bzl	Nps	DCC,ONp	HCl-Diox.	NH_3-MeOH-DMF
236	70	Phenol-trioxane	Phenyl	Aoc	DCC	HCl-MeOH	NH_3-MeOH
237	76	S-DVB-2%	Bzl	Boc	ONp,OSu-triazole	HCl-HOAc	NH_3-MeOH
238	42	S-DVB-2%	Bzl	Boc	DCC,ONp	HCl-HOAc,TFA	NH_3-MeOH
239	70	Phenol-trioxane	Phenyl	Aoc	DCC	HCl-MeOH	NH_3-MeOH
240	44	NO_2-S-DVB-2%	NO_2Bzl	Boc	DCC,ONp	HCl-HOAc,TFA	NH_3-MeOH
241	44	NO_2-S-DVB-2%	NO_2Bzl	Boc	DCC,ONp	HCl-HOAc,TFA	NH_3-MeOH
242	42	S-DVB-2%	Bzl	Boc	DCC,ONp	HCl-HOAc,TFA	NH_3-MeOH
243	23	S-DVB-2%	Bzl	Boc	DCC,ONp	HCl-HOAc	NH_3-EtOH-DMF
244	23	S-DVB-2%	Bzl	Boc	DCC,ONp	HCl-HOAc	NH_3-EtOH-DMF
245	42	S-DVB-2%	Bzl	Boc	DCC,ONp	HCl-HOAc-TFA	NH_3-MeOH
246	77	S-DVB-2%	Bzl	Boc	DCC,ONp	HCl-HOAc,TFA	NH_3-MeOH
247	77	S-DVB-2%	Bzl	Boc	DCC,ONp	HCl-HOAc,TFA	NH_3-MeOH
248	76	S-DVB-2%	Bzl	Boc	ONp,OSu-triazole	HCl-HOAc	NH_3-MeOH
249	38	S-DVB-2%	Bzl	Boc	DCC	HCl-HOAc	HBr-TFA
250	55	S-DVB-2%	Bzl	Boc,Nps	DCC	HCl-HOAc	HBr-TFA
251	112	S-DVB-2%	Bzl	Boc	DCC,ONp	HCl-HOAc,TFA	NH_3-MeOH
252	116	S-DVB-2%	Bzl	Boc	DCC,ONp	HCl-HOAc,TFA	NH_3-MeOH
253	72	S-DVB-2%	Bzl	Boc	ONp,OSu-triazole	HCl-HOAc,HCl	MeOH-*N*-Methyl piperidine
254	122	S-DVB-2%	Bzl	Boc	DCC,ONp	HCl-HOAc,TFA	NH_3-MeOH
255	116	S-DVB-2%	Bzl	Boc	DCC,ONp	HCl-HOAc,TFA	NH_3-MeOH
256	106	S-DVB-2%	Bzl	Boc	DCC	HCl-HOAc	HBr-TFA
257	58	S-DVB-2%	Bzl	Boc	ONp,OSu-triazole	HCl-HOAc	CH_3NH_2-MeOH
258	58	S-DVB-2%	Bzl	Boc	ONp,OSu-triazole	HCl-HOAc	NH_2NH_2-MeOH
259	58	S-DVB-2%	Bzl	Boc	ONp,OSu-triazole	HCl-HOAc	NH_2NHOH-MeOH
260	85	S-DVB-2%	Bzl	Boc,Nps	DCC,ONp, OPcp	HCl-HOAc,HCl-Diox. HCl-HOAc	NH_2NH_2-DMF
261	25	S-DVB-2%	Bzl	Boc	DCC	HCl-HOAc	HBr-TFA
262	19	S-DVB-2%	Bzl	Boc	DCC	TFA	HBr-HOAc
263	19	S-DVB-2%	Bzl	Boc	DCC	TFA	HBr-HOAc
264	18	S-DVB-2%	Bzl	Boc	DCC	HCl-Diox.	HBr-TFA
265	18	S-DVB-2%	Bzl	Boc	DCC	HCl-Diox.	HBr-TFA
266	18	S-DVB-2%	Bzl	Boc	DCC	HCl-Diox.	HBr-TFA
267	18	S-DVB-2%	Bzl	Boc	DCC	HCl-Diox.	HBr-TFA
268	18	S-DVB-2%	Bzl	Boc	DCC	HCl-Diox.	HBr-TFA
269	18	S-DVB-2%	Bzl	Boc	DCC	HCl-Diox.	HBr-TFA
270	38	S-DVB-2%	Bzl	Boc	DCC	HCl-Diox.	HBr-TFA
271	13	S-DVB-2%	Bzl	Boc	OSu,ONp	HCl-Diox.	HBr-TFA

PEPTIDES PREPARED BY SOLID PHASE PEPTIDE SYNTHESIS (continued)

Entry number	Number of residues	Trivial name or source	Sequence
272	10	**[Gly^5,Gly^{10}]-gramicidin S**	cyclo[-Val-Orn-Leu-D-Phe-Gly-Val-Orn-Leu-D-Phe-Gly-]
273	10	**[Gly^5,Gly^{10}]-gramacidin S**	cyclo[-Val-Orn-Leu-D-Phe-Gly-Val-Orn-Leu-D-Phe-Gly-]
274	10	**[Ile^5]-angiotensin I**	H-Asp-Arg-Val-Tyr-Ile-His-Pro-Phe-His-Leu-OH
275	10	**Alanylalanylalanylalanylalanylleucylaspartylalanythreonylarginine**	H-Ala-Ala-Ala-Ala-Ala-Leu-Asp-Ala-Thr-Arg-OH
276	10	**Oligolysine**	H-(Lys)$_n$-OH, n = 2–10
277	10	**Antamanide**	cyclo[-Pro-Phe-Phe-Val-Pro-Pro-Ala-Phe-Phe-Pro-]
278	11	**D-Alanyl-L-alanyl-D-alanyl-L-alanyl-D-alanyl-L-alanyl-D-alanyl-L-alanyl-D-alanyl-L-alanyl-2,4 dinitropheny-L-lysine**	H-D-Ala-Ala-D-Ala-Ala-D-Ala-Ala-D-Ala-Ala-D-Ala-Ala-Lys(Dnp)-OH
279	11	**Met-Lys-bradykinin**	H-Met-Lys-Arg-Pro-Pro-Gly-Phe-Ser-Pro-Phe-Arg-OH
280	11	**Casein peptide**	H-Val-Asp-Glu-Glu-Glu-DL-Ser-Ile-Ala-Met-Glu-Lys-OH
281	11	**Valylserylglutamylglutamylglutamylaspartylisoleucylalanylmethionylglutamyllysine**	H-Val-Ser-Glu-Glu-Glu-Asp-Ile-Ala-Met-Gly-Lys-OH
282	12	**Dansyl-oligoprolyl-hydrazide**	Dansyl-(Pro)$_n$-NHNH$_2$, n = 1–12
283	12	**Valinomycin**	cyclo-[L-Val-D-Hyv-D-Val-L-Lac]$_3$
284	12	**Oligo(glycylprolylprolyl)glycylprolylproline,** labelled	H-[Gly-Pro-(3,4-^{3}H)-Pro]$_3$-Gly-Pro-Pro-OH
285	12	**Oligo(glycylprolyl-3-methylprolyl)-glycylprolyl-3-methylproline**	H-[Gly-Pro-Pro(Me)]$_n$-OH n = 1, 2, 3 and 4
286	12	**Oligo(glycylprolyl-4-fluoroprolyl)-glycylprolyl-4-fluoroproline**	H-[Gly-Pro-Pro(F)]$_n$-OH n = 1, 2, 3 and 4
287	13	**Nylon 6 oligomers**	H-[Cap]$_n$-OH n = 2, 3, 4, 5, 7, 9, 11 and 13
288	14	**Glucagon (16–29)**	H-Ser-Arg-Arg-Ala-Gln-Asp-Phe-Val-Gln-Try-Leu-Met-Asn-Thr-OH
289	15	**Cysteinylalanylvalylcysteinyl-(im-benzyl)histidylalanylvalylalanyllysylalanyl-alanyl-alanylalanylalanylalanine**	H-Cys-Ala-Val-Cys-His(Bzl)-Ala-Val-Ala-Lys-Ala-Ala-Ala-Ala-Ala-Ala-OH
290	15	**Lysylalanylvalylcysteinylhistidylalanylvalylalanyllysylalanylalanylalanylalanylalanylalanine**	H-Lys-Ala-Val-Cys-His-Ala-Val-Ala-Lys-Ala-Ala-Ala-Ala-Ala-Ala-OH
291	15	**Ribonuclease—(1–15)**	H-Lys-Glu-Thr-Ala-Ala-Ala-Lys-Glu-Phe-Arg-Gln-His-Met-Asp-Ser-OH
292	16	**Tyrosylhistidylhistidylphenylalanylphenylalanylaspartylhistidylserylaspartylserylhistidylphenylalanylaspartylphenylalanylhistidylphenylalanine**	H-Tyr-His-His-Phe-Phe-Asp-His-Ser-Asp-Ser-His-Phe-Asp-Phe-His-Phe-OH
293	16	**[Ala^{16}]-ribonuclease—(1–16)**	H-Lys-Glu-Thr-Ala-Ala-Ala-Lys-Glu-Phe-Arg-Gln-His-Met-Asp-Ser-Ala-OH
294	17	**Angiotensinyl-bradykinin**	H-Asp-Arg-Val-Tyr-Ile-His-Pro-Phe-Arg-Pro-Pro-Gly-Phe-Ser-Pro-Phe-Arg-OH
295	18	**Bradykininyl-bradykinin**	H-Arg-Pro-Pro-Gly-Phe-Ser-Pro-Phe-Arg-Arg-Pro-Pro-Gly-Phe-Ser-Pro-Phe-Arg-OH
296	18	**Polistes kinin**	PCA-Tyr-Asn-Lys-Lys-Lys-Leu-Arg-Gly-Arg-Pro-Pro-Gly-Phe-Ser-Pro-Phe-Arg-OH
297	18	**Valylserylisoleucylglutamylglutamylglutamylserylserylprolylserylvalylglutamylglutamylglutamylaspartylserylisoleucylalanine**	H-Val-Ser-Ile-Glu-Glu-Glu-Ser-Ser-Pro-Ser-Val-Glu-Glu-Glu-Asp-Ser-Ile-Ala-OH
298	20	**Des Ile^2-insulin A-chain *S*-sulfonate (bovine)**	H-Gly-Val-Gly-Gln-Cys(SO$_3$)-Cys(SO$_3$)-Ala-Ser-Val-Cys(SO$_3$)-Ser-Leu-Tyr-Gln-Leu-Glu-Asn-Tyr-Cys(SO$_3$)-Asn-OH
299	20	**Lysylaspartylglutamylglutamylglutamylglutamylvalylglutamylserylphenylalanylserylglycylprolylaspartylalanylprolylleucylprolylalanylglycine**	H-Lys-Asp-Glu-Glu-Glu-Glu-Val-Glu-Ser-Phe-Ser-Gly-Pro-Asp-Ala-Pro-Leu-Pro-Ala-Gly-OH
300	21	**Insulin A-chain *S*-sulfonate (bovine)**	H-Gly-Ile-Val-Glu-Gln-Cys(SO$_3$)-Cys(SO$_3$)-Ala-Ser-Val-Cys(SO$_3$)-Ser-Leu-Tyr-Gln-Leu-Glu-Asn-Tyr-Cys(SO$_3$)-Asn-OH

PEPTIDES PREPARED BY SOLID PHASE PEPTIDE SYNTHESIS (continued)

Entry number	Reference	Solid support	Type of bond	Monomer protection	Coupling reagent	Deprotection	Cleavage
272	115	Styrene	Bzl	Boc	OSu	HCl-Diox.	HBr-HOAc
273	54, 80	S-DVB-2%	Bzl	Boc	DCC,OSu	HCl-HOAc	HBr-TFA
274	47	S-DVB-2%	Bzl	Boc	DCC	HCl-HOAc	HBr-TFA
275	110	S-DVB-2%	Bzl	Boc	DCC	HCl-HOAc	HBr-TFA
276	75	S-DVB-2%	Bzl	Boc	DCC	HCl-HOAc	HBr-TFA
277	130	S-DVB-2%	Bzl	Boc	DCC	HCl-HOAc	HBr-TFA
278	26	S-DVB-2%	Bzl	Boc	DCC	HCl-HOAc	HBr-TFA
279	27	S-DVB-2%	Bzl	Boc	DCC	HCl-HOAc	HBr-TFA
280	100	S-DVB-2%	Bzl	Boc	DCC	HCl-HOAc	HBr-TFA
281	109	S-DVB-2%	Bzl	Boc	DCC	HCl-HOAc	HBr-TFA
282	87	S-DVB-2%	Bzl	Boc	DCC	HCl-HOAc	NH_2NH_2-EtOH
283	74	S-DVB-2%	Bzl	Boc	DCC	HCl-Diox.	HBr-TFA
284	79	S-DVB-2%	Bzl	Boc	DCC	HCl-Diox.	HBr-TFA
285	79	S-DVB-2%	Bzl	Boc	DCC	HCl-Diox.	HBr-TFA
286	79	S-DVB-2%	Bzl	Boc	DCC	HCl-Diox.	HBr-TFA
287	19	S-DVB-2%	Bzl	Boc	Mixed anh.	HCl-HOAc	HBr-TFA
288	60	S-DVB-2%	Bzl	Boc	DCC,ONp	HCl-HOAc-BME	HF
289	14	S-DVB-2%	Bzl	Boc	DCC	HCl-HOAc	HF
290	101	S-DVB-2%	Bzl	Boc	DCC	HCl-HOAc	HF
291	124	S-DVB-2%	Bzl	Boc	DCC,ONp	HCl-Diox.	HF,HBr-TFA
292	29	S-DVB-2%	Bzl	Boc	DCC	HCl-Diox.	HBr-TFA
293	124	S-DVB-2%	Bzl	Boc	DCC,ONp	HCl-Diox.	HF,HBr-TFA
294	30, 31	S-DVB-1%	Bzl	Boc	DCC	HCl-Diox.	HBr-TFA
295	32	S-DVB-2%	Bzl	NPS	DCC	HCl	HBr-TFA
296	95	S-DVB-2%	Bzl	Boc	DCC,ONp	HCl-Diox.	HBr-TFA
297	109	S-DVB-2%	Bzl	Boc	DCC	HCl-HOAc	HBr-TFA
298	119	S-DVB-2%	Bzl	Boc	ONp	HCl-HOAc	HBr-TFA
299	109	S-DVB-2%	Bzl	Boc	DCC	HCl-HOAc	HBr-TFA
300	33	S-DVB-2%	Bzl	Boc	DCC	HCl-Diox.	HBr-TFA

PEPTIDES PREPARED BY SOLID PHASE PEPTIDE SYNTHESIS (continued)

Entry number	Number of residues	Trivial name or source	Sequence
301	21	**Insulin A-chain *S*-sulfonate (bovine)**	H-Gly-Ile-Val-Glu-Gln-Cys(SO_3)-Cys(SO_3)-Ala-Ser-Val-Cys(SO_3)-Ser-Leu-Tyr-Gln-Leu-Glu-Asn-Tyr-Cys(SO_3)-Asn-OH
302	21	**Insulin A-chain *S*-sulfonate (sheep)**	H-Glu-Ile-Val-Glu-Gln-Cys(SO_3)-Cys(SO_3)-Ala-Gly-Val-Cys(SO_3)-Ser-Leu-Tyr-Gln-Leu-Glu-Asn-Tyr-Cys(SO_3)-Asn-OH
303	21	**[Ala^{12},Ala^{14}]-insulin A-chain *S*-sulfonate (sheep)**	H-Gly-Ile-Val-Glu-Gln-Cys(SO_3)-Cys(SO_3)-Ala-Gly-Val-Cys(SO_3)-Ala-Leu-Ala-Gln-Leu-Glu-Asn-Tyr-Cys(SO_3)-Asn-OH
304	21	**[Ala^1]-insulin A-chain *S*-sulfonate (bovine)**	H-Ala-Ile-Val-Glu-Gln-Cys(SO_3)-Cys(SO_3)-Ala-Ser-Val-Cys(SO_3)-Ser-Leu-Tyr-Gln-Leu-Glu-Asn-Tyr-Cys(SO_3)-Asn-OH
305	21	**[Ala^7,Ala^{20}]-insulin A-chain (bovine)**	H-Gly-Ile-Val-Glu-Gln-Cys-Ala-Ala-Ser-Val-Cys-Ser-Leu-Tyr-Gln-Leu-Glu-Asn-Tyr-Ala-Asn-OH
306	21	**Insulin A-chain *S*-sulfonate (human)**	H-Gly-Ile-Val-Glu-Gln-Cys-Cys-Thr-Ser-Ile-Cys(SO_3)-Ser-Leu-Tyr-Gln-Leu-Glu-Asn-Tyr-Cys(SO_3)-Asn-OH
307	21	**[Val^2]-insulin A-chain *S*-sulfonate (sheep)**	H-Gly-Val-Val-Glu-Gln-Cys(SO_3)-Cys(SO_3)-Ala-Gly-Val-Cys(SO_3)-Ser-Leu-Tyr-Gln-Leu-Glu-Asn-Tyr-Cys(SO_3)-Asn-OH
308	21	**[Glu^5,Ala^{12},Ala^{14},Glu^{15},Ala^{18},Phe^{19}, Ala^{21}]-insulin A-chain *S*-sulfonate (sheep)**	H-Gly-Ile-Val-Glu-Glu-Cys(SO_3)-Cys(SO_3)-Ala-Gly-Val-Cys(SO_3)-Ala-Leu-Ala-Glu-Leu-Glu-Ala-Phe-Cys(SO_3)-Ala-OH
309	21	**[Val^2,Glu^5,Ala^{12}, Ala^{14}, Glu^{15}, Ala^{18}, Phe^{19}, Ala^{21}]-insulin A-chain *S*-sulfonate (sheep)**	H-Gly-Val-Val-Glu-Glu-Cys(SO_3)-Cys(SO_3)-Ala-Gly-Val-Cys(SO_3)-Ala-Leu-Ala-Glu-Leu-Glu-Ala-Phe-Cys(SO_3)-Ala-OH
310	21	**[Leu^2]-insulin A-chain *S*-sulfonate (sheep)**	H-Gly-Leu-Val-Glu-Gln-Cys(SO_3)-Ala-Gly-Val-Cys(SO_3)-Ser-Leu-Tyr-Gln-Leu-Gln-Asn-Tyr-Cys(SO_3)-Asn-OH
311	21	**[Ala^6,Ala^{11}]-insulin A-chain *S*-sulfonate (sheep)**	H-Gly-Ile-Val-Glu-Gln-Ala-Cys(SO_3)-Ala-Gly-Val-Ala-Ser-Leu-Tyr-Gln-Leu-Glu-Asn-Tyr-Cys(SO_3)-Asn-OH
312	23	**Hemoglobin B-chain (124–146) (human)**	H-Pro-Pro-Val-Gln-Ala-Ala-Tyr-Gln-Lys-Val-Val-Ala-Gly-Val-Ala-Asn-Ala-Leu-Ala-His-Lys-Tyr-His-OH
313	25	**Oligoaminocaproic acid amide**	H$[NH(CH_2)_5CO]_n$-OH n = 3,5,7,9,11,13,15,17,21,25
314	27	**Oligolysyl-decaalanyl-oligolysylglycine**	H-$[Lys]_n$-$[Ala]_{10}$-$[Lys]_n$-Gly-OH n = 7.9
315	28	**[des-Phe^1,des-Val^2,Ala^4]-insulin B-chain *S*-sulfonate (bovine)**	H-Asn-Ala-His-Leu-Cys(SO_3)-Gly-Ser-His-Leu-Val-Glu-Ala-Leu-Tyr-Leu-Val-Cys(SO_3)-Gly-Glu-Arg-Gly-Phe-Phe-Tyr-Thr-Pro-Lys-Ala-OH
316	29	**[des-Phe^1]-insulin B-chain *S*-sulfonate (bovine)**	H-Val-Asn-Gln-His-Leu-Cys(SO_3)-Gly-Ser-His-Leu-Val-Glu-Ala-Leu-Tyr-Leu-Val-Cys(SO_3)-Gly-Glu-Arg-Gly-Phe-Phe-Tyr-Thr-Pro-Lys-Ala-OH
317	30	**Insulin B-chain *S*-sulfonate (bovine)**	H-Phe-Val-Asn-Gln-His-Leu-Cys(SO_3)-Gly-Ser-His-Leu-Val-Glu-Ala-Leu-Tyr-Leu-Val-Cys(SO_3)-Gly-Glu-Arg-Gly-Phe-Phe-Tyr-Thr-Pro-Lys-Ala-OH
318	30	**Insulin B-chain *S*-sulfonate** (human)	H-Phe-Val-Asn-Gln-His-Leu-Cys(SO_3)-Gly-Ser-His-Leu-Val-Glu-Ala-Leu-Tyr-Leu-Val-Cys(SO_3)-Gly-Glu-Arg-Gly-Phe-Phe-Tyr-Thr-Pro-Lys-Ala-OH
319	30	**[Ala^3]-insulin B-chain *S*-sulfonate (bovine)**	H-Phe-Val-Ala-Gln-His-Leu-Cys(SO_3)-Gly-Ser-His-Leu-Val-Glu-Ala-Leu-Tyr-Leu-Val-Cys(SO_3)-Gly-Glu-Arg-Gly-Phe-Phe-Tyr-Thr-Pro-Lys-Ala-OH
320	30	**[Ala^4]-insulin B-chain *S*-sulfonate (bovine)**	H-Phe-Val-Asn-Ala-His-Leu-Cys(SO_3)-Gly-Ser-His-Leu-Val-Glu-Ala-Leu-Tyr-Leu-Val-Cys(SO_3)-Gly-Glu-Arg-Gly-Phe-Phe-Tyr-Thr-Pro-Lys-Ala-OH
321	30	**[Ala^9]-insulin B-chain *S*-sulfonate (bovine)**	H-Phe-Val-Asn-Gln-His-Leu-Cys(SO_3)-Gly-Ala-His-Leu-Val-Glu-Ala-Leu-Tyr-Leu-Val-Cys(SO_3)-Gly-Glu-Arg-Gly-Phe-Phe-Tyr-Thr-Pro-Lys-Ala-OH

PEPTIDES PREPARED BY SOLID PHASE PEPTIDE SYNTHESIS (continued)

Entry number	Reference	Solid support	Type of bond	Monomer protection	Coupling reagent	Deprotection	Cleavage
301	13	S-DVB-2%	Bzl	Boc	ONp	HCl-Diox.	HBr-TFA
302	78	S-DVB-2%	Bzl	Boc	DCC,ONp	HCl-HOAc	HBr-TFA
303	63	S-DVB-2%	Bzl	Boc	ONp	HCl-HOAc	HBr-TFA
304	65	S-DVB-2%	Bzl	Boc	ONp	HCl-HOAc	HBr-TFA
305	43	S-DVB-2%	Bzl	Boc	DCC,ONp	HCl-Diox.	HBr-TFA
306	86	S-DVB-2%	Bzl	Boc	DCC,ONp	HCl-HOAc	HBr-TFA
307	83	S-DVB-2%	Bzl	Boc	DCC,ONp	HCl-HOAc	HBr-TFA
308	83	S-DVB-2%	Bzl	Boc	DCC	HCl-HOAc	HBr-TFA
309	83	S-DVB-2%	Bzl	Boc	DCC	HCl-HOAc	HBr-TFA
310	83	S-DVB-2%	Bzl	Boc	DCC,ONp	HCl-HOAc	HBr-TFA
311	78	S-DVB-2%	Bzl	Boc	ONp	HCl-HOAc	HBr-TFA
312	120	S-DVB-2%	Bzl	Boc	DCC,ONp	HCl-HOAc	HBr-TFA
313	28, 102	S-DVB-2%	Bzl	Boc	Mixed anh.	HCl-HOAc	HBr-TFA
314	131	S-DVB-2%	Bzl	Boc	NCA,DCC	HCl-HOAc	HBr-TFA
315	81	S-DVB-1%	Bzl	Boc	DCC,ONp	HCl-Diox.	HBr-TFA
316	81	S-DVB-1%	Bzl	Boc	DCC,ONp	HCl-Diox.	HBr-TFA
317	33, 34	S-DVB-2%	Bzl	Boc	DCC,Np	HCl-HOAc	HBr-TFA
318	35, 84	S-DVB-2%	Bzl	Boc	DCC,ONp	HCl-HOAc	HBr-TFA
319	81	S-DVB-1%	Bzl	Boc	DCC,ONp	HCl-Diox.	HBr-TFA
320	81	S-DVB-1%	Bzl	Boc	DCC,ONp	HCl-Diox.	HBr-TFA
321	81	S-DVB-1%	Bzl	Boc	DCC,ONp	HCl-Diox.	HBr-TFA

PEPTIDES PREPARED BY SOLID PHASE PEPTIDE SYNTHESIS (continued)

Entry number	Number of residues	Trivial name or source	Sequence
322	42	**Staphylococcal nuclease (6–47)**	H-Lys-Leu-His-Lys-Glu-Pro-Ala-Thr-Leu-Ile-Lys-Ala-Ile-Asp-Gly-Asp-Thr-Val-Lys-Leu-Met-Tyr-Lys-Gly-Gln-Pro-Met-Thr-Phe-Arg-Leu-Leu-Leu-Val-Asp-Thr-Pro-Glu-Thr-Lys-His-Pro-OH
323	50	**Oligolysyl-oligovalyl-oligolysylglycine**	(DL-Lys)$_{18}$-(Val)$_{15}$-(DL-Lys)$_{16}$-Gly-OH
324	55	**Apoferrodoxin *S*-sulfonate** *C. pasteurianum*	H-Ala-Tyr-Lys-Ile-Ala-Asp-Ser-Cys(SO_3)-Val-Ser-Cys(SO_3)-Gly-Ala-Cys(SO_3)-Ala-Ser-Glu-Cys(SO_3)-Pro-Val-Asn-Ala-Ile-Ser-Gln-Gly-Asp-Ser-Ile-Phe-Val-Ile-Asp-Ala-Asp-Thr-Cys(SO_3)-Ile-Asp-Cys(SO_3)-Gly-Asn-Cys(SO_3)-Ala-Asn-Val-Cys(SO_3)-Pro-Val-Gly-Ala-Pro-Val-Gln-Glu-OH
325	55	**Polylysyloligoalanylglycine**	H(DL-Lys)$_{30}$-(Ala-d_4)$_{24}$-Gly-d_2-OH
326	60	**Poly(prolylprolylglycyl)prolylprolylglycine**	H-(Pro-Pro-Gly)$_{20}$OH
327	104	**[His(Bzl)^{26,33}, Phe^{59}]-apocytochrome *c*** horse heart	Ac-Gly-Asp-Val-Glu-Lys-Gly-Lys-Lys-Ile-Phe-Val-Gln-Lys-Cys-Ala-Glu-Cys-His-Thr-Val-Glu-Lys-Gly-Gly-Lys-His-Lys-Thr-Gly-Pro-Asn-Leu-His-Gly-Leu-Phe-Gly-Arg-Lys-Thr-Gly-Gln-Ala-Pro-Gly-Phe-Thr-Tyr-Thr-Asp-Ala-Asn-Lys-Asn-Lys-Gly-Ile-Thr-Trp-Lys-Glu-Glu-Thr-Leu-Met-Glu-Tyr-Leu-Glu-Asn-Pro-Lys-Lys-Tyr-Ile-Pro-Gly-Thr-Lys-Met-Ile-Phe-Ala-Gly-Ile-Lys-Lys-Lys-Thr-Glu-Arg-Glu-Asp-Leu-Ile-Ala-Tyr-Leu-Lys-Lys-Ala-Thr-Asn-Glu-OH
328	124	**Pancreatic Ribonuclease A *S*-sulfonate** (bovine)	H-Lys-Glu-Thr-Ala-Ala-Ala-Lys-Phe-Glu-Arg-Gln-His-Met-Asp-Ser-Ser-Thr-Ser-Ala-Ala-Ser-Ser-Ser-Asn-Tyr-Cys(SO_3)-Asn-Gln-Met-Met-Lys-Ser-Arg-Asn-Leu-Thr-Lys-Asp-Arg-Cys(SO_3)-Lys-Pro-Val-Asn-Thr-Phe-Val-His-Glu-Ser-Leu-Ala-Asp-Val-Gln-Ala-Val-Cys(SO_3)-Ser-Gln-Lys-Asn-Val-Ala-Cys(SO_3)-Lys-Asn-Gly-Gln-Thr-Asn-Cys(SO_3)-Tyr-Gln-Ser-Tyr-Ser-Thr-Met-Ser-Ile-Thr-Asp-Cys(SO_3)-Arg-Glu-Thr-Gly-Ser-Ser-Lys-Tyr-Pro-Asn-Cys(SO_3)-Ala-Tyr-Lys-Thr-Thr-Gln-Ala-Asn-Lys-His-Ile-Ile-Val-Ala-Cys(SO_3)-Glu-Gly-Asn-Pro-Tyr-Val-Pro-Val-His-Phe-Asp-Ala-Ser-Val-OH

PEPTIDES PREPARED BY SOLID PHASE PEPTIDE SYNTHESIS (continued)

Entry number	Reference	Solid support	Type of bond	Monomer protection	Coupling reagent	Deprotection	Cleavage
322	126	S-DVB-2%	Bzl	Boc	DCC,ONp	HCl-Diox.	HF
323	56	S-DVB-2%	Bzl	Boc	NCA,DCC	HCl-HOAc	HBr-TFA
324	52	S-DVB-2%	Bzl	Boc	DCC,ONp	HCl-HOAc	HBr-TFA
325	62	S-DVB-2%	Bzl	Boc	OSu	HCl-HOAc	HBr-TFA
326	73	S-DVB-2%	Bzl	Aoc	DCC	HCl-HOAc	HF
327	136	S-DVB-2%	Bzl	Boc	DCC,ONp, OSu	HCl-Diox.	HF
328	66	S-DVB-1%	Bzl	Boc	DCC,ONp	TFA-CH_2Cl_2	HF

Compiled by G. R. Marshall and R. B. Merrifield.

REFERENCES

1. **Merrifield, R. B.,** unpublished results (1961).
2. **Letsinger and Kornet,** *J. Amer. Chem. Soc.,* 85, 3045 (1963).
3. **Letsinger, Kornet, Mahadevan, and Jerina,** *J. Amer. Chem. Soc.,* 86, 5163 (1964).
4. **Weygand and Ragnarsson,** *Z. Naturforsch.,* 21B, 1141 (1966).
5. **Deer,** *Angnew Chem.,* 78, 1064 (1966).
6. **Bodanszky and Sheehan,** *Chem. Ind.,* 1423 (1964).
7. **Marshall and Merrifield,** *Biochemistry,* 4, 2394 (1965).
8. **Kessler and Iselin,** *Helv. Chim. Acta,* 49, 1330 (1966).
9. **Smith and Dunn,** *Fed. Proc.,* 25, 801 (1966).
10. **Merrifield,** *Fed. Proc.,* 21, 412 (1962).
11. **Merrifield,** *J. Amer. Chem. Soc.,* 85, 2149 (1963).
12. **Shemyakin, Ovchinnikov, Kinyushkin, and Kozhevnikova,** *Tetrahedron Lett.,* 27, 2323 (1965).
13. **Merrifield and Corigliano,** *Biochem. Prep.,* 12, 98 (1968).
14. **Lenard and Robinson,** *J. Amer. Chem. Soc.,* 89, 181 (1967).
15. **Bodansky and Sheehan,** *Chem. Ind.,* 1597 (1966).
16. **Shchukina and Sklyarov,** *Khim. Prirodn. Soedin.,* 2, 200 (1966).
17. **Khosla, Smeby, and Bumpus,** *Biochemistry,* 6, 754 (1967).
18. **Stewart and Woolley,** *Hypotensive Peptides,* Erdos, Back, and Sicuteri, Eds., Springer-Verlag, 1966.
19. **Kusch,** *Kolloid-Z. and Z. Polymer.,* 208, 138 (1966).
20. **Merrifield,** *Biochemistry,* 3, 1385 (1964).
21. **Merrifield and Stewart,** *Nature,* 207, 522 (1965).
22. **Stewart and Woolley,** *Nature,* 206, 619 (1965).
23. **Meienhofer and Sano,** *J. Amer. Chem. Soc.,* 90, 2996 (1968).
24. **Mizoguchi and Woolley,** *J. Med. Chem.,* 10, 251 (1967).
25. **Stewart, Young, Benjamini, Shimizu, and Leung,** *Biochemistry,* 5, 3396 (1966).
26. **Richards, Sloane, Jr., and Haber,** *Biochemistry,* 6, 476 (1967).
27. **Merrifield,** *J. Org. Chem.,* 29, 3100 (1964).

PEPTIDES PREPARED BY SOLID PHASE PEPTIDE SYNTHESIS (continued)

28. **Rothe, Schneider, and Dunkel,** *Makromol. Chem.*, 96, 290 (1966).
29. **Woolley,** *J. Amer. Chem. Soc.*, 88, 2309 (1966).
30. **Merrifield,** *Hypotensive Peptides,* Erdos, Back and Sicuteri, Eds., Springer-Verlag, 1966.
31. **Merrifield,** *Recent Progress in Hormone Research,* Academic Press, New York, 1967.
32. **Najjar and Merrifield,** *Biochemistry,* 5, 3765 (1966).
33. **Marglin and Merrifield,** *J. Amer. Chem. Soc.*, 88, 5051 (1966).
34. **Merrifield and Marglin,** *Peptides, Proc. Eur. Peptide Symp., 8th,* Noordwijk, Netherlands, 1966, Beyerman, Van de Linde, and Maassen van den Brink, Eds., North-Holland Publishing Company, Amsterdam, 1967, 85.
35. **Zahn, Okuda, and Shimonishi,** *Peptides, Proc. Eur. Peptide Symp., 8th,* Noordwijk, Netherlands, 1966, Beyerman, Van de Linde, and Maassen van den Brink, Eds., North-Holland Publishing Company, Amsterdam, 1967, 108.
36. **Klostermeyer, Halstrom, Kusch, Fohles, and Lunkenheimer,** *Peptides, Proc. Eur. Peptide Symp., 8th,* Noordwijk, Netherlands, 1966, Beyerman, Van de Linde, and Maassen van den Brink, Eds., North-Holland Publishing Company, Amsterdam, 1967, 113.
37. **Beyerman, Van Zoest, and Boers-Boonekamp,** *Peptides, Proc. Eur. Peptide Symp., 8th,* Noordwijk, Netherlands, 1966, Beyerman, Van de Linde, and Maassen van den Brink, Eds., North-Holland Publishing Company, Amsterdam, 1967, 117.
38. **Young, Benjamini, Stewart, and Leung,** *Biochemistry,* 6, 1455 (1967).
39. **Semkin, Gafurova, and Shchukina,** *Khim. Prir. Soedin.*, 3, 220 (1967).
40. **Semkin, Smirnova, and Shchukina,** *Zh. Obshch. Khim.*, 37, 1169 (1967).
41. **Shchukina, Semkin, and Smirnova,** *Khim. Prir. Soedin.*, 3, 358 (1967).
42. **Manning,** *J. Amer. Chem. Soc.*, 90, 1348 (1968).
43. **Marglin and Cushman,** *Biochem. Biophys. Res. Commun.*, 29, 710 (1967).
44. **Takashima, duVigneaud, and Merrifield,** *J. Amer. Chem. Soc.*, 90, 1323 (1968).
45. **Losse, Grenzer, and Neubert,** *Z. Chem.*, 8, 21 (1968).
46. **Losse, Madlung, and Lorenz,** *Chem. Ber.*, 101, 1257 (1968).
47. **Thampi, Schoellmann, Hurst, and Huggins,** *Life Sci. (Oxford),* 7, Part II, 641 (1968).
48. **Tilak and Hollinden,** *Tetrahedron Lett.*, 1297 (1968).
49. **Young, Leung, and Rombauts,** *Biochemistry,* 7, 2475 (1968).
50. **Bayer and Hagenmaier,** *Tetrahedron Lett.*, 2037 (1968).
51. **Ives,** *Can. J. Chem.*, 46, 2318 (1968).
52. **Bayer, Jung, and Hagenmaier,** *Tetrahedron,* 24, 4833 (1968).
53. **Sieber and Iselin,** *Helv. Chim. Acta,* 51, 622 (1968).
54. **Klostermeyer,** *Chem. Ber.*, 101, 2823 (1968).
55. **Yaron and Schlossman,** *Biochemistry,* 7, J2673 (1968).
56. **Epand and Scheraga,** *Biopolymers,* 6, 1551 (1968).
57. **Blake and Li,** *J. Amer. Chem. Soc.*, 90, 5882 (1968).
58. **Beyerman and Maassen Van den Brink-Zimmermannova,** *Rec. Trav. Chim.*, 87, 1196 (1968).
59. **Kettman, Benjamini, Michaeli, and Leung,** *Biochem. Biophys. Res. Commun.*, 29, 623 (1967).
60. **Marshall,** *Advan. Exp. Med. Biol.*, 2, 48 (1967).
61. **Shchukina, Sklyarov, and Gorbunov,** *Khim. Prir. Soedin,* 4, 120 (1968).
62. **Tripp,** *Diss. Abstr.*, B28, 3661 (1968).
63. **Weitzel, Weber, Hornle, and Schneider,** in *Peptides, Proc. Eur. Peptide Symp.*, 9th, Orsay, France, April 1968, Bricas, Ed., North-Holland Publishing Company, Amsterdam, 1968, 171.
64. **Weygand,** *Peptides, Proc. Eur. Peptide Symp.*, 9th, Orsay, France, April 1968, Bricas, Ed., North-Holland Publishing Company, Amsterdam, 1968, 171.
65. **Weitzel, Weber, Hornle, and Schneider,** in *Peptides, Proc. Eur. Peptide Symp.*, 9th, Orsay, France, April 1968, Bricas, Ed., North-Holland Publishing Company, Amsterdam, 1968, 222.
66. **Gutte and Merrifield,** *J. Amer. Chem. Soc.*, 91, 501 (1969).
67. **Garson,** *Diss. Abstr.*, B28, 3219 (1968).
68. **Polzhofer,** *Tetrahedron,* 25, 4127 (1969).
69. *Omenn and Anfinsen,* J. Amer. Chem. Soc., 90, 6572 (1969).
69. **Omenn and Anfinsen,** *J. Amer. Chem. Soc.*, 90, 6572 (1969).
70. **Inukai, Nakano, and Murakami,** *Bull. Chem. Soc. (Japan),* 41, 182 (1968).
71. **Khosla, Chaturvedi, Smeby, and Bumpus,** *Biochemistry,* 7, 3417 (1968).
72. **Beyerman, Hindriks, and deLeer,** *Chem. Commun.*, 1668 (1968).
73. **Sakakibara, Kishida, Kikuchi, Sakai, and Kakiuchi,** *Bull. Chem. Soc.(Japan),* 41, 1273 (1968).
74. **Gisin, Merrifield, and Tosteson,** *J. Amer. Chem. Soc.*, 91, 2691 (1969).
75. **Grahl-Nielsen and Tritsch,** *Biochemistry,* 8, 187 (1969).
76. **Beyerman, Boers-Boonekamp, Maasen, and van den Brink-Zimmermannova,** *Rec. Trav. Chem.*, 87, 257 (1968).
77. **Manning, Wuu, Baxter, and Sawyer,** *Experientia,* 24, 659 (1968).
78. **Weber, Hornle, Kohler, Nagelschneider, Eisele, and Weitze,** *Hoppe-Seyler's Z. Physiol. Chem.*, 349, 512 (1968).

PEPTIDES PREPARED BY SOLID PHASE PEPTIDE SYNTHESIS (continued)

79. Hutton, Marglin, Witkop, Kurtz, Berger, and Udenfriend, *Arch. Biochem. Biophys.*, 125, 799 (1968).
80. Halstrom and Klostermeyer, *Justus Liebigs Ann. Chem.*, 715, 208 (1968).
81. Weber and Weitzel, *Hoppe-Seyler's Z. Physiol. Chem.*, 349, 1431 (1968).
82. Schoellmann, *Hoppe-Seyler's Z. Physiol. Chem.*, 348, 1629 (1967).
83. Hornle, Weber, and Weitzel, *Hoppe-Seyler's Z. Physiol. Chem.*, 349, 1428 (1968).
84. Okuda and Zahn *Makromol. Chem.*, 121, 87 (1969).
85. Visser, Roeloffs, Kerling, and Havinga, *Rec. Trav. Chim.*, 87, 559 (1968).
86. Zahn, Okuda, and Shimonishi, *Angew. Chem.*, 78, 424 (1967).
87. Stryer and Haugland, *Proc. Natl. Acad. Sci. USA*, 58, 719 (1967).
88. Stewart, Mizoguchi, and Woolley, *Abstr. 153rd Meet. Amer. Chem. Soc.*, (1967).
89. Givas, Centeno, Manning, and Sehon, *Immunochemistry*, 5, 314 (1968).
90. Loffet, *Experientia*, 23, 406 (1967).
91. Park, Smeby, and Bumpus, *Biochemistry*, 6, 3458 (1967).
92. Marshall, in *Proc. 1st Amer. Peptide Symp.*, Weinstein and Lande, Eds., Dekker, New York, 1968.
93. Robinson, Thesis, University of California, San Diego, (1967).
94. Ryle, Leclerc and Falla, *Biochem. J.*, 4P (1969).
95. Stewart, *Fed. Proc.*, 27, 534 (1968).
96. Shapiro, Leng, and Felsenfeld, *Biochemistry*, 8, 3219 (1969).
97. Semkin, Smirnova, and Shchukina, *Zh. Obshch. Khim.*, 38, 2884 (1968).
98. Pereira, Close, Jellum, Patton, and Halpern, *Aust. J. Chem.*, 22, 1337 (1969).
99. Tritsch and Grahl-Nielson, *Biochemistry*, 8, 1816 (1969).
100. Ney and Polzhofer, *Tetrahedron*, 24, 6619 (1968).
101. Robinson and Kamen, in *Structure and Function in Cytochromes*, Okunuki and Kamen, Eds., University Park Press, Baltimore, 1968.
102. Rothe and Dunkel, *J. Polymer. Sci.*, B5, 589 (1967).
103. Krumdieck and Baugh, *Biochemistry*, 8, 1568 (1969).
104. Wang and Merrifield, unpublished results.
105. Mizoguchi, Shigezane, and Takamura, *Chem. Pharm. Bull.*, 17, 411 (1969).
106. Miura, Toyama, and Seto, *Sci. R. Toh. A*, 20, 41 (1969); *Chem. Abstr.*, 70, 47832 (1969).
107. Jolles and Jolles, *Helv. Chim. Acta*, A51, 980 (1968).
108. Nanzyo and Sano, *J. Biol. Chem.*, 243, 3431 (1968).
109. Polzhofer and Ney, *J. Chromatogr.*, 43, 404 (1969).
110. Benjamini, Shimizu, Young, and Leung, *Biochemistry*, 7, 1261 (1968).
111. Bondi, Fridkin, and Patchornik, *Israel J. Chem.*, 6, 22 (1968).
112. Baxter, Manning, and Sawyer, *Biochemistry*, 8, 3592 (1969).
113. Southard, Brooke, and Pettee, *Abstr. 158th Meet. Amer. Chem. Soc. NY (1969).*
113. Southard, Brooke, and Pettee, *Abstr. 158th Meet. Amer. Chem. Soc.* NY (1969).
114. Ovchinnikov, Kiryushkin, and Kozhevnikova, *Zh. Obshch. Khim.*, 38, 2551 (1968).
115. Wang and Merrifield, *Abstr. 158th Meet. Amer. Chem. Soc.* NY (1969).
116. Sawyer, Wuu, Baxter, and Manning, *Endocrinology*, 85, 385 (1969).
117. Mitchell, Gupta, and Roeske, *Abstr. 158th Meet. Amer. Chem. Soc.* New York (1969).
118. Green and Garson, *J. Chem. Soc.*, 401 (1969).
119. Weber, Hornle, Grieser, Herzog, and Weitzel, *Hoppe-Seyler's Z. Physiol. Chem.*, 348, 1715 (1967).
120. Chillemi and Merrifield, *Biochemistry*, 8, 4344 (1969).
121. Wang and Merrifield, *Int. J. Protein. Res.*, 1, 235 (1969).
122. Wang and Merrifield, *J. Amer. Chem. Soc.*, 91, 6488 (1969).
123. Felix and Merrifield, *J. Amer. Chem. Soc.*, 92, 1385 (1970).
124. Kato and Anfinsen, *J. Biol. Chem.*, 244, 5849 (1969).
125. Jorgensen, Windridge, Patton, and Lee, *J. Med. Chem.*, 12, 733 (1969).
126. Ontjes and Anfinsen, *Proc. Natl. Acad. Sci. USA*, unpublished results.
127. Southard, Brooke, and Pettee, *Tetrahedron Lett.*, 3505 (1969).
128. Ohno, Eastlake, Ontjes, and Anfinsen, *J. Amer. Chem. Soc.*, 91, 6842 (1969).
129. Vlasov and Bilibin, *Izv. Akad. Nauk. SSSR Ser. Khim.*, 1400 (1969).
130. Wieland, Birr, and Flor, *Justus Liebigs Ann. Chem.*, 727, 130 (1969).
131. Ingwall, Scheraga, Lotan, Berger, and Katchalski, *Biopolymers*, 6, 331 (1969).
132. Anfinsen, Ontjes, Ohno, Corley, and Eastlake, *Proc. Natl. Acad. Sci. USA*, 58, 1806 (1967).
133. Hornle, *Hoppe-Seyler's Z. Physiol. Chem.*, 348, 1355 (1967).
134. Ohno and Anfinsen, *J. Amer. Chem. Soc.*, 89, 5994 (1967).
135. Young, Benjamin and Leung, *Biochemistry*, 7, 3113 (1968).
136. Sano and Kurihara, *Hoppe-Seyler's Z. Physiol. Chem.*, 350, 1183 (1969).

This table originally appeared in Sober, *Handbook of Biochemistry and selected data for Molecular Biology*, 2nd ed., Chemical Rubber Co., Cleveland, 1970.

POLY(α-AMINO ACIDS), THEIR SOLUBILITY AND SUSCEPTIBILITY TO ENZYMATIC ACTIVITIES[a]

Substrate[b,c]	Soluble in[c]	Enzyme	pH	Reacted (+) or not (–)	Major products[c]	Reference[a]
Ala_n	Cl_2AcOH, F_3AcOH					
$(DL\text{-}Ala)_n$	H_2O, HCO_2H, Cl_2AcOH	Aminopeptidase		+		1
		Carboxypeptidase		+		1
		Chymotrypsin	7.7, 8.0	slow		2
		Papain	3.7, 7.0	+	Ala, oligomers	2, 3
		Pepsin	2.3, 4.2	–		2, 4
		"Pronase"	7.6, 8.0	+		2
		Trypsin	7.7, 8.0	–		2
$(Ala\text{-}Pro\text{-}Gly)_n$[d]	H_2O	Prolyl hydroxylase	7.8	+		59
$[Cys(Aet)]_n$	H_2O	Trypsin	7.4	+	$[Cys(Aet)]_2$, $[Cys(Aet)]_3$	5
$[DL\text{-}Phe(NH_2)]_n$	H_2O < pH 5					
Arg_n (7)	H_2O					
		Histone acetylation system		+	Acetylated polyarginine	6
		Protein (histone) methylase	8.5	+	Methylated polyarginine	6
		Trypsin	7.6	+	Arg, Arg_2	7
Arg-Gly-(Leu-Pro-Gly)$_5$	H_2O	Prolyl hydroxylase	7.8	+		53
Arg-Gly-(Pro-Pro-Gly)$_5$	H_2O	Prolyl hydroxylase	7.8	+		53
Asp_n	H_2O > pH 4	Acid protease from germinated sorghum	3.6 susp.	+	Asp, oligopeptides	8
		Chymotrypsin	7.7, 8.0	–		2
		Keratinase	9.0	–		9
		Pepsin	2.3, 4.2	–		2, 4, 10
		Penicillium cyaneofulvum protease	4.3	+		11
		"Pronase"	7.6, 8.0	–		2
		Trypsin	7.7, 8.0	–		2
		Yeast protease C	4.2	+	Asp	12
Cys_n	H_2O > pH 9					
$[Tyr(I_2)]_n$	H_2O > pH7					
$[Lys(Me_2)]_n$	H_2O	Carboxypeptidase B	7.5	+	$Lys(Me_2)$	13
Glu_n	H_2O > pH 4, $HCONMe_2$	Acid protease from germinated sorghum	3.6	+	Glu, oligopeptides	8
		Aspergillus fumigatus	5.1	+		14, 15
		Carboxypeptidase-A	5.0, 5.3	+	Glu	16, 17
		Carboxypeptidase-A (DFP-treated)	5.0–7.7	–		18
		Carboxypeptidase B	4.6, 4.9, 5.2	+	Glu	18
		Cathepsin D_2[e]	4.6	+		14
		Chymotrypsin	7.7, 8.0	–		2, 16
		Chymotrypsin	4.5–7.0	+		18
		Chymotrypsin 1	5.3	+		52
		Chymotrypsin 2	5.3	+		52
		Chymotrypsin C	5.3	+	Oligopeptides	19
		Elastase	5.0–7.0	+		18
		Elastase 1	8.8	–		52
		Elastase 2	8.8	–		52
		Esteroproteolytic enzyme from porcine pancrease	4.6–5.6	+	Oligopeptides	20

[a]Taken in part from Katchalski et al.[62] The reader is referred to this source if no other reference is given. References 33 and 63 to 82 also deal with properties of polyamino acids.
[b]Only homopolymers and sequence ordered copolymers are listed in this table.
[c]Abbreviations: AcOH, acetic acid; Cys(Aet), *S*-β-amino-ethyl-cysteine; DL-phe(NH_2), *p*-amino-DL-phenylalanine; Cl_2AcOH, dichloroacetic acid; $Tyr(I_2)$, 3,5-diiodotyrosine; $Lys(Me_2)$, N^ϵ,N^ϵ-dimethyllysine; $HCONMe_2$, dimethyl formamide; Me_2SO, dimethyl sulfoxide; HCO_2H, formic acid; Har, homoarginine; Hse, homoserine; Lys(Me), ε,N-methyl lysine; Pyr, pyridine; susp., suspension; F_3AcOH, trifluoroacetic acid; H_2O, water.
[d]The *N*-tert-butyloxycarbonyl and methyl ester blocked oligopeptides, n = 2, 3, 4, 5, and 6, were also hydroxylated.
[e]Crude enzymatic preparation has been used.

POLY(α-AMINO ACIDS), THEIR SOLUBILITY AND SUSCEPTIBILITY TO ENZYMATIC ACTIVITIES (continued)

Substrate[b,c]	Soluble in[c]	Enzyme	pH	Reacted (+) or not (–)	Major products[c]	Reference[a]
Glu_n (continued)	H_2O > pH 4, $HCONMe_2$ (continued)	Ficin	4.0–7.5	+	Glu, Glu_3	10
		Human thyroid proteinase[e]	4.3–7.7	–		14
		Keratinase	9.0	–		9
		Leucine aminopeptidase	5.0–7.7	–		18
		Leucine aminopeptidase I from *Aspergillus oryzae*	6.5	+	Glu	21
		Leucine aminopeptidase II from *Aspergillus oryzae*	6.5	slow		22
		Leucine aminopeptidase III from *Aspergillus oryzae*	6.0	slow		23
		Pancreatic protease 1	5.3	+		52
		Pancreatic protease 2	5.3	+		52
		Papain	4.0–8.0	+		16, 18, 24
		Penicillium cyaneofulvum protease	4.5	+		11
		Pepsin	2.0–5.0	+	Glu_3, Glu_4	2, 4, 25
		Pronase	7.6, 8.0	+		2
		Rennin	2.0–5.0	+		10
		Staphylococcus aureus[e]	4.8	+		14
		Subtilisin	3.5–7.5	+		18
		Trypsin	5.0–8.6	–		2, 18
		Yeast proteinase C	4.2	+	Glu	12
$(D\text{-}Glu)_n$	H_2O > pH 4	Chymotrypsin	7.7, 8.0	–		2
		Pepsin	2.3, 4.2	–		2, 4
		"Pronase"	7.6, 8.0	–		2
Glu-Gly(Pro-Pro-Gly)$_5$	H_2O	Prolyl hydroxylase	7.8	+		53
Gly_n	Cl_2AcOH, F_3AcOH, Conc. Li^+ and NH_4^+ halides	Keratinase	9.0	–		9
(Gly-Ala-Pro)$_4$	H_2O	Prolyl hydroxylase	7.2	+		55
(Gly-Pro-Ala)$_n$	H_2O	Collagenase	7.0	+	Gly-Pro-Ala	26
		Prolyl hydroxylase	7.8	very slow		59
(Gly-Pro-Gly)$_4$	H_2O if n < 4	Prolyl hydroxylase	7.8	–		59
(Gly-Pro-Pro)$_{\sim 16}$	H_2O	Prolyl hydroxylase	7.8	+		59
Har_n		Chymotrypsin	8.0	slow		61
		Trypsin	8.0	slow	Har_2, Har_3	61
His_n	H_2O < pH 6	Chymotrypsin	7.7, 8.0	–		2
		Lactoperoxidase	6.0	+	Iodinated polyhistidine	27
		Pepsin	2.3, 4.2	–		2
		Takadiastase[e]	5.3	+	His	28
$(DL\text{-}HSe)_n$		Chymotrypsin	7.7, 8.0	slow		2
		Pepsin	2.3, 4.2	slow		2
		"Pronase"	7.6, 8.0	+		2
		Trypsin	7.7, 8.0	–		2
Hyp_n	H_2O	Carboxypeptidase C	5.3	–		29
		Pepsin	2.3	–		4
Lys_n	H_2O	Acid protease from germinated sorghum	3.6	–		8
		Arthrobacter proteinase	8.0	+	Oligopeptides	30
		Carboxypeptidase-A	5.6–9.5	–		31, 32
		Carboxypeptidase-B	7.9–9.3	+	Lys	31, 32
		Cathepsin D_2[e]	4.7	+		14
		Chymotrypsin	7.7, 8.0	+		2, 14, 32, 33
		Elastase,	7.0–10	+		32
		Fibrinolysin (bovine)	7.4	+	Lys_2, Lys_3, Lys_4	28
		Ficin	7.0–12	+	Lys_3, Lys_4, Lys_5	10, 32
		Human thyroid proteinase	4.3–9.7	–		14
		Keratinase	9.0	+	Lys_2, Lys_3, Lys_4	9

POLY(α-AMINO ACIDS), THEIR SOLUBILITY AND SUSCEPTIBILITY TO ENZYMATIC ACTIVITIES (continued)

Substrate[b,c]	Soluble in[c]	Enzyme	pH	Reacted (+) or not (–)	Major products[c]	Reference[a]
Lys_n (continued)	H_2O (continued)	Leucine aminopeptidase	4.3–9.7	+	Lys	32
		Leucine aminopeptidase I from *Aspergillus oryzae*	6.5	+		21
		Leucine aminopeptidase II from *Aspergillus oryzae*	6.5	+		22
		Leucine aminopeptidase III from *Aspergillus oryzae*	6.0	slow		23
		Pancreas-protease	7.65	+	Lys, Lys_2, Lys_3	60
		Papain	7.0–12	+	Oligopeptides	10, 32
		Penicillium cyaneofulvum protease	10.7	+	Oligopeptides	11
		Pepsin	2.3–4.6	–		2, 32, 34
		"Pronase"	7.6, 8.0	+		2, 14
		Pronase E	7.0, 8.3	+	Lys	60
		Staphylococcus aureus[e]	3.2–6.1, 7.1–9.5	–		14
		Subtilisin	6.5–10	+		32
		Takadiastate	4.85, 5.0	+	Lys	28, 60
		Thrombin	7.8	–		28
		Trypsin	6.0–10	+	Lys_2, Lys_3	32, 34, 35
$(DLys)_n$	H_2O	Carboxypeptidase-B	7.65	–		60
		α-Chymotrypsin	7.65	–		60
		Pancreas powder extract	6.0	+		36
		Pancreas-protease	7.65	–		60
		Pronase E	7.00, 8.3	–		60
		Takadiastase	4.85	–		60
		Trypsin	8.0	–		36
$(Lys\text{-}Ala\text{-}Ala)_n$	H_2O	Elastase	8.6	+	Lys-Ala-Ala, $(Lys\text{-}Ala\text{-}Ala)_2$	37
		Trypsin	7.5	+	Ala-Ala-Lys	37
Met_n	$CHCl_3$, Cl_2AcOH					
$[Lys(Me)]_n$	H_2O	Carboxypeptidase B	7.5	+	Lys(Me)	13
		Trypsin	7.2	+		51
Orn_n(38)	H_2O	Trypsin		–		39
Phe_n	33% HBr/AcOH, Hot AcOH	Chymotrypsin	susp.	–		40
		Keratinase	9.0	–		9
Pro_n	H_2O, AcOH	Acid protease from germinated sorghum	3.6	–		8
		Aminopeptidase P	8.6	+	Pro	41
		Carboxypeptidase C	5.3	+		42
		Chymotrypsin	7.7, 8.0	–		2
		Clostridial aminopeptidase	8.6	–		43
		Dipeptido carboxypeptidase from *E. coli*	8.1	–		44
		Prolyl hydroxylase	7.5	–		48, 58
		Leucine aminopeptidase from bovine lens	9.1	–		45
		Leucine aminopeptidase I from *Aspergillus oryzae*	6.5	very slow		21
		Leucine aminopeptidase II from *Aspergillus oryzae*	6.5	very slow		22
		Leucine aminopeptidase III from *Aspergillus oryzae*	6.0	very slow		23
		Pepsin	2.3, 4.2	–		2, 4
		Penicillium cyaneofulvum protease	4.0–10.7	–		11
		Proline iminopeptidase	7.8–9.5	+	Pro	46, 47
		"Pronase"	7.6, 8.0	+		2
		Prolyl hydroxylase	7.5	–		48
		Trypsin	7.7, 8.0	–		2
		X-Prolyl aminopeptidase	7.7	very slow		49
		Yeast proteinase C	5.0–8.0	–		12
$(Pro\text{-}Gly\text{-}Ala)_n$	H_2O	Collagenase	7.0	–		26
$(Pro\text{-}Gly\text{-}Gly)_n$	H_2O	Collagenase	7.0	+	Pro-Gly, Gly-Pro-Gly, Gly-Pro-Gly-Gly	26

POLY(α-AMINO ACIDS), THEIR SOLUBILITY AND SUSCEPTIBILITY TO ENZYMATIC ACTIVITIES (continued)

Substrate[b,c]	Soluble in[c]	Enzyme	pH	Reacted (+) or not (–)	Major products[c]	Reference[a]
$(Pro\text{-}Gly\text{-}Pro)_n$	H_2O	Carboxypeptidase C	5.3	–		29
		Clostridial aminopeptidase	8.6	+	Pro, Gly-Pro-$(Pro\text{-}Gly\text{-}Pro)_{n-1}$	43
		Collagenase	7.0	+	Gly-Pro, Gly-Pro-Pro, Pro-Gly-Pro-Pro	26
		Dipeptido carboxypeptidase from *E. coli*	8.1	+	Gly-Pro, $(Pro\text{-}Gly\text{-}Pro)_{n-1}$-Pro	44
		Prolyl hydroxylase	7.2, 7.5, 7.6, 7.8	+	Hydroxylated poly-(Pro-Gly-Pro)	48, 54, 55, 58
$(Pro\text{-}Pro\text{-}Gly)_n$[f]	H_2O for $n < 10$, 10% AcOH or 50% EtOH for n = 20	Prolyl hydroxylase	7.5, 7.8	+		56, 57
Ser_n	Conc. LiBr in H_2O	Acid protease from germinated sorghum	3.6, susp.	–		8
Trp_n	Pyr, $HCONMe_2$, Cl_2AcOH	Chymotrypsin	susp.	–		40
Tyr_n	H_2O > pH 9, Pyr, $HCONMe_2$, Me_2SO	Chymotrypsin	7.7, 8.0	+		2
		Chymotrypsin	7.3–7.5, 8.3	–		50
		Keratinase	9.0	–		9
		Pepsin	2.3, 4.2	–		2, 4
		Trypsin	7.7, 8.0	–		2
$(Tyr\text{-}Ala\text{-}Glu)_n$	H_2O > pH 5.5	Acid proteinase from germinated sorghum	3.6	+	Oligopeptides with N-terminal tyrosine	8
Val_n	F_3AcOH					2

[f]The nonapeptides, n = 3 with *N*-tert-pentyloxycarbonyl and benzyl ester blocked N- and C-terminal groups, respectively, were also hydroxylated.

Compiled by Arieh Yaron.

POLY(α-AMINO ACIDS), THEIR SOLUBILITY AND SUSCEPTIBILITY TO ENZYMATIC ACTIVITIES (continued)

REFERENCES

1. Lindestrom-Lang, *Acta Chem. Scand.*, 12, 851 (1958).
2. Simons and Blout, *Biochim. Biophys. Acta,* 92, 197 (1964).
3. Katchalski, Sela, Silman, and Berger, in *The Proteins,* Vol. II, 2nd ed., Neurath, Ed., Academic Press, New York, 1964, 524.
4. Neumann, Sharon, and Katchalski, *Nature,* 195, 1002 (1962).
5. Lindley, *Nature,* 178, 647 (1956).
6. Kaye and Sheratzky, *Biochim. Biophys. Acta,* 190, 527 (1969).
7. Ariely, Wilchek, and Patchornik, *Biopolymers,* 4, 91 (1966).
8. Garg and Virupaksha, *Eur. J. Biochem.*, 17, 13 (1970).
9. Nickerson and Durand, *Biochim. Biophys. Acta,* 77, 87 (1963).
10. Katchalski, Levin, Neumann, Riesel, and Sharon, *Bull. Res. Counc. Isr. Sect. A,* 10, 159 (1961).
11. Ankel and Martin, *Biochem. J.*, 91, 431 (1964).
12. Hayashi and Hata, *Biochem. Biophys. Acta,* 263, 673 (1972).
13. Seely and Benoiton, *Biochem. Biophys. Res. Commun.*, 37, 771 (1969).
14. Lundblad and Johansson, *Acta Chem. Scand.*, 22, 662 (1968).
15. Martin and Jonsson, *Can. J. Biochem.*, 43, 1745 (1965).
16. Green and Stahmann, *J. Biol. Chem.*, 197, 771 (1952).
17. Avrameas and Uriel, *Biochemistry,* 4, 1750 (1965).
18. Miller, *J. Am. Chem. Soc.*, 86, 3913 (1964).
19. Folk and Schirmer, *J. Biol. Chem.*, 240, 181 (1965).
20. Gjessing and Hartnett, *J. Biol. Chem.*, 237, 2201 (1962).
21. Nakadai, Nasuno, and Iguchi, *Agric. Biol. Chem.*, 37, 757 (1973).
22. Nakadai, Nasuno, and Iguchi, *Agric. Biol. Chem.*, 37, 767 (1973).
23. Nakadai, Nasuno, and Iguchi, *Agric. Biol. Chem.*, 37, 775 (1973).
24. Miller, *J. Am. Chem. Soc.*, 83, 259 (1961).
25. Simons, Fasman, and Blout, *J. Biol. Chem.*, 236, PC64 (1961).
26. Harper, Berger, and Katchalski, *Biopolymers,* 11, 1607 (1972).
27. Holohan, Murphy, Flanagan, Buchanan, and Elmore, *Biochim. Biophys. Acta,* 322, 178 (1973).
28. Rigbi, Ph.D. thesis, Hebrew University, Jerusalem, 1957.
29. Nordwig, *Hoppe-Seyler's Z. Physiol. Chem.*, 349, 1353 (1968).
30. Hofsten and Reinhammar, *Biochim. Biophys. Acta,* 110, 599 (1965).
31. Gladner and Folk, *J. Biol. Chem.*, 231, 393 (1958).
32. Miller, *J. Am. Chem. Soc.*, 86, 3918 (1964).
33. Katchalski, *Adv. Protein Chem.*, 6, 123 (1951).
34. Waley and Watson, *Biochem. J.*, 55, 328 (1953).
35. Katchalski, Grossfeld, and Frankel, *J. Am. Chem. Soc.*, 70, 2094 (1948).
36. Tsuyki, Tsuyuki, and Stahmann, *J. Biol. Chem.*, 222, 479 (1956).
37. Yaron, Tal, and Berger, *Biopolymers,* 11, 2461 (1972).
38. Debabov, Davidov, and Morozkin, *Izvest. Akad. Nauk. SSR SER Khim.*, p. 2153 (1966).
39. Katchalski, Sela, Silman, and Berger, in *The Proteins,* Vol. II, 2nd ed., Neurath, Ed., Academic Press, New York, 1964, 521.
40. Rigbi and Gros, *Bull. Res. Counc. Isr.*, 11A, 44 (1962).
41. Yaron and Mlynar, *Biochem. Biophys. Res. Commun.* 32, 658 (1968); Yaron and Berger, *Methods in Enzymology,* Perlman and Lorand, Eds., Vol. 19, Academic Press, New York, 1970, 521.
42. Nordwig, *Hoppe-Seyler's Z. Physiol. Chem.*, 349, 1353 (1968).
43. Kessler and Yaron, *Biochem. Biophys. Res. Commun.*, 50, 405 (1973); *Methods in Enzymology,* Perlman and Lorand, Eds., Academic Press, New York, in press.
44. Yaron, Mlynar, and Berger, *Biochem. Biophys. Res. Commun.*, 47, 897 (1972).
45. Wiederanders, Lasch, Kirschke, Bohley, Ansorge, and Hanson, *Eur. J. Biochem.*, 36, 504 (1973).
46. Sarid, Berger, and Katchalski, *J. Biol. Chem.*, 234, 1740 (1959).
47. Sarid, Berger, and Katchalski, *J. Biol. Chem.*, 237, 2207 (1962).
48. Kivirikko and Prockop, *J. Biol. Chem.*, 242, 4007 (1967).
49. Dehm and Nordwig, *Eur. J. Biochem.*, 17, 364 (1970).
50. Rigbi, Seliktar, and Katchalski, *Bull. Res. Counc. Isr.*, 6A, 313 (1957).
51. Paik and Kim, *Biochemistry,* 11, 2589 (1972).
52. Uriel and Avrameas, *Biochemistry,* 4, 1740 (1965).

POLY(α-AMINO ACIDS), THEIR SOLUBILITY AND SUSCEPTIBILITY TO ENZYMATIC ACTIVITIES (continued)

53. **Kivirikko, Kishida, Sakakibara, and Prockop,** *Biochim. Biophys. Acta,* 271, 347 (1972).
54. **Prockop, Juva, and Engel,** *Hoppe-Seyler's Z. Physiol. Chem.,* 348, 553 (1967).
55. **Rhoads and Udenfriend,** *Arch. Biochem. Biophys.,* 133, 108 (1969).
56. **Kikuchi, Fujimoto, and Tamiya,** *Biochem. J.,* 115, 569 (1969).
57. **Suzuki and Koyama,** *Biochim. Biophys. Acta,* 177, 154 (1969).
58. **Hutton, Jr., Marglin, Witkop, Kurtz, Berger, and Udenfriend,** *Arch. Biochem. Biophys.,* 125, 779 (1968).
59. **Kivirikko and Prockop,** *J. Biol. Chem.,* 244, 2755 (1969).
60. **Darge, Sass, and Thiemann,** *Z. Naturforsch.* 28c, 116 (1973).
61. **Rigbi and Elzab,** *6th FEBS Meet.,* Madrid, Abstr. No. 730, 1969; Rigbi, Segal, Kliger, and Schwartz, *Bayer Symp. V,* Fritz, Tschesche, Greene, and Truscheit, Eds., Springer-Verlag, Berlin, 1974, 541.
62. **Katchalski, Sela, Silman, and Berger,** in *The Proteins,* Vol. II, 2nd ed., Neurath, Ed., Academic Press, New York, 1964, 436.
63. **Bamford, Elliott, and Hanby,** *Synthetic Polypeptides,* Academic Press, New York, 1956.
64. **Katchalski and Sela,** *Adv. Protein Chem.,* 13, 243 (1958).
65. **Szwarc,** *Adv. Polymer Sci.,* 4, 1 (1965).
66. **Stahmann,** *Polyamino Acids, Polypeptides and Proteins,* Wisconsin Press, Madison, 1962.
67. **Sela and Katchalski,** *Adv. Protein. Chem.,* 14, 391 (1959).
68. **Katchalski,** *Proc. VI Int. Congr. Biochem.,* p. 80 (1965).
69. **Katchalski,** *Harvey Lect.,* 59, 243 (1965).
70. **Katchalski,** in *New Perspectives in Biology,* Sela, Ed., Elsevier, New York, 1964, 67.
71. **Kauzmann,** *Annu. Rev. Phys. Chem.,* 8, 413 (1957).
72. **Scheraga,** *Annu. Rev. Phys. Chem.,* 10, 191 (1959).
73. **Leach,** *Rev. Pure Appl. Chem.,* 9, 1 (1959).
74. **Urnes and Doty,** *Adv. Protein Chem.,* 16, 401 (1961).
75. **Harrap, Gratzer, and Doty,** *Annu. Rev. Biochem.,* 30, 269 (1961).
76. **Schellman and Schellman,** in *The Proteins,* Vol. II, 2nd ed., Neurath, Ed., Academic Press, New York, 1964, 1.
77. **Fasman, Tooney, and Shalitin,** in *Encyclopedia of Polymer Science and Technology,* Bikales, Ed., John Wiley & Sons, New York, 1965, 837.
78. **Fasman,** in *Biological Macromolecules, Poly-α-Amino Acids, Protein Models for Conformational Studies,* Timasheff and Fasman, Eds., Marcel Dekker, New York.
79. **Goodman, Verdini, Choi, and Masuda,** *Top. Stereochem.,* 5, 69 (1970).
80. **Scheraga,** *Chem. Rev.,* 71, 195 (1971).
81. **Johnson,** *J. Pharm. Sci.,* 63, 313 (1974).
82. **Blaut, Bovey, Goodman, and Lotan,** *Peptides, Polypeptides and Proteins,* John Wiley & Sons, New York, 1974.

Index

INDEX

A

B

C

D

E

F

G

H

I

J

K

L

M

N

O

P

Q

R

S

T

U

V

W